NASA SP-2001-4227

UPLINK-DOWNLINK
A HISTORY OF THE DEEP SPACE NETWORK
1957–1997

Douglas J. Mudgway

The NASA History Series

National Aeronautics and Space Administration
Office of External Relations
Washington, DC
2001

Library of Congress Cataloging-in-Publication Data

Mudgway, Douglas J. 1923–
    Uplink-Downlink: A History of the NASA Deep Space Network , 1957–1997 / Douglas J. Mudgway.
       p. cm. – (NASA SP ; 2001-4227) (The NASA history series)
    Includes bibliographical references and index.
      1. Deep Space Network–History. I. Title. II. Series. III. Series: The NASA history series

TL3026 .M84 2000
629.47'43'0973--dc21
                                                                                                    00-058220

This book is dedicated to the memory of

Nicholas A. Renzetti

1914–1998

Throughout his entire career at JPL (1959-1996), Nicholas A. Renzetti brought the strength and dynamism of his own character to the Deep Space Network. Some of it passed to all of those he touched and to every task he undertook. Due in no small part to his continued advocacy, there exists a continuous record of the technical achievements of the Network from its inception to 1997, in the form of the "TDA Progress Reports" series. He gave rise, also, to many other publications that reflected the breadth of activity in which the Network engaged.

For many years, Renzetti envisioned a project that would distill the essence of materials collected into a single book-length narrative covering the full story of the Deep Space Network—scientific, technical, operational, personal, and political. My pleasure at having undertaken the task is tempered only by my regret that he was unable to see it completed.

Douglas J. Mudgway
Sonoma, California

ISBN 0-16-066599-X

For sale by the Superintendent of Documents, U.S. Government Printing Office
Internet: bookstore.gpo.gov   Phone: toll free (866) 512-1800;   DC area (202) 512-1800
Fax: (202) 512-2250 Mail: Stop SSOP, Washington, DC 20402-0001

ISBN 0-16-066599-X

Contents

iii	**Dedication**
xv	**Foreword**
xix	**Acknowledgments**
xxi	**Preface**
xxvii	**Introduction**
xxix	The Solar System
xxix	Size and Composition
xxxi	Terrestrial Planets
xxxi	Jovian Planets
xxxii	Inner and Outer Planets
xxxii	Asteroids
xxxii	Comets
xxxiii	Earth and Its Reference Systems
xxxiv	Earth Motions
xxxv	Time Conventions
xxxv	Interplanetary Trajectories
xxxvii	Deep Space Communications
xxxvii	Bands and Frequencies
xxxviii	Doppler Effect
xxxviii	Cassegrain Focus Antennas
xxxix	Uplink and Downlink
xl	Signal Power
xl	Coherence
xli	Decibels
xliii	Modulation and Demodulation
xliv	Spacecraft Radio System
xliv	An Essential Part of the Answer

Chapter 1—Genesis: 1957–1961

2	**Goldstone, California**
7	The Pioneer Lunar Probes
12	The Echo Balloon Experiment
15	The Venus Radar Experiment
16	Research and Development Antenna
18	**Washington, DC**
20	**Woomera, Australia**

| 23 | Johannesburg, South Africa |
| 26 | Endnotes |

Chapter 2—The Mariner Era: 1961–1974

30	**The Mariner Era Mission Set**
33	The Ranger Lunar Missions
36	Mariner Planetary Missions
37	Mariner 1962
39	Mariner 1964
41	Mariner 1967
42	Mariner 1969
45	Mariner 1971
46	Mariner 1973
48	Pioneer Interplanetary Missions
50	Pioneer Jupiter Missions
52	Surveyor Lunar Lander Missions
54	Lunar Orbiter Missions
56	Apollo Missions
59	**The Deep Space Instrumentation Facility (DSIF)**
61	The Facility (DSIF) Becomes a Network (DSN)
61	L-band to S-band
62	Improvements for the Mariner Mars Missions
64	The Need for a Second Network
66	Larger Antennas are Needed
70	Incremental Improvements in the Network
72	Flight Project Requirements Become Formalized
73	The DSN Becomes a Multimission Network
75	Openings and Closings
78	**Looking Back**
79	**Endnotes**

Chapter 3—The Viking Era: 1974–1978

82	**The Viking Era Mission Set**
84	**The Network**
87	**Network Operations**
87	The Helios Mission
91	The Viking Mission

Contents

107	The Voyager Mission
113	*Pioneers 6–11*
116	Pioneer Venus
119	**Network Engineering and Implementation**
119	The DSN Mark III System Design
121	Implementation for Viking
123	DSN Mark III-75 Model
125	The 64-meter Stations
129	The 26-meter Stations
130	High-Power Transmitter
132	Mars Radar
133	The Introduction of X-band
135	The 64-meter Antenna Problems
137	DSN Mark III Data System Project
141	Mark III Data System (MDS) Implementation
145	DSN Mark III-77 Model
146	DSN Mark III-77 Telemetry System
148	DSN Mark III-77 Command System
150	DSN Mark III-77 Tracking System
151	DSN Mark III-77 Ground Communications
151	DSN Mark III-77 Arraying Capability
153	DSN Mark III-77 Decoding Capability
155	The 26-meter Antenna S-X Conversion Project
159	**Overview**
160	**Endnotes**

Chapter 4—The Voyager Era: 1977–1986

166	**The Voyager Era**
166	Deep Space Missions
169	Earth-Orbiting Missions
171	**Network Operations**
171	Prime Missions
171	Voyager at Jupiter
177	Voyager at Saturn
190	Venus Balloon/Pathfinder
193	Voyager at Uranus
202	Uranus Encounter
206	**Extended Missions**

206	Helios
207	Viking at Mars
211	*Pioneers 6–9*
212	*Pioneers 10, 11*
215	*Pioneer 12* (Venus Orbiter)
216	**Highly-elliptical Orbiters**
216	ICE
219	AMPTE
221	**Reimbursable Missions**
224	**Network Engineering and Implementation**
224	New 34-meter High-Efficiency Antennas
226	The 64-meter to 70-meter Antenna Extension
230	Rehabilitation of DSS 14 Pedestal
233	DSS 14 Azimuth Radial Bearing
233	DSS 43 Pedestal
234	Antenna Performance Upgrade
236	Microwave Design Considerations
237	Kicker Braces
238	Subreflector
238	Surface Panels
239	Madrid Site Implementation
240	Canberra Site Implementation
240	Goldstone Site Implementation
242	**Networks Consolidation**
246	**Mark IVA Implementation**
251	Signal Processing Centers
253	Mark IVA Network Operations Control Center
255	Mark IVA Ground Communications Facility
255	DSN Mark IVA System Upgrades
255	Mark IVA Telemetry
258	Mark IVA Command
263	Mark IVA Tracking
265	**An Abrupt Transition**
266	**Endnotes**

Chapter 5—The Galileo Era: 1986–1996

272	**The Galileo Era**
272	A Defining Moment

Contents

273	Mission Set
274	Overview
276	**Network Operations**
276	Deep Space Prime Missions
276	The Comet Halley Missions
277	Giotto
279	Phobos
283	Background for Magellan
285	Magellan Mission to Venus
288	Voyager at Neptune
294	Background for Galileo
299	Galileo Mission to Jupiter
319	Background for Ulysses
324	Ulysses Solar Mission
326	Background for Mars Observer
327	Mars Observer Mission
330	Deep Space Extended Missions
330	The Pioneer Missions
331	International Cometary Explorer (ICE)
331	Voyager Interstellar Mission
333	Earth Orbiter Missions
333	General
335	Emergency Mission Support
337	**Network Engineering and Implementation**
337	The 70-meter Antennas
338	Elevation Bearing Failure
340	Gearbox Rehabilitation
341	Subreflector Drive Problems
343	Interagency Arraying
343	Parkes-Canberra Telemetry Array
348	VLA-Goldstone Telemetry Array
354	The X-band Uplink
358	Block V Receiver
359	The DSN Galileo Telemetry (DGT) Subsystem
371	Beam Waveguide Antennas
383	**Signal Processing Center Upgrade Task**
383	Background
383	Signal Processing Centers Upgrade
385	SPC Telemetry

388	SPC Command
390	SPC Tracking
393	Network Operations Control Center
394	Ground Communications Facility
398	**A Successful Conclusion**
399	**Endnotes**

Chapter 6—The Cassini Era: 1996–1997

408	**The Cassini Era**
408	Winds of Change
409	The Cassini Era Mission Set
410	**Deep Space Missions**
410	General
410	Galileo
410	Ulysses
410	*Voyagers 1 and 2*
412	*Pioneer 10*
412	Near-Earth Asteroid Rendezvous
413	Mars Global Surveyor
421	Mars Pathfinder
428	Cassini
434	**Earth-Orbiting Missions**
434	General
435	Space VLBI Observatory Program
437	**The Network**
437	Complexes and Antennas
440	Ka-band Downlink
445	Orbiting VLBI Subnetwork
450	Emergency Control Center
453	**Other Aspects**
454	**Endnotes**

Chapter 7—The Advance of Technology in the Deep Space Network

458	**The DSN Technology Program**
459	The Great Antennas of the DSN
467	Forward Command/Data Link (Uplink)
468	Return Telemetry/Data Link (Downlink)

469	Low-Noise Amplifiers
472	Phase-Lock Tracking
472	Synchronization and Detection
473	A Digital Receiver
474	Encoding and Decoding
477	Data Compression
478	Arraying of Antennas
483	Radio Metric Techniques
483	Doppler and Range Data
485	Timing Standards
486	Earth Rotation and Propagation Media
486	Radio Science
487	VLBI and Radio Astronomy
489	The Global Positioning System
489	Goldstone Solar System Radar
492	Telecommunications Performance of the Network
494	Cost-Reduction Initiatives
496	**Ka-band Development**
498	Other Technologies
498	Optical Communication Development
499	DSN Science

Chapter 8—The Deep Space Network as a Scientific Instrument

502	**The TDA Science Office**
507	**Radio Science**
511	Celestial Mechanics
513	Solar Corona and Solar Wind
514	Radio Propagation and Occultation
517	Relativistic Time Delay
518	Gravitational Waves
521	**Radio Astronomy**
523	International Cooperation
524	Tidbinbilla Interferometer
525	Orbiting VLBI
525	Host Country Programs
526	Cross-Support Agreements
526	Guest Observer Program
526	Antenna Utilization for Radio Astronomy

528	The DSN Also Benefits
529	Significant Advances
533	**Search for Extraterrestrial Intelligence (SETI)**
541	**Crustal Dynamics**
550	**Planetary Radar**
564	Asteroids
566	Radar Astrometry
567	Renaissance
569	Significant Events
572	**Endnotes**

Chapter 9—The Deep Space Network as an Organization in Change

582	**In the Beginning (1958 to 1963)**
582	Background
584	The Rechtin Years
594	**The Formative Years (1964 to 1994)**
594	A DSN Manager for Flight Projects
596	The Space Flight Operations Facility
601	The Bayley/Lyman Years
606	The Dumas/Haynes Years
610	The JPL Strategic Plan: 1994
612	Birth of the TMOD
615	**Reinventing the Future (1994 into the New Millennium)**
615	Reengineering TMOD
616	Data Capture Process
618	Activity Plan Process
619	Reinventing the Future
621	"Bridging the Space Frontier"
624	TMOD Primary Challenge: 1997
627	**Endnotes**

629	**List of Figures**
641	**Appendix**
656	**About the Author**
657	**Index**
669	**Other Books in the NASA History Series**

The Sun sitting on his throne commands all things
To tend downward toward himself, and does not
   allow the chariots of the heavenly bodies to move
Through the immense void in a straight path, but
   hastens them all along
In unmoving circles around himself as center.

*From "Ode on This Splendid
Ornament of Our Time and
Our Nation, the Mathematico–
Physical Treatise by the Eminent
Isaac Newton"
by Edmond Halley, Astronomer Royal
London, 1684*

FOREWORD

From the very beginning of its association with NASA in 1958, the Jet Propulsion Laboratory (JPL) received its fair share of public recognition for its successes and failures in pursuing the exploration of deep space. It started with the Explorers, the first American satellites to orbit Earth. Later there came the Rangers, the first spacecraft to reach the surface of the Moon; the Mariner spacecraft, first to visit Venus and Mars; and the Voyagers that pushed the boundaries of deep space communication further out to Jupiter and Saturn, and eventually to Uranus and Neptune. There were other spacecraft that put landers, probes, or orbiters into planetary orbits or atmospheres, or onto planetary surfaces. There were probes whose mission was to explore the composition and dynamics of the interplanetary medium, and probes to observe the physics of the Sun. There were the huge missions, such as Viking to Mars, Galileo to Jupiter, and Cassini to Saturn, and there were small missions like Pathfinder to Mars and the New Millennium missions to asteroids and comets. There was also science that did not require a spacecraft for its experiments such as radio astronomy, radar astronomy, and the search for extraterrestrial intelligence.

The public accolades that were engendered by the bountiful science returned from all of these NASA projects were shared by NASA and the scientists whose exquisite instruments and innovative interpretation of the data produced the new knowledge reflected in their results. However, what the press conferences, news releases, and media coverage did not reveal was the incredibly complex infrastructure that made each of these marvelous deep space missions possible. This infrastructure, which had been built over the years at JPL, included the Deep Space Network (DSN), an essential, integral part of every mission. There was, in effect, a relationship between the planetary missions, the spacecraft that carried them out, and the Deep Space Network that enabled such missions to be planned in the first place.

Without the remarkable improvement in performance of the DSN, scientific missions to the distant planets would have been impossible. In 1964, when Mariner IV flew past Mars and took a few photographs, the limitation of the communication link meant that it took eight hours to return to Earth a single photograph from the Red Planet. By 1989, when Voyager observed Neptune, the DSN capability had increased so much that almost real-time video could be received from the much more distant planet, Neptune.

It is timely that, some 40 years after its inception, the Deep Space Network should be recognized for its remarkable litany of progress in radio communications over vast distances, thereby allowing planetary scientists to collect data from sites throughout the solar system. This book succeeds in bringing the history of the DSN forward for the attention of curious, generally informed, or technical specialist readers.

NASA is to be commended for commissioning this book as part of its History Series, for the NASA/JPL Deep Space Network is the world's largest and most advanced facility for tracking, navigating, and acquiring data from interplanetary spacecraft. Worldwide in scope, and international in concept, the DSN has supported not only NASA space missions, but also those of the space agencies of Japan, Germany, Russia, France, and Canada. Also, in concert with a NASA policy of international cooperation, the resources of the Network have been made available to support qualified enterprises from all nations.

In planning this book, New Zealand-born author Douglas Mudgway was faced with the formidable problem of making an extremely complex technology comprehensible to the curious or generally informed reader, while at the same time presenting for the specialist reader a historically accurate account of the advance of the technology that made possible more sophisticated planetary missions. The task was further complicated by the fact that the DSN was in a constant state of change, changes that were in consonance with the requirements of the space missions it was supporting.

For the former type of reader, Mudgway describes what the DSN actually did, within the framework of several overarching eras, each corresponding to a period of time during which a major NASA deep space mission dominated the public scene. The Mariner, Viking, Voyager, and Galileo Eras are examples. He provides an inside view of what it took to design, build, and operate those tenuous radio communication links between the controllers, engineers, and scientists at computer terminals at JPL and billion dollar spacecraft about to land on Mars, orbit around Jupiter, or fly through the rings of Saturn.

For the latter reader, the author provides an excellent review of the growth of the specialized technology that underlaid the remarkable expansion of Network capability that enabled the design of increasingly ambitious planetary missions. A comprehensive appendix provides help for the technical researcher.

The unique capabilities of the great antennas of the DSN, together with a generous NASA policy of making them available for non-NASA scientific research, attracted radio and radar astronomers from the United States and many other countries to the extent that the requests for observing time far exceeded the time that could be made available for these ground-based scientific purposes. Chapter 8, "The Deep Space Network as a Scientific

Foreword

Instrument," brings this important aspect of the work of the DSN to the attention of the reader, and provides a basic review of the published work that resulted.

Uplink-Downlink transforms the technical records of a major NASA facility, unique in the world, into a viable historical narrative covering 40 years of its critical involvement in the United States space program. The Deep Space Network emerges from this study not only as a complex, human-machine system of worldwide dimensions, but also, more convincingly, as a focus for the aspirations of the NASA scientists for ever-bigger science, and of the JPL engineers for ever-greater innovation and enterprise in navigating to distant targets and communicating at ever-greater data-rates, in spite of the fluctuations in available NASA funding for both, driven in some measure by the conflicting priorities of the piloted versus unpiloted programs within NASA itself.

Throughout the narrative we observe the interaction of these powerful currents, not from the lofty heights of a dispassionate historical observer, but from the eye-level of a dedicated participant, for the author's long career at JPL was played out at the vortex of these often conflicting currents of self-interest. As a consequence, the engineers, technicians, scientists, and managers at NASA, JPL, and its partner institutions in Spain and Australia, whose names appear from time to time in this book, are presented as real-life people contributing their various talents to the milieu in which they found themselves. It was the totality of these individual efforts, driven by a common inquiring interest in space and focused toward the common goal of its exploration, that produced the remarkable entity known throughout the world as the NASA Deep Space Network.

I am confident that this work will be a valuable addition to the documented history of the United States space program.

W. H. Pickering

ACKNOWLEDGMENTS

To have been part of the history recorded here is sufficient reason in itself to acknowledge my indebtedness to many colleagues in the DSN who supported and assisted me throughout my long and rewarding career at the Jet Propulsion Laboratory (JPL). That said, there remains the need to recognize the important contributions that were made to the writing of this history by many of those engineers and scientists, and by other persons less directly associated with my life in the Deep Space Network.

I must begin with Nicholas A. Renzetti, for it was he who brought me from Australia to the United States in 1962 to begin a career at the Deep Space Instrumentation Facility (DSIF), and it was he who, in 1996, after I had retired from JPL, stimulated my personal interest in producing a history of the Deep Space Network. When the project began to falter for lack of funding support, MacGregor Reid provided much needed encouragement, and Paul Westmoreland responded with the limited resources available to him to keep it going.

Later, when those resources expired, the project came to the attention of the NASA Chief Historian, Roger Launius. His encouragement, backed with adequate resources, moved the project rapidly forward to completion. Without his enthusiastic support it is unlikely this book would have been published. Along the way, I was ably assisted by Louise Alstork and other members of the NASA History Office staff.

As the work got under way, my access to historical documents, files, and photographs was eased immeasurably by the generous help of members of the JPL Archives and Records section, notably John Bluth, Elizabeth Moorthy, and Robin Morris.

At all times, Shirley Wolff of the Telecommunications and Missions Operations (TMO) Outreach Office was my lifeline to the daily pulse of the Network. I came to depend on her patience and energy for transmitting documents and other technical material provided, at my request, by various engineers and scientists associated with the Network. She, too, helped in bringing this project to life.

Last and longest, but by no means least, were the contributions in the form of interviews, discussions, technical briefings and materials, narrative reviews, and encouragement on various DSN-related topics that were provided by: Catherine Thornton on geodesy; Michael Klein on the search for extraterrestrial intelligence; Martin Slade on radar astronomy; James Hodder on network operations; Thomas Kuiper, Marvin Wick, and Pamela Wolken on radio astronomy; George Textor on Voyager; Leslie Deutsch on DSN telemetry for Galileo; Joseph Wackley for DSN systems; Joseph Statman on the Big Viterbi Decoder; Robert Wallace on 34-m antennas; Dan Bathker on microwaves; Robert Clauss on masers; Charles Stelzried on system noise temperature, Venus Balloon, and Giotto; Bob Preston and John Ovnick on orbiting VLBI; Fred McLaughlin on the 70-m antennas; Dale Wells on 70-m antenna maintenance; James Layland on coding and arraying; Patrick Beyer on Galileo; Dennis Enari on Ulysses and Mars Pathfinder; Marvin Traxler on Mars Observer; Allen Berman on Magellan; Robert White on 34-m antennas; Thomas Wynne for photographs; Ronald Gillette on Cassini; John McKinney on Mars Missions; Ed Massey on Voyager and Ulysses; Nick Fanelli and Joe Goodwin on the Earth orbiters and reimbursables; Bob Ryan and David Lozier (Ames Research Center) on the Pioneers; Michael Stewart on Magellan; Don Mischel, Tom Reid, Richard Mallis, and Robertson Stevens for early background material; George Schultz for an early draft; and finally, Olivia Tyler, Bobby Buckmaster, and Lynda McKinley for miscellaneous but nevertheless indispensable help.

The onerous task of reviewing the draft version of the book was undertaken by Larry Dumas, Michael Hooks, Roger Launius, MacGregor Reid, Gael Squibb, and Jose Urech. Their insightful comments and suggestions greatly enhanced the accuracy, consistency, and quality of the narrative.

The families of Nicholas Renzetti and William Merrick kindly provided background material on the personal lives of these two important figures in the history of the Deep Space Network. Their contributions are gratefully acknowledged.

Douglas J. Mudgway
Sonoma, California

PREFACE

Although the subtitle for this book, "A History of the Deep Space Network" has an air of finality about it, the suggestion of a task ended or a work completed, that does not represent the true state of affairs. The Deep Space Network (DSN) is really a work in progress. It had a beginning, of course, and a life that, at the time this book was written, had spanned four decades. That is what this book is about. The task of recording "what happened next" will fall to future historians.

Eloquent histories of NASA's planetary explorers have been written by others, and it was not my intent to revisit those magnificent enterprises. To them belonged the high drama associated with spacecraft encounters and landings on distant planets, spectacular launches, startling new science, and stunning color images from distant corners of the solar system.

As a key element in all of those dramatic spacecraft events, the DSN shared their excitement, but saw a quiet, unreported drama of its own. There were occasions when the success or failure of a multimillion dollar spacecraft, the reputation of NASA, or the recovery of critical science data from a far planet, lay in the hands of the DSN and, not infrequently, those of a crew member at a distant tracking station. Such events appear throughout the narrative. More often though, the determined, sometimes heroic, efforts of engineers and technicians in laboratories at JPL, and in control rooms and antennas at remote tracking stations, provided drama enough for those of us who were aware of it. Struggling to meet the seemingly insatiable demands of the planetary space missions for more, bigger, or better capability, and all of it sooner, those individuals bore the brunt of the burden of change that characterized progress in the DSN from its beginning. These events, too, are identified in the narrative.

In these pages, the informed reader will discover a simple description of what the DSN is about, and how it works—an aspect of NASA's spectacular planetary program that seldom found its way into the popular media coverage of those major events. Future historical researchers will find a complete record of the origin and birth of the DSN, its subsequent development and expansion over the ensuing four decades, and a description of the way in which the DSN was used to fulfill the purpose for which it was created. At the same time, the specialist reader is provided with an abundant source of technical references that address every aspect of the advanced telecommunications technology on which the success of the DSN depended. And finally, archivists, educators, outreach managers, and article writers will have ready recourse to the inner workings of the DSN and how they related to the more publicly visible events of the planetary space program.

Despite the way it might have appeared to many of us within it, the DSN was not created as "an end unto itself;" it was created by NASA to serve the needs of the NASA planetary spaceflight program. The spaceflight projects like Mariner, Surveyor, Pioneer, Viking, Ulysses, Galileo etc., were its "customers." It seemed appropriate, therefore, to record the history of the DSN against a timeline based on the great planetary explorers. Indeed it was their requirements for tracking and data acquisition services, to support more ambitious flight missions, that drove the DSN to the heights of excellence and achievement described herein.

At some point in the future it will be timely to produce another history of the DSN. The present work will serve as a clear point of departure for what follows, for by that time the rapidly diminishing "corporate" memory reflected herein will have surely been lost. There is a final point to be made. At a conservative estimate, NASA invested $4–5 billion over 40 years in developing and operating the Deep Space Network as a major NASA facility. An investment of public money of that magnitude deserves a public record of how the money was spent and an account of the return on the investment. Hopefully, this history will suffice for that purpose also. Those were my objectives in writing this book.

Before proceeding further, a few words about internal communications practices within the DSN organization at the Jet Propulsion Laboratory (JPL) during the period covered in this history is in order. Perhaps in keeping with the very nature of JPL's academic associations, communications within the various departments, offices, laboratories, and individuals that constituted the DSN at JPL were always very informal. Most interactions between engineers and managers and their people were conducted "eyeball to eyeball" in offices, corridors, bathrooms, or cafeterias, as the need arose. When two or more people were together, a vacant office was used as an ad hoc conference room. Larger groups met in a designated conference room, where the overhead projector and transparencies were the communications medium, and a handout, consisting of hard copies of the transparencies, on which the attendee made notes, was the only record of the material presented and the decisions reached. Interoffice memos were generally used to address issues that went beyond the writer's immediate organizational boundary. Anyone could write a memo on any subject, and did, when a meeting would not suffice. There never was a centralized, or even unified, filing system within the DSN organization. Everybody was responsible for keeping their own files in any way they chose. However, throughout the laboratory, there was always a paper reduction or "records roundup" program in effect. People were continually encouraged to discard inactive files in the interests of saving space. Furthermore, when people retired, they took files with them or discarded

Preface

them. So, for one reason or another, all DSN interoffice memos and personnel files, with few exceptions, have disappeared.

Fortunately, however, this loose approach did not apply to documents that were tightly controlled by the formal DSN Documentation System. It was generally considered that anything worth preserving should be published in a formal document, and personnel were strongly encouraged to do so. As a result, a most complete record of DSN activity in research, engineering, and deep space tracking "operations" is to be found in the series of documents called "Progress Reports." Such reports are referenced extensively in the chapters that follow, and are readily available from the JPL Archives Section.

The regrettable absence of interoffice memos, meeting minutes, verbatim records of discussions, letters, personnel records, etc., relating to DSN activity through the years accounts for the paucity of anecdotal material carried in the narrative.

The NASA Deep Space Network is a crucial element in the mission to explore our solar system. Its main purpose is to provide a high-quality communications link between an Earth-based mission control center and the many spacecraft of the United States and several foreign countries that are, or were, engaged in that mission. Over the forty years since the inception of the Deep Space Network, its people have been engaged in a continuous, pervasive drive to refine communications links. The remarkable record of achievement described herein is the result of their effort.

In choosing *Uplink-Downlink* as the primary title of this book, I endeavor to convey to the reader the key characteristic of the Deep Space Network and the essential focus of the people who make it work.

This theme, *Uplink-Downlink*, runs through the book and is used to show how the DSN has developed from a single-mission network of limited capability to a giant multiple-mission network. Together with a series of innovative improvements in spacecraft communications systems, the Network provides a deep space communications capability which now exceeds its modest beginnings by 12 orders of magnitude.

While the DSN is a prime example of exquisite telecommunications technology and engineering, it is a fully operational, global system of people and machines supporting NASA and other programs of planetary exploration. The way in which the DSN has supported these programs through the years, and the part it has played in all of the momentous and historical scientific discoveries resulting from this endeavor, forms the timeline against which the history of the DSN is described.

The spectacular successes, and occasional failures, of NASA's ventures into space are well covered by the media in various forms. We are quite accustomed to live television coverage of new spacecraft launches; scientists presenting their latest "amazing" facts or pictures; images from remote spacecraft flying by, onorbit about or on the surface of distant planets; and NASA animation movies of complex inflight or orbital maneuvers. The complex infrastructure of tracking stations, intercontinental communication links, data processing services, and highly skilled engineers and technicians that make it all possible are seldom mentioned, much less shown, on screen.

That infrastructure, so essential to the NASA program of Planetary Exploration, is known as the Deep Space Network. The fascinating history of its evolution, from a simple, ground-based antenna and receiver recording the signal from a telemetry transmitter onboard the first simple spacecraft attempting to reach the Moon, to the present worldwide network of giant antennas keeping track of spacecraft at the edge of the solar system, is the subject of this book.

Like the NASA space program itself, the history of the Deep Space Network is one of continual change, of performance enhancement, of striving for excellence and, of going where none has gone before.

To provide the general reader with a basis for understanding the significance of these changes in the context of the Deep Space Network, we begin with a simple discussion of the dynamics of spaceflight and why a tracking station is needed to maintain contact with an artificial object in space. The function of a deep space station in keeping track of one or more spacecraft is addressed, and is followed by an explanation of how a tracking station does what it does. Then, since the course of history in the Deep Space Network was determined at various times by decisions based on radio frequency technology, it seemed that a short discourse on the principles involved would be in order.

The early phase of DSN history, from its inception in 1958 to the mid-1970s, was associated with NASA's first attempts to fly spacecraft to the Moon, Venus, Mars, and Jupiter. I have chosen to call this the Mariner Era because most of the major DSN development was driven by the requirements of the Mariner missions. By 1974, the Mariner Era was drawing to a close and the Viking Era of the bigger, more complex, more costly missions to revisit the planets, was beginning. The Viking Era opened in 1975 with the Viking mission to Mars.

Dependent on the NASA planetary exploration program for its funding, the DSN continued to evolve in size, capability and cost at a rate commensurate with the level of

Preface

NASA interest in unmanned exploration of the solar system. For the next twenty-five years, the Voyager, Galileo, and Cassini programs were the principal drivers for expansion and development in the Deep Space Network. Each program effectively determined a unique era of influence in the overall story of the DSN.

From the outset, the esoteric technologies of deep space telecommunications and information processing were dominant features of research and development programs in the DSN. Key characteristics of this work are described in chapter 7, "The Advance of Technology in the Deep Space Network."

Much of the technology in the Deep Space Network shares a lot in common with that of the giant radio telescopes of the world. It should not be surprising that the antennas of the Deep Space Network were used, both individually and in collaboration with various radio astronomy observatories, for radio astronomy investigations as an adjunct to their principal task of spacecraft tracking. These investigations, and others related to scientific experiments that involved the DSN but did not involve a spacecraft, are described in chapter 8, "The Deep Space Network as a Scientific Instrument."

To organizations born of the space age, and particularly those involved directly with the NASA space program, growth and change were almost synonymous. Through the years covered here, organizational changes at NASA Headquarters, at Caltech's Jet Propulsion Laboratory (parent body of the DSN), and within its own management structure, were a continual and essential part of its evolution. The changes that took place and their impact on the DSN ability to cope with conflicting challenges for new planetary missions, while continuing support for ongoing missions within a fluctuating level of NASA funding support, is the substance of chapter 9, "The Deep Space Network as an Organization in Change."

I have elected to address the general reader in describing the history of the DSN, with the expectation that a specialist reader will take advantage of the references I have given to follow-up in specific areas of interest using the rich repository of related technical material contained in the series of Tracking and Data Acquisition (TDA) Progress Reports now available on the World Wide Web at *http://tda.jpl.nasa.gov/progress_report*.

INTRODUCTION

Surely there can be few people in this day and age who are not familiar with the television images of a NASA spacecraft blasting off from Cape Kennedy in a splendid gout of smoke, fire, and steam to a rendezvous with an asteroid, comet, or planet in some remote corner of the solar system. From the comfort of our living rooms we watch the launch rocket rise majestically off the pad, pass through the local cloud cover at the launch site and soon become an ever-diminishing point of light in the center of the television screen. Sometimes we observe the booster rockets burn out and drop away before the main rocket stage engines accelerate the vehicle and its delicate planetary payload out of sight. The Titan-Centaur launch of *Voyager 1*, shown in Figure Intro-1, was a typical example. What happened next? you may ask and, Where did that whole thing go?

Figure Intro-1. *Voyager 1* **outward bound for Jupiter, September 1977.** *Voyager 1* was launched by a Titan-Centaur vehicle from Cape Kennedy on 5 September 1979. After encountering Jupiter in February 1979, *Voyager 1* used a gravity-assist maneuver to reach Saturn in November 1980. It then continued on a trajectory which eventually took it beyond the boundary of the solar system. Twenty years later it had become the remotest of all spacecraft and was still being tracked by the Deep Space Network in 1998.

Months or even years after a launch, an item on the evening television news covering a NASA press release about an impending rendezvous of a NASA spacecraft with Mars, Jupiter, Saturn, or an asteroid, followed by a video clip or the latest set of pictures from the planetary object, might have caught the attention of viewers. For certain, the pictures showed areas and details never before seen and will be said to have risen new questions about origins and development of the solar system, and possibilities for water, maybe even life, on the planet or one of its satellites. The *Voyager 1* view of Jupiter as the spacecraft approached the

xxvii

Figure Intro-2. *Voyager 1* views Jupiter close up, February 1979. One of the most spectacular planetary photographs ever taken, this close-up picture of Jupiter was taken by *Voyager 1* on its approach to Jupiter in February 1979. Passing in front of the planet are two of the Galilean satellites, Io and Europa. Both satellites are larger than Earth's Moon, and the Great Red Spot seen on Jupiter's surface is larger than Earth itself.

planet, shown in Figure Intro-2, is a prime example of science data return with high value for public interest.

There will be some with enquiring minds who may connect new scientific evidence with a spacecraft launch, recalled from the distant past (see Figure Intro-1), and ask the question, How did "it" get from there to what I am seeing now? (See Figure Intro-2.)

Well, the answer depends on what is meant by "it," but an essential part of the answer is the subject of this book, namely, the Deep Space Network. To properly understand that part of the answer, some explanation of the nature of the problem and the terminology used to describe it will be necessary. The discussion that follows is directed to

Introduction

that end, to readers who have a general interest in the subject but whose field of professional interest lies elsewhere. For the expert reader, numerous technical references are provided throughout the book for followup on topics of specific interest.

Before beginning a discussion of radio communication with spacecraft traveling in deep space, we shall briefly review the environment of space and the motions of planetary spacecraft within that environment. This review and the discussion of space communications that follows it will explain various technical terms commonly used in describing the environment of space, the motions of Earth and spacecraft, and space communications. It is hoped that the general reader, who may not be familiar with many of these terms, will find this helpful in understanding the terminology used in the main chapters of the book.

The Solar System

NOTE: The following discussion makes extensive use of material contained in "The Basics of Space Flight," a learning document produced by the Jet Propulsion Laboratory for use in its spaceflight operations training program.

The solar system has been studied for religious or scientific reasons from the very earliest times. For most of that time, studies of the solar system have had to rely on indirect measurements of the various objects in the solar system such as the visible light emitted by, or reflected from the objects, or later, by the radio waves emitted by the Sun. However, with the emergence of space flight, instruments can be sent to many objects in the solar system to make direct measurements of their physical properties and dynamics, at close range. Data collected from such measurements have resulted in an unprecedented increase in our knowledge of the solar system.

Size and Composition

The solar system consists of an "average" star we call the Sun and the planets Mercury, Venus, Earth, Mars, Jupiter, Saturn, Uranus, Neptune, and Pluto. It also includes numerous satellites of the planets, comets, asteroids, meteoroids and of course, the interplanetary medium. The planets, most of the satellites of the planets, and the asteroids, revolve around the Sun in the same direction, in nearly circular orbits. The Sun and the planets rotate on their individual axes. Except for Pluto, all the planets orbit the Sun in or near the same plane, called the plane of the ecliptic.

The most common unit of measurement for distances within the solar system is the astronomical unit (AU). One AU equals the mean distance of the Sun from Earth, about

150,000,000 km. Prompted by the need for a more accurate figure for spacecraft navigation purposes, the DSN refined the value of the AU in the 1960s using radar echoes from Venus. Distances within the solar system, the distance between a DSN antenna on Earth and a planetary spacecraft for instance, are often indicated in terms of the distance light travels in a unit of time at the speed of 300,000 km per second.

For example:

Light Time	Light Travels	Approx. Distance
1 second	300,000 km	0.75 Earth-Moon
1 minute	18,000,000 km	0.125 AU
8.3 minutes	150,000,000 km	1 AU, Earth-Sun
1 hour	1,000,000,000 km	1.5x Sun-Jupiter

Although the Sun is characterized as a typical star, it dominates the gravitational field of the solar system; it contains 99.85 percent of the mass of the solar system. All the planets combined contain only 0.135 percent of the total mass with the satellites of the planets, asteroids, meteoroids and interplanetary medium making up the balance of 0.015 percent. Even though the planets account for only a small portion of the total mass of the solar system, they retain the greater part of the angular momentum of the solar system. This storehouse of energy can be utilized by interplanetary spacecraft to make so-called "gravity assist" changes in their interplanetary trajectories.

In terms of volume, nearly all of the solar system appears to be an empty void. Far from being just nothingness, however, this void comprises the interplanetary medium and includes various forms of energy and at least two material components: interplanetary dust and interplanetary gas. The dust consists of microscopic particles which can be measured by special instruments carried by interplanetary spacecraft. Interplanetary gas is a tenuous flow of gas and charged particles called plasma, which when streaming from the Sun is called solar wind. The solar wind can also be measured by interplanetary spacecraft. The point at which the solar wind meets the interstellar medium, that is the stellar wind streaming from other stars, is called the heliopause. This boundary, which marks the edge of the Sun's influence, is theorized to lie perhaps 100 AU from the Sun. At the time of this writing, the Voyager and Pioneer spacecraft were making their way toward this remote limit of the solar system. The magnetic field of the Sun extends outward into interplanetary space and it, too, can be measured by instruments carried on planetary spacecraft.

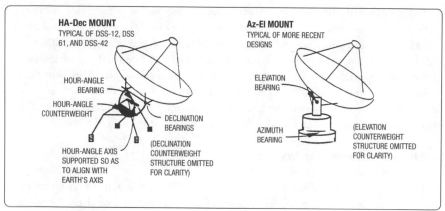

Figure Intro-5. HA-Dec and Az-El mounted antennas of the DSN.

that it requires an asymmetrical design, which is unsuited to the support of very heavy structures. By contrast, Az-El mounted antennas are basically symmetrical structures which can support heavy weights, but to track celestial objects they require driving in two axes, AZ and EL simultaneously. However, the advent of high speed computers has obviated the problems formerly associated with converting coordinates from one system to the other. The essential differences between the HA-Dec and Az-El types of mount for DSN antennas are illustrated in Figure Intro-5.

Earth Motions

Earth rotates on an axis inclined at 23.5 degrees to the plane of its orbit around the Sun. Its period of rotation is 24.0 hours mean solar time. This motion is of prime importance to the configuration of the Deep Space Network. Due to the rotation of Earth, a spacecraft moving away from Earth along its trajectory in space appears to an observer on Earth to set toward the west. To an observer further west, the spacecraft will appear to rise in the east and travel across the sky until, 8 to 10 hours later, it sets. The process repeats until 24 hours later the spacecraft appears, rising in the East for the first observer. It follows that three observers, each located 120 degrees of longitude apart on the surface of Earth, would have the spacecraft continuously in view as the view period passed from one to the other. This is the basic idea underlying the location of the three Deep Space Communication Complexes of the Network at Goldstone, California, near Madrid, Spain, and near Canberra, Australia. From these locations, approximately 120 degrees of longitude apart, their great antennas, some HA-Dec, others Az-El, are able to maintain continuous radio view of planetary spacecraft as Earth performs its daily rotations.

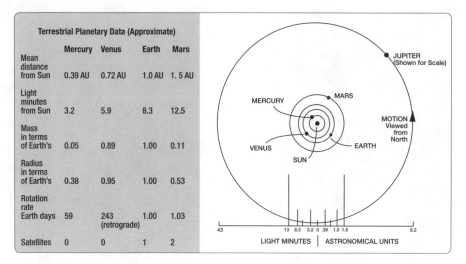

Figure Intro-3. Mean distances of the terrestrial planets from the Sun. The orbits in this figure are drawn approximately to scale. The orbit of Jupiter is shown for scale reference only.

The Terrestrial Planets

The four terrestrial planets of the solar system—Mercury, Venus, Earth, and Mars—have a firm rocky surface similar to that of Earth. The orbits and mean distances of the terrestrial planets from the Sun are shown in Figure Intro-3.

None of the terrestrial planets has rings. Earth has a layer of rapidly moving charged particles known as the Van Allen Belt, first detected by the JPL Explorer 1 space probe in 1958.

The Jovian Planets

The four Jovian planets—Jupiter, Saturn, Uranus, and Neptune—are all much larger than the terrestrial planets and have gaseous natures like that of Jupiter. They all have satellites and rings, although the size and number of each varies considerably between them. The orbits and mean distances of the Jovian planets from the Sun are shown in Figure Intro-4.

Jovian Planetary Data (Approximate)				
	Jupiter	Saturn	Uranus	Neptune
Mean distance from Sun	5.2 AU	9.5 AU	19.2 AU	30.1 AU
Light hours from Sun	0.72	1.3	2.7	4.2
Mass in terms of Earth's	318	95	15	17
Radius in terms of Earth's	11	9	4	4
Rotation rate in hours	9	10.5	15.6	18.4
Number of known satellites (1993)	16	14	15	7

Figure Intro-4. Mean distances of the Jovian planets from the Sun. The orbits in this figure are drawn approximately to scale. Pluto, at a mean distance of 39.5 AU from the Sun, is not discussed in this book and is omitted from the figure to accommodate the scale of the other planets.

Inner and Outer Planets

Frequently, the planets whose orbits lie inside Earth's orbit—namely, Venus and Mercury—are referred to as the inner planets. Mars, Jupiter, Saturn, Uranus, Neptune, and Pluto are generally known as the outer planets.

Asteroids

Asteroids are rocky objects orbiting the Sun at a distance of about 2.7 AU, between the orbits of Mars and Jupiter, and moving in the same direction as the planets. They vary from the size of pebbles to objects measured in hundreds of kilometers. Some have orbits which cross the orbit of Earth from time to time. The DSN Solar System Radar at Goldstone is used to investigate the properties of asteroids with Earth crossing orbits.

Comets

Comets are believed to be composed of rocky material and water ice. Their highly elliptical orbits bring them very close to the Sun and swing them deeply into space often beyond the orbit of Pluto. Comet structures are diverse and very dynamic, but they a[ll] develop a cloud of diffuse material called a coma, that usually grows in size and brigh[t]ness as the comet approaches the Sun. A small bright nucleus is usually visible in t[he] middle of the coma. The coma and nucleus together constitute the head of the come[t]. As a comet approaches the sun it develops an enormous tail of luminous dust mater[i]al that extends for millions of kilometers from the head away from the sun. It al[so] develops a tail of charged particles and an envelope of hydrogen. Several comets ha[ve] been investigated by planetary spacecraft flying close to or through the coma.

Earth and Its Reference Systems

Without a system of coordinates to consistently identify the positions of observers, pla[n]ets, and interplanetary spacecraft, exploration of the solar system would not be possibl[e]. Because space is observed from an Earth platform, a system of Earth coordinates [is] required to establish the position of the observer. The locations of DSN antennas o[n] Earth's surface are specified in terms of latitude and longitude.

To establish a coordinate system for the sky, the concept of a celestial sphere whose cent[er] is at the center of the Earth, is used. The celestial sphere has an imaginary radius larg[er] than the distance to the farthest observable object in the sky, that is to say, it extends f[ar] beyond the limits of the solar system. The extended axis of Earth intersects the north an[d] south poles of the celestial sphere. The direction of a spacecraft, planet, or star, or any oth[er] celestial object, can be specified in two dimensions on the inside of this sphere using a sy[s]tem of coordinates analogous to Earth's latitude and longitude system. Although the referen[ce] origins are different, the analogous terms latitude and longitude on the celestial sphere a[re] declination (DEC, latitude) and right ascension (RA, longitude). When used in connectio[n] with a specific location such as a DSN antenna, at a particular time of day, the term R[A] is replaced by a different term called hour angle or HA.

A somewhat simpler system for describing the position of a distant spacecraft relative t[o] a particular antenna and time of day uses the local horizon and true north as its refe[r]ences. Its measurements are azimuth (AZ), measured in degrees clockwise around the horizo[n] from true north and elevation (EL), measured in degrees above the local horizontal datum. Optical telescopes, radio telescopes, and the DSN antennas are designed with mounting[s] that make best advantage of either the HA-Dec or Az-El coordinate systems.

In a HA-Dec system the HA axis is parallel to Earth's axis of rotation. Thus an anten[n]na built on a HA-Dec mount has the advantage that motion is required mostly in onl[y] one axis, HA, to track an object like a spacecraft, as Earth rotates. The disadvantage i[s]

Introduction

The ability of the DSN antennas to view the planetary spacecraft is not changed by the annual rotation of Earth around the Sun. It does, however, introduce a cyclic change to the distance between Earth and spacecraft depending on the orientation of the spacecraft trajectory relative to Earth's orbit.

Time Conventions

In addition to local time and Greenwich Mean Time, there are several other systems for measurement of time that are used for tracking planetary spacecraft in the Deep Space Network. Three of the most important are Universal Time Coordinated (UTC), Earth Received Time (ERT), and Round-Trip Light Time (RTLT).

UTC is a worldwide scientific standard of timekeeping based upon carefully maintained atomic clocks that are accurate to a few microseconds via the addition or subtraction of leap seconds as necessary at two opportunities every year. Its reference point lies on the Earth prime meridian at Greenwich, England. All spacecraft operations in the DSN are conducted on the basis of UTC.

The time (in UTC) at which a DSN tracking station observes an event associated with a planetary spacecraft is called the Earth Received Time (ERT). The time (in UTC) that a tracking station observed the loss of signal due to a spacecraft occultation by a planet would be an example of such an event.

The elapsed time that it takes a radio signal (traveling at the speed of light) to travel from a DSN station to a spacecraft or planetary body, and after retransmission by the spacecraft or reflection from the body, return to the tracking station, is known as the Round-Trip Light Time (RTLT). It is used by the DSN to measure spacecraft range and other navigational and scientific parameters. The RTLT from Earth to the Moon is about 3 seconds; to the Sun, about 17 minutes. In 1993, the RTLT for the *Voyager 1* spacecraft was 14 hours, 13 minutes. By the end of 1998, this had increased to 20 hours, 18 minutes, making *Voyager 1* the most distant spacecraft in the solar system.

Interplanetary Trajectories

A spacecraft sitting atop a launch vehicle at Cape Canaveral can, in one sense, be considered as already in orbit around the Sun by virtue of the motion of the Earth orbit around the Sun. To send such a spacecraft to an outer planet, Mars for example, the spacecraft's existing orbit must be adjusted to cause it to intercept the orbit of Mars at a single point. The portion of the new orbit between Earth and Mars is called the interplanetary trajectory.

To achieve such a trajectory, the spacecraft is lifted off the pad by the launch vehicle, rises above Earth's atmosphere, orients itself to the right attitude and is accelerated in the direction of Earth's orbit around the Sun to the extent that it becomes free of Earth's gravitational effect. The magnitude and direction of the accelerating force is carefully calculated to achieve a new orbit which will have an aphelion (farthest distance) equal to that of the orbit of Mars. After injection into its new orbit, the spacecraft simply "coasts" the rest of the way to its destination. Of course, to get to Mars itself, rather than just to Mars' orbit, the spacecraft must be inserted into its interplanetary trajectory at precisely the right time to reach the orbit of Mars at the same time that the planet itself reaches the point where the spacecraft will intercept the orbit of Mars. This calls for very precise timing indeed, and results in constrained opportunities for launch called "launch windows." A spacecraft interplanetary trajectory can be adjusted or trimmed, to a limited degree, by means of midcourse maneuvers which use the onboard thrusters to change the spacecraft speed and direction, or by means of a "gravity assist," which uses the gravitational force of a nearby planet for the same purpose. The motions of Earth that play an important part in the interplanetary orbit insertion process are shown in Figure Intro-6.

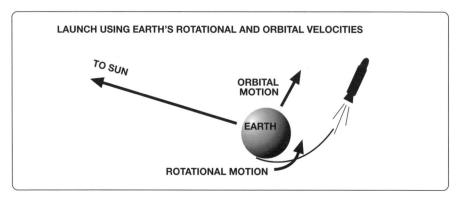

Figure Intro-6. Insertion into interplanetary orbit. For interplanetary launches, the spacecraft takes advantage of Earth's orbital motion to supplement the limited energy available from the launch vehicle itself. In the diagram, the launch vehicle is in addition to using Earth's rotational speed, accelerating generally in the direction of Earth's orbital motion, which has an average velocity of approximately 100,000 km per hour along its orbital path. The launch vehicle is used to increase or decrease this velocity in the tangential direction to insert the spacecraft into the desired transfer orbit.

Introduction

The vital DSN operation of "initial acquisition" is performed during, or as close as possible to, the interplanetary trajectory injection maneuver right after launch. After that, spacecraft and mission controllers are entirely dependent upon the DSN for communications with the spacecraft throughout the life of the mission.

The foregoing paragraphs have provided a greatly simplified answer to the question of what happens to a spacecraft after it leaves the launch pad. They have also described the environment in which it will move for the rest of its life. An answer to the next question, How do spacecraft and mission controllers communicate with their spacecraft during its long passage through the environment of deep space? requires some understanding of the nature of deep space communications.

Deep Space Communications

The Deep Space Network makes use of electromagnetic radiation for communicating with interplanetary spacecraft. All planetary spacecraft are equipped with radio transmitters and receivers for sending signals to and receiving signals from the Earth-based antennas of the DSN. In addition to signals from spacecraft, DSN antennas and receivers are capable of detecting signals from natural emitters of electromagnetic radiation, such as the stars, the Sun, molecular clouds, and giant gas planets such as Jupiter. The "signals" from these sources appear as random noise to the sensitive receivers of the DSN, and their study and interpretation is the field of radio astronomy. The DSN supports many research projects in radio astronomy.

Bands and Frequencies

Electromagnetic radiation from the natural emitters combines with radiation from artificial sources to create a background level of electromagnetic noise from which the spacecraft signals must be detected. The ratio of the signal level to the noise level is known as the signal-to-noise ratio (SNR). SNR is one of most common terms used in the DSN to describe the quality of a communication link.

Electromagnetic radiation with frequencies between about 10 kHz and 100 GHz are referred to as radio frequency (RF) radiation. For convenience in managing its use, RF radiation is divided into groups called bands, such as S-band and X-band. The bands are further divided into small ranges of frequencies, or channels, some of which are allocated by international agreement, for the use of deep space telecommunications. The following table shows the approximate range of frequency and wavelength for each band.

Radio Communication Bands vs. Wavelength and Frequency

Band Name	Wavelength (cm)	Frequency (GHz)
L	30–15	1–2
S	15–7.5	2–4
C	7.5–3.75	4–8
X	3.75–2.4	8–12
K	2.4–0.75	12–40

Early spacecraft used L-band for deep space communications. Within a few years, S-band replaced L-band, and more recently X-band came into general use for deep space communications. Experiments to demonstrate the advantages of telecommunications systems using K-band were in progress at the time of this writing (1997). Generally speaking, the higher frequency bands offer greater advantages for space communications, although these tend to be offset by other factors, such as increasing losses at the highest bands.

Doppler Effect

The Doppler effect is routinely observed as changes in the frequency of spacecraft radio signals received by a DSN tracking station. This is caused by the relative motion between a spacecraft and the tracking station due to the spacecraft trajectory, its orbit around a planet, Earth's orbit around the Sun, or the daily rotation of Earth about its axis. When the distance decreases, the frequency decreases proportionally to the rate of change of distance, and vice versa. If two widely-separated tracking stations observe a single spacecraft, they will have slightly different Doppler signatures. This information, described as the Doppler type of radio metric data, is generated at the DSN tracking stations and used by spacecraft navigators to describe the path of the spacecraft through deep space, very precisely and in three dimensions.

Cassegrain Focus Antennas

Whether the mount is of the HA-Dec or Az-El type, all DSN antennas, irrespective of their size, use the Cassegrain focus system to concentrate the electromagnetic energy incident upon their surfaces. The incident energy may originate from a distant spacecraft, a celestial body, or the station's own uplink transmitter. It works equally well for either uplink or downlink. A diagram of a cassegrain focus antenna is shown in Figure Intro-7.

In a Cassegrain antenna, a secondary reflector is added to the structure to fold the electromagnetic beam back to a prime focus near the primary reflector in the manner shown

Introduction

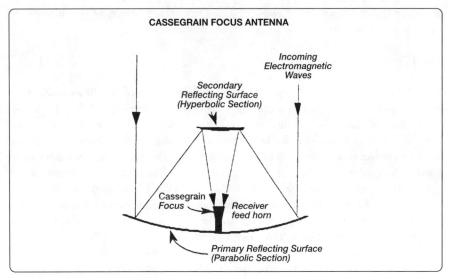

Figure Intro-7. Cassegrain focus antenna.

in the figure. Incoming electromagnetic waves are focused by the prime reflector on to the secondary reflector, or hyperboloid, which refocuses them into the receiver feed horn located at the prime focus. Usually, several feed horns are mounted on a single cone structure, and by rotating the slightly offset secondary reflector, the main beam can be directed to any receiver horn as required. For transmitting, the system works in the reverse way.

This design accommodates very large-diameter antennas and allows bulky, heavy receiving and transmitting equipment to be located near the center of gravity of the composite structure.

Uplink and Downlink

The radio signal transmitted from a DSN antenna to a distant spacecraft is known as an uplink. The radio signal transmitted by the spacecraft to the DSN is known as a downlink. Uplinks or downlinks may consist of a pure RF tone called a carrier, or a carrier which has been modulated to carry information in each direction, including commands to the spacecraft or telemetry data from the spacecraft. A spacecraft communications link that involves only a downlink is called a one-way link. When an uplink is being received by a spacecraft at the same time that a downlink is being received, the communications mode is said to be two-way. These two distinct modes of operation play a significant part in the operation of deep space communications.

Signal Power

Typically, a local broadcasting station with a transmitting power of 50,000 watts will deliver an acceptable radio signal to a portable receiver at a distance of 100 km. How then can a spacecraft transmitter, limited to a power of 20 watts, deliver an acceptable signal to a DSN receiver across hundreds of millions of kilometers of interplanetary space? The first step involves concentrating all of the available energy into an extremely narrow beam pointed in one direction, rather than spreading it in all directions, as in a broadcast station. At the spacecraft, this is done with a small parabolic dish antenna, typically one to five meters in diameter. Even so, when these concentrated signals reach Earth they have vanishingly small power, perhaps as small as 1×10^{-20} watts. The rest of the solution is provided by the receiving power of the large aperture antennas of the DSN and the extraordinary sensitivity of its cryogenically-cooled low-noise receivers. Aided by special coding and decoding schemes to discriminate against radio noise, the DSN can extract the science and engineering data from these unimaginably weak signals and deliver it in real time to the intended users. As this history will show, the ability of the DSN to do this has improved by many orders of magnitude over the forty or so years since its inception.

Coherence

In addition to its use as a conveyor of modulated telemetry data, the downlink carrier is also used by the DSN for tracking the spacecraft and for carrying out some types of radio science experiments. Each of these applications requires the detection of minute changes (fractions of 1 Hz) in many GHz of carrier frequency over many hours. This can only be accomplished when the frequency of the downlink carrier itself is extremely stable and is known with very great precision. Since the spacecraft itself could not carry the massive equipment needed to do this, it makes use of the uplink, which does have the requisite stability, as a frequency reference for the downlink. In effect, the spacecraft simply retransmits the DSN uplink after modulating part of it with the desired telemetry data. When this is done, the downlink is said to be two-way and coherent, that is, in-phase with the uplink.

At each Deep Space Complex, a hydrogen maser-based frequency standard provides the reference for generating an extremely stable uplink frequency for transmitting to the spacecraft. The resulting spacecraft downlink, based on and coherent with the uplink, has practically the same high frequency stability as the original reference frequency standard. Comparison of one with the other at the phase level of individual cycles produces the desired tracking

Introduction

or scientific data. The entire process ultimately depends for its success on the "phase coherence" of the uplink and downlink carriers, and the accuracy with which corrections can be made for the many sources of error in the end-to-end system.

Under some operational conditions the spacecraft may not have an uplink. To cover such cases, all spacecraft carry a small auxiliary oscillator, which serves as a reference for generating a non-coherent downlink in the absence of a DSN-generated uplink. It is not highly stable and its frequency is affected by temperature variations on the spacecraft. Nevertheless, there have been numerous instances in the long history of the DSN when a spacecraft's local oscillator played a major role in saving an otherwise doomed spacecraft. Some spacecraft also carry a thermally stabilized ultra-stable oscillator (USO) which is used for precision radio science experiments involving planetary occultations. Because of stringent frequency requirements for spacecraft tracking navigation and radio science, the DSN remains at the forefront of the technology for frequency and timing standards.

Decibels

Perhaps the most commonly used technical term in the DSN is the ubiquitous decibel. Since the term decibel (or dB) appears frequently in the following history, a simplified, non-rigorous explanation is in order. The decibel is a unit of measure used to describe the ratio between two power levels. For example, the power delivered by a 100-watt audio amplifier is said to be 10 dB higher than the power delivered by a small 10-watt amplifier. The ratio of the powers is 10 to 1. Inversely, where the ratio is 1 to 10 (1/10) the equivalent term would be -10 dB.

Because of its mathematical origin, the decibel scale is not linear, it is logarithmic; the 10 dB corresponds to a ratio of 10, 20 dB corresponds to a ratio of 100, 30 dB to a ratio of 1000 and so on. For ratios less than 10, 3 dB corresponds to a 2 to 1 power ratio, 6 dB to a ratio of 4, etc. When the ratios are inverted they are represented by a negative sign before the corresponding dB value, so that a ratio of 1 to 2 or 1/2 is represented by -3 dB.

Much of the technical progress in the DSN is reflected in terms of dB values for high power transmitters, gain of large antennas, threshold of receiving systems, gain of maser amplifiers, signal-to-noise ratios, bit-error rates for decoding systems, etc.

The remarkable improvement in downlink performance that took place during the Mariner, Viking, and Voyager Eras, the first three great eras of DSN growth, is shown graphically in Figure Intro-8.

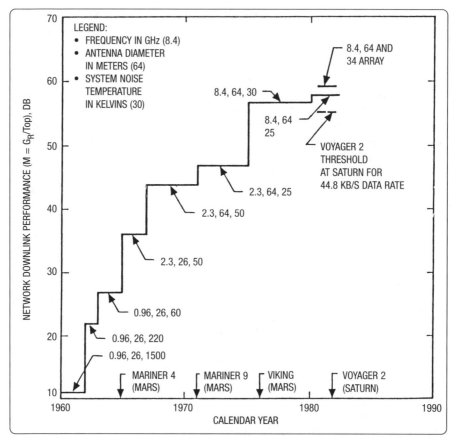

Figure Intro-8. Improvement in DSN downlink performance during the first twenty years. In this graphic, published by C. T. Stelzried et al. in 1982, downlink performance is stated in terms of the parameter M which relates the two key elements of downlink performance: the gain (size) of the receiving antenna G_R, and the noise temperature (sensitivity) of the receiving system, Top. It is expressed in dB as described above.

In the context of this figure, improved downlink performance implies an increased capability for a DSN tracking station to return more data at the same range, the same data at a greater range, or some other enhanced combination of data and range from a planetary spacecraft. The effect of changes in operating frequency, from L-band to S-band to X-band, is evident in the stepwise improvements, as is the contribution of the larger diameter antennas both individually and in array with

xlii

Introduction

other antennas. The parameter called system noise temperature is a measure of the sensitivity of the maser receiving systems that are a key element in the overall performance of the downlink.

After the early 1980s, improvements in terms of dB were less dramatic, not because of any less incentive for further improvement, but because by that time the technology of deep space communications was approaching the practical limits of physical realization for that era. More esoteric approaches, such as advanced data coding methods, were called upon, and although significant improvements continued to be made, they were much smaller in terms of dB. This issue, and how the DSN dealt with it, is discussed in chapter 7, "The Advance of Technology in the Deep Space Network."

Modulation and Demodulation

Scientific and engineering data is generated by a planetary spacecraft in digital form and modulated to an S-band or X-band carrier signal for transmission to the antennas of the DSN. Modulation is a two step process in which the bits of raw data are first applied to a subcarrier of much lower frequency. The data-modulated subcarrier is then applied to the RF carrier for radiation, via the spacecraft's transmitter and antenna, to Earth. At the DSN receiving station, the process is reversed. Sensitive receivers detect the carrier signal, extract the subcarrier and pass it to a subcarrier demodulator. The subcarrier demodulator extracts the data and conditions it for recording and forwarding to JPL. Most spacecraft apply a complex code to the raw data to protect it against noise-induced errors during transmission. Special equipment at the tracking stations performs the extra step of decoding the data before it is processed for recording and delivery. This end-to-end function of a deep space communication system, and the data associated with it, is called telemetry.

There is an analogous function that uses the S-band or X-band uplink for transmission of command data to a spacecraft. It is called command. In this case, the subcarrier modulated with a bit-stream of command data. At the spacecraft, the carrier is detected, the subcarrier is demodulated and the command data bits are eventually passed to the command system for immediate execution or for storage for later action.

The distance from Earth, or range, of a distant spacecraft can be measured by modulating the uplink with a special code or sequence of digital characters, which, when turned around by the spacecraft and remodulated on the downlink, allow the time delay between transmission and reception to be determined with great precision. The

Earth/spacecraft range is then calculated from knowledge of the time delay and the propagation velocity of the modulated radio signal.

Spacecraft Radio System

Intentionally, the spacecraft itself was not among the topics discussed up to this point. While a planetary spacecraft is a miracle of modern technology, the history of planetary spacecraft development is beyond the scope of this book. It is important to note, however, that the spacecraft radio system is an integral part of a deep space communications system although not an integral part of the DSN organization during the period of time covered here. The spacecraft radio system was provided by the flight project organization as part of the spacecraft. Nevertheless, responsibility for RF compatibility between the spacecraft and the DSN was always a DSN responsibility. In the ensuing chapters, where a spacecraft radio system is discussed, the general explanations given above will suffice for an understanding of the matter.

An Essential Part of the Answer

In the preceding discussion, I elected to use generalities in describing some of the basic concepts of deep space communications. The reader will encounter all of these terms and topics many times as the history of the Deep Space Network unfolds. These concepts also allow us to answer the questions prompted by the launch pictures, and by the remarkable new images shown at the press conference. An essential part of the answer is provided by NASA's Deep Space Network. Worldwide in concept, continuous in operation, the DSN is the link between the two images. The global scale of the Network is illustrated by the map shown in Figure Intro-9.

In terms of the foregoing discussion, the Deep Space Network may be characterized as a scientific instrument of worldwide proportions that uses a single up- and downlink radio signal, in combination with a coherent spacecraft transponder, to perform three basic functions. Interplanetary spacecraft depend on the first two for their communications with Earth. The third and no less significant function brings the DSN into the science community in a more direct way. Each of these functions is associated with one of the unique types of data carried on the up- and downlinks.

The first, and perhaps the most important, function is that of generating radio metric data. Consisting of Doppler data, ranging data, and interferometric comparisons between two tracking stations, radio metric data are used by spacecraft navigators to determine the precise location of a spacecraft along its trajectory at all times. Radio metric data

INTRODUCTION

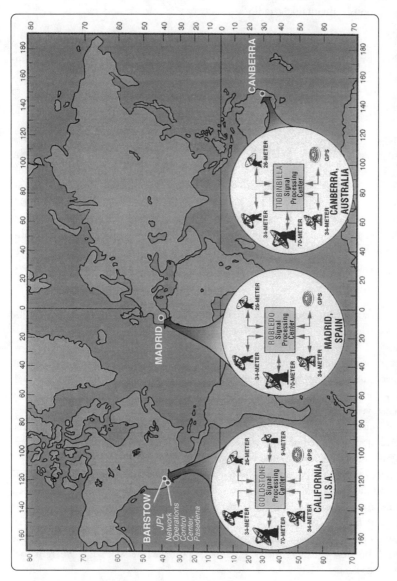

Figure Intro-9. An essential part of the answer; global map of the Deep Space Network, 1992. The three Deep Space Communication Complexes are located at intervals of 120 degrees of longitude around Earth, and are identified by the local name of the area and the generic name of the Complex. The antennas at each Complex are shown as they were in 1992; more have been added since then. For the sake of clarity, the NASA communications system linking each Complex to the Network Operations Control Center have been omitted.

xlv

are also key to the task of pointing the DSN antennas in the right direction to establish, and continuously maintain, a radio link between spacecraft and Earth.

The second function, telecommunications, makes use of appropriate modulation impressed on the uplink (by a tracking station), and on the downlink (by a spacecraft), to connect a spacecraft with its Earth-based controllers. Command instructions are sent to the spacecraft on the uplink, while engineering and science data are telemetered back to Earth, on the downlink.

A third function relates to the use of the DSN as a scientific instrument for research in radio and radar astronomy. Scientists were quick to recognize the potential of the state-of-the-art capabilities of the DSN as a powerful new resource for advancing their experiments. Encouraged by a generous NASA policy for making time observation available, the large antennas that could be pointed with great precision, sensitive receivers, and extremely stable timing systems of the DSN soon attracted the attention of many of the world's leading researchers. Using spacecraft radio signals, Extra-Galactic Radio Sources or echoes from powerful radar transmitters at Goldstone, scientists continue to use the DSN as a scientific instrument to widen our knowledge of distant regions of the universe.

What follows is an end-to-end scenario that brings all the pieces together to describe the deep space communications process for a typical planetary mission.

Soon after the launch vehicle was lost to view on the television screen, the first stage booster burned out and fell away, leaving the injection vehicle, with the spacecraft attached, to coast for a short time until it reached the exact position and time for the trajectory injection maneuver to begin. At about that same time, the injection vehicle had moved far enough around Earth to be in radio view for the DSN tracking stations in Australia. The Canberra stations tuned their receivers to the spacecraft's auxiliary oscillator frequency, making allowance for the one-way Doppler effect, and locked-on their receivers to automatically follow the spacecraft. The telemetry data on the one-way downlink carried engineering data that confirmed that the spacecraft survived the stresses of the launch events. At the appointed time, the injection vehicle engine was fired to accelerate the spacecraft free of Earth gravity and on to its interplanetary trajectory. The small initial acquisition antenna at Canberra had been following the spacecraft and now steered the more powerful, but less agile, 34-meter antenna to point at the spacecraft where it too, locked up its receivers.

Introduction

Telemetry data, demodulated and decoded at the tracking station, continued to flow to mission controllers at JPL, but they needed to send commands to the spacecraft immediately to orient it to the Sun and to reconfigure it for its long cruise to the planet. Besides, spacecraft navigators were anxious to have the better, more-stable Doppler data that was associated with the two-way, coherent mode. The transmitter at the tracking station was turned on, and making allowance for the uplink Doppler offset, its ultra-stable S-band (or X-band) uplink was tuned across the frequency of the spacecraft receiver. On detecting the uplink, the spacecraft automatically switched over to the coherent mode of operation and began retransmitting the uplink back to the tracking station. The DSN receivers were then in the two-way, coherent mode, the two-way Doppler was reaching the navigators, and commands could be transmitted at any time as required.

As the spacecraft continued to move away from Earth on its new trajectory, Earth rotation brought the other DSN Complexes into view, and the stations there passed the spacecraft in turn, from one to the other. This involved a complicated operational procedure to avoid losing telemetry data, but the net result was that communication with the spacecraft continued, uninterrupted by Earth's rotation, for the approximately two remaining years of the mission.

During the long cruise period, the Navigation Team used two-way Doppler and ranging data generated at the tracking stations to verify, and when necessary adjust, the trajectory of the spacecraft to ensure that it reached the planet at the correct position and time for a successful encounter. Because the distance between the spacecraft and Earth had been steadily increasing, the power level of the carrier signal received by the 34-meter antennas gradually dropped to the point where the signal-to-noise ratio was too low for the DSN receivers to operate properly. At that point, the tracking stations changed over to the big 70-m antennas for tracking this spacecraft. The 70-m antennas allowed them to receive telemetry data at a higher data rate, a capability that would be needed at the greater range of the forthcoming planetary encounter.

Encounter day finally arrived. All of the complex sequences of operation that the spacecraft would carry out during encounter had been transmitted to the spacecraft from the DSN and were stored in its computer memories. For some time, the spacecraft cameras and other scientific instruments had been trained on the approaching planet; engineering and the telemetry data rate had been increased to the maximum possible. All the terrestrial communications circuits linking the three Deep Space Complexes to JPL were on high alert. The Deep Space Network was then doing what it does best: providing a real-time, two-way, coherent, digital data communications link between NASA's Jet

Propulsion Laboratory in Pasadena, California, and a planetary spacecraft encountering one of the Jovian planets.

Responding to its stored commands, the spacecraft executed the encounter sequence of events according to a pre-determined plan that included all of the onboard science instruments. As the spacecraft entered occultation (was obscured by the planet), radio science experiments were carried out using the spacecraft USO. The immense amount of data collected in the short time of its actual encounter was stored by the spacecraft on its tape recorder for later playback via the downlink. Only the most important data could be sent back in real time, but this was sufficient to give anxious mission scientists their first glimpse of the planet, up close. Even after demodulating and decoding the downlink data stream that contained the images, the tracking stations could not view them. They had to be separated from the other science data and further processed by computers at JPL before they could be analyzed by the imaging team and presented for public viewing.

It was those pictures that the public first saw on television at the early NASA press conferences from JPL. Later, as the remainder of the encounter data were retrieved by the DSN, scientists conducted much more detailed analyses of their data and their findings were ultimately presented to the world at NASA press conferences in Washington. The DSN stations continued to track the spacecraft as it pursued its orbit around the Sun, returning science data at an ever decreasing rate and conducting radio science experiments as the opportunity arose. So long as the spacecraft continued to function properly, scientists continued to press for more data from deep space, and for the support of the Deep Space Network to obtain it.

The foregoing end-to-end scenario established the DSN as an essential part of the answer to the questions invoked by a search for a less obvious connection between the launch of a spacecraft and science data presented at a NASA press conference. Perhaps that answer begets further the question of how the DSN came into being and how it developed. With that question in mind, what a timely open to the history of the Deep Space Network.

CHAPTER 1

GENESIS: 1957–1961

GOLDSTONE, CALIFORNIA

A fine, wide, blacktop road runs past the salt-encrusted bed of dry lake toward the Mars site at Goldstone, deep in California's Mojave Desert. Opposite the lake bed off to the right, a disused minor road leads through a narrow cleft in the desiccated, brown and brick red-colored landscape to another dry lake bed, surrounded by low, barren hills and also encrusted with dried salt. There is nothing else here except for a small brick building and an unusual, spidery, three-legged structure supporting a large, perforated metal dish. The open side of the dish faces the sky, and high above its center, a smaller dish is suspended at the apex of four tall struts protruding from the surface of the large dish. Although the immensity of the totally barren landscape dwarfs the whole structure, it is, in fact, well over thirty meters to its highest point. A chain link fence encloses the area, and a bronze tablet on the nearby stone marker informs the occasional visitor that this is the site of the NASA Pioneer Deep Space Station. The inscription on the plaque, overlaid on the photograph of the Pioneer site shown in Figure 1-1, identifies the station's association with the NASA space program.

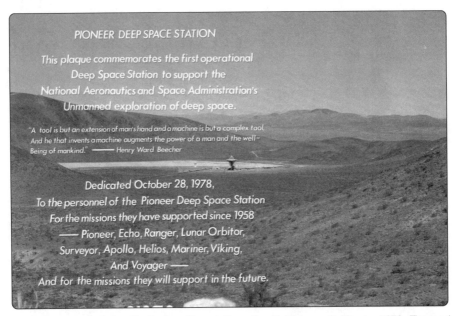

Figure 1-1. View of the Pioneer tracking station site, Goldstone, California, 1978. The road entrance to the site is on the far side of the picture, while the antenna, seemingly diminished by the immensity of surrounding landscape, stands in the center near the edge of the lake bed.

Figure 1-2. Prominent personalities of the Deep Space Network unveiling the commemorative plaque at the Goldstone Pioneer Site, 28 October 1978. Eberhardt Rechtin, principal founder of the Deep Space Network, is on the right. William H. Bayley, the original General Manager of the Network, is on the left. Richard K. Mallis, DSIF Operations Manager, is standing next to Bayley. Charles Koscielski, Director of the Goldstone Complex at the time of the dedication ceremony, is next to Rechtin.

Several of the people who played a prominent part in the establishment of the original Network attended the dedication ceremony. They appear in the unveiling picture shown in Figure 1-2. Their role in the chronicle of the Deep Space Network (DSN) will become apparent in later chapters.

This antenna, built in 1958, known through most of its working life as Deep Space Station 11 (DSS 11), was the first antenna in what eventually became the NASA Deep Space Network. As the inscription of the plaque implies, it supported all of the NASA deep space missions during the twenty years of its operational life.

At deactivation in 1981, Station Director Lou Butcher sent a message to his colleagues throughout the Network which neatly summarized the Station's long association with the NASA space program: "The closing of the Pioneer (DSS 11) station marks the end of 23 years of spaceflight tracking. DSS 11 was built when the Jet Propulsion Laboratory

(JPL) was under contract to the Army and was the first 26-meter antenna to be built in the DSN. DSS 11 was the leader for all DSN stations to follow in tracking the first JPL space mission (*Pioneer 3*) in December 1958. It was the station that tracked *Surveyor 1* onto the Lunar surface, and the station that tracked the lunar module (*Apollo 11*) onto the Lunar surface. It has played a major role in most of the NASA Lunar and Planetary spaceflight achievements." The United States Department of the Interior designated the Pioneer site as a National Historic Landmark on 27 December 1985. In a very literal sense, this is where the history of the Deep Space Network began.

The International Geophysical Year (IGY)[1], 1957, a cooperative international enterprise to advance the state of scientific knowledge about Earth and its environment, commenced in July and was running its course. In October, the Russians launched the world's first Earth-orbiting satellite, called *Sputnik*, as part of their contribution to the IGY activities. About a month later, Soviet scientists launched a larger version of *Sputnik* containing a live dog, named Laika, into a high elliptical orbit around the Earth. Both in and out of government, scientific, and engineering circles, the effect of the *Sputnik* launch was one of complete surprise, followed by damaged pride and embarrassment. From a military viewpoint, the implications were obvious: the Soviets demonstrated an operational, intercontinental ballistic missile (ICBM) capability ahead of the United States. While to the scientific world at large, Soviet science was clearly leading the "race to space," the hopes of U.S. scientists were pinned on Project Vanguard, a U.S. Navy project to place a grapefruit-sized scientific package in Earth orbit. In full view of a national television audience, the first Vanguard launch at Cape Canaveral in December 1951 was a complete failure, enhancing even more the prestige of the Soviet accomplishment with Sputnik.

The U.S. government was engaged in a desperate "catch-up" game with its Soviet counterpart in the newly emerging field of space research.

The events that followed the December Vanguard debacle are the stuff of which the history of JPL is made, but to properly appreciate what happened next, one needs to make a short backtrack to 1955.

As a consequence of a lengthy experience with the U.S. Army as a contractor on various guided missile development programs at the White Sands Missile Range, JPL had already developed a great deal of expertise in the techniques and technology of guidance and tracking systems for large rockets and ICBMs. While working for the Army as long ago as 1955, JPL, in competition with the Navy and the Air Force, submitted a proposal for an Earth-orbiting satellite for the U.S. contribution to the IGY. JPL's satellite was to be called Orbiter. The Department of Defense (DOD), however, selected the Navy's proposal called Vanguard, and JPL's Orbiter pro-

GENESIS: 1957–1961

posal was shelved. Perhaps it was serendipity, but the U.S. Army encouraged JPL to continue with development of Orbiter, at a low level, for the next couple of years. High-speed upper stages for the booster rockets were developed and guidance and telemetry systems were designed and tested. Of particular significance to the then-distant future of a deep space tracking network was the tracking system designed by JPL to track the high altitude test rockets. It was called Microlock, and working in conjunction with a minimum weight radio transmitter carried by the flight test unit, could provide telemetry and positional data to a range of several thousand kilometers. A phase-lock tracking receiver, a key feature developed at JPL by Eberhardt Rechtin and Richard Jaffe some years earlier, was operational at each of the Microlock stations.

Toward the end of 1957, perhaps sensing that the Vanguard program was in trouble, Secretary of Defense Neil H. McElroy authorized the Army to reactivate the Orbiter program and proceed with all deliberate speed to an earliest possible launch. The Army provided a launch vehicle based on an existing Redstone rocket design and JPL provided a suitable satellite carrying a small scientific payload, including a radiation-measuring instrument designed by Dr. James A. Van Allen of the University of Iowa. The satellite was renamed Explorer and the Army-JPL team was committed to launch in 90 days.

To track the new satellite and to receive its downlink telemetry signals, JPL expanded its existing Microlock ground-based tracking facilities to include stations at Cape Canaveral; Singapore; Nigeria; and San Diego, California. Primitive communications services to the overseas sites sometimes relied on native "runners" to transport messages and tapes to the nearest telegraph office. Although the U.S. stations had interferometric tracking antennas, a single, helical antenna at each overseas site provided only telemetry and Doppler data. Primitive though it was, JPL had acquired its first taste of worldwide network development and operations.[2]

On 31 January 1958, less than 60 days after the Vanguard explosion on the launch pad, and just 84 days after receiving approval from Secretary McElroy, *Explorer I* lifted off the launch pad at Cape Canaveral atop a Juno I launch vehicle and, to the acclaim of a national television audience, became America's first Earth-orbiting satellite. Telemetry data received by the Microlock ground stations from the Van Allen Geiger-counters revealed the presence of a high altitude band of radiation encircling Earth. Eventually named for Van Allen, this became one of the most important discoveries of the IGY.

Suddenly, the "catch-up" roles of the previous year were reversed and JPL never looked back. Henceforth, attention was focused not on the near-Earth region but outward, toward deep space, and its chief advocate was the ambitious and far-sighted director of JPL, William H. Pickering.

Knight Commander of the British Empire, conferred by England's Queen Elizabeth II in 1975, is among the many honors and citations bestowed upon William Pickering by prestigious scientific and technical organizations throughout the world. Knighthood not only recognized his scientific achievements, but also symbolized his British heritage and association with his native country of New Zealand. Pickering received his formal high school education in Wellington, the capital city of New Zealand, prior to embarking on a career in electrical engineering at California Institute of Technology (Caltech) in Pasadena, California, in 1929. Seven years later, he had earned a Ph.D. in physics and an appointment to the Caltech faculty, where, in addition to teaching, he was engaged in cosmic ray research with Robert A. Milikan and H. Victor Neher.

During World War II, Pickering organized and taught electronics courses for military personnel at Caltech, and became acquainted with the Radiation Laboratory at MIT and its director, Lee A. Dubridge. By the time Dubridge was named to the presidency of Caltech in 1946, Pickering was working at Caltech's Jet Propulsion Laboratory on the design and development of telemetering systems for rocket research vehicles. Pickering was appointed director of JPL in 1954 and immediately began moving the laboratory toward the forefront of applied engineering research and development. Within a few short years, the results of this move became apparent to the world-at-large in a spectacular way. The *Explorer I* Earth-orbiting satellite, the United States' initial response to the Soviets' *Sputnik*, made its appearance in January 1958, thanks to the combined efforts of the teams led by Pickering (satellite), von Braun (launch rocket), and Van Allen (radiation measuring experiment).

Following this spectacular success, historian Cargill Hall described Pickering as, "spare, intense, reserved, and in a quiet way, implacable . . . determined to mount a JPL program of lunar flights. To his mind," he said, "lunar flights might once have been a subject fit only for science fiction, but now they were on the reachable frontier of engineering science, exactly the frontier where Pickering wanted JPL to be." Prophetic words indeed. As director of what soon became a NASA Field Center, Pickering led JPL through the highs and lows of the first U.S. spaceflights to the Moon, and later, to Venus and Mars. Under his leadership, NASA/JPL, part of which included the Deep Space Network, quickly established and maintained a position of pre-eminence in the emerging, new art of science and technology for deep space planetary exploration.[3]

He retired from JPL in 1976 and served for two years as director of the research institute of the University of Petroleum and Minerals in Saudi Arabia. He later returned to Pasadena to enter private consulting practice.

GENESIS: 1957–1961

THE PIONEER LUNAR PROBES

On 7 February 1958, responding to President Eisenhower's initiative for a United States Space Program, the Department of Defense established an organization called the Advanced Research Projects Agency (ARPA) to promote, coordinate, and manage all existing military and civilian space activities. ARPA was to function only as an interim organization pending congressional establishment of a civilian space management agency.[4]

Toward the end of March 1958, the Secretary of Defense announced approval of a lunar space program, to be directed by ARPA as part of the United States' participation in the International Geophysical Year (IGY) activities. The program was named Pioneer.

The Pioneer lunar program involved three launches by the U.S. Air Force, using an existing Thor-Able booster rocket combination, and two U.S. Army launches utilizing the new Jupiter Intermediate Range Ballistic Missile (IRBM), designated as Juno II. The scientific objectives of the Pioneer program were to measure cosmic radiation in the region between Earth and the Moon, and to make a more accurate determination of the mass of the Moon. At the same time, valuable experience in the design of a lunar probe trajectory and a tracking and telecommunications system could be obtained. Because of its long prior association with the Army's missile development programs at White Sands Missile Range, JPL was to be involved in the two Army launches, *Pioneers 3* and *4*. The target launch dates were set for 11 November and 14 December 1958.

Research into the tracking and communications systems required to support a space program of lunar and planetary spacecraft had been in progress at JPL for some time.

Confronted with the problem of tracking and communicating with the two Pioneer probes, JPL put that fund of knowledge to good account.

At that time, JPL was fortunate to have, as the head of its small missile guidance research division, a visionary whose advocacy of the long term approach determined the course of U.S. solar system exploration far into the future.

Tall, imposing in appearance, of quiet disposition and authoritative manner, Eberhardt Rechtin was the epitome of the top-level executive. Articulate, with a brilliant engineering mind, he seemed to be way ahead of other the speaker in a conversation, or the presenter at a conference. He was patient with those less acute in their thinking, and courteous in pointing out errors of fact or judgement. A dark suit and tie completed his picture. He was also, not surprisingly, an accomplished violinist.

Eberhardt Rechtin brought a B.Sc. degree in electrical engineering (1946) from the California Institute of Technology with him when he joined JPL in 1946 as a member of the very talented communications and radio guidance team on the Corporal and Sergeant programs. He completed a Ph.D. in electrical engineering at Caltech in 1950 and along with Richard Jaffe, went on to study the theory and design of "phase-lock" circuits. With Walt Victor's help, the basis of the "phase-lock" receivers became an integral part of the DSN.[5]

As the first head of the Tracking and Data Acquisition Office at JPL, it was Rechtin's vision of a worldwide network of tracking stations, and his considerable powers of persuasion and tenacity in pursuing his ideas with NASA, that gave impetus and direction to the DSN in the early formative years. It was his endorsement that gave substance to the far-reaching proposals of Victor, Stevens, and Merrick for the new 64-meter antennas that would be much larger and perform better than any existing antennas designed for tracking distant spacecraft. He introduced the concept of the "sum of the negative tolerances" as the standard criterion for the margin of safety in the design of uplinks and downlinks for planetary spacecraft. To control the electrical interfaces (uplink and downlink) between the DSN and the many different JPL and non-JPL designed spacecraft that he perceived would require tracking support from the DSN in the years ahead, he mandated fixed, definable specification for the DSN radio links. This would become a permanent and vital feature of DSN management in the years ahead.

Father of the phase-lock loop and architect of the DSN were his legacies. Rechtin retired from JPL in 1967 to head the Advanced Research Projects Agency of the U.S. Department of Defense.

It was the perception of Eberhardt Rechtin that solar system exploration should continue to evolve in coming years and eventually become a major part of the overall NASA space program. He envisioned not only a lunar program which included soft landings on the Moon's surface, but also a program that involved a photographic survey of Venus and Mars.

A long-range vision such as his required not one, but three antennas spaced at 120 degrees of longitude around Earth so that they could maintain continuous contact with the spacecraft as Earth rotated about its axis. Although time and funding precluded the full-scale plan, for the immediate task of tracking the Pioneer lunar probes, one antenna sufficed. It did not allow for continuous contact with the probe, but the time of launch and trajectory could be arranged so that the actual flyby occurred when the probe was in view of the single antenna site. Later, assuming approved appropriate funding, the additional antennas could be added to create the full network and give the U.S. the capability to communicate with future spacecraft travelling to the edge of the solar system.

Genesis: 1957–1961

To lend credence to his ideas, Rechtin gathered experts in all the fields of technology necessary to accept a challenge of those dimensions.

Victor, Stevens, Merrick, and Bell, each an expert in his own field, had the knowledge and drive to lead the way in developing and building the super-sensitive receivers, telemetry and guidance systems, high power transmitters and large precision antenna structures, all needed on a very short time scale.

Immediate decisions were needed on basic communications parameters, such as operating frequency, antenna gain, antenna diameter, beamwidth, angular pointing accuracy, slew rate, receiver sensitivity, Doppler tracking rates, bandwidth, transmitter power, and signal-to-noise ratios. Similar parameters were needed for the probe radio system so that the probe and Earth communications systems continued to work together as the distance between them stretched as far as the Moon and beyond.

The team had already determined that the existing ground-based tracking systems, which were being used by JPL for the Earth-orbiting Explorer flights in progress at that time, were not suitable for tracking lunar probes. A more efficient, high-performance, Earth-based antenna was needed. After considerations of antenna gain, automatic tracking performance, minimum galactic radio noise and the state-of-the-art radio communications at that time, a frequency of 960 MHz was chosen as near-optimum for the communications link between the lunar probe and Earth. The Pioneer radio transmitter transmitted at 960 MHz and the Earth-based tracking station received on that frequency. A large-diameter, steerable, parabolic dish antenna capable of operating at 960 MHz was needed to receive the transmissions from the Pioneer probes and be available for the support of possible follow-on missions, perhaps even for missions beyond the Moon to the nearest planets. The launch dates were set for yearend. Cost and construction time were vital factors in the search for a suitable antenna. While incorporating advanced and reliable principles of design, the antenna was to be built and operating in approximately six months.

The choice of 960 MHz as the operating frequency for the Pioneer probe brought with it another most significant advantage. It allowed JPL to make use of an existing design for a large radio astronomy antenna developed some years earlier for the burgeoning science of radio astronomy by the Naval Research Laboratory, with assistance from the Carnegie Institute, Associated Universities, and the Blaw-Knox Company. As a result of this prior work, the Blaw-Knox Company could provide a large parabolic antenna that would meet the fixed price and construction time constraint of six months. The antenna, 26 meters (85 feet) in diameter, was designed for radio astronomy applications, and required substantial modifications to the antenna drive system to make it suitable for precision tracking of lunar space probes.

Confident in JPL's ability to make the necessary changes and get the antenna built and operating in time for the Pioneer 3 launch, ARPA decided to purchase not one, but three, of the 26-meter antennas. One of the antennas would be used for the imminent Pioneer launches, while the other two were intended for ARPA longer range plans for what it termed its World Net. It was a calculated risk on the part of ARPA, for even though the design was complete, none of the antennas had yet been built. With the pressure of the Pioneer launch schedule, however, there was little choice in the matter.

With the operating frequency and antenna decisions already made, attention turned to a location for the antenna site. To avoid contaminating or obscuring the very weak radio signals received from the distant spacecraft with artificial radio interference, a location remote from any metropolitan area was desired. At the same time it needed to be close enough to an established community to be practical for the staff that would be required to operate the equipment. The combined constraints of funding and schedule strongly influenced a search for a suitable site on Government-owned property. A convenient location that met all of these criteria was found in a natural bowl-shaped area surrounded by low hills near the Goldstone Dry Lake on the Fort Irwin military reservation, 72 kilometers northwest of the city of Barstow, in the Mojave Desert, California.

In March 1958, a JPL radio interference survey team certified the area as free from radio interference. A month later a construction company began work on access roads, facilities, services, and buildings. While work was in progress at Goldstone, a mobile tracking station, using a three-meter diameter antenna, was erected near Mayaguez, Puerto Rico to obtain initial trajectory data downrange from the Florida launch site.

In early June, the steel components for the antenna started to arrive onsite at Goldstone. Assembly of the antenna commenced in mid-August and was completed in November 1958. By that time, all the receiving, recording, communications, servo, and antenna drive modifications and other electronics had been installed and tested. After a short period of crew training, the station was ready for its first operational mission, but there had been no time to spare. A photograph of the first Goldstone antenna, as it looked shortly after beginning operation in 1958, is shown in Figure 1-3.

While the antenna construction work was proceeding on an accelerated schedule at Goldstone, the direction of the space program was changing rapidly in Washington. President Eisenhower's Executive Order No. 10783 officially established the National

Figure 1-3. The 26-meter antenna and tracking station (DSIF 11) at Pioneer site, Goldstone, California, 1958. Pioneer, the first Goldstone antenna as it looked shortly after beginning operation in 1958. The antenna was deactivated in 1981 and has been designated a National Historic Landmark by the U.S. Department of the Interior.

Aeronautics and Space Administration (NASA) on 1 October 1958. Discussions regarding the status of the Jet Propulsion Laboratory and its possible acquisition by NASA as a component Field Center quickly followed.

NASA outlined its proposal for the transfer of JPL from the Army on 15 October 1958. A few weeks later, the Department of Defense indicated acceptance of the proposal and on 3 December 1958, Presidential Order No. 10793 transferred the functions and facilities of the Jet Propulsion Laboratory of the California Institute of Technology from the Department of the Army to NASA.

NASA launched *Pioneer 3* on a trajectory toward the Moon three days later, on 6 December 1958. Due to a rocket booster problem, the probe only reached an altitude of 63,500 miles before falling back to Earth, somewhere in central Africa. The small antenna in Puerto Rico tracked it for 14 hours (round-trip) and was able to maintain telemetry contact to at least 60,000 kilometers. The much larger Goldstone antenna was able to acquire telemetry for the entire time that the *Pioneer 3* probe was above its horizon. The telemetry received from the Puerto Rico station proved to have significant scientific value because it contained the data from two passes through the then-mysterious Van Allen radiation belts.

Although the *Pioneer 3* flight was a disappointment, three months later, the next lunar probe, *Pioneer 4*, became the first United States spacecraft to leave Earth's gravitational field. Launched on 3 March 1959, on a trajectory similar to that of *Pioneer 3*, the probe traveled 435,000 miles toward destination before its batteries became depleted and its transmissions to Earth ceased. After 41 hours of flight, *Pioneer 4* passed 37,300 miles from the Moon's surface on a flight path that was a good deal ahead of and below the planned trajectory. Although one of the key experiments to observe the hidden side of the Moon was defeated by the perturbed trajectory, data from the onboard radiation counters identified a third belt of radiation and sensed significant changes in the intensity of the Van Allen radiation belts discovered by *Pioneer 3*. With its large antenna working well, the Goldstone station recorded over 24 hours of telemetry data from three successive 10-hour passes over the site, before the signal ceased during the fourth pass on 6 March 1959.

While the *Pioneer 3* and *4* missions had not been entirely successful, the performance of the new antenna exceeded all expectations. It demonstrated the efficacy of its design and fully vindicated the confidence that ARPA had shown in JPL's expertise and ability to accomplish a challenging task. It would henceforth be known as the Goldstone Pioneer tracking station, identified as DSIF 11.

The Echo Balloon Experiment

Important and highly visible as they were, the *Pioneer 3* and *4* lunar probes were not the only projects that engaged the attention of Rechtin and his engineering team at JPL in the spring of 1959. In January, JPL had agreed to cooperate with NASA and Bell Telephone Laboratories (BTL) of Holmdel, New Jersey, in an experiment to test the feasibility of long-range communications between two distant points on the surface of the Earth by means of a reflected signal from the surface of a large orbiting balloon. Based on separate

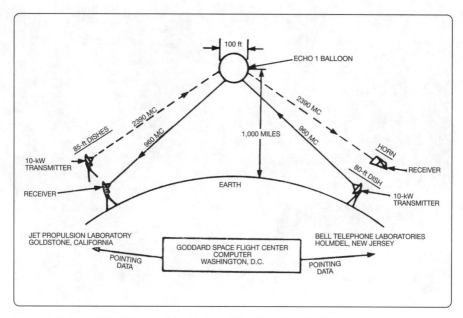

Figure 1-4. Essential features of the Project Echo Experiment, 1960.

and independent proposals from scientists at BTL, who were interested in studying the use of "passive repeaters" for radio communications, and scientists at the Langley Aeronautical Laboratory in Hampton, Virginia, who were interested in air density measurements in Earth's upper atmosphere, this experiment, named Project Echo, was endorsed several years earlier by the United States IGY Committee. Scientists had agreed on a design for an experiment to satisfy both scientific objectives. A radio signal would be transmitted from the east coast of the United States, reflected off the metalized surface of a large (100-foot diameter) Earth-orbiting balloon, and received on the west coast. A similar link, using a much higher frequency, would be set up in the west-to-east direction. The plan called for setting up separate links at 960 MHz (east to west) and 2,390 MHz (west to east). The essential features of the experiment are shown in Figure 1-4.

In the fall of 1958, while considering how the Echo balloon experiment might be implemented, it became apparent to engineers at JPL that the new antenna, then under construction at Goldstone, could serve as the west coast receiving terminal for the 960-MHz, east-to-west leg of the experiment. For its part, BTL would provide a 960-MHz transmitting station and a 2,390-MHz receiving station at its east coast facilities in New Jersey. However, one major problem remained unsolved. There was no transmitter for

Figure 1-5. Az-El transmitting antenna for the Echo Balloon Experiment, Goldstone, 1960.
The antenna was constructed between July and December 1959, and the complete station became operational in July 1960.

the 2,390-MHz for the west-to-east leg of the experiment. The new Goldstone antenna was designed as a receiving station only. It had no transmit capability.

To solve that problem, a second 85-foot antenna was built at Goldstone in 1959 using funds provided by the NASA Communications Satellite program. To avoid mutual radio interference between the two antennas, a site some miles away but similar to the Pioneer site was selected. A low range of hills separated the two sites and provided the necessary level of radio isolation between them. JPL named the site "Echo" for its obvious association with the Echo experiment.

An early photograph of the transmitting antenna at the Goldstone Echo site is shown in Figure 1-5.

Unlike the Pioneer antenna, which used a polar or hour angle-declination (HA-Dec) drive system, the Echo antenna used a high-speed, azimuth-elevation (Az-El) drive which was more suitable for tracking Earth satellites, such as Echo, than the more astronomy-oriented polar drive used for deep space probes. The Echo station also included a 10-kilowatt transmitter operating at 2,390 MHz, which at the time represented a very advanced, state-of-the-art development. In addition to supporting the Echo balloon experiment, the advanced features of the Echo station allowed JPL engineers to evaluate the relative merits of large antenna drive systems and the performance of high-power transmitters and low-noise receivers, operating at S-band, for future applications in deep space tracking stations.

Less than a month after the full station was completed, on 12 August 1960, *Echo 1* was launched. Within two hours of launch, a recording of President Eisenhower's voice was transmitted from the Echo site at Goldstone to BTL in New Jersey. Echo had scored a "first" in long distance passive reflector communications. Ironically, this distinction was almost immediately eclipsed by the informal Goldstone to Woomera "moon-bounce" experiment described later in this chapter. Experiments in long distance communications and upper atmospheric air density soon followed.

At the conclusion of the Echo experiment, ownership of the Echo antenna was transferred to the new NASA Office of Tracking and Data Acquisition and became an official part of the Deep Space Instrumentation Facility. Its ultimate destiny, however, lay in a different direction from that of Pioneer, the other DSIF antenna at Goldstone. The first step along this path began with the Venus Radar Experiment in March 1961.

The Venus Radar Experiment

The possibility of detecting an Earth-based radio signal reflected from the surface of Venus had intrigued radio astronomers through most of the 1950s. In fact, investigators at Lincoln Laboratory, Massachusetts Institute of Technology, and at Jodrell Bank Experimental Station, University of Manchester, had attempted to detect echoes from Venus in 1958 and 1959 without success.

A successful experiment of this kind could offer great potential for a plethora of new scientific knowledge about Venus and its physical properties, and the possibility, for example, to determine the reflective properties of the surface for electromagnetic radiation, the orientation of the planet's axis of rotation, and its rate of rotation. Also, much could be learned about the characteristics of the interplanetary medium along the transmission path between Earth and Venus. Most importantly though, it could offer the possibility to independently determine a value for the astronomical unit (AU) with much greater accuracy than had been previously possible.

The astronomical unit (AU), used by astronomers as a measure of the mean distance of Earth from the Sun, is a vital parameter in the calculation of the ephemerides, or the paths of planets around the solar system. It was therefore of immense significance to JPL's long-range plans for sending spacecraft on missions to close encounters with distant planets. JPL engineers viewed the Venus radar experiment as a technological challenge for the type of equipment required for future missions to the planets and for personnel to operate it. Without the very highest level of performance in both areas, the hope for success would be minimal, and would establish a standard against which the performance of future operational DSIF stations could be judged.

During the experiment, a 13-kilowatt, continuous wave, S-band signal was transmitted from the new Echo antenna toward the planet, while the reflected signal was received by the Pioneer antenna fitted with a Maser receiver, specially designed for reception of very weak signals. The first undisputable radar returns from the planet Venus were obtained with the Goldstone radar on 10 March 1961. Using advanced, spectrum analysis techniques, the radar team, led by Richard Goldstein, measured the delay to, and changes in, the fundamental character of the returned signal due to its round-trip to Venus and back to Earth. These data were recorded for subsequent scientific analysis. During the time that Venus remained within 50 and 75 million miles from Earth, over 200 hours of good radar data were obtained.

After analysis, this impressive bank of scientific data yielded the much sought after "improved value for the AU" in addition to satisfying the several other scientific objectives of the experiment. The new value for the AU was determined to be 149,598,500 ± 500 km, a most significant improvement in the accuracy of previous values.[6] Subsequently, JPL estimated that *Mariner 2*, the first mission to Venus the following year, would have missed the planet by too great a distance to have been of any scientific value, had the old value of the AU been used in the trajectory computations.

The Venus radar experiment thus represented the first use of the DSN (DSIF) as a scientific instrument. Its immediate success marked the appearance of a significant astronomical research tool at Goldstone, whose stature in the scientific community would continue to rise as the years went by. The technical details of the Venus radar experiment, and the history of what evolved into the Goldstone Solar System Radar, is described in chapter 8, "The Deep Space Network as a Scientific Instrument."

Research and Development Antenna

By the time the Venus radar experiments ended in May 1961, it had become apparent that JPL communications engineers needed a dedicated antenna research and development facility at Goldstone, where new components used in the network could be tested, measured, and evaluated. With the pressing needs of future missions in mind, and the Ranger and Mariner missions about to start, it was simply not possible to devote one of the two existing Goldstone antennas (Pioneer or Echo) solely to research work, despite the vital importance to the technological future of the network. Predictions of future mission requirements mandated a need for two operational antennas, not only at Goldstone, but also at the other two longitudes.

Furthermore, until a way could be found to simultaneously transmit and receive on a single antenna without severely degrading the downlink signals, two antennas, one for transmitting and the other for receiving, would be required for deep space communications.

On this basis, the case was made for yet another antenna at Goldstone. The new antenna, which would be of the same type as the Pioneer and two overseas antennas, would be erected at the Echo site where it could make use of the existing support facilities. When that was completed, the original Echo Az-El antenna would be moved to another site a few miles away, and established as the basis for a new facility dedicated to deep space communications research and development (R&D).

The plan was rapidly approved by NASA and immediately put into effect at Goldstone. Between February and May 1992, a new HA-Dec antenna was erected near the Az-El antenna at the Echo site. It was put into service two months later, just in time to support the first Mariner missions to Venus.

With a second operational deep space station now in place at Goldstone, the Az-El antenna could be released from network support and prepared for the complex task of relocation.[7]

In June 1962, the entire Az-El antenna—mounted on its pedestal—was jacked up, loaded on a supporting undercarriage, and moved approximately four miles across the desert to a new location that became known as the Venus site, after the now famous Venus radar experiment. There, the antenna began a brilliant new life as a test facility for communications research and development, and over the next 40 years, proved vital to the continued preservation of JPL pre-eminence in the field of deep space communications.

The research work that was conducted at that facility, known through the DSN R&D Station 13, is described in chapter 7, "The Advance of Technology in the Deep Space Network."

WASHINGTON, DC

Before proceeding further with the history of the DSN, it will be instructive to review the events occurring simultaneously in a related but quite different arena from that just described. The focus of that activity was in Washington, DC.

Before NASA was created, ARPA had planned to develop a worldwide network of tracking stations to support its forthcoming programs for the exploration of deep space. The network was to be called Tracking and Communications Extraterrestrial Network (TRACE). Starting with the first three TRACE stations—Goldstone, Puerto Rico, and another small station at Cape Canaveral—ARPA planned to expand TRACE into a full-scale worldwide network by adding two additional overseas stations, similar to the one it had funded for Goldstone. This expansion was the rationale for the ARPA decision to purchase three of the 85-foot-diameter antennas from Blaw-Knox in 1957.

Plans for the worldwide network were well under way in early 1958. ARPA had already approved a JPL proposal for a three-station network which would provide optimum coverage for tracking deep space probes. One of the sites would be sites at Goldstone, California. The two overseas stations were to be located at Luzon, the Philippines, and in Nigeria.[8]

Taking a broader view, however, the Department of Defense expressed some reservations about the proposed sites in terms of their utility to all U.S. space vehicles, rather than the specific, deep space probes considered by JPL. With this in mind, ARPA asked JPL to reconsider its proposal for the station sites with the object of improving coverage for a broader range of space vehicle trajectories. The resulting study showed that better orbital coverage for all planned U.S. space missions would be possible if the Nigerian site was moved north to southern Portugal or Spain, and the Philippines site moved south to central Australia. The coverage of orbits with an inclination of 34 degrees to 51 degrees would be much improved. Future, piloted, Earth-orbiting flights were very much in the minds of the planners at the time.[9] In retrospect, their foresight was well justified since these locations effectively served the needs of the major tracking stations of the Deep Space Network for over forty years.

In the turmoil of space-related activity in that tumultuous year, 1958, ARPA never did get the opportunity to deploy the overseas stations of its World Net. NASA was created in October of that year, JPL was transferred to NASA in December, Pioneer 3 was launched, and by the opening of the new year, 1959, it was NASA and not ARPA setting the course for JPL and the future shape of the Deep Space Network. It was NASA that ultimately acquired and erected the two remaining antennas procured by ARPA. To the nation's space program, and to the work at JPL, the formation of NASA was transparent. It was mainly a transfer of authority and funding for the existing programs. Instead of working under an Army Contract, JPL now worked under a NASA contract. NASA inherited JPL's experienced personnel its facilities, including the new antenna at Goldstone and, perhaps most important of all, the JPL vision of a worldwide network of tracking stations for deep space probes. ARPA, representing military interests in space applications, pressed forward with its own programs under the direction of DOD.

As soon as NASA began to make its long-range plans for a civilian space program known, the need for completion of what ARPA had called the World Net immediately became obvious. One station already existed at Goldstone and sites had been proposed for two overseas stations. The antenna at Goldstone now belonged to NASA, but ARPA still owned the remaining two of the three originally ordered from Blaw-Knox. In January 1959, NASA and ARPA representatives decided each agency would get one antenna. The NASA antenna would be shipped to Australia as part of a three-station NASA operated network with sites in Goldstone (existing), South Africa, and Australia. ARPA would use its antenna as part of a deep space network for military purposes with stations in Japan and Spain. However, the military interest in deep space soon faded and NASA was eventually able to purchase the third antenna from ARPA to form the basis for its station in South Africa.

In moving toward the ultimate creation of the worldwide network, NASA directed its principal efforts in 1959 toward onsite surveys of the locations that had been proposed for Australia and South Africa and the diplomatic negotiations that were necessary to secure approval for their use.[10] In both countries, NASA was able to secure the interest of the government agencies—Australian Department of Supply (DOS)/Weapons Research Establishment (WRE) and the South African Council for Scientific and Industrial Research (CSIR)/National Institute of Telecommunications Research (NITR), respectively—in cooperating in the operation of the proposed deep space stations. Both WRE and NITR assisted the NASA/JPL survey team in identifying sites for the stations which met the three major criteria; a stable land area capable of supporting large antenna structures, surrounding hilly terrain to provide natural shielding against electrical interference and a surrounding area relatively free of radio interference in the frequency region useful for space communications.

WOOMERA, AUSTRALIA

The favored Australian site lay on the southern edge of the great inland desert region at a place called Woomera, about 350 kilometers north of the city of Adelaide, capital of the state of South Australia. Approximately 110 degrees west of the longitude of Goldstone, Woomera was already a missile and long-range rocket test center operated by WRE. The local language was English, and the nearby rocket and missile test activities would provide a pool of technical expertise and facilities. Furthermore, in the same area, the Australians had already installed, and were operating, a U.S. Navy minitrack station and a Smithsonian Baker-Nunn tracking camera as part of its participation in the IGY.

The signing of a "construction and operation" contract for the antenna around early April 1960 allowed both JPL and WRE to begin making major moves toward construction of a NASA deep space tracking station at Woomera. WRE initiated the road, buildings, power generation, and foundation work. JPL began shipping antenna components and the electronics for the station.

The antenna was finally built at a site known locally as "Island Lagoon," so named for the nearby dry lake which appeared to have an island at the center. Working under a JPL contract with supervisor Floyd W. Stoller, Blaw-Knox began assembling and erecting the antenna in May 1960. By August, the antenna was complete and an electronics team began installing the radio and tracking equipment, most of which had been supplied by Collins Radio Company. When NASA built the second and third 26-m antennas in Woomera, Australia, and Johannesburg, South Africa, the task of integrating the new antennas with their electronics equipment and bringing the two new stations into operation fell to Richard "Dick" Mallis.

Richard Mallis was an outgoing individual, easy to work with, sociable, and much respected by his colleagues. He was an excellent manager with good communications and technical skills and an appreciation for the different institutional environments at all three antenna locations. When he went to JPL in 1955 to work on radio guidance systems for the Army's Sergeant missile program, native Californian Richard K. Mallis took with him a degree in mechanical engineering from the University of Southern California and a Navy background. Caught up in the changes that swept JPL into the space program in 1958, he assisted with the construction of the first 26-m antenna at Goldstone, and later implemented the down-range tracking station in Puerto Rico to cover the launches of the Army's two Pioneer lunar probes.

Genesis: 1957–1961

Together with Goldstone, the three stations of the Deep Space Instrumentation Facility finally formed a worldwide network. With these completed in time to support JPL's first Ranger lunar missions, Mallis returned to JPL to take up a staff position in Renzetti's new Communications Engineering and Operations Section. He was responsible for Operations, regulating the way the Network carried out its day-to-day tracking functions. In this role, he set-up a Network-wide logistics and repair program, a frequency and timing standards program, a documentation system, and a training program for operations and maintenance personnel. This essential infrastructure remained the basis for all operations, maintenance, quality control, and configuration management processes as the worldwide Network expanded in size and capability through the years. He integrated the first commercial contractor, Bendix Field Engineering Corporation, into the DSIF as the operations and maintenance service provider for the Goldstone facility. In later years, as his responsibilities expanded to include the Space Flight Operations Facility at JPL in addition to the DSN, he became Manager of the Operations Division. Eventually he transferred elsewhere in JPL to further his professional career. However, because of his unique experience with service contract management, he was frequently called upon to assist the DSN in evaluating new contract proposals when existing service contracts expired. He retired in 1993 after 37 years of service at JPL and later took up residence in Australia.

In a final spectacular exercise on 3 November 1960, the Woomera station demonstrated its operational status by receiving voice and teletype messages transmitted from Goldstone via reflection from the Moon. The JPL onsite manager, Richard K. Mallis, departed Woomera four days later, after turning the new facility over to WRE for its future management and operation. The Australian engineers soon demonstrated their ability to handle the technical complexities of the new "space age" facility for which they had accepted operational responsibility. In a repeat of the "Moon bounce" experiment on 10 February 1961, during the official opening formalities, the station passed a congratulatory message from NASA Deputy Administrator Hugh Dryden in Washington to Australian Minister of Supply Alan Hume at Woomera over a "Moon bounce" communications link.

The photograph of the completed Woomera Tracking Station in Figure 1-6 shows the 26-meter-diameter antenna, the electronics equipment building, and service and facility structures. Island Lagoon is visible at the horizon to the left of the antenna structure.

The first overseas station of the Deep Space Instrumentation Facility (DSIF) was ready to enter operational service. Designated DSIF 41, Woomera, it would see eleven years of valuable service before being superseded by new stations at Tidbinbilla, near Canberra,

Figure 1-6. The 26-meter antenna and tracking station (DSIF 41), Woomera, Australia, 1961.

in southeast Australia. The Woomera station ceased operations on 22 December 1972, as part of a NASA program to consolidate overseas station facilities. Initial proposals to move the antenna to a new, more accessible location where it could be used for Australian radio astronomy purposes were not successful because the Department of Supply determined that the cost of transporting the antenna was excessive. Eventually, it was dismantled and sold for scrap despite the vigorous protests of several prominent members of the Australian scientific community.

Genesis: 1957–1961

JOHANNESBURG, SOUTH AFRICA

The requirements for continuous tracking coverage of deep space probes outlined in the JPL proposals for a worldwide network dictated that the third element in the network should be located in the longitude band which included Spain and South Africa. Although a site in Spain was preferred by NASA/JPL, the political complexities associated with existing international treaties with Spain caused NASA/JPL to look to South Africa for a suitable site. South Africa also offered many advantages. The South African government was anxious to participate with NASA in the new space venture, the language was familiar, and a great deal of technical expertise was already available in that country. Also, and most importantly, the flight path of all deep space probes launched from Cape Canaveral would pass over or near the station within an hour or so of launch. This meant that the downlink from the space probe could be first "acquired" close to the point at which the spacecraft would be injected into its planetary trajectory. During this period, mission controllers depended on telemetry and navigation data from the first acquiring tracking station to make critical decisions very early in the mission, when corrective action is necessary. Supplemented with the mobile tracking station that would be moved from Puerto Rico, an 85-foot antenna at this site would provide the DSIF with an excellent "initial acquisition" capability in addition to the normal tracking capability of the station.

The site that was chosen lay near the Hartebeestpoort Dam about 40 miles north of Johannesburg, on government-owned property intended for use as a radio research facility. As had been the case at Woomera, a NASA Minitrack station and a Baker-Nunn camera were already operating in the vicinity. An earlier site survey confirmed that the Hartebeestpoort site also satisfied the antenna site-selection criteria.

Although site surveys conducted in mid-1958 had confirmed the suitability of the site for the antenna, it was not until September 1960 that an agreement between the governments of the United States and the Union of South Africa permitted NASA to issue a contract to the South African Council for Scientific and Industrial Research for the construction, management, and operation of the station. NASA wanted the station to be completed by July 1961 in time to support the first launch of new Ranger missions to the Moon. Time was truly "of the essence." To prepare the site, erect the antenna, install and test the equipment, and train personnel in its operation, under the pressure of a high profile mission like Ranger, posed a formidable challenge for everyone involved.

Work began in earnest in mid-January 1961, as soon as the foundation for the antenna was ready. Antenna structural components travelling by ship, train, and truck from

Figure 1-7. The 26-meter antenna and tracking station (DSIF 51), Hartebeestpoort, South Africa, 1961.

Philadelphia had arrived a few days earlier. The erection team arrived from JPL a few days later. Erection of the antenna finally began on 16 January and was completed 69 days later on 25 March. The JPL team—Don Meyer, Dick McKee, and Howard Olsen—completed installation of the electronics in early June and spent the next few weeks conducting the necessary tests, checkouts, and calibrations. The station was ready for operation by 1 July 1961. Provided with teletype equipment and communication circuits by the South African Department of the Postmaster General, the new station was able to participate in the first operational readiness test for the Ranger missions later in July, as originally planned.

A photograph of the Johannesburg Tracking Station soon after its completion in 1961 is shown in Figure 1-7.

JPL and South African officials discussed the possibility of holding a public event to mark the formal opening of the station as had been done for the Woomera station. There was, however, some reluctance on the part of NASA Headquarters to publicly recognize a cooperative project with South Africa in light of the United States' attitude toward the political situation then emerging in that country. Eventually, the South African government held its own opening ceremony without formal NASA representation. Some years later, this issue would intensify and ultimately lead to the closure of the station as a part of the Deep Space Network.

Nevertheless, the new station, designated DSIF 51, Johannesburg, played a vital role in the early NASA missions to the Moon and planets. In that role, it served the DSN with distinction for 13 years until DSN operations at that site were terminated in June 1974. Soon after, NASA transferred the antenna and equipment to the South African National Institute for Telecommunications Research (NITR) where it was used successfully for at least another 25 years, as an instrument for radio astronomy research.[11]

The completion of the Johannesburg station marked the fulfillment of the world network initiatives proposed years earlier by JPL, carried forward under ARPA, and brought to fruition under NASA. Together with a small launch monitoring facility at Cape Canaveral, the three 85-foot antennas at Goldstone, Woomera, and Johannesburg became known as the Deep Space Instrumentation Facility (DSIF). The DSIF had become a separate facility of the NASA Office of Space Operations. It was managed, technically directed, and operated by the Jet Propulsion Laboratory (JPL) for the California Institute of Technology (Caltech), a prime contractor for NASA's Solar System Exploration program, in Pasadena, California. Within a few years, the DSIF would change its name to the Deep Space Network (DSN) and rapidly increase in size, complexity, and capability to a level unimaginable to its founders. But that lay in the distant future. In the immediate future lay the first operational challenge for the new network—the Ranger missions to the Moon and the first Mariner missions to Venus. It was to be a busy future. The Mariner Era was about to begin.

Endnotes

1. N. A. Renzetti et al., "A History of the Deep Space Network from Inception to 1 January 1969." JPL Technical Report TR (September 1971): 32–1533.

2. William R. Corliss, "A History of the Deep Space Network" (Washington, DC: National Aeronautics and Space Administration CR-151915, 1976).

3. R. Cargill Hall, *Lunar Impact: A History of Project Ranger* (Washington, DC: National Aeronautics and Space Administration, 1977).

4. Homer E. Newell, *Beyond the Atmosphere; Early Years of Space Science* (Washington, DC: National Aeronautics and Space Administration SP-4211, 1980).

5. E. Rechtin and R. Jaffe, "The Design and Performance of Phase-Lock Circuits," *IRE Transactions on Information Theory* (March 1955).

6. Andrew J. Butrica, *To See the Unseen, A History of Planetary Radar Astronomy* (Washington, DC: National Aeronautics and Space Administration SP-4218, 1996): 42.

7. Jet Propulsion Laboratory, "Az-El Move and Reinstallation," *JPL Space Programs Summary 37-13* (Vol. III, 1 October 1962): 9–13.

8. Jet Propulsion Laboratory, *Description of World Network for Radio Tracking of Space Vehicles* (Pasadena, CA: JPL Publication No. 135, 1 July 1958).

9. E. Rechtin, H. L. Richter, Jr., and W. K. Victor, "National Ground-Based Surveillance Complex." JPL Technical Memorandum TM 39-9 (15 December 1958).

10. Support for NASA facilities in Australia, South Africa, and Spain was provided under cooperative international agreements between the United States and those sovereign countries. The diplomatic, political, and administrative complexities of the negotiations that culminated in those agreements is beyond the scope of this history of the Deep Space Network. Readers interested in pursuing those details will find useful material in the JPL and NASA Archives.

Genesis: 1957–1961

11. A more detailed account of many of the topics discussed in this chapter is contained in the unfinished and unpublished notes on the early (prior to 1962) history of the Deep Space Network compiled by Craig Waff at JPL in 1993. The Waff notes are held in the JPL Archives.

CHAPTER 2

THE MARINER ERA: 1961–1974

THE MARINER ERA MISSION SET

The worldwide network, envisioned by Rechtin in 1958, became a reality between July 1959 and July 1961. In that hectic two-year period, funding was proposed and approved, international agreements were negotiated, suitable sites were found, equipment was procured and shipped overseas, roads and facilities were built, and JPL contractors erected antennas and installed equipment. The Deep Space Instrumentation Facility now consisted of a real network of 26-meter-diameter antennas located at intervals of approximately 120 degrees of longitude around the globe. Two such antennas were located at Goldstone, California, and one each at Johannesburg, South Africa, and Woomera, Australia. A world view of the DSIF as it existed at that time is shown in the figure below.

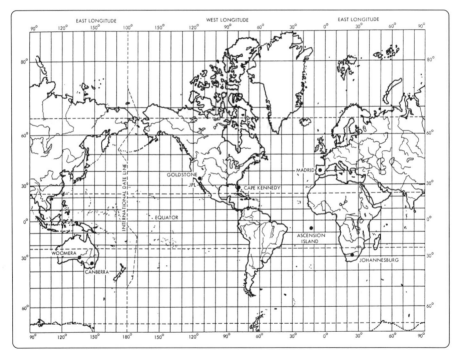

Figure 2-1. World view of the Deep Space Instrumentation Facility, 1961. Although the locations of future sites, near Canberra, Australia, and near Madrid, Spain, are also shown on this diagram, they had not been established at that time.

The Mariner Era: 1961–1974

Teletype and telephone circuits linked the stations to a rudimentary flight operations control center at JPL. In the 1961 photograph of the operations control center at JPL shown in Figure 2-2, the presence of rotary telephones, mechanical calculating machines, wall boards, and mechanical status displays is a poignant reminder of the pre-digital age.

At JPL and at each of the tracking stations, highly motivated crews of operators and engineers were trained to operate and maintain the complex, state-of-the-art equipment that JPL had designed and tested, but not yet demonstrated, to communicate with distant spacecraft.

At the same time, DSIF planners knew that in the years ahead, the network would be called upon to support much more complex missions to the Moon, missions to fly by and possibly even orbit Venus and Mars, and heliocentric probes to measure fields and particles and the solar wind in interplanetary space. At the extreme limit of possibility lay spacecraft encounters with Jupiter and Saturn.

Figure 2-2. Flight operations control center, JPL, 1961.

To appreciate the magnitude of the challenge facing the embryonic DSIF at that time, it is instructive to arrange this inventory of future missions into a mission set in the following way:

Deep Space Mission Set for the Mariner Era

Program	Type	Launches	Missions	First Launch	Last Launch	Span
Ranger	Lunar Photo	9	3	Aug 61	Mar 65	4 yrs
Mariner	Venus, Mars flyby	0	7	Jul 62	Nov 73	11 yrs
Pioneer	Interplanetary	4	4	Dec 65	Nov 68	3 yrs
	Jupiter	2	2	Mar 72	Apr 73	1 yr
Apollo	Piloted	2	7 Test	Nov 67	Dec 72	5 yrs
	Lunar	14	6 Landers			
Surveyor	Lunar Lander	7	5	May 66	Jan 68	2 yrs
Lunar Orbiter	Lunar Orbiter	5	5	Aug 66	Aug 67	1 yr

As this table makes clear, all launches did not necessarily result in successful missions. Failures were a common feature of the early programs but the failure rate greatly improved as time went on. The Ranger program is an example. The launch span, the time between the first and last launches of individual programs, is an indicator of the program's impact on the DSIF. Many launches in a short span affected the level of tracking activity for the DSIF but did not require significant new technology between launches. Lunar Orbiter and Surveyor were such programs. The reverse was true for programs with a long launch span like Pioneer and Mariner.

In terms of influence on the growth and technical development of the network, the Mariner program exceeded all of the other programs in both complexity and launch span. It did not produce the longest duration missions, however. The Pioneer program would still have viable missions in flight long after the last Mariner mission ended in 1974. Nevertheless, the Mariner program drove the early development of new capabil-

The Mariner Era: 1961–1974

ity for the DSIF to a greater extent than any other program and, at a time before the public and scientific palate had become jaded with a surfeit of "solar system science," it produced many of NASA's most spectacular scientific results. For that reason, the period of DSIF history that began with the disastrous launch of the first *Ranger* in 1961 and ended with the hugely successful *Mariner 10* spacecraft in orbit around Mercury and Venus in 1974, is identified here as the "Mariner Era."

Of the five projects included in the Mariner Era mission set, three, the Ranger, Mariner, and Surveyor projects, were fully managed by JPL. The Ames Research Center in Mountain View, California, directed the Pioneer Program, while Lunar Orbiter was directed by the Langley Research Center in Hampton, Virginia.

DSN participation in the Apollo program was directed by the Goddard Space Flight Center, Greenbelt, Maryland. Because these later projects were not controlled by JPL, the different management styles and demands for new standards of performance that they brought to the DSIF led to far-reaching changes in the DSIF arrangements for supporting non-JPL or "outside" flight projects. It was the sum total of these often conflicting influences that shaped the character and capability of DSIF during the Mariner Era.

In a real sense, the DSIF began its effective operational life as a complete operational network with the Ranger program, an ambitious JPL mission to land a spacecraft on the surface of the Moon, in August 1961.

The Ranger Lunar Missions[1]

While one stream of activity in JPL had, over the past several years, been focused on the problems associated with approving, designing, and implementing the DSIF, the principal effort had been devoted to developing the first spacecraft specifically designed to reach the surface of the Moon. Designated the Ranger Program, this project had been assigned to JPL by NASA in December 1959 and was to be completed within 36 months.

The Ranger spacecraft were much more complex than the earlier Pioneer-type probes. In particular, the Rangers were to be attitude stabilized to keep their high-gain radio antennas pointed toward Earth and the three antennas of the DSIF. This also allowed the solar panels, which provided power for the spacecraft electronics, to remain pointed at the Sun during flight.

The battery powered Pioneers had been simply spun like an artillery shell to maintain their spatial orientation during flight. In addition to attitude stabilization and more advanced radio communication systems, some of the spacecraft would carry an ejectable capsule. This capsule would descend to the lunar surface on a small retrorocket and thereafter right itself to transmit data on local radio activity and moonquakes. Later models were to carry high definition video systems to return detailed pictures of the lunar surface prior to impact.

The original Ranger program comprised five spacecraft, the first two of which were to be test flights ejected into an elliptical orbit around Earth. These would be essentially engineering flights to check out the new capabilities of attitude stabilization and solar power for the spacecraft and, for the DSIF, the all important telecommunication design. The remaining three spacecraft would consist of a basic spacecraft "bus" with an ejectable capsule containing a science instrument package and the retrorocket to set it on the lunar surface. During the flight to the Moon, these spacecraft were to make fields and particle science measurements and, during the final moments of approach before impact, take close-up vidicon pictures of the lunar surface.

In retrospect, this appeared to be an ambitious plan indeed for a fledgling technology, as later events proved. However, at the time, it was perceived as the appropriate major step forward that was needed to reestablish the prestige and forefront position of American scientific endeavor, so adversely affected by the Russian success with *Sputnik* and *Lunik* some years previously.

Although all five of the first Ranger flights were unsuccessful due to launch vehicle and spacecraft system failures, the excellent performance of the DSIF proved that the world-wide network concept of tracking stations managed and controlled from a central location was not only feasible but would be an essential element of future space missions to other planets and the Moon.

The Ranger flights demonstrated the value of continuous communication between the spacecraft and the network stations. This provided the Mission Operation teams back in the Flight Operations Control Center with immediate information about the status and condition of spacecraft and operating systems during the entire flight. This included engineering and scientific data. Ironically, this capability allowed the spacecraft controllers to estimate the reason for all of the Ranger failures attributed to the spacecraft.

The Mariner Era: 1961–1974

Following a soul-searching evaluation of the causes of failure of the first five Ranger missions, and some consequent reorganization of technical responsibilities at JPL, the program resumed in January 1964. The Ranger team redefined its objectives to the sole task of securing close-up television pictures of the lunar surface, ostensibly to aid in the design of two more sophisticated lunar missions then under study by NASA, namely the unpiloted Surveyor landers and the piloted Apollo program. Although the instrumented science capsule had been removed, the mission remained very ambitious for those days, as it included a launch into Earth-parking orbit followed by a second burn of the launch vehicle to transfer the spacecraft into a lunar trajectory. Attitude control maneuvers then followed to orient the solar panels to the Sun and the spacecraft antenna toward Earth. Further midcourse and terminal maneuvers were required to be correctly executed before the Ranger vidicon driven cameras could view the lunar surface for the brief period before final impact. During that time, the precious video data would be transmitted to the DSN stations where it would be processed and delivered to the Ranger team in the Operations Control Center at JPL.

With the reputation of JPL on the line, the first of the final group of four Rangers was launched in January 1964. Despite the political pressure to demonstrate success after the poor results from the first group of five launches, *Ranger 6* was also a failure. It performed perfectly until the last 10 minutes before lunar impact, at which point the spacecraft television system should have turned on to start transmitting pictures to the Goldstone station. There was no indication that this happened, and no pictures were sent. The cause of this failure was attributed to high voltage "corona discharge" in the camera's electrical insulation and appropriate changes were made to the TV subsystem on the remaining three spacecraft.

Six months later, the fortunes of the Ranger program changed dramatically. *Ranger 7*, launched in July 1964, was a resounding success and radioed 4,300 high-resolution pictures of the lunar surface back to Earth. Immediately after lunar impact terminated the flight, President Johnson called JPL and told Director William Pickering that *Ranger 7* was "a magnificent achievement." On behalf of the whole country he congratulated NASA, JPL, and NASA contractors, saying, "This is a basic step forward in our orderly program to assemble the scientific knowledge necessary for man's trip to the Moon." News conferences and presentations at the White House followed as soon as the films arrived at JPL from Goldstone and could be processed. The mood at NASA was elation, the Ranger team at JPL was ecstatic.

The final flights of the Ranger program in February and March 1965 added to the success of *Ranger 7*. Thousands more pictures of the lunar surface were returned by missions 8 and 9, and the reputation of JPL as a leader in the field of space exploration was established to worldwide acclaim.

No less delighted than those who were directly involved in the mission was the lunar science community, which, for the first time, had over 17,000 close-up pictures of the lunar surface for study and analysis. In addition, engineers and scientists designing the Surveyor and Apollo missions, soon to follow Ranger to the Moon, had a greatly improved basis for their designs.

In his History of Ranger project, Cargill Hall wrote, "No longer a liability, the Ranger program had vindicated American space policies and presaged accomplishments yet to come." It had indeed, and they would not be long in coming.

Mariner Planetary Missions[2]

Important as it was, Ranger was not the only program demanding the attention of the DSIF in 1962. By midyear, the first Mariner spacecraft, designed by JPL for planetary missions as distinct from lunar missions, were on their way to the launch pads at Cape Canaveral, Florida (later renamed Cape Kennedy).

Unlike lunar missions, planetary missions have only a very limited opportunity for launch, or "launch window," which occurs every few years. For missions to Venus, 1962 was one such year, and the next opportunity for launch would not occur until 1964. NASA and JPL could not wait that long for a first planetary mission. The attempt to send a spacecraft to Venus would have to be made in 1962.

Since the time between the NASA go-ahead and the opening of the Venus launch window was less than a year, much of the Ranger technology, including the radio subsystem, was utilized for the Mariner spacecraft. For the DSIF, this meant a continuation of the L-band uplink and downlink support and the possibility of competition with Ranger for the attention of DSIF tracking stations when both Ranger and Mariner spacecraft would be in view at the same time.

This situation, which first appeared in the Ranger/Mariner Era, was to become of immense importance in the years ahead as the number of flight missions increased. Conflicts for use of the DSN antennas when view periods of different spacecraft overlapped were a

The Mariner Era: 1961–1974

frequent occurrence. In the onrush of preparing the Mariners for launch the problem was deferred.

In addition to the existing 26-meter stations at Goldstone, Woomera, and Johannesburg, the DSN had, at the time of the first Mariner launches, grown to include a Launch Station at Cape Canaveral identified as DSIF 0, and a Mobile Tracking Station (DSIF 1), which was located near the Johannesburg station in South Africa.

It was the purpose of DSIF 0 to track the spacecraft as it lifted off the launch pad atop the launch vehicle, by manually pointing a small two-meter diameter dish antenna at the rapidly disappearing launch vehicle/spacecraft combination. This was not a job for the fainthearted, since the distance between the tracking antenna and the launch pad was necessarily short to ensure that the spacecraft signal was received continuously for as long as possible after liftoff. This period, seldom more than a minute or two, provided vital engineering information for the DSN regarding the status of the downlink as the spacecraft started its long journey around the world to reappear over South Africa at the next tracking station, DSIF 5.

There, the small 3-meter antenna of the Mobile Tracking Station rapidly acquired the spacecraft and used its pointing information to direct the large 26-meter antenna to the spacecraft. Once acquired, the spacecraft downlink was tracked and the data it carried was processed, recorded, and returned to JPL by airmail, a process which at that time could, and usually did, take seven days.

Mariner 1962

The first two Mariner spacecraft were essentially long-range versions of the Ranger spacecraft. They carried instruments to make scientific measurements in interplanetary space while moving along its trajectory between Earth and Venus. On arrival, they would measure the radiation and magnetic fields while the microwave and infrared radiometers analyzed Venus' atmosphere. The spacecraft used the same type of radio system (L-band) as had been carried by the Ranger spacecraft.

Two complicated, inflight operations were to be carried out after launch. The first of these was the midcourse maneuver to correct the trajectory to aim it more precisely at Venus. The second maneuver would point the science instruments to scan the Venus surface as the spacecraft flew by. The commands to carry out the maneuvers would be sent from the tracking stations. The science data would be transmitted by the spacecraft's three-watt transmitter to the DSN antennas, by then 58 million kilometers from the spacecraft. Data would "trickle" across those millions of miles of interplanetary space at just over eight data bits per second. Apart from the tremendous scientific value of the science data itself, this was regarded as a major technological achievement at that time.

Mariner 1 was launched in July 1962, but was destroyed by the Range Safety when it ran off course shortly after liftoff. However, *Mariner 2* injected into a Venus trajectory by an Atlas-Agena launch vehicle the following month, performed exactly as intended. Following a perfect liftoff, DSIF 0 followed the spacecraft until it disappeared over the horizon. The spacecraft appeared over South Africa some 28 minutes later and was acquired by DSIF 1 with no problem. Within three minutes, the big, 26-meter Johannesburg antenna had found the spacecraft and began receiving engineering data to verify the spacecraft had survived the stresses of the launch environment.

The midcourse maneuver was carried out successfully and, after some further commanding to correct a temperature problem, reached Venus in mid-December. After passing Venus at a range of 40,000 miles, the spacecraft continued in orbit around the Sun until it fell silent on 2 January 1963. The DSN stations searched the empty airwaves for several days trying to find some sign of life from the first planetary spacecraft but none was found. In its brief lifetime, Mariner had transmitted 11 million data bits of science and engineering data back to Earth.

As a consequence of experience with the first Ranger and Mariner missions, the DSIF began work to improve its capability so that engineering and science data could be relayed back to JPL as it was received from the spacecraft, that is to say in real time. Automatic monitoring of the DSIF stations would be added to permit the engineering experts at JPL who had designed the station equipment originally to aid the remote station staff in troubleshooting and problem identification. Even greater emphasis was to be placed on good engineering practices, conservative design, and thorough testing to minimize equipment failures in the field. Techniques to acquire the spacecraft downlink as early in flight as possible were studied. Small wide-beam antennas for "initial acquisition" were added to the network antennas and the accuracy of flight path predictions was improved.

The experience gained from these early flights also led to a new approach to DSN telecommunications link design, in terms of receiver sensitivity and transmission power of the ground tracking station. Combined with similar considerations for the spacecraft radio system, the basic concepts of uplink and downlink performance became firmly established as the methodology for all future deep space communications engineering design.

Mariner 1964

A "launch window" for Mars only opens every 25 months, and such an opportunity occurred in late November 1964. To take advantage of this situation, NASA chose a Mars-1964 project based on JPL's experience with the earlier Venus missions. Three spacecraft were to be built, two for flight, with one as backup that operated at the then-new S-band frequency.

The *Mariner 3* and *Mariner 4* missions were to be supported with the existing DSIF 26-m stations at Goldstone, Johannesburg, and Woomera, plus a new station in Australia which had been built at Tidbinbilla near Canberra. The stations at Woomera and Johannesburg that retained the old L-band receiving equipment were fitted with conversion equipment, which enabled them to receive the S-band signals from the Mariner spacecraft. In addition, the R&D station at the Goldstone Venus site, which by then had acquired a high-power, 100-kW transmitter, was to provide backup for the 10-kW transmitters at the other stations in the event spacecraft problems required a more powerful uplink for commanding purposes.

These first Mars flights also coincided with very significant improvements in the capability and operational management structure of the Ground Communications Facility (GCF) and Space Flight Operations Facility (SFOF) at JPL. The SFOF was a large data-handling machine, albeit housed in a barely finished new building at JPL, where data arriving from the tracking stations was processed in real time and distributed to the DSN and Flight Project Operations Control Groups for action and to Science Groups for analysis. The simple control room of the Ranger and early Pioneer days with a few desks, teletype machines, and display boards was "history."

Although the first of the Mariner 1964 missions failed due to a launch vehicle problem, the second, *Mariner 4*, was launched successfully on 8 November 1964. A little more than seven months later, it reached Mars. On 14 July 1965, in response to commands from the Johannesburg tracking station, the *Mariner 4* spacecraft turned on its cameras, tape recorders, and scientific instruments in preparation for the first close-up look at Mars. As the encounter point approached, the *Mariner 4* cameras focused on Mars and took a 25-minute sequence of pictures of the Martian surface before the spacecraft trajectory took it behind the planet. The pictures were stored on the spacecraft tape recorder. Playback to Earth began when the spacecraft came into view of the tracking stations, about an hour later.

It must be pointed out here that although the many millions of data bits in the picture images could be recorded almost instantly by the spacecraft, it would take many days to return the recorded data back to Earth, due to the limited capability of the downlink. The first pictures of Mars took ten days to playback. This would change radically in the years ahead, but at the time of *Mariner 4*, the downlink capability between Mars and Earth for the 26-m antennas of the DSIF was limited to only 8.33 bits per second.

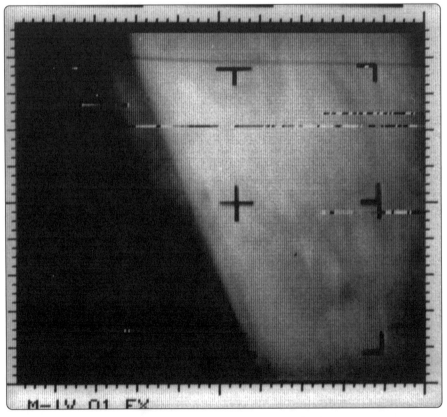

Figure 2-3. First close-up picture of Mars. Returned from *Mariner 4* on 14 July 1965. The photograph shows the Phelgra region and Mars horizon from a distance of approximately 10,500 miles.

The Mariner Era: 1961–1974

After a second playback of images and science data, *Mariner 4* continued along its orbit around the Sun, sending back more science and engineering data until the downlink was inadvertently "lost" about two and a half months later.

Scientifically, the mission was judged to be a great success, affording humankind the first close-up views of the Martian surface. Crude as they appear now, those first images, transmitted laboriously at about 8 bits per second over millions of miles of deep space, were hailed with tremendous enthusiasm by the public at large. A composite image of the Phelgra region of Mars, taken by the *Mariner 4* spacecraft, was proudly displayed in the Main Lobby of the Engineering Building at JPL, where for many years, it served both as a reminder of the humble beginnings of the planetary program and as the point from which to gauge the astonishing progress that would take place over the next few years. The original picture is reproduced in Figure 2-3.

For some time after the downlink signal was lost by the DSIF stations, the R&D Venus station at Goldstone continued to search for it. Eventually the signal was recovered and used to check out and calibrate the new 64-meter antenna at Goldstone. It was also planned to use *Mariner 4* in conjunction with *Mariner 5* to perform radio science experiments in interplanetary space from three vantage points: two spacecraft and Earth. However, the spacecraft supply of attitude control gas was exhausted during engineering tests in this extended phase of the *Mariner 4* mission. After three years of operation in space, and 1,119 passes by the DSN tracking stations, the mission was terminated.

Mariner 1967

NASA chose to use the 1967 planetary launch window for a second mission to Venus. The spacecraft for this mission would be the one remaining spacecraft from the three that had been built for the 1964 Mars missions. Fitted with a different set of science instruments, including a photometer, radiometer, and plasma probe, the spacecraft would further explore the dense, optically opaque Venusian atmosphere.

The tracking and data acquisition requirements for support of *Mariner 5* were complicated by the DSN desire to involve *Mariner 4*, which had since been recovered, in simultaneous radio science experiments in interplanetary space. In addition, new telemetry and command processing equipment added to the Network, between the *Mariner 4* and *Mariner 5* launches, further complicated the technical and operational interfaces between the DSN stations and the flight project organization at the SFOF. The spacecraft was more complex, too. It carried, in addition to the standard lunar ranging system, a newly developed system of ranging for its first flight evaluation. The Tau system, as it was called, was designed to extend the DSN ranging capability beyond the limit of the lunar distance.

At Goldstone, *Mariner 5* would be supported by the two 26-m stations (Echo and Pioneer) and the new 64-m Mars station. In Australia, it would be Woomera and the new 26-m station at Tidbinbilla that would be used. Johannesburg and another new 26-m station at Robledo, near Madrid, Spain, would complete the DSN support for this mission. At the SFOF, a new 7044 computer system would see its first operational use on *Mariner 5* and demonstrate the capability to process *Mariner 4* and *Mariner 5* data simultaneously, even though the data streams came from different DSN stations. To handle so many data streams from the increasing number of tracking stations, a newly installed communications processor in the SFOF used computer switching to direct all communications throughout the DSN.

Launched on 14 June, the spacecraft was first placed in an Earth parking orbit before being injected into a final trajectory to Venus. Before reaching the limit of its capability, the DSN lunar ranging system followed the spacecraft to a distance of 10 million km, making *Mariner 5* the first spacecraft to be ranged beyond lunar distance. The new Tau ranging system was activated about two weeks before success in tracking the spacecraft to the Venus encounter and beyond, to about 75 million km. Together with Doppler and angle data, ranging data became a standard type of radiometric data generated by the DSN and greatly enhanced the orbit determination process for the navigation of future spacecraft. With the downlink running at 8.33 bits per second and encounter activities proceeding smoothly, *Mariner 5* made its closest approach to Venus on 18 October 1967, at a range of 3,946 km. The science return satisfied the mission objectives and the DSN continued to track the spacecraft until it passed out of radio range at about 160 million km.

Mariner 1969

In discussing the background to the Mariner 1969 missions Corliss wrote, "Earlier probes to Venus and Mars had indicated that Mars was the most likely planet in the solar system to support life. The richly detailed and cratered surface of the planet revealed by *Mariner 4* had surprised planetologists and made them anxious for more photos. Mars thus became NASA's prime planetary target. The Mariner Mars 1969 mission, therefore, concentrated on TV imaging of the planet's surface, and experiments that might aid the design of future missions, particularly those looking for life. Besides the TV cameras, the two new Mariner-type spacecraft assigned to the mission carried an ultraviolet spectrometer, an infrared spectrometer, and an infrared radiometer."

The Mariner Era: 1961–1974

For the DSN, the Mariner 1969 missions were of special importance because they included the first flight demonstrations of the new DSN high rate telemetry system. Previously, with downlinks limited to just over 8 bits per second, playback of telemetry data from spacecraft at the range of Mars or Venus was measured in days or even weeks. With the DSN's new block-coded telemetry system running at 16,200 bits per second, roughly two thousand times faster than the former system, low resolution pictures could be sent in real time from Mars. The new high rate telemetry did not replace the existing low rate system. It was an addition to the DSN's rapidly increasing capability for planetary exploration. Corliss continued, "The DSN was ready for the much greater distances and higher rates required (for planetary exploration). The various technical advances and new facilities just described had begun years before the new missions left their launch pads. As usual though, the new DSN capabilities, which had seemed perhaps unnecessarily ambitious when proposed, were quickly absorbed by the new mission designers."

Tracking and data requirements for the Mariner 1969 missions (*Mariners* 6 and 7), included not only the existing 26-m stations, but the new 64-m antenna at Goldstone as well. The downlink would carry telemetry data at either 8.33 or 33.33 bits per second for normal spacecraft operations during the cruise period. When the spacecraft reached Mars, the new high rate telemetry experiment would be turned on. Of course telemetry reception at 16,200 bits per second would be possible only when the spacecraft was in view of Goldstone, where the big 64-m antenna was required to enhance the downlink sufficiently to carry data at this rate.

Orbit determination requirements for the very precise navigation needed to fly the spacecraft close enough for a Mars encounter would be based on standard radiometric data, including an improved version of the Tau planetary ranging system.

Both *Mariners* 6 and 7 were launched successfully in February and March 1969. For the first time, NASA had successfully launched a pair of spacecraft. Both spacecraft were on target and, from a DSN viewpoint, both were in the same part of the sky. This meant that both spacecraft would be in view of a given DSN site, although not within a single antenna beamwidth, simultaneously. Tracking both spacecraft simultaneously required two antennas and two telemetry data processors, one for each downlink. At the same time, there were the Pioneer spacecraft and backup support for the first Apollo missions to be considered as well. The "loading" on the DSN antennas had begun to reach the point of saturation, and the resource allocation system was hard pressed to service all of its customers.

As Mars began to draw near towards the end of July, encounter operations began with *Mariner 7* only five days behind *Mariner 6*. Corliss describes what happened next. "All seemed to be going well until about six hours before the *Mariner 6* encounter, when Johannesburg reported that the signal from *Mariner 7* had disappeared. It was an emergency that came at the worst possible time. The Robledo, Spain[,] antenna discontinued its tracking of *Pioneer 8* and began to search for the lost spacecraft. When Mars came into view for Goldstone, the Pioneer 26-m antenna joined the search, while the Echo 26-m antenna continued tracking *Mariner 6*. It was decided to send a command to *Mariner 7* to switch from the highly directional high-gain antenna to its omnidirectional low-gain antenna. The spacecraft responded correctly, and suddenly both the Pioneer station and the Tidbinbilla station began receiving low-rate telemetry from the recovered spacecraft. Something had happened to the spacecraft but no one knew just what."

While the DSN was committed to support one *Mariner* at a time in a mission-critical phase, this situation presented one spacecraft approaching encounter and a second one with a serious and unknown problem. To deal with it, the DSN applied its main effort to the ongoing *Mariner 6* encounter, while a special team at JPL studied the *Mariner 7* anomaly.

Fortunately, the *Mariner 6* encounter events executed without any problems. Many pictures of Mars were taken and successfully returned to Earth using both the high-rate and normal low-rate telemetry systems. The special "Tiger Team" at JPL was able to overcome the *Mariner 7* attitude problem by using the real-time high-rate telemetry sight, the TV cameras on Mars, in time to carry out a very successful encounter.

For both encounters, the new High-Rate Telemetry System (HRT) proved its worth, not only in recovering from the *Mariner 7* emergency, but also in providing a much faster channel for playing back TV and other high-rate science from Mars to Earth. In all, 202 pictures covering almost 20 times the area covered by *Mariner 4* were returned to Earth by the *Mariner 6* and *7* missions.

Following the formal end of the mission in November 1969, both spacecraft, still operating perfectly, were used to perform radio science experiments in relativity, astronomical constants, and electron densities in interplanetary space.

The Mariner Era: 1961–1974

Mariner 1971

We turn again to Corliss for a view of these missions: "The logical follow-on mission to Mariner-Mars 1969, using the lunar program analogy, would be picture-taking orbiters around Mars. The next Mars opportunity was in the spring of 1971, and two Mariner-class spacecraft were prepared accordingly. The new Mariners drew heavily upon the technology of the 1969 mission. Instrumentation was very similar, with emphasis again on photography. The primary objectives were the search for evidence of life and the gathering of data that would aid Mars landers. After mapping as much of the surface as possible, scientists wanted more data on the density and composition of the Martian atmosphere. A 90-day orbital mission for each spacecraft was planned. These spacecraft would be the first terrestrial satellites of another planet."

The requirements levied upon the DSN for these missions were even more extensive than those of the 1969 missions. This was due to the precise navigation and maneuvers needed to inject the spacecraft into orbit around Mars and to the DSN antenna coverage required for 90 days of orbital operations for two spacecraft simultaneously.

During the intervening two years between the Mars opportunities, DSN capabilities expanded considerably. The Multimission Telemetry System had been inaugurated on the 1969 flights, and for the 1971 mission. The experimental HRT, which had been demonstrated so successfully in 1969, was fully operational. Pictures and science could be transmitted back from Mars to Goldstone much more rapidly. A 50,000 bit per second wideband communications link was available to carry two 16,200 bits per second (16.2 kbps) data streams simultaneously between Goldstone and JPL. In effect, the DSN now had a downlink capable of delivering two simultaneous 16.2 kbps data streams directly from Mars to JPL in real time. Of course the Mars station at Goldstone was the only 64-m antenna in operation at that time, but the high speed readouts of the Mariner 71 tape recorders were planned to take place only when DSS 14 had Mars in its view. Another important capability of the DSN, which was planned for use on these missions, was the ability of a single DSN antenna to simultaneously handle two spacecraft located in the same beamwidth.

The Mariner 71 flights began inauspiciously with a failure of the upper stage (Centaur) of the Atlas-Centaur launch vehicle which resulted in the loss of *Mariner 8*. Investigation of the Centaur problem delayed the next launch to 30 May when *Mariner 9* lifted off on a direct ascent trajectory for Mars with an arrival in mid-November 1971. Although the loss of one spacecraft placed the burden of responsibility for the entire mission on *Mariner 9*, it reduced the burden of tracking for the DSN to a single spacecraft.

After a perfect midcourse maneuver in June and a nominal cruise to the planet, *Mariner 9* was injected into Mars orbit on 14 November. Unfortunately, just at the time of *Mariner 9*'s arrival, the planet was covered by a huge dust storm, which precluded any useful photography. *Mariner 9* continued to make its regular 12.567 hour orbits while the scientists waited for the dust to settle. While the spacecraft waited in orbit, the other instruments were busy and pictures of Mars's moons Phobos and Deimos were taken.

By 3 January 1972, the dust storm had cleared sufficiently for the 90-day mission to begin. Soon, the pictures came by the thousands to reveal, for the first time, deep fluvial channels, possibly cut by water, and evidence of ice action in the polar regions. Detailed maps of the Martian surface, needed to plan the Viking landings to come a few years later, were being drawn even as the *Mariner 9* cameras covered more and more of the surface of the planet.

Because the spacecraft was still operating so well at its nominal end date in April, NASA extended the mission to re-examine some specially interesting areas and observe solar occultations as the spacecraft moved through the shadow of Mars twice per day. Despite its rigorous and longer than planned mission, *Mariner 9* survived until 27 October 1972.

Mariner 1973

The launching of *Mariner 10* in November 1973, on the first mission to Venus and Mercury, marked the end of the Mariner spacecraft era. Immensely successful as it was, the Mariner 1973 Venus-Mercury mission was the last to use the so-called Mark III-73 network model which had supported the previous Mariner missions to Mars in 1969 and 1971 and the later Pioneer missions to Jupiter.

High-rate telemetry, wideband communication circuits connecting all complexes to JPL, and 64-m antennas at all three sites formed the basis for the DSN support of the Mariner 10 mission. However, the spacecraft carried a redesigned imaging system that required the DSN to provide a downlink capability from Mercury of 117.6 kbps to fully exploit the new imaging system capability. This data rate greatly exceeded the operational capability of the existing DSN high rate telemetry system of 16 kbps.

Responding to pressure from the Project, the DSN had proposed a substantial number of improvements to the existing DSN downlink capability to increase the mission data return and enhance its science value. These improvements included 1) the installation of a developmental supercooled maser and ultra cone at DSS 43; 2) the installation of an S/X-band dichroic plate and feed cones at DSS 14; 3) the implementation of a special 230-kbps wideband data transmission circuit from Goldstone to JPL for real-time picture transmission and a 28.5-kbps circuit from Australia and Spain to JPL for transmission of recorded data; 4) the redesign of the telemetry and command processor (TCP) computer software to handle the data rates as high as 117 kbps; and 5) the installation of a Block IV receiver-exciter at DSS 14 for the S/X-band Radio Science Experiment.

As the mission extended into a second encounter with Mercury in 1974, the increasing Earth-spacecraft range introduced further losses into the telecommunication path. To compensate for this additional loss, the DSN implemented a new signal enhancement technique, previously developed and demonstrated by Spanish engineers at the Madrid Complex in 1969 and 1970. In this technique, the spacecraft signals from several DSN stations were added together at a central point so that their combined signal strength would be greater than that of a single station. The new "antenna arraying and signal combining" technique was implemented, this time at Goldstone, using stations DSS 14, DSS 12, and DSS 13. Tests with the spacecraft demonstrated an improvement in downlink performance of 0.7 dB. In September 1974, the new technique was used for the first time by the DSN, with good results, as a fully operational capability to support the second *Mariner 10* Encounter of Mercury.

The *Mariner 10* data return statistics were impressive. At its conclusion, it had returned to Earth over 12,000 images of the planets Venus and Mercury. In terms of scientifically useful pictures, this exceeded the combined total of the *Mariner 6* and *7* missions to Mars by a factor of almost 15.

The time *Mariner 10* had spent in deep space, short though it was, had been very eventful. Years later, Bruce Murray, who at the time was leader of the *Mariner 10* Imaging Team, expressed his view of the scientific worth of the mission, "The economy class Mariner flyby of Venus and Mercury was one of the most productive space science experiments ever carried out."[3]

On 24 March 1975, shortly after its third flyby of the planet Mercury and exactly 506 days after its launch in November 1973, the last of the Mariner series of planetary spacecraft fell silent. With its attitude control gas supply exhausted, *Mariner 10* started to slowly tumble out of control, and controllers at the Jet Propulsion Laboratory in Pasadena, California, sent a command to turn off its transmitter to avoid contaminating the deep space environment with an unwanted source of radio emissions.

From the DSN point of view, the *Mariner 10* Mission to Venus and Mercury was associated with a long list of "firsts."

The *Mariner 10* mission was the
- first multiplanet gravity assist mission,
- first spacecraft to photograph Venus,
- first spacecraft to fly by and photograph Mercury,
- first spacecraft to have multiple encounters with a target planet,
- first JPL spacecraft to transmit full-resolution pictures in real time from planetary distances,
- first mission to use dual-frequency radio transmission, and
- first mission to use arrayed ground station antennas to improve signal-to-noise ratio.

For the DSN, the *Mariner 10* mission to Venus-Mercury provided a fitting conclusion to the Mariner Era. During that era, there had been many other important missions sharing the attention of the DSN, which will be discussed later. Nevertheless, it was the Mariners that influenced the future capabilities of the DSN for the longest time and in the most significant way, as is evident in the summary of the Mariner missions contained in the following table (on the following page).

Pioneer Interplanetary Missions[4]

Unlike the short missions of the five earlier Pioneer lunar probes, the second generation of *Pioneer* spacecraft consisted of long duration missions designed to measure the interplanetary fields and particles environment along trajectories that placed the spacecraft in heliocentric orbits ranging between 0.7 and 1.2 AU. They were sometimes described as interplanetary weather stations. The four spacecraft, *Pioneers 6–9*, were each launched approximately one year apart, starting in December 1965 with *Pioneer 6* and ending in November 1968 with *Pioneer 9*.

All four spacecraft transmitted continuous streams of scientific data from various orbits around the Sun. This meant that the DSIF was required to provide almost continuous tracking for first one, and eventually four Pioneers, to ensure that all of the rapidly changing science data were captured. Furthermore, instead of dying after the first year or so of operation as was expected, all four Pioneers continued to operate actively in deep space for many years. At the end of the Mariner Era they were still going strong and had become a permanent feature of the DSIF routine tracking schedules.

The Mariner Era: 1961–1974

Summary of the Mariner Planetary Missions: 1962–73

Mission; Launch Date	Objective Result	Science Return	Max. Data Rate at Encounter
1, Venus; Jul 62	Flyby, Failure	None	None
2, Venus; Aug 62	Flyby, Success	Scanned Venus surface at encounter	8.33 bits/sec
3, Mars; Nov 64	Flyby, Failure	None	None
4, Mars; Nov 64	Flyby, Success	First close-up photos of Mars surface	8.33 bits/sec
5, Venus; Jun 67	Flyby, Success	Science data on Venus surface environment	16.33 or 8.33 bits/sec
6, Mars; Feb 69	Flyby, Success	High-res. photos of Mars' equatorial region	16,200 bits/sec
7, Mars; Mar 69	Flyby, Success	High-res. photos of Mars southern hemisphere	16,200 bits/sec
8, Mars; May 71	Orbiter, Failure	None	None
9, Mars; May 71	Orbiter, Success	Mapped whole Mars surface photos and data from Mercury	16,200 bits/sec
10, Venus/Mercury; Nov 73	Flyby, Orbiter, Success	UV photos of Venus, close-up photos and data from Mercury	117,600 bits/sec

The new Pioneer program was directed by the NASA Ames Research Center (ARC) at Mountain View, California, and thus it became the first major non-JPL or "off-Lab" flight project to require support from the DSIF. In addition, ARC wanted to exercise mission control functions from Mountain View rather than from the mission control center at JPL in Pasadena. These and other requirements, related to enhanced downlink capability, presented a new situation for the DSN and forced it into broadening its management structure to deal with flight projects other than those directed by JPL. In due course, special positions were created in the DSN Office to deal with the unique requirements of each new flight project. This unique position was given the title of "DSN Manager for Project . . . ."

Like other flight projects in these early years, the Pioneer Project provided the tracking stations with a set of unique Ground Operational Equipment (GOE) for use when the station was tracking its particular spacecraft. The stations were obliged to take on the onerous tasks of accommodation, installation, maintenance and operation. The GOE provided real-time telemetry data processing and spacecraft commanding functions, which could not be handled by the existing DSIF equipment because the Pioneer spacecraft design differed from that of the current JPL spacecraft. Eventually, the Madrid engineering staff developed new software for the DSN's Multiple Mission Telemetry and Command Processor that emulated the GOE functions and interface signals, thereby avoiding the need for the special Pioneer-unique equipment in the Network.

For the most part, the DSIF used the network's 26-m antennas to support the Pioneer missions. However, as the Pioneers reached the threshold of detection capability for the 26-m antennas, improvements in downlink sensitivity became necessary to the continued life of the missions. DSN engineers made systematic improvements in the S-band maser sensitivity, microwave equipment, and receiver "tracking loop" performance and advanced demodulation hardware, while Madrid engineering staff developed special convolutional decoding software for *Pioneer 9*. As a result of these improvements, the threshold of performance for Pioneer on the 26-m antennas was extended from 0.4 AU in 1965, to 1.5 AU by 1969. Although it was driven by the enduring performance of the Pioneers, the improved downlink capability was, of course, of great benefit to later missions and the to the DSN in general.

Inevitably, as the Pioneers moved steadily deeper into space, the strength of their downlink signals dropped below the detection capability of the enhanced 26-m antennas. Now, fully compatible with the Network's multiple mission capabilities, the Pioneer spacecraft were transferred to the new 64-m antennas which were, fortuitously, just coming into service in the network. With the great increase in downlink capability that the 64-m antennas brought to the Network, the communication range of the Pioneers was extended significantly into, and even beyond, the next era.

PIONEER JUPITER MISSIONS

Much smaller than the JPL-designed Mariner-class spacecraft, the interplanetary Pioneers were simple and rugged and, being spin-stabilized, avoided the complexities of three-axis attitude control used by the Mariners. The efficacy of their design had been amply demonstrated by four years of continuous deep space operation when, in 1969, NASA selected the "third generation" Pioneers for the first missions to Jupiter.

The Mariner Era: 1961–1974

Although the Jupiter Pioneers were technically more advanced than the previous spacecraft, the design philosophy was much the same. This project was directed as before, by ARC.

The spacecraft were instrumented to carry out fields and particles measurements throughout the entire mission. At Jupiter they were to explore the Jovian system, obtain the first spin scan images of the planet, and investigate its enormous magnetosphere. To test new technologies for future missions into this previously unexplored region of deep space, where sunlight was too weak for satisfactory operation of solar cells, the spacecraft were powered by Radioisotope Thermal Generators (RTGs). Instead of a mast-type of antenna, the new Pioneers carried 2.75-m parabolic dish antenna for communications with the DSN. They were, for their time, very advanced spacecraft.

Pioneer 10 finally left the launch pad on 3 March 1972, and made its closest approach to Jupiter 21 months later, on 4 December 1973, at a distance equal to 2.86 Jupiter radii, 302,250 km from the center of the planet. Science data from the encounter instruments confirmed the complexity of the Jupiter system. For the first time, close-up photos of the planet surface were obtained and an atmosphere was detected on the satellite Io. The giant planet's magnetosphere was found to be disk shaped and bigger than the Sun itself.

DSN support for the *Pioneer 10* Jupiter mission was complicated by serious conflicts for the 26-m and 64-m antennas by the *Mariner 10* mission to Mercury and Venus and by the need for Goldstone 64-m radar surveillance of possible Viking-Mars landing sites. The multiplicity and complex nature of these overlapping view periods presented the DSN with a problem it had not encountered up to this time, but which loomed ominously in the years ahead. A Network Allocation Working Group was established to resolve the conflicts by mutual compromise between the conflicted projects. The DSN had moved from an age when it could assign antennas for the exclusive use of a single flight project to an age when it would be necessary to assign missions to antennas on a day to day, or even hour to hour, basis. Thus was born the DSN scheduling system, which soon became a permanent feature of daily DSN activity.

By the close of the Mariner Era, eleven DSN stations had provided 2000 tracking passes for *Pioneer 10* between April 1972 and January 1974, the equivalent of 21,000 hours of tracking time. During this time the spacecraft distance from Earth varied from 22 million to 890 million kilometers. During the 60-day Jupiter encounter period, over 17,000 commands were sent to control the complex spacecraft sequences.

Except for a sling-shot (later called a gravity-assist) encounter with Jupiter, the *Pioneer 11* mission was essentially the same as that of *Pioneer 10*. It was launched on 6 April 1973 and reached Jupiter at the end of the year.

Building on the success of the *Pioneer 10* encounter with Jupiter, the ARC mission controllers retargeted *Pioneer 11* in flight in such a way that, at its encounter with Jupiter in December 1973, trajectory was altered to eventually intercept the ringed planet Saturn. *Pioneer 11* went on to pass within 20,000 kilometers of Saturn's main outer ring system on 1 September 1979, and became the first spacecraft to do so. The remarkable success of *Pioneer 11*, discussed further in the chapter on the Voyager Era, paved the way for the Voyager spacecraft to later pass Saturn.

Surveyor Lunar Lander Missions[5]

Originally the objectives of the Surveyor Lunar Lander program, complemented by Lunar Orbiters, were primarily scientific in nature. After Apollo became a national goal, the objectives of the Surveyor program were redirected somewhat towards determining whether a human could land safely on the surface of the Moon. The Surveyor program was managed by JPL and controlled from the Space Flight Operations Facility (SFOF) which had replaced the simple mission control center of the early days.

The DSN felt the impact of Surveyor in a number of ways. These stemmed from a new (to the DSN) concept that Surveyor planned to use to control its spacecraft once they landed. The spacecraft had few automatic features and depended almost entirely on real-time commands for its operation on the lunar surface. This concept of interactive control of a distant spacecraft was based on real-time video and other data displays of the current state of the spacecraft. These data would be used by spacecraft controllers to make decisions regarding the desired future state of the spacecraft. Commands transmitted from the DSN stations would control the actions of the spacecraft accordingly. To make this possible, video data streams from the tracking stations, 4,400 bits per second from Goldstone, 1,100 bps from the overseas stations, were first transmitted via the Ground Communications circuits to the SFOF. There the data was converted to television displays for operational decisions by the spacecraft controllers, and to photographic images for scientific interpretation by the project scientists.

Obviously this level of operational decision making was completely dependent upon the reliability of the communications circuits between the DSN sites and JPL. To ensure the integrity of the overseas communication circuits for Surveyor, all DSN traffic was, for the first time, carried on a high-quality satellite channel provided by NASCOM.

The Mariner Era: 1961–1974

As was the case for Pioneer, the Surveyor Project provided special mission-dependent equipment in the form of a Command and Data Console (CDC), for installation at each of the DSN stations, that would support the missions. The CDCs were operated and maintained by project people who were resident at the stations, pending the transfer of responsibility to the DSN. In the event of a loss of communications between a DSN station and mission control at the SFOF, these personnel would be capable of controlling the spacecraft to ensure its safety until the situation was corrected.

In the period prior to launch of the first Surveyor, lengthy and elaborate simulations were carried out to prepare the DSN and Surveyor operations teams for the scenarios they would likely face in interacting with the spacecraft on the lunar surface.

Using its new Atlas-Centaur launch vehicle, NASA launched seven Surveyors between May 1966 and January 1968. The success of all seven launches demonstrated the reliability of the new high-energy Centaur upper stage, a potent liquid-hydrogen-fuelled booster rocket which had recently been developed at NASA's Lewis Research Center. The Centaur subsequently came into general use as an upper stage for deep space mission launches, until the program was discontinued in 1986 as a consequence of the Space Shuttle Challenger disaster.

Landing sites for the Surveyor spacecraft had been chosen primarily for their interest to the future Apollo program and four of the spacecraft soft-landed successfully at, or near, these sites and returned large quantities of engineering and scientific data. Two of the spacecraft failed to reach the lunar surface. *Surveyor 7*, the final spacecraft of the series, was targeted to land in the more scientifically interesting lunar highlands. After a successful landing it demonstrated the advantages of real-time control and manipulatory capability by carrying out a number of experiments that involved digging trenches, moving rocks, and even recovering an alpha-scattering experiment that had failed to deploy correctly after touchdown. The effort that went into the pre-launch simulation efforts clearly proved to be worthwhile.

Despite the complexity of the Surveyor program, DSN support for all seven flights was deemed to be excellent due in no small measure to the exhaustive testing and training program that preceded the first launch. The table below, which summarizes the data returned from each of the Surveyor missions, attests to the quality of the DSN data retrieval support.

Summary of Surveyor Lunar Lander Missions

Mission; Launched	Spacecraft Operations on the Lunar Surface
1; 1966; 30 Mar	Soft landing, operated for 2 lunar days. Returned 13,000 pictures, received 108 commands.
2; 1966; 20 Sep	Spacecraft became unstable, destroyed itself in flight.
3; 1967; 17 Apr	Soft landing, operated for 1 lunar day. Returned 10,000 pictures, received 63,000 commands.
4; 1967; 14 Jul	Lost radio contact touchdown.
5; 1967; 11 Sep	Soft landing, operated for 2 lunar days, 118 hours of alpha-scatter experiments. Returned 27,000 pictures, received 123,000 commands.
6; 1967; 7 Nov	Soft landing, operated for 1 lunar day, 30 hours of alpha-scatter experiments. Returned 45,000 pictures, received 170,000 commands.
7; 1968; 7 Jan	Soft landing, operated for 1 lunar day, 100 hours of alpha-scatter, many surface sampler experiments. Returned 28,000 pictures, received 150,000 commands.

Note: One lunar day is equivalent to approximately 27 Earth days.

Lunar Orbiter Missions[6]

Although an orbital lunar mapping mission had been originally conceived by JPL as an adjunct to the Surveyor lander program, NASA eventually assigned the development and management of the Lunar Orbiter project to the Langley Research Center in Hampton, Virginia.

The main objective of the Lunar Orbiter project was to obtain high-resolution photographs of potential Apollo landing sites on the Moon to complement those made by Surveyor on the actual surface. From its continuous orbit about the Moon, it would also make a detailed photographic survey of the entire lunar surface. To accomplish this task, the spacecraft carried a complex wide-angle camera and film processing package which employed facsimile transmission techniques to return the images to Earth via the tracking stations of the Deep Space Network. There were five identical spacecraft in the Lunar Orbiter program, all launched on Atlas-Agena vehicles in one year, beginning August 1966. All five missions were successful. Each of the missions orbited the Moon at a different inclination to the lunar equator and was therefore able to scan a different section of the lunar surface. The combined result of all five missions amounted to a survey of almost 99 percent of the entire lunar surface. By the time the project ended, Lunar Orbiter had surveyed the entire front surface of the Moon and most of the back surface, in some 2,000 individual high-definition, photographic images.

A standard S-band network of three deep space tracking stations was used to support the five Lunar Orbiter missions. Like the Pioneer and Surveyor projects, the Lunar Orbiter project provided these stations with special purpose mission-dependent ground reconstruction equipment (GRE), as well as personnel to operate them. The Lunar Orbiter missions were conducted concurrently with the Surveyor and Mariner-Venus missions. This multiple-mission environment heavily taxed the resources of the DSN and, by requiring precise scheduling of the activities of all three projects and a high degree of cooperation and coordination between them, presaged the later establishment of the Network Resources Allocation Working Group.

The Lunar Orbiter missions provided the DSN with the first opportunity to evaluate its new "turnaround" lunar ranging system in actual flight operations. Prior to this time, Earth to spacecraft range had been estimated by electronically processing the Doppler tracking data, a method that was cumbersome and prone to errors from "slipped cycles." The Lunar Orbiter spacecraft carried a ranging transponder, which, after detecting a special code carried on the uplink, would retransmit the code along with telemetry data on the downlink back to Earth. At the DSN tracking station, special ranging equipment compared the downlink code with uplink code to calculate the time it took the range code to travel out to the spacecraft and back. After proper calibrations and corrections were applied, spacecraft navigators used these data to obtain a very precise value for the spacecraft range (distance between tracking station and spacecraft). So successful were these inflight demonstrations that turnaround ranging, in various more refined forms, soon came into regular use along with Doppler data as the standard types of radiometric data generated by the DSN for all lunar spacecraft navigation purposes.

In addition to refining the new turnaround ranging system, DSN engineers extended its use to improve the time synchronization between DSN stations throughout the world. This enabled a more accurate determination of the spacecraft orbit about the Moon, which in turn yielded more and better information about the geophysical properties of the Moon itself. In addition to purely scientific interest, these data also provided valuable background information for the design of the Apollo piloted missions to the Moon, which were about to begin.

After the Lunar Orbiter program ended, scientists at JPL continued to analyze the wealth of DSN ranging and other radiometric data that had been generated during the Lunar

Orbiter missions. The outcome of this work greatly improved our understanding of the radius, orbit, and gravitational field of the Moon. It also led to the discovery of hitherto unknown anomalies in the lunar gravitational field due to unexpected concentrations of mass, or "mascons" as they came to be called, lying beneath the lunar surface.

Apollo Missions[7]

Just as NASA had assigned tracking and data acquisition responsibility for the planetary program to JPL, it assigned tracking and data acquisition responsibility for the Apollo program to the Goddard Space Flight Center (GSFC), Greenbelt, Maryland. And, just as JPL built the DSN for its planetary program, GSFC built the Manned Space Flight Network (MSFN) for supporting Apollo. The MSFN comprised two types of tracking networks. The first consisted of a large number of 9-m stations to handle the Apollo spacecraft during Earth-orbit operations, while the second network consisted of three 26-m stations to handle the translunar and lunar phases of the flights. For obvious reasons, the 26-m stations of the MSFN were located near the DSN stations at Goldstone, Canberra, and Madrid.

Although the MSFN provided the prime network for tracking and data acquisition support for the Apollo flights, the DSN also contributed a great deal of technology and facility support to Apollo. The DSN provided the MSFN with S-band receiving, transmitting and ranging equipment and computer software for lunar trajectory orbit determination purposes. Also, because the MSFN stations were (intentionally) very similar in design to the DSN stations, it was decided to equip one DSN station at each longitude (DSS 11, DSS 61 and DSS 42) with sufficient MSFN equipment to act as backup for the prime MSFN 26-m stations at the same site, or one nearby. These became known as the "mutual" stations. Corliss has described this aspect of the JPL/DSN role in Apollo in his account of the history of the Piloted Space Flight Network (MSFN).

A second control room called the MSFN wing was added to each DSN station and connected via a microwave link to the nearby prime Apollo station. The MSFN wing housed the special Apollo transmitting and receiving equipment, the MSFN electronics and recorders, and the switching connections that allowed the single DSN 26-m antenna to be used for tracking either deep space or Apollo spacecraft.

By mutual agreement between JPL and GSFC, Apollo mission operations at both the DSN mutual station and the MSFN prime station were directed by a single coordinator provided by the MSFN at each complex. This operational arrangement proved to

The Mariner Era: 1961–1974

be so effective that it was used without change for all the Apollo flights, beginning with the Apollo 4 test flights in 1967 until the program concluded with Apollo 17 in 1972.

Following the success of the early test flights, DSN support for the Apollo program expanded to include the Goldstone 64-m antenna as backup for special events and emergencies. To make this possible, equipment was added to the receivers at the 64-m antenna to allow the Apollo downlink signals to be transferred several miles via microwave link to the prime MSFN station for processing and onward transmission to the Apollo Mission Control Center in Houston. Furthermore, during Apollo tracks, the 64-m antenna would have to be driven by computer generated pointing data derived from Apollo spacecraft trajectory information provided by the MSFN.

These new interfaces, implemented and tested in great haste, were completed just in time for the first piloted lunar orbit flight on Apollo 8 in 1968. The then new 64-m antenna at Goldstone was called upon to provide backup support for the 26-m MSFN antennas for receiving television signals for public broadcast while the spacecraft was en route to and from the Moon and in orbit around the Moon. The arrangement worked perfectly, and from then on the Goldstone 64-m antenna was used on a regular basis to provide critical and, in some cases, emergency (Apollo 13) support for all Apollo missions.

Apollo 13 was the third piloted mission with the Moon as its destination, and a landing and return to Earth as its objective. The mutual stations of the DSN, together with the big Mars antenna (DSS 14) at Goldstone, had been scheduled to provide the backup coverage for the MSFN prime stations as was normal practice by the time of Apollo 13.

After the Saturn V launch from Cape Kennedy on 11 April 1970, the mission proceeded "nominally." The DSN stations commenced to follow the detailed tracking schedule developed to minimize impact to the Mariner 1969 Mars spacecraft, then in its extended mission. All went well until Apollo 13 was making its third pass over Goldstone on 14 April. At that point, about 55 hours into the mission, the unthinkable happened—an explosion in an onboard oxygen tank rendered the Lunar Service Module, with all its redundant systems, useless. All mission objectives, except return to Earth, were abandoned. While the mission controllers at Houston set about trying to recover the crew, all tracking stations were placed on high alert for the spacecraft emergency. The Goldstone Mars antenna was immediately requested to extend its tracking coverage of the current and all subsequent Apollo 13 passes, to the maximum possible, horizon to horizon.

The unique capabilities of the 64-meter antenna were about to be demonstrated once again. To conserve electrical power, the spacecraft began transmitting with low power through its low-gain omnidirectional antenna, an emergency procedure which resulted in a much weakened downlink. Because of its great receiving capability, DSS 14 was the only tracking station able to maintain contact with the disabled spacecraft and its crew during many parts of the aborted mission. Even when the 26-meter stations were able to receive the downlink, the improved clarity of the astronauts' voice channels when DSS 14 came into view was of great assistance and encouragement to those in danger during the subsequent recovery efforts.

After the astronauts were safe and the mission concluded, the DSN received the following appreciative message from the director of flight operations at Houston. "We wish to commend the entire Network for their superior performance in support of *Apollo 13*. In the midst of this most difficult and critical mission, it was extremely reassuring to have a Network with so few anomalies and one which provided us with urgently needed voice and data to bring the crew back safely. We thank you for your outstanding support." Cognizance of the mutual stations passed to the DSN at completion of the Apollo program in 1972, and much of the electronics technology from the wing stations was subsequently incorporated into DSN systems.

The Mariner Era: 1961–1974

THE DEEP SPACE INSTRUMENTATION FACILITY (DSIF)

When the Mariner Era opened in 1961, the network, then known as the Deep Space Instrumentation Facility (DSIF), consisted of three stations—one at Goldstone, California; another at Woomera, Australia, in the desert 320 kilometers north of Adelaide; and a third near Johannesburg, South Africa. In addition to its 26-m diameter dish antenna, each station had receive and transmit capabilities at a frequency of 960 MHz (L-band), and could process and record telemetry data at low data rates from the JPL-designed space probes. Telephone and teletype circuits linked each site to a simple control and coordination center at JPL. These circuits were carried between continents via undersea cable, and by land line, or in the case of Johannesburg, by a high-frequency radio link, to their ultimate destination.

At that time, the loose organizational structure of JPL allowed for a great deal of interaction between the technical groups responsible for the spacecraft and its operation, and the groups responsible for the development, implementation, and operation of the DSIF. This was of prime importance in the area of telecommunications, where it was essential that the spacecraft radio system was designed to be compatible with the radio systems installed at the tracking stations. While the spacecraft and tracking station design groups each focused on their individual design issues, important questions of "telecommunication compatibility" were responded to, informally, by the engineers involved. It was a carryover from earlier times, when the spacecraft radio and DSIF radio systems had been designed and built in the same technical division at JPL. This situation soon changed, however, and the question of telecommunications compatibility, with all its ramifications, became a major concern at JPL as the NASA space program gathered momentum.

Mission Operations was another factor that dominated the planning processes for early deep space missions. Determining and controlling the flight path of a spacecraft, displaying and analyzing the engineering and scientific data received from it, selecting and issuing commands to the spacecraft, and coordinating the activities of the worldwide stations of the DSIF and the communications links connecting them to JPL, are all part of Mission Operations. This specialized activity is conducted in a central location called the Flight Operations Control Center.

As early as 1960, JPL had completed studies for a facility that could handle mission operations not only for the forthcoming Ranger mission, but also for all future space missions. Plans called for a Space Flight Operations Facility consisting of the equipment, computer programs, and groups of technical and operations personnel that would carry out these and all other functions required to fly the Ranger missions. The facility would

Figure 2-4. Network Operations Control Center (NOCC) at JPL, 1969. Refurbished many times since then, and organized in several different ways to meet the expanding needs of the DSN, the NOCC continued to perform that vital function through 1997.

be located in a single building at JPL, and the same teams would work each of the missions to ensure continuity of experience and expertise.

In the short time available before the first Ranger launch, a temporary facility was established next to the existing computer room. Blackboards and pinboards to post current flight status lined the walls, desks, phones, calculating machines, and teletype machines stood wherever space permitted. Later, a few console displays were added for closed circuit television display of printouts from the large mainframe IBM 7040 computers in the adjacent computer room. A large static display board showed the status of the DSIF stations and the communications links between them.

When it became obvious that the makeshift facility would be inadequate even for missions in the immediate future, planning and funding for the much larger permanent facility was accelerated. Designed specifically for spaceflight operations, the Space Flight Operations Facility (SFOF) was completed and began supporting flight missions in 1964. A separate control center for directing the expanded operational functions of the Network occupied a large portion of the new SFOF as shown in Figure 2-4.

The Mariner Era: 1961–1974

The Facility (DSIF) Becomes the Network (DSN)

In a memo to senior staff dated 24 December 1963, the Director of JPL, William H. Pickering, redefined the responsibilities of the DSIF to include "all mission-independent portions of the Space Flight Operations Facility," in addition to the existing responsibility for the tracking stations, the communications system linking them to JPL. The combined organizational structure was named the Deep Space Network (DSN) and would be directed by Eberhardt Rechtin, with the functional title of assistant laboratory director for tracking and data acquisition.

L-band to S-band

At this time also, a most significant engineering change was being made in the DSN. It concerned the change to a higher operating frequency for the uplinks and downlinks. Mainly as a matter of convenience and expediency, the original receivers and transmitters in the DSN had been designed in 1957 and 1958 to operate at a frequency of 960 MHz. In telecommunications terminology, 960 MHz lies within a narrow band of frequencies identified as L-band. The Ranger spacecraft radio system had to match that frequency to be compatible so that it too operated at L-band. However, there are significant advantages to the uplink and downlink performance to be had from operating at a much higher frequency of 2,200 MHz, or S-band.

By 1963, the availability of new radio frequency amplifiers and transmitters, which would operate well at S-band, allowed the DSN to take advantage of the better uplink and downlink performance at the higher frequencies. With an eye to the requirements of future missions, the DSN set about converting all the tracking stations from L-band to S-band. But where would that leave the Ranger program about to restart its lunar flights, and how would it affect the recently approved Mariner missions to Mars?

The solution devised by DSN engineers was to install S-band to L-band conversion equipment at the stations in parallel with the older L-band equipment. This would accommodate the remaining four Rangers and the first two Mariner missions to Venus on L-band. It could also provide for the later Mariner missions to Mars and the proposed Surveyor missions, both of which would use the more efficient S-band uplinks and downlinks. The L/S-band converter would remain in place until the end of the L-band missions, by which time a new, fully S-band system would be in place. The L/S-band converters would then be removed and the conversion to the more efficient S-band operation would be complete. All missions from then on would be on S-band. Of all the improvements in the early years of the DSN, the move to S-band was probably the most significant.

Improvements for the Mariner Mars Missions

The Mariner 1964 Mars missions were the first in which the DSN would use the newly implemented L/S-band capability to adapt the spacecraft S-band downlink to suit the existing DSN L-band receivers. The capability to send commands to the spacecraft would be provided by 10-kW transmitters recently installed at Goldstone-Pioneer, Johannesburg, and Woomera stations.

In addition to the uplink/downlink improvements accruing from the change to S-band, significant improvements to navigation accuracy were expected from the introduction of atomic clocks throughout the DSN to replace the less stable crystal controlled oscillators. These rubidium frequency standards improved the quality of the radio Doppler data provided to the spacecraft navigators by the DSN stations, enabling them in turn to improve the trajectory determination process necessary to deliver the spacecraft to a small aim point in the vicinity of Mars.

Improvements in the NASA ground communications system (NASCOM), which connected the far-flung stations of the DSN to the Flight Control Center at JPL, had already taken place. Teletype links, voice, and high speed data circuits using a worldwide network of microwave links, and undersea cables and radio circuits were connected to a central communications center at the NASA Goddard Space Flight Center in Greenbelt, Maryland. From there the circuits could be distributed to JPL and other NASA Centers scattered throughout USA, as required. The DSN could call on NASCOM to bring up the communications necessary to support a particular mission whenever needed. When the mission was completed, the circuits were turned back to NASCOM to be used for some other NASA space mission, maybe at a different NASA Center.

By the end of 1964, the primitive Flight Operations Control Center used for the first Ranger missions had been superseded by the newly completed Space Flight Operations Facility (SFOF). In a dual string computer arrangement, new IBM 7094 computers performed the data processing for the *Mariner 3* and *4* missions to Mars, while the Ranger processing was still carried out on a later version of the original 7040 machines. This arrangement minimized the need for changes to the existing Ranger software.

In the spacious facilities of the new SFOF building, a large high ceiling room in the middle of the first floor housed the new Mission Control Center. An attractive entrance lobby with reception desk and space-related displays occupied the front of the building, while offices and conference rooms for the various teams of flight project personnel surrounded a DSN Control Room on the other three sides. In the huge basement below the Control Room, engineers in the DSN Communications Terminal monitored, coordinated, and routed the flow of voices and data between the stations, the DSN Controllers above them, and the data processing computers on the second and third floors of the SFOF. An elaborate internal communications system enabled all the users of the SFOF to access and direct data to and from sources and destinations as their duties and authority required during the mission. The Facility was designed to run 24 hours a day, and incorporated a generator-driven "uninterruptible power supply" that would supply primary power if the commercial power failed for any reason. During critical parts of a mission, the generators would be turned on anyway to ensure that the SFOF would not suffer any kind of a "glitch" while critical mission operations were in progress. The muffled roar of the JPL generators always informed the quiet neighborhood that something important was going on at the Lab.

Over the course of thirty years, 1964 to 1994, the management responsibility for the large data processing facilities in the SFOF passed back and forth several times between the DSN and the Flight Project offices at JPL, depending on the current "financial climate" at NASA Headquarters.

However, responsibility for the two DSN control functions, Network Control and Ground Communications Control, always remained in the hands of the DSN. The consolidation of these vital functions, together with expansion of the Network itself, benefited from a continuum of long-term planning and management. As a result, the DSN was perceived from time to time to have a "monolithic structure" by other organizations within the JPL and, indeed, even by NASA Headquarters. Good relations with NASA were essential to the well-being of continued JPL involvement in the planetary space program. Nowhere was this recognized more acutely than in the Tracking and Data Acquisition Office at JPL, where the encouragement of frequent and open channels of communication between DSN personnel and NASA on DSN-related matters did much to improve working relationships and dispel the "monolithic" image of the DSN in those years. With the beginning of the S-band missions and the advent of the SFOF containing the DSN Network and Communications Control Centers, as well as the Mission Operations Control Center and its data processing facilities, the DSN began to assume the form, and much of the substance, of the modern DSN.

The Need for a Second Network

To support the more sophisticated missions of the 1965 to 1968 period, the DSN recognized the need to expand and improve its communications, mission, and network control capabilities. The two major lunar missions nearing launch readiness, Lunar Orbiter and Surveyor, would pave the way for the start of the Apollo program and would transmit data streams at thousands of data bits per second rather than the tens or hundreds of bits per second received from the Mariners and Rangers. The increased complexity of the spacecraft would require expanded and faster monitor, control, and display facilities.

For the first time, the DSN began to find that the simultaneous presence of several spacecraft on missions to different destinations created new problems in network and mission control. The vexing problem of DSN "antenna scheduling" began to arise as several spacecraft began to demand tracking coverage from the single DSN antenna available at each longitude. The difficulty of assigning priority among competing spacecraft whose view periods overlapped at a particular antenna site was to prove intractable for many years. The problem was exacerbated by competition between flight projects from NASA Centers other than JPL, each of which felt entitled to equal consideration, for the limited DSN resources. The DSN was placed in the impossible situation of arbitrating the claims for priority consideration. The regular "Network Scheduling" meetings conducted by the DSN often resulted in the establishment of priorities that were determined more by the dominant personalities in the group than by the real needs of the projects.

With all of these imminent new requirements in mind, NASA decided to embark on a program to construct a second network of DSN stations. Arguments as to where the stations were to be located were complicated not only by technical considerations, but by political and international considerations.

There were already two stations at Goldstone, one at the Pioneer site and a second at the Echo site. Eventually, NASA decided to build two new stations, one at Robledo, about 65 kilometers west of Madrid, Spain, and the other at Tidbinbilla, about 16 kilometers from Canberra, Australia.

NASA looked to the Spanish Navy's Bureau of Yards and Docks to design and construct the Robledo station. For its new facilities in Australia, NASA dealt with the Australian Government Department of Supply through its representative, Robert A. Leslie.

The Mariner Era: 1961–1974

As an Australian foreign national, Robert A. Leslie played a major role in shaping the relationship between NASA-JPL and the Australian government, on whose good offices NASA depended for support of its several tracking stations in that country. With family origins in the state of Victoria, Australia, and an honors degree in electrical engineering from the University of Melbourne (1947), Leslie had worked on radio controlled pilotless aircraft for the military in both England and Australia for fifteen years before he encountered NASA. He was a high-ranking officer with the Australian Public (Civil) Service (an affiliation that he retained throughout his career) when, in 1963, he became the Australian government's representative for NASA's new deep space tracking facility being built at Tidbinbilla, near Canberra in southeastern Australia.

As might be expected, the success of a NASA venture in a foreign country depended to a large extent on the personalities of the people who were directly involved on each side of the international interface. The foundation for the success of what later became the Canberra Deep Space Communications Complex (CDSCC) was, in no small part, due to Bob Leslie's personal ability to "get along" with people at all levels. In representing the Australian side of negotiations between NASA and JPL, Bob Leslie was firm but gracious, capable, and friendly. His unassuming "paternal" manner endeared him alike to counterparts at NASA, his colleagues at JPL, and his staff in Australia.

Along with a few key Australian technical staff members, Leslie spent a year at Goldstone assembling and testing the electronic equipment that would subsequently be reassembled at Tidbinbilla to complete the first 26-meter tracking station (DSS 42) at the new site. He was the first director of the new Complex when it began service in the Network in 1965. It was there that he established the procedures and protocols on which all future DSN operational interactions between JPL and the Australian stations would be based.

A few years later, in 1969, Leslie left the "hands-on" environment of the deep space tracking station to head the Australian Space Office, a branch of the Australian Government that, under various names and government administrations, would guide future expansion and consolidation of all NASA facilities in Australia. In that capacity, his charm, experience, and wisdom served the DSN well.

Leslie's build was stocky and solid, his appearance craggy, his attitude "laid back." Cheerful, sociable, easy to talk to, and blessed with a good sense of humor, he was held in high regard by everyone he met, Australian or American. Tennis was his sport, fishing his hobby, and "do-it-yourself" home building his passion. In his younger days, he actually excavated the ground with shovel and wheelbarrow and single-handedly built the family swimming pool at his home in Canberra. Many a JPL engineer enjoyed a poolside barbecue at the Leslie home in the course of a technical visit to the station.

Robert Leslie retired in 1983 and died in Canberra, Australia, in 1996.

By mid-1965, the two new stations were completed and declared operational. The DSN then had two stations in Australia, (Woomera and Tidbinbilla), one in Spain, and one in South Africa. In addition, a permanent spacecraft monitoring station had been built at Cape Canaveral to replace the temporary facility with its hand-steered tracking antenna. Impressive as this growth was, still greater changes were in progress.

Larger Antennas Are Needed

Consistent with the JPL vision for missions to more distant planets, and more powerful communications links to support them, the DSN had long recognized a need for larger antennas, that is, larger than the existing 26-m antennas. Studies had shown that a diameter of 64 meters was about the maximum practical limit to an antenna which would have sufficient stability and structural integrity for DSN purposes. Interest in building a giant new antenna of that size, employing radical new design and construction techniques, became a reality as early as 1962 at NASA and JPL. Feasibility study contracts for an advanced antenna had already been issued to numerous U.S. corporations and, finally in January 1963, the Rohr Corporation was selected to build the antenna at a suitable site a few miles from the existing Echo and Pioneer locations at Goldstone. Since the advanced antenna was originally intended to support the first missions to Mars, the site was appropriately called "the Mars site."

Design work began immediately (1963), and procurement and fabrication of steel components started a year later. Roads, concrete foundations, alidade structure, and diesel generator buildings were completed in 1964, followed by the control room and elevation bearings. The supporting framework for the antenna panels followed next, and by mid-1966, the 64-meter Mars Antenna was ready for service. Although the first signals from Mars—transmitted by *Mariner 4*—were detected in March 1966, a great deal of performance testing, personnel training, and calibration remained to be completed before the Mars antenna was considered a fully operational addition to the DSN.

The Mariner Era: 1961–1974

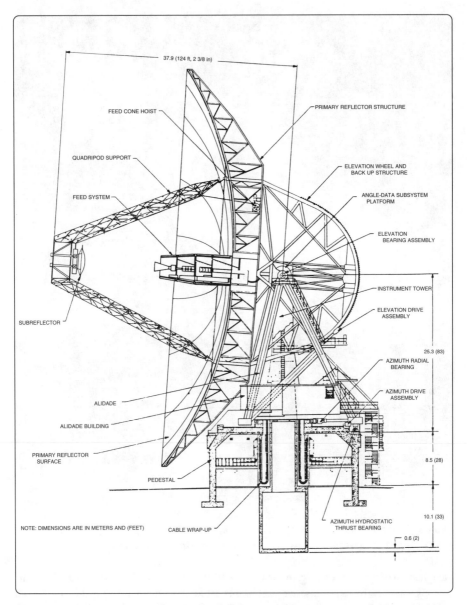

Figure 2-5. Side elevation of the 64-meter azimuth-elevation antenna at Goldstone, 1966.

Figure 2-6. The 64-meter antenna at Goldstone, 1966. The pickup truck parked to the right side of the pedestal gives an impression of the immense size of the antenna.

Designed to remain operating in wind speeds as high as 80 km per hour, the Mars antenna was constructed of massive steel beams, which, together with its pedestal, weighed about 33 million kilograms. At wind speeds exceeding 80 km per hour, the antenna was stowed in a fixed position to protect it from permanent damage. In that position it could withstand hurricane force winds. The dish, and its azimuth-elevation mounting atop the pedestal, weighed nearly 13 million kg. The structure rotated in azimuth on three flat bearing surfaces that floated on a pressurized film of oil, about the thickness of a sheet of paper. The reflecting area of the dish surface was 3,850 square meters. It could be pointed to a given position in space with an accuracy of 0.006 degree. The major structural components and dimensions of the 64-m antenna are evident in the side elevation shown in Figure 2-5.

In a single, major step forward, the Mars antenna provided the DSN with more than six times the transmitting power and receiving sensitivity of the 26-meter antennas, and more than doubled their tracking range. This was a significant new capability indeed. But, important as it was, it was only one in a series of stepwise improvements that took place in the DSN in the mid-1960s. A photograph of the new 64-meter antenna at Goldstone, taken soon after its completion in mid-1966, is shown in Figure 2-6.

Most of the credit for the conceptual design of the Goldstone 64-m antenna, the first of its kind ever built for tracking planetary spacecraft, went to a brilliant electrical engineer who had come to JPL originally in 1951 to work on missile tracking systems during the Corporal and Sergeant tests at White Sands. His name was William D. Merrick. In 1958, it was Bill Merrick who masterminded the design and construction of the first two 26-m antennas at Goldstone before taking up the job of project manager for the Advanced Antenna System. The "Hard Core" Design Team that Merrick formed to carry out the complex design and analysis task for the new 64-m antenna contained the best engineering talent in the fields of servos, microwaves, mechanical, optical, structural, civil, electrical, and hydraulic engineering, stress analysis and contract management that could be found at JPL or elsewhere.

He managed his "Hard Core" team, and the contractors who eventually built the antenna, in a somewhat unorthodox manner, but the on-time, in-budget and to-specification end result demonstrated the efficacy of his management style.

Although he insisted on high standards of performance from his people, his propensity for practical jokes and his sense of humor endeared him to those who worked on his team. Those who needed to direct or limit his activities found him difficult. He thought of himself as a person who got things done regardless of obstacles such as budget constraints, administrative orders, or alternative opinions that stood in his path. He could always find a way around them.

When the Goldstone 64-m antenna was completed in 1966, it was regarded by many who appreciated the effort that he put into it as a testament to Bill Merrick. After the 64-m antennas in Spain and Australia were completed, Merrick moved from the "hands-on" engineering, at which he excelled, to a staff position which offered little challenge for his special talents.

Bill Merrick retired from JPL in 1984 to take up a position in industry, and died at his home in Ventura, California, in 1997.

Incremental Improvements in the Network

Important as it was, the new 64-m antenna at Goldstone was only one of many improvements in technology and management that were added to the Network in the mid-1960s. Like the 64-m antenna, each change represented an incremental increase in the overall capability and performance of the Network. As Corliss put it, "The DSN did not suddenly change from a lunar to a planetary network, or a low-bit-rate network to a high-bit-rate network, or a network burdened with mission-dependent equipment to a multi-mission network. The DSN was always being upgraded, some of the steps were small, some of the others were big." Although there were many changes in many areas, some small, some large, among the most important at this time was the improvement in the DSN ranging system.

The technique of measuring the range (distance) from a reference point on Earth, to a spacecraft at lunar distance, to an accuracy of a few centimeters was one thing. To do the same thing for a spacecraft at planetary distance required a significant new technique for measurement of range. A Lunar Ranging system, known as the Mark I, was already in place in the mid-1960s. Although it was designed for the lunar missions, it had successfully tracked *Mariner 5* out to nearly ten times the distance of the Moon.

Planetary ranging, however, would require a hundredfold increase in capability. This could be achieved with new and more complex range codes transmitted to the spacecraft on the uplink, and new methods of detecting and decoding the range code retransmitted from the spacecraft on the downlink. More precise methods for calibrating the system, and for measuring time, would also be required. Two planetary ranging systems were developed in this period, one by Robert C. Tausworthe, the other by Warren L. Martin. The two systems were aptly named the "Tau" and the "Mu." In Tausworthe's design, the time delay between a pseudo-random code transmitted on the uplink and returned on the downlink was used to measure the range of the spacecraft. Martin's design used a sequential binary code, assisted by a Doppler rate-aided function to carry out the decoding, for range measurement. This design also automatically produced a type of data called DRVID, for Differenced Range versus Integrated Doppler, which allowed the range data to be corrected for the effects of charged particles along the path between spacecraft and tracking station. Together these improvements extended the performance of the Mu system to the point where high-precision ranging measurements were possible out to a range of 2.6 AU.

The Mark I, and a later Mark IA version, was used for the Lunar Orbiter and Surveyor missions in 1966 and 1967. The Tau planetary ranging system saw its first operational use on *Mariner 5* in 1967 and 1968, and an improved and more stable version of the Tau system was used from 1969 to 1971 for *Mariners 6* and *7*. Later still, the Mu ranging system appeared in the Network and was used for relativity experiments on the extended missions of *Mariners 6* and *7* in 1970.

For the accurate ranging required by the Lunar Orbiter mission, the electronic clocks at the various tracking stations had to be synchronized to within 50 microseconds of the master clock at Goldstone. In the early days, the DSIF had to rely on radio signals from WWV for station time-keeping purposes. The most accurate measurement at the time was in milliseconds rather than microseconds. Later, transportable cesium atomic clocks from the National Bureau of Standards offered improved accuracy for station time synchronization. However, this method proved to be too expensive and inconvenient for operational use. A more operationally convenient method of time synchronization using the spacecraft itself was eventually implemented for Lunar Orbiter and improved the accuracy to 20 microseconds.

As usual though, the DSN was looking for something better. It showed up in 1966 as the "Moon-bounce Time Sync" scheme. This depended upon the propagation delay of a precision timed X-band signal which was transmitted from the Goldstone Venus station to each of the overseas stations during their mutual lunar-view periods. The Goldstone Venus station acted as master timekeeper and timing signals reached the desired stations via the Moon. By 1968, the DSN was setting its clocks to an accuracy of 5 microseconds using this technique, which had been adopted for operational use throughout the Network.[8]

Driven by the rapidly increasing data rates being transmitted by the new spacecraft coming into the DSN, the capability of the ground communications system had to be expanded to permit the tracking stations to return data to JPL as rapidly as possible. More stations needed more operational voice and teletype traffic to control them. Consequently, in 1967, a new computer-based teletype communications switcher was installed in the Com Center in the SFOF and the high-speed data circuits were upgraded to carry 2,400 bits per second.

The SFOF, too, felt the pressure to keep up with the ever-increasing requirements for higher data rates on the uplinks and downlinks for the new missions. The SFOF data processing capability was expanded to three computer strings, each comprising an IBM 7044 and large disk memory for input/output processor functions, followed by an IBM 7094. The 7094 was used as the primary processor for the complex calculations related to orbit determination; it generated antenna pointing and receiver tuning predictions used for initial acquisition, calculated the parameters needed for spacecraft maneuvers, and manipulated the spacecraft tracking data generated by the DSN stations. In addition, a special system was developed to display monitor and performance data for the DSN station controllers.

Flight Project Requirements Become Formalized

By 1967, the interactions between the DSN and new flight projects needing DSN support had become very formalized. No longer were all the missions managed by JPL, where loose interdepartmental agreements would suffice to commit funds and resources to meet a specific JPL objective. Now, other NASA Centers were responsible for the spacecraft and its scientific payload, and the function of JPL was to provide the tracking and data acquisition support using the resources of the DSN. Even within the JPL institution, the changing management organization at NASA Headquarters was reflected in entirely separate funding channels for the flight project and the DSN organizations. This situation resulted in a very formal process for the presentation of flight project requirements to the DSN, and the acceptance (or rejection) of these requirements by the DSN.

NASA Headquarters insisted that these negotiations be set out in detail in two formal documents, the Support Instrumentation Requirements Document (SIRD) for the flight projects, and the NASA Support Plan (NSP) for the DSN. Before any flight project could begin to effectively design its mission, it not only had to have a signed SIRD from one office of NASA Headquarters but a signed NSP from another. With these in hand, a newly approved flight project could truly begin. In the years ahead, the relationship between the DSN and all flight projects would always be a reflection, for better or for worse, of the SIRD/NSP negotiations.

Finally, the DSN was beginning to use computer assistance to resolve mounting conflicts for station time, not only for spacecraft tracking purposes but also for station maintenance, implementation of new hardware and software, and testing time. Also, the stations were becoming much more complex, and consequently required more time for calibration, con-

The Mariner Era: 1961–1974

figuration and check-out prior to the start of each tracking pass. In addition, the flight projects were becoming less tolerant of station outages which depleted their hard won tracking time allocation. The persistent issue of tracking time allocations would never go away; it was ameliorated somewhat as techniques to manage it improved, but it never went away.

This was the state of the DSN when, within a very short period it found itself dealing with more spacecraft than any of its people could have imagined a few short years earlier.

The DSN Becomes a Multimission Network

Referring to this period in the history of the DSN, William Corliss wrote, "The NASA lunar exploration program absorbed the bulk of the DSN support capability during the 1966-1968 period. Surveyor, Lunar Orbiter, and the backup support provided to the first Apollo flights combined to utilize the DSN almost fully. The lunar program at this time consisted of short-lived missions, a few days long, and the DSN was able to divert some of the support necessary for the *Mariner 5* shot to Venus and also accord some support for *Pioneers 6–9*, which kept on operating long after the ends of their nominal missions. *Mariner 4* was also picked up again and became another example of an extended mission. The new 64-meter Mars antenna was called upon to support almost all of these missions, although not always as a prime station. The DSN during this period was not yet a multimission network, although it was supporting many missions simultaneously. The DSN stations were still crowded with mission-dependent equipment and this situation was the very antithesis of a multimission philosophy."

These and other issues related to the general problem of data return from deep space are discussed by Hall, Linnes, Mudgway, Siegmeth, and Thatcher[9] and an excellent description of the growth of the DSN from inception through 1969 is given by Renzetti et al.[10]

To simplify the increasing problem and cost of accommodating the many different kinds of special telemetry and command processing equipment appearing at the tracking stations to support the various missions, the DSN developed a "multi-mission" philosophy.

Instead of each flight project bringing its own equipment (and people to operate it) to the stations, the DSN would provide a generic set of equipment capable of operating over a wide range of parameters at each station. Future projects would have to design their uplinks and downlinks to fall within the capability of the equipment provided by the DSN. That way, all flight projects could use the same set of ground support equipment, and operations costs and complexity could be reduced. With some modification, this concept has been preserved to the present time.

Its success has been largely due to the DSN policy of advanced development, in keeping the DSN capability well ahead of the flight projects' demands to use it. The DSN was able to do this by basing its long term planning on judicious forecasting of future mission requirements.

By the early part of 1968, the Lunar Orbiter and Surveyor Lander missions had been concluded. Spectacularly successful, they had returned an avalanche of lunar surface and other science data, which was assimilated into the design studies for the Apollo piloted missions.

It was not realized at the time that a DSN mission would not view the Moon again for nearly twenty-five years. In December 1992, a remarkable picture of the Moon in orbit around Earth would be captured by a spacecraft called *Galileo*, outbound on a mission to Jupiter and its satellites. By then, the DSN and the environment in which it functioned would be very different indeed.

The missions to follow over the next six years (1968–1974) would focus the DSN's attention outward to even more distant planets and to the Sun. During these years, the DSN made enormous improvements in its operational capability to meet the requirements of the new planetary missions.

New telemetry and command systems with true multimission features were added throughout the Network. Not only were these additions multimission in nature but their ability to run at higher data rates on the uplink and downlink had been increased to 16 kilobits per second for telemetry and 32 bits per second for the command link. New forms of coding the downlink data to improve the quality of the science data delivered to the flight project had also been developed.

The speed of identification and correction of problems in the Network was improved by giving station controllers better tools to monitor and control the configuration, status, and performance of remote tracking stations.

The Doppler and ranging data generated by the tracking stations and used by the flight projects for spacecraft navigation purposes was improved in accuracy and extended in range. Studies were made to better understand the disturbing effect of charged particles on the radio path between the spacecraft and the DSN antennas. Thrown off by the Sun, these

The Mariner Era: 1961–1974

particles introduce delays in the radio path and consequent inaccuracies in the navigation data. Ways of calibrating out these undesirable effects were found and put into effect.

To create a realistic operational environment to train DSN operators prior to the start of a real mission, the DSN uses a Simulation System. Electrically generated signals can be programmed to simulate a real spacecraft under a variety of flight conditions. Artificial faults can be introduced and the operations personnel can be trained in the proper reactions. With an increasing number of more complex spacecraft coming into the Network in the early 1970s, a Simulation Center was established at JPL to carry out the simulation tasks for all missions. Conversion assemblies at the stations allowed the individual stations to interact with the "SIM Center" to suit their individual training needs.

The addition of "wideband" circuits that could handle data at 50 kilobits per second between the Goldstone Mars station and the SFOF, and 28.5 kilobits per second to Spain and Australia, gave the Ground Communications Facility (GCF) as it was now called, a capability to deliver the increased volume of data from the now enhanced DSN to the SFOF. Because the existing high speed circuits, operating at 4800 bits per second, could handle much of the traffic formerly handled by teletype, the teletype services were phased out.

It had become obvious that the SFOF itself would not be able to cope with the demands of the future flight projects for data processing and handling. The data streams planned for the future Mariner and Viking missions to Mars would overwhelm the data-handling capability of the old IBM 7040/7044 generation of computers. An Advanced Data System study group recommended a completely new design for the new Mark III SFOF, which involved, among other things, replacing the old machines with new IBM 360/75s.

No longer an appendage to the DSN, the SFOF became an autonomous organization with full responsibility for supporting the flight missions. DSN responsibility in the SFOF was redefined to cover only Network control and the data processing necessary to carry out that function.

Openings and Closings

The tremendous improvement (a factor of six times over a 26-meter station) in uplink and downlink performance, afforded by the 64-meter antenna at Goldstone, was offset somewhat by the limitation of a single antenna at only one longitude. To provide the continuous coverage with high performance required by the new missions, similar antennas in Spain and Australia would be needed. The viability of the 64-meter design had already been established by the Mars station experience, and by June 1969, the Collins

DSCC	Location	DSS	DSS serial designation	Antenna Diameter, m (ft)	Antenna Type of mounting	Year of initial operation
Goldstone	California	Pioneer	11	26(85)	Polar	1958
		Echo	12	26(85)	Polar	1962
		(Venus)[a]	13	26(85)	Az-El	1962
		Mars	14	64(210)	Az-El	1966
Tidbinbilla	Australia	Weemala	42	26(85)	Polar	1965
		Ballima	43	64(210)	Az-El	1973
—	Australia	Honeysuckle Creek	44	26(85)	X-Y	1973
Madrid	Spain	Robledo	61	26(85)	Polar	1965
		Cebreros	62	26(85)	Polar	1967
		Robledo	63	64(210)	Az-El	1973

[a] A maintenance facility. Besides the 26-m (85-ft) diam Az-El mounted antenna, DSS 13 has a 9-m (30-ft) diam Az-El mounted antenna that is used for interstation time correlation using lunar reflection techniques, for testing the design of new equipment, and for support of ground-based radio science.

Figure 2-7. Composition of the Deep Space Network, 1974.

The Mariner Era: 1961–1974

Radio Company of Richardson, Texas, had been selected to build two more 64-meter antennas. To simplify logistics and support facilities, the new antennas would be built near the existing 26-meter sites at Canberra, Australia, and Madrid, Spain.

Construction of the Australian antenna began soon after and was completed in three years, without significant delays, in mid-1972. Electronics were installed and tested and, after a period of calibration, operator training, and performance demonstration tracking, the station reached full operational status in April 1973.

Although work on the Spanish antenna encountered some problems along the way, it followed a similar pattern and was declared operational in September 1973. By this time, a second 26-meter antenna had been added to each of the complexes at Madrid and Canberra.

The DSN now had two networks of 26-meter antennas plus a complete network of 64-meter antennas. They would be numbered so as to indicate the areas in which the antennas were located; 11 to 19 for Goldstone, 41 to 49 for Australia and 61 to 69 for Spain. Thus the DSN would consist of Deep Space Station 14 (DSS 14) at Goldstone, DSS 63 at Madrid, and DSS 43 at Canberra, plus the six 26-meter stations, which were similarly identified.

At each location, the number and types of antennas continued to increase. Larger buildings were built to house the hundreds of racks of electronic equipment required to control the antennas, and to transmit, receive, record, and process the spacecraft data before it could be fed to the NASCOM communication circuits for transmission back to JPL. Instead of a single antenna and a modest Control Room at each longitude, there were now several antennas, a Signal Processing Center, power station, cafeteria, laboratory, workshops, offices, and supporting facilities. The entire establishment had become a Deep Space Communications Complex (DSCC)—GDSCC for Goldstone, CDSCC for Canberra, and MDSCC for Madrid.

The new complexes at Canberra and Madrid now provided the support formerly given by Woomera and Johannesburg, both closed by NASA: Woomera for economic reasons, Johannesburg for diplomatic reasons. The composition of the Network as it existed in 1974 after these closings had taken place is tabulated in Figure 2-7.

LOOKING BACK

Born out of the challenge to the prestige of American technology created by the appearance of the Soviet satellite *Sputnik* in October of 1957, by 1974 the DSN had evolved into the world's largest and most sensitive radio communications and navigation network for unpiloted interplanetary spacecraft engaged in the exploration of the solar system.

From a loose collection of remote transmitting and receiving stations maintained and operated by essentially dedicated engineering personnel from JPL, the DSN had matured into a fully integrated global network of operational tracking stations, maintained and operated by competent nationals of the cooperating countries in which the sites were located.

It had become a worldwide organization of tracking stations, located on three continents and connected by high quality communications satellite and cable links to a Control Center in Pasadena, California. From there, a web of phone, teletype, and modem circuits distributed scientific data in various forms to the science community in research centers and universities throughout the U.S. and to several experimenters at locations in other countries.

Fifteen or so years after inception, the Pioneer, Ranger, Surveyor, Lunar Orbiter, and Mariner spacecraft had carried out scientific missions to the Moon, Sun, Mercury, Venus, Mars, and Jupiter with great success. These spacecraft were spread out in all directions and distances throughout the solar system. During these years, the DSN was frequently called upon to track as many as six such spacecraft simultaneously, the limit to the requirements being set by the number of antennas in the DSN itself.

Two big Voyager spacecraft were being built for 1977 missions to Jupiter. Two ambitious Viking missions to Mars, even though they would not be launched for another year, were demanding and getting most of the DSN attention.

Yet, this was the era of the Mariners. As it drew to a close, the DSN was moving toward a new phase of development. It would be simultaneously involved with several large flight projects directed not by JPL, as had been the case in the Mariner program, but by other NASA Centers or by foreign space agencies. We will refer to this period of the DSN history as the Viking Era.

The Mariner Era: 1961–1974

Endnotes

1. R. Cargill Hall, *Lunar Impact: A History of Project Ranger*, NASA History Series (Washington, DC: NASA SP-4210, National Aeronautics and Space Administration, Scientific and Technical Information Branch, 1977).

2. William R. Corliss, "A History of the Deep Space Network," NASA CR-151915, 1976.

   N. A. Renzetti, "Tracking and Data Acquisition Support for the Mariner Venus 67 Mission," JPL Technical Memorandum TM 33-85, Vols. I–III, September 1969.

   N. A. Renzetti, "Tracking and Data Acquisition Support for the Mariner Mars 1969 Mission," JPL Technical Memorandum TM 33-474, Vols. I–III, September 1971.

   R. P. Laeser, et al., "Tracking and Data Acquisition Support for the Mariner Mars 1971 Mission," JPL Technical Memorandum TM 33-523, Vols. I–IV, March 1972.

   E. K. Davis, "Mariner 10 Mission Support," Deep Space Network Progress Report PR 42-27, March/April 1975, 15 June 1975, pp. 5–9.

3. Bruce C. Murray, *Journey Into Space* (New York: W. W. Norton, 1989).

4. Richard O. Fimmel and Earl J. Montoya, "Space Pioneers and Where Are They Now" (Washington, DC: NASA EP-264, Educational Affairs Division, National Aeronautics and Space Administration, 1987).

   N. A. Renzetti, "Tracking and Data Acquisition Support for Pioneer Project," JPL Technical Memorandum TM 33-426, Vols. I–IV, November 1970.

   A. J. Siegmeth, "Pioneer Mission Support," JPL Technical Report TR 13-1526, Vol. III, 15 April 1972, pp. 8–15.

5. N. A. Renzetti, "Tracking and Data Acquisition Support for Surveyor Project," JPL Technical Memorandum TM 33-301, Vols. I–V, December 1969.

6. J. R. Hall, "Tracking and Data System Support for the Lunar Orbiter (Project)," JPL Technical Memorandum TM 33-450, Vol. I, 15 April 1970.

7. William R. Corliss, "Histories of the Space Tracking and Data Network (STADAN), Piloted Space Flight Network (MSFN) and NASA Communications Network (NASCOM)," NASA CR-140390, June 1974.

P. S. Goodwin, "Apollo Mission Support," The Deep Space Network, Space Programs Summary SPS 37-56, Vol. II, 31 March 1969, pp. 35–42.

R. B. Hartley, "Apollo Mission Support," JPL Technical Report TR 32-1526, Vol. X, 15 August 1972, pp. 41–48.

R. B. Hartley, "Apollo 13 Mission Support," The Deep Space Network, Space Programs Summary, SPS 37-64, Vol. II, 31 August 1971, pp. 7–11.

8. Jet Propulsion Laboratory, "The Deep Space Network," Space Program Summary, Vol. III, SPS 37-43, SPS 37-45, SPS 37-53, 1966–1968.

9. J. R. Hall, K. W. Linnes, D. J. Mudgway, A. J. Siegmeth, J. W. Thatcher, "The General Problem of Data Return From Deep Space," *Space Science Reviews* 8 (1968): 595–664; Dordrecht-Holland: D. Reidel Publishing Co.

10. N. A. Renzetti, "A History of the Deep Space Network From Inception to January 1969," JPL Technical Report TR 32-1533, Jet Propulsion Laboratory, Pasadena, California, September 1971.

CHAPTER 3

THE VIKING ERA: 1974–1978

THE VIKING ERA MISSION SET

The tracking and data acquisition requirements of the Viking Mission to Mars dominated the engineering and operations resources of the DSN during the years 1974 through 1978. Obviously, there were also many other missions during this period, all of them demanding and receiving the attention of the DSN as they moved through the various phases of their respective missions. The Deep Space mission set for that period is shown in Figure 3-1 with the relative importance of each noted.

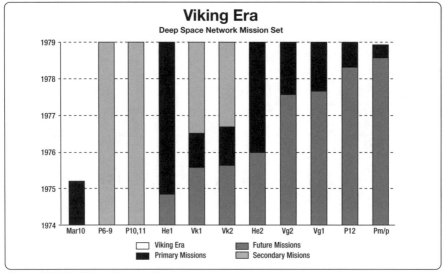

Figure 3-1. The Viking Era mission set.

Primary Missions are those in the prime mission phase of their project. Secondary Missions are those that have completed their prime mission objectives and are in an extended mission phase. Future Missions are those which, although not yet launched, have significant prelaunch support demands on DSN attention and resources. Figure 3-1 also shows the spacecraft that were moving through these mission phases during this arbitrarily defined Viking Era. *Mariner 10*, for example, was engaged in multiple encounters with Venus and Mercury. *Pioneers 6-9* were observing interplanetary fields and charged particle phenomena from solar orbits inward and outward of Earth orbit. *Pioneers 10 and 11* were making the first encounters with the planet Jupiter.

The Viking Era: 1974–1978

Also during this era, two Helios spacecraft from the Federal Republic of Germany (West Germany), the first of the non-NASA spacecraft to be supported by the DSN, were launched, then traveled on helio-centric orbits around the Sun. The long-awaited grand tour of the solar system was also inaugurated with the launch of the two JPL Voyager spacecraft. *Voyagers 1* and *2* would also become dominating influences on DSN operations and engineering and establish an era of their own. Toward the end of the Viking Era, two other missions would revisit Venus: the *Pioneer Venus Orbiter* for remote sensing, and the *Pioneer Multiprobe* to determine the composition and structure of the Venusian atmosphere.

As the Viking Era opened, the Viking project was replanning its previously cancelled 1973 Mars Lander missions for two launches in 1976. Strongly managed by the Langley Research Center in Hampton, Virginia, Viking was to dominate the DSN scene before, during, and after the rather short Prime Mission, and for that reason it seems appropriate to consider these years as the Viking Era.

It would be the tracking and data acquisition requirements for the new missions in the Viking Era Mission Set that would provide the rationale for DSN Operations and Engineering activity in the years 1974 through 1978.

THE NETWORK

The DSN moved from the Mariner Era into the Viking Era with one complete subnetwork of 64-meter antennas and two complete subnetworks of 26-meter antennas. The 64-meter antennas at DSS 14 Goldstone (Mars), DSS 43 Canberra (Tidbinbilla), and DSS 63 Madrid (Robledo) had come into operational service in 1966, 1973, and 1974, respectively. Each station could transmit and receive on S-band. In addition to S-band, DSS 14 could receive on X-band.

At the Canberra and Madrid Complexes, one of the 26-meter antennas was colocated with the 64-meter antennas at the Tidbinbilla and Robledo sites, while the second 26-meter antennas were located at the Honeysuckle and Cebreros sites. Goldstone 26-meter antennas were at separate sites (Pioneer and Echo). The 26-meter antennas at Goldstone, Canberra, and Madrid (DSS 11, 42, and 61) became operational on S-band in 1964, 1965, and 1965, respectively, while DSS 12, 44, and 62 became operational in 1965, 1974, and 1966, respectively. All 26-meter antennas had S-band receive-and-transmit capability.

Operations control and signal processing for DSS 43/42 and DSS 63/61 were conducted from a single control facility at each site. For this reason, these stations were called conjoint stations. The remaining stations were operated and controlled by individual control centers.

By the end of the Viking Era in 1978, work was in progress to convert one 26-meter subnet (DSS 12, 42, 61) to 34-meter, S/X-band capability. The first station in the series, DSS 12, had been completed in October and was supporting inflight operations with its new capability in November. A block diagram of the DSN as it appeared toward the end of 1978 is shown in Figure 3-2.

The Viking Era would see many significant changes in the structure, technical capabilities, and operational management of the DSN.

The spacecraft and mission control functions for the inflight projects would be conducted from a facility now called the Mission Control and Computing Center (MCCC). The MCCC would be a completely separate entity from the DSN. Located in Building 230 at JPL, the former Space Flight Operations Facility (SFOF), it would house the Mission Director and his mission operations teams, as well as the computers and data processing facilities needed to process, distribute, and display the real-time mission data as they arrived from the DSN tracking stations. A clear, manageable interface for the transfer of data between the DSN and MCCC had been defined, and all DSN data were delivered by the Ground Communications Facility (GCF) to that interface in accordance with pre-established specifications regarding the quality, quantity, and validity of the data.

The Viking Era: 1974–1978

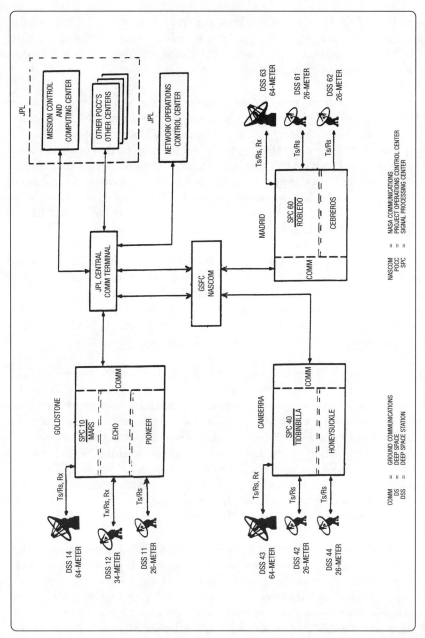

Figure 3-2. Functional block diagram of the Deep Space Network, 1978.

As individual flight missions were completed, they moved out of the MCCC and new missions took their place. Some flight projects, notably Pioneer and Helios, elected to establish Project Operations Control Centers (POCCs) at their own institutions. In these cases, the GCF routed the DSN data directly to the POCCs.

The operations control and management functions for the Network were conducted from a facility now called the Network Operations Control Center (NOCC). It, too, was located in the SFOF Building at JPL and housed the DSN Operations Teams and the hardware and software necessary to generate and transmit the data products such as predicts, sequence of events, configurations, and schedules needed by the stations to conduct real-time spacecraft tracking operations. At the NOCC, station controllers coordinated activities with the remote site operations personnel. The controllers acted as a single point of contact for the flight Project Mission Operations people in the MCCC (or POCC) and the DSN on matters concerning real-time flight operations.

All data passing to and from the Deep Space Stations (DSSs) and the MCCC also passed in parallel through the NOCC. There it was monitored, validated for quality and completeness, recorded, and eventually transferred to the flight project in the form of a digital tape called an Intermediate Data Record (IDR). A very extensive communications, monitoring, and display capability was provided in the NOCC to allow the DSN real-time operations staff to perform all of these functions.

The operations control and communications circuit coordination functions for the GCF were carried out in the Central Communications Terminal (CCT), also located in the SFOF Building at JPL. In the basement of the SFOF resided the complex terminal and switching equipment for the worldwide network of communications circuits on which the DSN depended for its voice, high-speed, and wide-band data communications with the remote tracking stations of the DSN. Working closely with the NOCC real-time operations teams, the GCF Operations Teams performed similar functions of monitoring and coordinating the activities of the NASA Communications Division (NASCOM) of the Goddard Space Flight Center, which provided the intercontinental circuits, the local short-haul line carriers, and the communications supervisors at the DSSs.

All communications traffic between the DSSs and the NOCC/MCCC first passed through the CCT, where it was routed to its prescribed destination. This function, originally carried out by established line-switching methods, would be replaced by more flexible and reliable message-switching techniques by the end of the Viking Era. By that time also, new capabilities for automatic detection and correction of short line outages and data block errors would ensure continuous and error free communications support for real-time flight operations.

The Viking Era: 1974–1978

NETWORK OPERATIONS

The Helios Mission

As the urgency of the NASA space program of the mid-1970s carried the DSN forward into the Viking Era of 1975–1977, it was not an American but a German spacecraft that vied with the NASA missions for DSN attention.

Conceived in 1966 as a cooperative project between NASA and the Federal Republic of Germany (West Germany), the Helios Mission would eventually consist of two solar probes launched toward the Sun—one in each of calendar years 1974 and 1975—to achieve a perihelion distance (point of closest approach) from the Sun of approximately 0.3 AU (Astronomical Unit; the mean distance between Earth and Sun, approximately 150,000,000 kilometers or 93,000,000 miles).

Each spacecraft would carry ten scientific experiments to perform field and particle measurements in the region between Earth and the Sun. The target perihelion distance of 0.3 AU was chosen because it bounded a previously unexplored region of interplanetary space and was within the estimated limits of the high-temperature capability of the solar cells necessary to generate power for the spacecraft. As viewed from Earth, the Helios trajectory took the form shown in Figure 3-3.

It would provide unique opportunities for high-value scientific observations during the repeated periods of perihelion (closest approach), aphelion (farthest approach), and solar conjunctions (spacecraft obscured by Sun).

The spacecraft carried two 20-watt S-band transponders, both of which operated on the same uplink and downlink frequencies. The transponders provided a capability for coherent or non-coherent operation, turnaround ranging, and ground command of the spacecraft.

The Helios spacecraft employed one telemetry channel to transmit both science and engineering data back to Earth. Both data types were convolutionally encoded and modulated on to a single 32.768-kilohertz subcarrier, which in turn was phase-modulated onto the S-band downlink. The combined science and engineering information data rate could be varied from 8 bits per second to 4,096 bits per second in steps of two.

Ground command capability was provided by two actively redundant receiver/command detector chains; one connected permanently to the low-gain (omni) antenna, the other to the medium-gain antenna system. Separate command subcarrier frequencies permit-

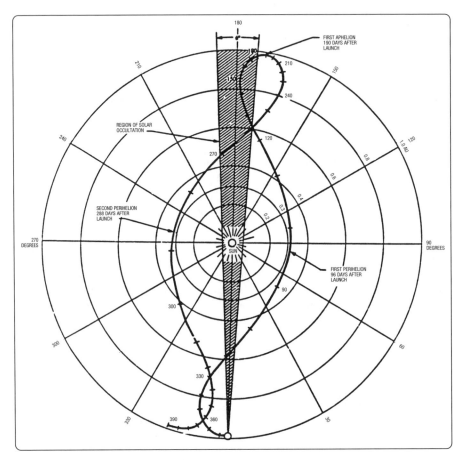

Figure 3-3. Heliocentric orbit of *Helios 1*.

ted ground control to select either chain, regardless of spacecraft orientation. The command data rate was Manchester-coded 8 symbols/second, which phase-modulated either the 448- or the 512-hertz subcarrier.

The first Helios spacecraft (*Helios 1*) was launched on a Titan-Centaur launch vehicle from Cape Canaveral, Florida, on 10 December 1974.

On 15 March 1975, 96 days after launch, *Helios 1* reached its first perihelion at 0.31 AU (see Figure 3-3). Although the perihelion was the focal point of scientific interest for the project, the period of 25 days about perihelion was also fascinating to investigators. During this period, the spacecraft would traverse hitherto unexplored regions of the inner solar system and collect scientific data that would increase our understanding of the Sun's influence on Earth and its surroundings.

Planning and scheduling the supporting stations (DSS 12, 14, 42, 44, and 62) to cover this event were enormously complicated by the conflicting requirements for DSN support generated by the *Mariner 10* mission (then making its third Mercury Encounter on 16 March 1975) and support for *Pioneer 11*, which was completing encounter operations at Jupiter at the same time. To further complicate the problem, the Viking and the Pioneer Venus projects were also demanding attention for pre-launch testing and training activities. Because of this situation, Helios was unable to obtain some of the prime 64-meter antenna support and much of the redundant 26-meter station coverage it desired. Nevertheless, by a process of pass-by-pass negotiation, and a willingness by all flight projects involved to adapt tracking schedules to real-time changes, the DSN commitment to provide continuous tracking coverage throughout this phase of the Helios mission was largely satisfied.

Throughout 1975, the *Helios 1* spacecraft continued along its predetermined path around the Sun, as shown in Figure 3-3. Perihelion, aphelion, and superior and inferior conjunctions continued to afford unique opportunities for science data observations, which in turn led to special requirements for DSN support by the 64-meter stations when the Sun-Earth-Probe (SEP) angle was small (less than 7 degrees).

Plans for launching a second Helios spacecraft in December 1975 were formally completed by the project at a Joint Working Group meeting in Munich in May 1975. The second Helios mission (*Helios 2*) would closely parallel that of *Helios 1* and provide the scientists with three high-interest, perihelion crossings during the course of the mission. However, initial launch and early phases of the new mission would be controlled in a markedly different manner.

The *Helios 2* launch phase would be controlled from the German Space Operations Center (GSOC) at Oberpfaffenhofen near Munich rather than from the Mission Operations Control Center at JPL, as it was for *Helios 1*. This decision, which marked a significant advance in the confidence and capability of the newly established German space agency to manage deep space mission operations, would eventually create a resource of great assistance to NASA in general and the DSN in particular. However, a backup spacecraft operations team would be located at JPL during the critical launch period to deal with any unforeseen emergency. The DSN would continue to provide standard tracking station support, subject to availability of resources, throughout the mission.

Realizing that it would be unable to meet all of the demands for tracking station support in 1976, the DSN managers had, in 1975, negotiated an agreement with the Spaceflight Tracking and Data Network (STDN) for its stations in Goldstone and Madrid to provide a limited receiving-and-analog tape recording capability for the *Helios 1* and *2* spacecraft telemetry signals. These tapes were to be shipped to the STDN station at Merritt Island, Florida (MIL 71), where they would be converted to digital recordings, which could then be replayed to JPL via the NASCOM communications circuits. At JPL, the digital telemetry data were merged into the Helios Master Data Record and delivered to the Helios Mission Control Center in Germany in the same manner used for a normal DSN telemetry record.

Beginning in January 1976, this rather cumbersome operational arrangement supported *Helios 1*, while the DSN focused on the launch of *Helios 2*. Throughout 1976 and 1977, "STDN cross-support," as it was called, was used extensively to recover Helios scientific data that would otherwise have been lost due to the over-subscription of DSN tracking resources in those years.

The *Helios 2* spacecraft was launched on 15 January 1976, with initial acquisition occurring over DSSs 42 and 44. Primary control of the *Helios 2* postlaunch activities was carried out by the GSOC with a backup team at JPL. The spacecraft successfully carried out the maneuvers necessary to establish its correct orientation for the life of the mission and began its long journey around the Sun. During its perihelion passage in mid-April 1976, it would pass three million kilometers closer to the Sun than did *Helios 1* (0.29 AU versus 0.31 AU). *Helios 1* and *2* traveled similar trajectories, and the observations from *Helios 2* could be correlated with those of *Helios 1* to improve the detail and precision of the data.

The primary mission for the *Helios 2* spacecraft ended in May 1976 as the first solar conjunction approached. Despite the degraded downlink as the SEP angle fell below 5 degrees, the solar conjunction period continued to be of great interest to the Radio Science Team. They continued to gather Faraday Rotation and Celestial Mechanics data from the spacecraft downlink as the radio path from the spacecraft to the Earth approached ever closer to the solar disk and eventually entered solar occultation. In this period (of small SEP angles), the sequential ranging technique of the Mu-2 ranging machine proved superior to the standard DSN planetary ranging system in generating good ranging data in the presence of increasing solar noise.

The Viking Era: 1974–1978

In May 1976, the Helios project announced a plan to modify, by the end of the year, the German Telecommand Station at Weilheim to include a telemetry data receiving capability for Helios. The additional coverage afforded by a dedicated German station, coupled with the spacecraft capability to store data for transmission during a later tracking pass, lessened the advantages of the cross-support mode with its attendant losses and time delays. The Helios requirement for STDN cross-support was therefore withdrawn at the end of the *Helios 2* prime mission.

The DSN managers decided, however, to continue engineering tests to evaluate the use of a microwave link between a STDN and a DSN station for real-time telemetry processing and commanding.

Helios 1 and *Helios 2* spacecraft, while in their extended mission phase, rated equal priority with *Pioneer 10* and *11* for DSN support. With Viking preparations for the first landing on Mars approaching their climax, only limited DSN tracking support for Helios could be expected. Unlike the JPL spacecraft whose lifetime depended critically on a supply of inert gas to power the attitude stabilization systems, the Helios spacecraft were spin-stabilized like the early Pioneers and not dependent on a limited gas supply. They could therefore continue to orbit the Sun and return science data as long as the uplink and downlink with the Earth tracking stations (DSN and Weilheim) remained operable.

Supported by the DSN at a low level of effort, the Helios mission continued to return a wealth of science data relative to the Sun and its environment until the mission was terminated in 1985.

Long before that point was reached, however, the focus of DSN attention had shifted to other missions.

It was June 1976, and the Viking Mars Lander Mission was about to move into the spotlight and demand the full resources of the DSN and a great deal of attention from the media.

The Viking Mission

Any way you looked at it, the Viking Mars Lander project was big. Big in concept, scale, price, success, and big in impact on the DSN.

Like the successful Lunar Orbiter project some years earlier, the Viking project was managed for NASA by the Langley Research Center (LaRC) in Hampton, Virginia. The Project Manager was James S. Martin, Jr. However, JPL had two roles to play in the Viking project. One was to design and build the Orbiter that would carry the Lander into Mars Orbit and the other was to provide the tracking and data acquisition support required by both Orbiters and both Landers, simultaneously. Langley would provide the Lander. The launch and mission operations would be conducted by a joint LaRC/JPL mission operations team led by a Mission Director appointed by LaRC.

The concept was to deliver an instrumented space vehicle to Mars orbit from where it would be released to descend softly to a preselected landing site, there to search for signs of extraterrestrial life, and make in situ observations of the local surface features and weather. Two such vehicles would be built and launched in the 1975 launch opportunity. The cost of the Landers and Orbiters was estimated at about $620 million (1977),[1] an enormous amount for a single project in the 1970s. The heavy combined Orbiter/Lander spacecraft weighed 3,500 kilograms and required the most powerful launch vehicle combination then available, the Titan/Centaur, to boost it to the Red Planet.

Viking held the unique promise of finding an answer to a question that had tantalized the thoughts of humans throughout the ages: "Does life, in any form, exist anywhere outside our own planet Earth?" At that time, the science community, although not unanimous in its opinion, held that the most likely candidate was the planet Mars.

Viking was a very high profile mission which demanded, and expected to get, all of the DSN resources that were available. It also demanded significant new capability that the DSN did not have, but was requested to implement specifically for Viking. The higher levels of DSN management that had for so long been used to accommodating JPL missions, did not at first respond well to this off-lab method of doing business. They were not ready to accept requirements for support, particularly those requiring new developments, with equanimity. They felt entitled to challenge the need for the stated support, conduct an independent ground/space telecom link trade-off analysis and evaluate alternate approaches. This was the very antithesis of the kind of response that the Viking Project Management Team could accept, and it took a long time and many harsh words, with appeals to higher authority, before the two sides found a workable solution. Part of the problem arose from the fact that the Viking Project Office was on the other side of the country from California.

The Viking Era: 1974–1978

DSN representatives became involved in negotiations for DSN services and resources far from their home office, with a project office that expected on-the-spot agreements. The Viking people could not understand why this was not always possible. The reason lay in the difference between the single-minded management style of the Viking Project Office, where everything was focused on and controlled by the project, and the Tracking and Data Acquisition Office at JPL, where the DSN was managed somewhat along the lines of "the greatest good for the greatest number." At that stage, despite a plethora of formal NASA documentation, the DSN process for working with off-lab projects had not matured to the point where there was well established procedure for negotiating and approving new flight project requests for DSN support and resources.

Eventually, the DSN office produced a large, new document that specified all of its capabilities and the technical details associated with them. It was titled the *Deep Space Network to Flight Project Interface Design Handbook* and was assigned the number 810-5.[2] In the future, new flight projects seeking DSN support for their missions would be expected to use this technical interface document as a "shopping list" to identify their requirements on the DSN for support. To a large extent, this avoided the difficult bargaining experienced in the Viking negotiations.

The scientific background to Viking stemmed from the Mariner missions to Mars in 1964–65, 1969, and 1971–73. In 1965, *Mariner 4* was the pioneering flight to Mars, providing a few television pictures of its cratered surface and defining the physical parameters of its atmosphere in a fast flyby mission. In 1969, *Mariners 6* and *7* conducted more extensive approach and flyby imaging, acquired spectral data on atmospheric composition, and did initial thermal mapping. *Mariner 9*, placed in orbit about Mars in 1972, surveyed the entire surface at various image resolutions and performed extensive spectral, thermal, and gravitational studies. A landing mission was the next logical step.

The concept of the Viking project involved two Viking spacecraft, each consisting of an Orbiter and a Lander, boosted by a Titan/Centaur launch Vehicle into a Mars trajectory at the 1975 launch opportunity. On arrival at the planet, each spacecraft was to be placed in orbit. Then, following a mapping survey and landing-site verification, the Lander capsule was to separate from its Orbiter, and descend through a complex landing sequence to the Mars surface. The configurations of the Viking Orbiter and Viking Lander are shown in Figure 3-4. Figure 3-5 depicts uplink/downlink telecommunications among the DSN, the Viking Orbiter, and the Viking Lander on Mars.

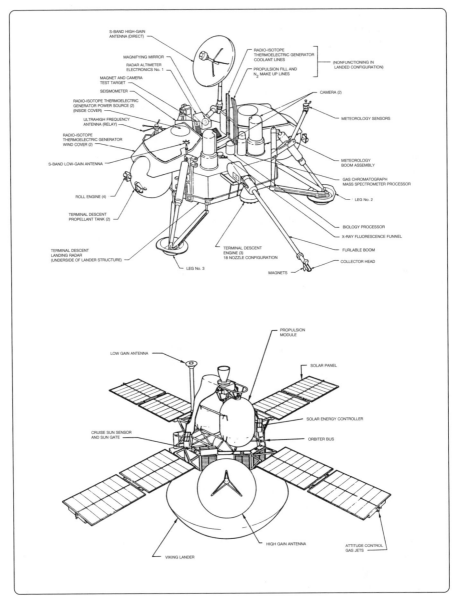

Figure 3-4. Configurations of the Viking Orbiter (top) and Viking Lander (bottom).

The Viking Era: 1974–1978

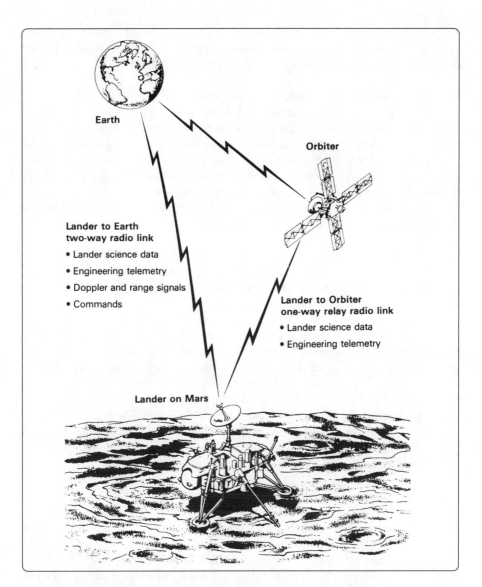

Figure 3-5. Viking Orbiter, Lander, Earth Telecommunications System. This diagram, by Ezell and Ezell, shows the uplink and downlink between the Deep Space Network and the Viking Orbiter, the uplink and downlink between the Deep Space Network and the Lander on Mars, and the uplink data relay from the Lander to the Orbiter.

Although the primary science investigations would be carried out by the Lander on the Mars surface, a significant science return would be obtained from the experiments conducted by the Orbiter and by the Lander during its descent through the Mars atmosphere.

At its inception in the late 1960s, the Viking project was intended to be launched in the 1973 opportunity, making use of the *Mariner 69* and *Mariner 71* Orbiter design and operations experience, and their data on the planet. *Viking 73* would complete the Mariner-Mars orderly sequence of planetary exploration by soft landing a package of scientific instruments on the surface of the planet. To increase mission reliability and scientific scope, two launches were planned. Precise navigation data would be required to achieve the desired landing accuracy, and large quantities of scientific and engineering data were to be telemetered from the two orbiting and two Lander craft. Preliminary Tracking and Data System planning was undertaken on this basis.

Construction of the 64-meter antenna subnet of the DSN was just then getting underway after a few years of successful operation of the initial advanced antenna system at DSS 14 at Goldstone, California. A contractor had been selected to build and install the 64-meter antennas at the overseas sites in Australia and Spain, and these facilities were scheduled to be ready to support the 1973 Viking missions. In addition, during this period, other new elements of the Tracking and Data Acquisition (TDA) system destined for Viking support were being demonstrated, such as high-rate telemetry, high-power transmitters, and the Differenced Range versus Integrated Doppler (DRVID) navigation technique.

At the beginning of 1970, funding considerations created far-reaching changes in the Viking project and resulted in an alteration of the flight schedule from a 1973 to a 1975 launch opportunity. The revised launch date permitted a more thorough assessment of the TDA support requirements and capabilities, and an increased level of commitment from what would have been possible for a 1973 mission.

By the end of 1973, the planning phase was essentially complete. The design had matured, and key capabilities and interfaces had been defined. In addition, the network of 64-meter antennas was operational, and the implementation of new capabilities needed for Viking had begun. Initial design compatibility testing with breadboard models of the Orbiter and Lander radio systems had begun in the Compatibility Test Area 21 (CTA21) at JPL. Known as "DSN Compatibility Testing," this practice had become standard by 1977 for all spacecraft with which the DSN was involved, and was indeed part of the agreement with the flight projects seeking tracking and data acquisition support from the DSN. The DSN had established two compatibility test facilities, one at JPL called Compatibility Test Area 21 (CTA 21), and the other on Merritt Island at Cape Canaveral

The Viking Era: 1974–1978

(MIL 71). Except for the actual large DSN antennas, both facilities were equipped with hardware and software representative of, and generally identical with, that which was installed at the DSN tracking stations around the world. Radio frequency uplinks and downlinks at S-band, X-band, and small (2-meter) calibrated antennas and radio frequency (RF) attenuators were used to accurately simulate the signal paths between the inflight spacecraft antennas and Earth station antennas. Following these tests, simulated spacecraft data were input to CTA 21 to verify the design of the end-to-end ground data link, beginning with the radio frequency data from the spacecraft and ending with data displays in the Mission Control Center. In progressive stages, these simulated operations were built to represent the maximum mission load conditions; system integration and data compatibility were demonstrated during mid-1974.

A succeeding series of ground data system tests, using actual tracking stations and test models of the Viking Orbiter and Lander provided training for the personnel of the Flight Operations System who would actually conduct the mission, as well as the DSN and Mission Control Center personnel who would support it.

Because there were to be two Viking launches close together, the DSN would be called upon to track two Viking spacecraft en route to Mars simultaneously. When they reached Mars, each spacecraft would separate into an Orbiter and a Lander, so that then four Viking spacecraft would be supported at various times by the DSN stations. Although only one Lander at a time would require a direct link back to Earth, this multiple spacecraft task would refire all the tracking stations in the DSN to handle it.

One set of 26-meter stations was required to provide support for the launch and cruise period of both Vikings until limiting uplink and downlink conditions necessitated the use of the large 64-meter stations. The second set of 26-meter stations was required to back up the first set if necessary, and to take over some of the tracking load when the Helios solar mission needed the services of the first set during its launch period later in the year. All four uplinks and downlinks of the Viking Orbiters and Landers would use the standard DSN S-band frequency channels, except that the Orbiters also carried X-band downlinks. These were intended to demonstrate the advantages of the higher frequency X-band channels for deep space communications, an advance in technology that the DSN proposed to introduce into the Network in the years ahead. It was expected that the increase in DSN capability would be even greater than that which resulted from the change to S-band several years before.

The DSN capabilities required to support the 1975 Viking Missions to Mars are shown in the following table.

Deep Space Station Capabilities for Viking (1975)

26-Meter Antenna Stations
 Deep Space Stations (DSS) 11, 42, 61, 12, 44, and 62
 One low-rate, plus one medium-rate telemetry data stream
 S-band ranging, shared at DSS 11, 42, 43, 61, and 63
 S-band Doppler at each station
 One high-speed ground communications channel
 Digital original data record with automatic total recall
 Analog original data record with station replay

64-Meter Antenna Stations
 Deep Space Stations (DSS) 14, 43, and 63
 Single command uplink, 20 kilowatt prime, 100 kilowatt/400 kilowatt backup
 Up to six low-, medium-, and high-rate telemetry data streams
 S- and X-band ranging
 S- and X-band Doppler
 One high-speed and one wideband ground communications channel
 Digital original data record with automatic total recall
 Analog original data record with station replay
 S- and X-band occultation data, receive and record
 S-band very long baseline interferometry data receive and record at DSS 14 and 42 only

Ground Communications Facility
 50 kilobit per second wideband channel DSS 14 to JPL (GCF Central Communications Terminal)
 27.6 kilobit per second wideband channel DSS 43 and 63 to JPL
 4800 bits per second high-speed channels to/from all stations and JPL
 Circuit performance monitoring and error detection

Network Operations Control Center
 Automatic total recall capability
 Intermediate data record production
 Network system performance monitoring, display, and data system validation
 Tracking performance monitoring
 Network operations control

Mars Planetary Radar (Goldstone)
X-band transmit and receive
Up to 400-kilowatt transmit power desired
Use of developmental (rather than operational) hardware and software

Individual uplinks and downlinks, each with its own particular parameters and critical data path, would present an operational challenge at every level to all elements of the DSN. The complexity of the mission is reflected in the downlink telemetry capabilities of each of the two simultaneous Viking spacecraft (see the following two tables).

Viking Orbiter Telemetry Channels

Channel	Data Type	Data Rate (bits per second)	Subcarrier (kilohertz)
Low	Engineering (uncoded)	8 1/3 or 33 1/3	24
High	Science (block-coded, 32, 6)	1k, 2k, 4k, 8k, 16k	240
	Science (uncoded)	1k, 2k, 4k	240

Viking Lander Telemetry Channels

Channel	Data Type	Data Rate (bits per second)	Subcarrier (kilohertz)
Low	Engineering (uncoded)	8 1/3 or 33 1/3	12
High	Science (block-coded, 32, 6)	1k, 2k, 4k, 8k, 16k	72
	Science (uncoded)	250, 500, 1000	72

Four Vikings (two Orbiters and two Landers), together with the other inflight missions being supported at the same time, were about to stress the DSN multimission concept to its limits. At the outset, it was recognized by both the DSN and the project that the most severe operational challenge lay in the successful establishment and manipulation of multiple communication and data links with the four spacecraft, two Orbiters and two Landers. A very extensive series of engineering "compatibility" tests were planned to demonstrate the integrity of the uplinks and downlinks. Realistic mission sequences were simulated to train the operations personnel, and comprehensive exercises of increasing complexity were used to measure the proficiency of the combined DSN and Viking flight operations teams. At the same time, similar test and training exercises were being carried out within the huge organization associated with the actual Titan-Centaur launch itself.

Eventually, all the prelaunch readiness criteria were met, the final launch readiness review was held and the approval for launch was given, and the countdown began. Unfortunately, both launches were delayed by technical problems that could not be corrected on the launch pad and that necessitated removal of the spacecraft from the Titan-Centaur launch vehicle. Reassembly and testing further delayed the liftoff. Finally, *Viking 1* was launched on 20 August, followed by *Viking 2* on 9 September 1975. Both launch vehicles performed perfectly to inject their payloads into Mars trajectories with arrival times at Mars of 19 June and 7 August 1976, respectively.

In both cases, the Australian 26-meter stations performed the initial acquisition sequences precisely as planned—a tribute to the careful planning and intensive preflight testing and personnel training that preceded it. Both Viking spacecraft survived the stressful launch conditions without incident, and the "cruise" phase of the Viking mission to Mars began.

During the next few months, each spacecraft would move continuously along its individual trajectory, gradually crossing the region of deep space between the injection point on the orbit of Earth (around the Sun) to its final injection point on the orbit of Mars (around the Sun). It would be performing maneuvers to correct its trajectory for small errors, carrying out scientific experiments in deep space, checking and calibrating the science instruments that the Orbiter would use to make observations from Mars orbit, and checking the condition of the all important Lander and the instruments it would use to carry out experiments on the Mars surface.

The DSN would maintain continuous uplinks and downlinks to communicate with both Vikings, generate radiometric tracking data for spacecraft navigation purposes,

The Viking Era: 1974–1978

and evaluate the performance of the new X-band downlink. The DSN would also be completing implementation of the new and extended capabilities required for the Mars "Orbiter operations" phase and "Lander operations" phase. Expansion was required in almost every area. The quality of the Doppler data used for spacecraft navigation had to be improved, and the data rates for the Orbiter telemetry had to be increased, as did the capability of the tape recorders used to record it at the Complexes. New hardware and software was needed to provide monitor and control functions for the new equipment. Earlier, the Viking project determined that in the event of adverse landing attitude conditions, the standard DSN 20-kilowatt transmitter power would not be sufficient to uplink commands to the lander to work around the situation. It demanded the newly installed 100-kilowatt transmitters which had not been proven for support of regular network operations, let alone critical support of the type suggested. Furthermore, because of greatly increased emphasis on the evaluation of alternate Mars landing sites, a request was made for the assistance of an experimental 400-kilowatt X-band transmitter that was being used at Goldstone for the study of planetary surfaces using radar techniques. At one stage, over 400 major change items were working their way through the DSN change control system to meet the Viking schedule. Engineering and operations staff were augmented to cope with the work and ongoing operations-everyone seemed to be working on Viking.

Nevertheless, the changes and additions were made and tested, and the DSN was ready when the first Viking arrived at Mars.

On 19 June 1976, Viking mission controllers started the 28-minute retro-engine burn that slowed *Viking 1* down from its approach speed of 14,400 kilometers per hour to 10,400 kilometers per hour to allow it to be captured into Mars orbit. Later, after evaluation of the navigation data, the retro-engine was burned again, this time for 132 seconds, to adjust the orbit slightly to bring it over the prime landing site. The spacecraft was now in a 24.6-hour orbit, corresponding to the length of a Martian day. The landing site would be viewed by the Orbiter on each subsequent revolution near the lowest part of the orbit. With the completion of orbit insertion and trim, the "cruise" phase of the *Viking 1* mission had been completed.

In the days that followed, the analysis team evaluated the Orbiter pictures and the Goldstone radar data to the last possible detail to arrive at consensus regarding the best (or rather, least risky) place to land consistent with meeting the principal science objectives of the Mission. The range of possible landing sites was constrained not only by the surface features, which could pose a threat to the safety of the lander as it descended, but also by the need for a biologically promising area, which would be indicated by the possible presence of water in some form.

The science teams searched long and hard for an agreement, and finally selected "alpha site" in the large central basin, at 22.5 degrees north latitude and 47.5 degrees west longitude. Once the decision to land had been made, the necessary planning followed rapidly. At 10:30 P.M. on 19 July, after calling for readiness reports from each of the major elements of the project, the project manager announced the "Go" for separation of the lander from the Orbiter. The separation and descent sequences were initiated and all events followed exactly as planned.

With the Orbiter low on its radio horizon, DSS 43 transmitted the Lander separation "Go" command to the Orbiter at 07:47:18. The Orbiter released the Lander about an hour later, and began sending relay data from the Lander as it descended through the Mars atmosphere to the surface. By 09:20, the Orbiter was beginning to set over the Australian Complex, and control was passed from Canberra to Madrid where DSS 63 continued to receive the Lander descent data and forward it to JPL. Back in the Mission Control Center at JPL, the Viking flight team anxiously watched the Lander telemetry data as the tension mounted. The Lander Team Leader called off the radar-altimeter readings and the critical events as they were executed in perfect sequence by the Lander. Finally, at 05:11:43 Greenwich Mean Time (GMT), came the announcement, "We have touchdown." Knowing that several critical housekeeping chores remained to be carried out before the first picture taking could begin, the jubilation in the Mission Control areas was rather restrained. The Lander appeared to be in a nearly nominal attitude, positioned in a stable, level landed configuration. The shutdown and reconfiguration tasks were accomplished without incident, and the first real-time imaging sequence began. The first picture (Figure 3-6) appeared on the monitors in Mission Control half an hour later. It showed the circular footpad at the end of one of the three legs of the Lander, neatly settled on the Martian soil. A few light-colored deeply pitted rocks lay in the foreground.

A second picture (Figure 3-7) soon followed. It showed a 300-degree panorama of what appeared to be a sandy plain covered with rocks of various sizes.

The quality of the first pictures from the surface of Mars astonished the flight team, the hundreds of guests gathered at JPL to observe the event, and the national television audience that saw the pictures a short while later. These images were humankind's first close-up view of Mars. Congratulations from the project manager, the NASA Administrator, and President Gerald Ford soon followed.

The Viking Era: 1974–1978

Figure 3-6. First Viking close-up picture of Mars surface. Viking Lander 1 took this close-up photograph of the Martian surface just minutes after its successful touchdown on the Plain of Chryse on 20 July 1976. The center of the image is about 1.4 meters (5 feet) from the spacecraft camera. A portion of the Lander footpad is visible in the lower right corner of the picture. The spacecraft continued to observe surface features and carry out numerous other scientific experiments on the Martian surface and environment until it ceased operating on 13 November 1983.

Figure 3-7. First panoramic picture of the surface of Mars.

With one Lander on the surface, one Orbiter in orbit, and the *Viking 2* spacecraft rapidly approaching its Mars orbit insertion point, the tempo of activity by the Viking flight team and in the DSN reached an unprecedented level. While some of the scientists were anxious to initiate the biology experiments related to the search for life, others were faced with the task of selecting the second landing site. The spacecraft controllers were involved with managing *Viking 2*, then in Mars orbit, as well as operating *Lander 1* and *Orbiter 1*. The DSN was doing its best to keep up with the demands for more and more tracking time, as well as with the avalanche of data from both Orbiters, now running at 16,000 bits per second.

Responding to commands from its controllers at JPL, *Lander 1* scooped up the first samples of Mars soil on 28 July and carefully transferred them to its various science instruments for biological, chemical, and molecular analyses.

By mid-August, *Orbiter 1* had been in orbit around Mars for 55 days making a comprehensive photo reconnaissance of possible landing sites for *Lander 2*. *Lander 1* had been on the surface for 24 days obtaining striking pictures of the Martian landscape, and making continuous observations of Mars weather, surface features, and composition, as well as analysis of the Mars atmosphere. For the first time, the DSN could establish a direct radio link between its antennas on Earth and an antenna on the surface of another planet. Radio scientists used this capability as a highly precise instrument to refine existing knowledge about the physical properties of Mars and its atmosphere.

The data from the *Lander 1* biology instrument, which was analyzing the Martian soil samples, was indicating the possibility of a biological response. However, other possible causes of the response were being considered and tested as the experiment continued. It was still much too soon to reach any conclusions regarding the existence (or nonexistence) of Martian life.

Late in August 1976, a decision was made regarding the second landing site. *Lander 2* would land in the eastern end of the Plains of Utopia at latitude 48 degrees north, longitude 226 degrees west. Once that decision was made, events followed a similar course to those of *Viking 1*. Touchdown would occur on 3 September. At 9 A.M. that day, the Viking Flight team met for the "Go/no go" for separation decision. All systems reported "Go" for separation. The commands were sent from DSS 14, and the separation sequence began to execute.

But there were some heart stopping moments to come before Viking mission control would know that *Lander 2* was on the surface. Shortly after separation, *Orbiter 2* ceased sending the telemetry data it was relaying from the descending Lander. Swift action by the spacecraft controllers at Mission Control and the station controllers at Goldstone determined that the spacecraft attitude stabilization system, on which it depended to

The Viking Era: 1974–1978

keep its main antenna pointed toward Earth, had failed. Without the main antenna, the Orbiter would not be able to transmit the Lander images to Earth.

In anticipation of such an eventuality, the Orbiter had been programmed to switch over to a small backup antenna and begin sending basic engineering data to aid in troubleshooting. Immediately, commands were sent to turn on the Orbiter tape recorder so that the pictures from the Lander would be saved while the engineers figured out how to reestablish the main Orbiter to Earth radio link.

Using the downlink from the small backup antenna on the Orbiter to transmit a star field map for reference, the spacecraft was commanded to slowly roll back to its original position and lock on to the celestial reference star, Vega. The main antenna was activated, and to the great relief of all concerned, the downlink signal returned to normal. All subsequent sequences were executed as planned, and the first pictures from *Lander 2* were received early in the morning of 4 September, nine hours later than originally scheduled.

The first two pictures of the Utopia site were essentially similar to those taken by *Lander 1* in Planitia six weeks earlier. The first showed a footpad slightly embedded in sandy material with some debris flung around from the exhaust of the Lander retro-engine. The second picture revealed a barren rock-strewn plain, somewhat less rocky than the corresponding picture from *Lander 1*. Following the landing, the science instruments began to gather data to complement the similar observations being made by *Lander 1* at Planitia.

For the next two months planetary operations continued at a furious pace. The flight teams and DSN personnel and resources were fully extended. Problems arose, and were corrected. Failures occurred and were worked around or backup systems were activated. Some data were lost, but very little. A few mistakes were made, but none which harmed any of the four active spacecraft. Within the Network, new operational procedures were developed to overcome the difficulties of working with multiple spacecraft simultaneously, and the multimission hardware and software held up to its arduous task of running continuously at top speed.

Although the other experiments delivered a wealth of fascinating new scientific data about Mars, the biology experiments continued to provide inconclusive results.

On 16 November 1976, the primary scientific objectives of the Viking Mission to Mars had been achieved and, as planned, the "Prime Mission" was declared complete.

The scientific community began the task of analyzing the new Mars data and publishing findings in science journals around the world. The findings of the biology experiment and its potential to reveal the presence of some form of life on Mars generated particular public interest. The investigators needed to exercise extreme caution in reaching a conclusion on these results. One of the investigators told a press conference, "Nobody wants to be wrong on a question as important as that of life on Mars." Efforts to explain the inconclusive results of the biology experiments continued, and hopes of a nice crisp answer faded. An enormous amount of scientific debate ensued over the following years, but eventually it was conceded that the three biology experiments did not provide conclusive evidence for the presence of living organisms on Mars. Neither did it rule out the possibility. A NASA spokesperson declared, "No clear, unambiguous evidence of life was detected in the soil near the landing sites." And added, "The question of life on Mars, according to the Viking biologists, remains an open one." So the question remained unanswered.

Although the Prime Mission had officially ended, NASA decided to extend Mars planetary operations with the four active spacecraft for 18 months. By that same time, interest and activity gradually moved to the events surrounding the rapidly approaching solar conjunction, the period of time when Mars moves behind the Sun as viewed from Earth. Radio scientists would observe changes to the S-band and X-band downlinks as they slowly became intercepted by the solar disk. These effects would be interpreted to yield new information about the structure and composition of the solar corona and the solar wind.

As Mars, with its four attendant spacecraft emerged from behind the Sun in late November 1976, the spacecraft-to-Earth telecommunications links began to return to normal, mission operations activity increased, and the Viking "extended mission" began.

No one realized it at the time, but the Viking mission would continue in various forms for the next six years. Although the Landers were designed to operate for three months in the Prime Mission, one of them, *Lander 1* would remain active until its last contact with DSS 14 in mid-November 1982. The two Orbiters eventually depleted their attitude stabilization gas supply and, without their antennas pointed toward Earth, were unable to maintain an uplink or downlink. *Lander 2* ceased operation in February 1980 when its battery failed.

The extension of the Viking mission into 1977 continued to place heavy demands on the DSN, not only because of the addition of radio science investigations to the original science experiments (except for biology, which had been completed in the Prime

The Viking Era: 1974–1978

Mission), but also because the DSN was now engaged in replacing all its obsolete SDS computers with new MODCOMP minicomputers. Each of the nine tracking stations, in turn, would have its old telemetry and command processors replaced with new machines and software, and then be requalified to resume inflight support. It would be called the Mark-III Data System (MDS) upgrade, and was to be accomplished without impact to any of the inflight missions, including Viking (then running at its peak).

It was under these difficult circumstances that the focus of DSN attention moved to another new flight project that would extend the capability of the DSN uplinks and downlinks to even further limits. This new flight project was called Voyager.

The Voyager Mission

In its time, the Voyager project was one of the most ambitious planetary missions ever undertaken. *Voyager 1*, which would encounter Jupiter in March 1979, was to investigate the planet and its large satellites; later, it would explore Saturn, its rings, and several of its satellites. *Voyager 2* would arrive at Jupiter in July 1979. This spacecraft would also examine Jupiter and Saturn and their satellites, then it would be hurled on toward an encounter with Uranus, which it would reach seven years later, in 1986. En route to these encounters, both spacecraft would make measurements of the interplanetary medium and its interactions with the solar wind.[4]

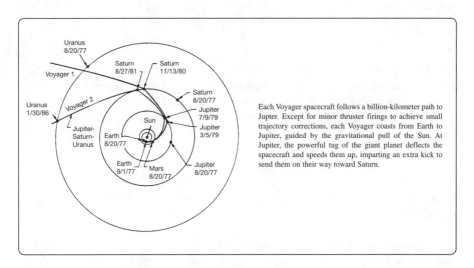

Figure 3-8. Voyager trajectories to Jupiter, Saturn, and Uranus.

Except for minor adjustments to the trajectories made by the spacecraft thrusters, the spacecraft would initially coast from Earth to Jupiter, guided by the powerful gravitational pull of the Sun. At Jupiter, the gravitational force of the planet itself would be used to redirect the spacecraft toward Saturn.

Due to a problem that required the removal of *Voyager 1* from the launch pad for repairs, *Voyager 2* was launched first. The launch from Cape Canaveral, Florida, in August 1977 utilized the most powerful launch vehicle then available, the mighty Titan/Centaur. *Voyager 1*, launched sixteen days later on 5 September, on a shorter, faster trajectory, soon overtook *Voyager 2* and eventually reached Jupiter four months earlier than *Voyager 2*.

To inject these heavy spacecraft into an Earth escape trajectory to Jupiter would require a five-stage rocket-motor burn sequence, two burns of the Titan, two burns of the Centaur and a final single burn of the solid rocket motor in the spacecraft's own propulsion module. All these complex events were executed flawlessly on both occasions. Sixty minutes after liftoff, the Voyager spacecraft were free of the influence of Earth's gravity, coasting on their way to Jupiter at a speed of more than 10 kilometers per second.

The first DSN involvement with the Voyager spacecraft had taken place in the early months of 1977 when the spacecraft uplinks and downlinks had been subjected to a rigorous series of radio "compatibility" tests to ensure that spacecraft and DSN would be able to communicate without problems after launch. For the Voyager tests, CTA 21 at JPL and MIL 71 at Cape Canaveral were each configured to simulate a 64-meter antenna station, and each included both Block-III and Block-IV Receiver-Exciter Subsystems and the new MDS for telemetry, command, and radiometric data. The early tests at CTA 21 in March and April were run with the spacecraft in the space simulator at JPL and established the validity of the radio system design as represented by the spacecraft components being tested. The tests with the actual flight configuration spacecraft took place at MIL-71 in July and August at Cape Canaveral. This was the last chance before launch to confirm beyond all reasonable doubt that the Voyager-to-DSN uplink and downlink would function in space "as designed." Both spacecraft passed the tests without reservation.

Sixty minutes after liftoff, separated from the Centaur and beginning its long journey to Jupiter, *Voyager 2* approached its second involvement with the DSN "initial acquisition."

"Initial acquisition" is the term used by the DSN to describe the chain of events that take place at the first tracking station to observe, lock onto, and track the downlink signal after the spacecraft has separated from the launch vehicle and leaves Earth orbit to

The Viking Era: 1974–1978

begin its planetary trajectory. A successful, two-way, initial acquisition is crucial to the safety of the spacecraft and subsequent mission events. When completed, the data it provides gives the mission controllers knowledge of the spacecraft condition via telemetry, the ability to control the spacecraft via the command uplink, and the capability to determine the spacecraft orbit from the Doppler radiometric data generated by the DSN from the interaction of the coherent uplink and downlink.

Based on long experience of many launches of different spacecraft and different launch vehicles, the DSN had refined the initial acquisition procedures by the time of the Voyager launches to the point where the probability of success was related to two main factors. The first of these, dispersion errors in the powered flight part of the launch sequence, affects exactly where and when the spacecraft will rise at the first viewing station and, therefore, whether or not it will rise within the beamwidth of the DSN antenna set to the predicted pointing angles. The second factor is the uncertainty in the predicted downlink frequency at the time the spacecraft appears above the radio horizon of the first acquiring station. This determines whether the spacecraft downlink will appear in the bandwidth of the DSN receiver set to the predicted downlink frequency. Time is of the essence. Because all the uncertainties grow with time, the longer it takes to acquire the spacecraft, the more difficult it becomes to accomplish.

When a spacecraft first appears over the radio horizon, its angular movement across the sky is too great for the large 64-meter antennas to track. For this reason, the spacecraft is first acquired by the smaller, more agile, wide-beam 26-meter antenna. Much later, when the trajectory has been well established and more accurate predicts become available, the spacecraft can be transferred to the large 64-meter antenna. To further increase the 26-meter antenna's ability to view the spacecraft in the event of significant trajectory dispersions, a small, wide-beam acquisition antenna is attached to the outer edge of the 26-meter antenna and fed to a separate receiving and display device. Station operators can use the acquisition antenna data to help point the 26-meter antenna to "lock on" to the rapidly moving spacecraft.

The designated initial acquisition station for *Voyager 2* was DSS 12 at Goldstone. The station had been configured according to the "Initial Acquisition Plan," the necessary prediction data for antenna pointing angles and rates had been loaded into the antenna pointing system (APS) and the receiver frequencies and Doppler offsets had been updated based on the latest spacecraft frequencies measured just before liftoff. Similar preparations had been made on board the NASA tracking ship *Vanguard*, which was on station in the Indian Ocean, and the Goddard STDN tracking stations on Guam, Kwajalein, and Hawaii.

All of the Titan and Centaur powered flight events observed by the downrange tracking stations appeared to be "nominal." About sixty minutes after liftoff, the spacecraft separated from the Centaur, burned its propulsion module engine for 45 seconds, and was finally injected into a direct orbit for Jupiter. Within the next several minutes the spacecraft carried out several pre-programmed maneuvers. After turning to point the roll axis toward Earth, the science and RTG booms were deployed, and the spent propulsion module was jettisoned. Then, with its S-band downlink transmitter turned "on" and its low-gain wide-beamwidth antenna pointed to Earth, it moved rapidly away from Earth orbit, passing high across the Indian and Pacific oceans toward the west coast of the United States. In rapid succession, each of the antennas on *Vanguard*, Guam, Kwajalein, and Hawaii briefly acquired the downlink signal as the spacecraft rapidly passed through their field of view.

Finally, approximately 1 hour 14 minutes after launch, the spacecraft signal appeared in the acquisition receiver at DSS 12. The receiver was immediately "locked up one-way" on the spacecraft signal and the 26-meter antenna put into the auto-track mode. A few minutes later, the DSS 12 transmitter was turned "on" and the uplink frequency swept across the spacecraft receiver rest frequency. Immediately, the spacecraft receiver locked up on the DSS transmitter signal and, after a small readjustment of the downlink frequency, the DSS established two-way acquisition of Voyager for the first time.

Once the first tracking station had established an uplink and downlink with the *Voyager* spacecraft, the telemetry, command, and radiometric data capabilities of the DSN became available to the Voyager Mission controllers. As the spacecraft continued to move away from Earth, its apparent eastward motion slowed and eventually reversed to the normal westerly motion as Earth rotated beneath it. A few hours later, Voyager set at Goldstone and the spacecraft was transferred to the Canberra stations without incident. DSN support for Voyager had begun.

The powered flight and initial acquisition sequences for the *Voyager 1* launch on 5 September 1977, followed essentially the same pattern of events as described above for *Voyager 2*, so that by early September 1977, the DSN had two additional planetary spacecraft to deal with, both of them bound for Jupiter.

DSN support on the 26-meter Network for both *Voyagers 1* and *2* during the last quarter of 1977 was fairly routine. The priority enjoyed by the project at that time was not curtailed by scheduling conflicts with other inflight missions. The mission controllers were involved with trajectory correction maneuvers, optical navigation imaging, analysis of spacecraft problems, and the calibration of various science instruments (the High-Gain Antenna

The Viking Era: 1974–1978

and the S- and X-band communication links. During this time, the telemetry data rate was running at 2,560 bits per second, and both spacecraft were operating well.

In November 1977, both spacecraft crossed the orbit of Mars to begin a nine-month crossing of the asteroid belt. By the end of the year on 15 December 1977, both spacecraft were more than 1 AU from Earth, with *Voyager 1* having overtaken *Voyager 2*. One-way radio communication time was 9 minutes 49 seconds for *Voyager 1*, and 9 minutes 40 seconds for *Voyager 2*. *Voyager 1* continued to increase its lead and would arrive at Jupiter in March 1979, four months ahead of *Voyager 2*.

In 1978, routine cruise operations continued normally on both spacecraft until early April. In the course of switching from primary to backup receivers on 5 April, it was discovered that the backup receiver in *Voyager 2* had developed a fault in its carrier tracking loop. Analysis by the project and DSN telecommunications engineers determined that a capacitor in the receiver carrier tracking loop had probably shorted. However, soon after switching back to the prime receiver, it was found that it also had developed a fault that proved to be permanent in nature. Since the spacecraft was unable to receive commands from Earth on the now-permanently failed primary receiver, intense efforts were made by the DSN to develop a strategy that would permit commanding the spacecraft using the partially disabled secondary receiver, despite its failed tracking loop.[5]

The carrier tracking loop enables the spacecraft receiver to follow changes in the uplink carrier frequency caused by the relative movement of the ground transmitter as Earth rotates. This is the radio telecommunications equivalent of the well known Doppler effect, and knowledge of its precise value is an essential element in the technique of spacecraft interplanetary navigation. When no uplink is present, the spacecraft receiver remains active at a frequency called its "rest frequency." To activate an uplink and send commands, the uplink transmission must first be tuned to this rest frequency, after which the spacecraft receiver tracking loop will lock-on and follow the uplink frequency changes due to Doppler throughout the tracking period.

The strategy that finally evolved from these efforts, to compensate for the inoperative tracking loop, required the DSN to slowly change the transmitter uplink frequency during a spacecraft tracking period in such a way that the frequency changes due to Doppler, as seen by the spacecraft receiver, would be cancelled out. The receiver could then lock on to the DSN uplink at its rest frequency. No tracking loop action would be necessary and commands would be received by the spacecraft in the normal way.

On 13 April, the *Voyager 2* recovery sequence, which was based on this strategy, was initiated by DSS 63. In less than an hour it was confirmed that the first commands had been received by the spacecraft receiver. Spacecraft controllers immediately sent a large sequence of commands to suspend all but essential spacecraft activity; these commands would minimize any risk of losing the vital uplink while they set about evaluating the spacecraft condition and determining how to deal with it in the future.

Throughout the emergency, *Voyager 2* had been tracked continuously by the 64-meter antennas with the cooperation of the Viking, Pioneer, and Helios projects, which were also dependent on those antennas for their mission support. Only the 64-meter antennas could receive the low-power S-band signals that *Voyager 2* had been transmitting when the emergency occurred. As soon as possible, commands were sent to turn the S-band transmitter to high power so that the spacecraft could be tracked by the 26-meter antennas, thus relieving the pressure for 64-meter antenna support.

The *Voyager 2* emergency placed an additional burden on the DSN operations organization in terms of planning, tracking analysis, and real-time operations. It occurred during a period when there was a large number of very significant changes in progress throughout the Network. These changes were associated with the MDS implementation, discussed in more detail later. Suffice it to say here that in 1978 new telemetry, command, tracking, radio science, and data delivery systems were installed at all sites. DSS 12 was withdrawn from operations and upgraded from a 26-meter, S-band station to a 34-meter S- and X-band station.

Apart from the actual installation and testing, all of these changes necessitated new procedures and retraining of the Network operating personnel, while maintaining a high level of support for all of the inflight projects. Nevertheless, by using innovative operational procedures to economize on station preparation time, and detailed negotiation and scheduling of the available resources, the DSN was able to maintain satisfactory tracking and data acquisition support for all ten interplanetary missions in 1978.

Over the next few months, the Tracking-Loop-Capacitor (TLC) tuning sequence was greatly refined and would eventually become a routine feature of DSN operational support for *Voyager 2*.

As a result of this timely recovery action, the *Voyager 2* mission was able to resume throughout 1978 with the high expectation of achieving all of its Jupiter mission objectives.

The Viking Era: 1974–1978

Both spacecraft continued to move through the 1978 cruise phase of the mission without incident. Voyager support in the last few months of 1978 was reduced somewhat to allow the DSN to divert resources to support the Pioneer Venus and Multiprobe missions. *Voyager 1* emerged from its crossing of the Asteroid belt in September followed in October by *Voyager 2*. No events of any significance were reported. By the end of the year, the project attention was focused on a new phase of the mission, due to begin on 4 January 1979. It was called the "Observatory Phase" and would be the prelude to the Jupiter encounters.

Pioneers 6–11

Since the start of the second-generation Pioneer program beginning with *Pioneer 6* in 1965, the DSN had introduced many modifications toward improving the threshold capability of the Network for Pioneer telemetry support. Seven 26-meter stations were equipped with improved masers, more efficient microwave equipment, linear antenna polarizers, 3-Hertz receiver carrier tracking loops, advanced demodulation hardware, and sequential decoding software (for support of *Pioneer 9*). By 1969, these systematic improvements had resulted in an extension of the Pioneer telemetry threshold on the 26-meter antennas from 0.4 AU (1965) to 1.5 AU (1969).[6] The availability of the 64-meter stations increased the telecommunications link margins for the Pioneers so much that they continued to be a significant presence in the DSN mission set throughout the Viking Era (Figure 3-1).

The *Pioneer 10* and *11* missions to Jupiter were designed to investigate the interplanetary medium beyond the orbit of Mars, the asteroid belt, and the near-Jupiter environment, hitherto unexplored regions of interplanetary space at that time. The spacecraft incorporated many technological advances over those in the interplanetary *Pioneers 6–9*, although the design philosophy was the same and the same design team (NASA's Ames Research Center and TRW Systems) developed and built the spacecraft.[7] The spacecraft were launched in 1972 and 1973, respectively, to solar system escape trajectories. *Pioneer 10* went on to become the first spacecraft ever to penetrate the asteroid belt and, in December 1973, to obtain close-up images of Jupiter. *Pioneer 11* reached Jupiter one year later and was successfully retargeted for an encounter with Saturn in September 1979.

The DSN supported both flights with the 26- and 64-meter stations, but there were serious conflicts with the overlapping *Mariner 10* mission to Mercury and Venus, during 1973 and 1974, and the Viking requirements for prelaunch test time, and for Mars radar observations from DSS 14 at Goldstone in 1975.

In March 1976, the Pioneer project was estimating a useful spacecraft life to 1983 for *Pioneer 10*.[8] With the then-existing 64-meter station, the downlink telecommunications limit at minimum bit rate (16 bits per second) would be reached in mid-1980 at a range of about 22 AU, which is 2 AU beyond the orbit of Uranus.

Pioneer was already pushing the outer limits of long range deep space communication links, but there was a strong motivation to push the uplinks and downlinks even further, hopefully to the point of matching the projected end of useful spacecraft life. Downlink performance improvements were also of interest for the *Pioneer 11* Saturn Encounter, where the highest data rate possible was desired to maximize the imaging science data return. The predicted downlink telecommunications performance for *Pioneers 10* and *11* with both 26- and 64-meter antennas is shown in Figure 3-9.

At the time, there were a number of possibilities for improving the existing downlink performance of the DSN antennas; these are listed in the table below. The table also gives the factor by which the resulting telecommunications downlink margin would be increased (in decibels) and the multiplying factor by which the maximum range would be increased.

Possible Downlink Performance Improvements for *Pioneers 10* and *11*

Item	Improvement	Downlink Margin Increase (decibel)	Range Increase Factor
(1)	Reduce receiver tracking loop (BW = 3 Hertz)	1.0	1.12
(2)	Improved low-noise feed-cone (18.5 kelvin)	0.7	1.09
(3)	Reshape the 64-meter subreflector	0.5	1.06
(4)	Increase 64-meter antenna diameter to 70 meters	0.8	1.10

With the 1976 configuration of the 64-meter antenna, *Pioneer 10* would reach the limit of its telecommunications downlink at about 22 AU in mid-1980. This limit could be extended to 24.6 AU, which the spacecraft would reach in early 1981, by using the 3-Hertz receiver tracking loop bandwidth then available in the new Block IV receivers. The improved S-band feed-cone would reduce the downlink system noise temperature

The Viking Era: 1974–1978

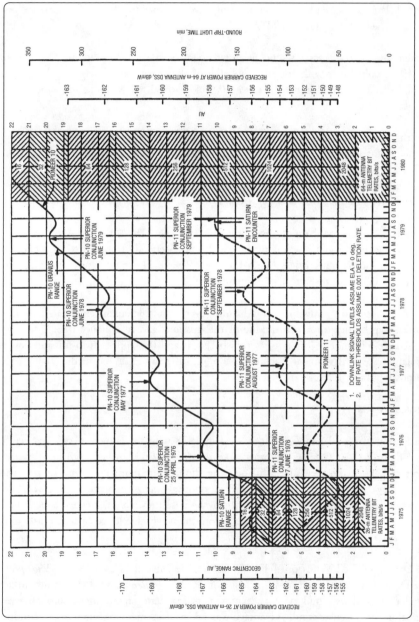

Figure 3-9. Downlink performance estimates for *Pioneers 10* and *11*.

to 18.5 kelvin and extend the potential range even further to 26.8 AU, which *Pioneer 10* would reach at the end of 1981.

Although the last two items would have extended the maximum range to 31.3 AU in early 1983, coincident with the projected life of the spacecraft, it was not thought at the time that they could benefit *Pioneer 10*, not planned for Network implementation until after 1983.

Over the course of time, these estimates turned out to be grossly pessimistic and, as subsequent chapters will show, *Pioneer 10* continued to maintain its downlink with the DSN until it finally vanished on 31 March 1997. *Pioneer 11* obtained some benefit from the low-noise feed-cone but not enough to enable it to achieve its desired downlink telemetry rate of 1,024 bits per second.

Following their successful Jupiter encounters in 1973 and 1974, the *Pioneer 10* and *11* spacecraft moved through the Viking Era, with two essentially trouble-free spacecraft on solar system escape trajectories. Along the way, *Pioneer 11* would return imaging data from an encounter with Saturn in September 1979.

The *Pioneer 10* and *11* missions would utilize every available enhancement to the DSN downlink capabilities to extend the life of the missions and to enhance their interplanetary science return. The longevity of the spacecraft and this approach to mission design and planning would yield far-reaching results into the future, much further than could have possibly been foreseen in that era.

Pioneer Venus

Investigation of the planet Venus continued in 1978 with the launch of two spacecraft, one of them a *Venus Orbiter*, the other a *Venus Multiprobe*. Managed by Ames Research Center as part of the very successful ongoing Pioneer program, the *Venus Orbiter* was launched by Atlas-Centaur in May 1978, and the *Venus Multiprobe* in August 1978. These two launches would also be noteworthy because they were last expendable launch vehicles to be used for planetary launches before the newly developed Space Shuttle took over all deep space launches in the 1980s.

The *Orbiter* reached Venus on 4 December, followed five days later by the *Multiprobe* on 9 December. After insertion into Venus orbit, the *Orbiter* mapped the Venus surface by radar, imaged its cloud systems, explored its magnetic environment, and observed interactions of the solar wind. Advantage was taken of its many occultations to conduct

The Viking Era: 1974–1978

Earth-based radio science experiments to determine the refractivity, turbulence, and ionospheric structure of the atmosphere.

On arrival at the planet, the *Multiprobe* spacecraft (bus) released four heat-resistant probes to penetrate the Venus atmosphere at widely separated locations. Each probe contained an S-band radio transmitter that continued to return telemetry measurements of atmospheric temperature, pressure, and density until it impacted the planet surface.

Both Pioneer Venus 1978 missions presented significant challenges to DSN engineering because they required a substantial amount of the new equipment and new technology then being installed as part of the MDS implementation program. In addition, the *Multiprobe* mission presented unique operational problems because of the need to rapidly acquire and simultaneously track all four probes as they were released into the Venus atmosphere from the bus.

Shortly after arrival and insertion into Venus orbit, the *Orbiter* began a long series of Venus occultations, during which the DSN conducted ninety S- and X-band radio science observations as the spacecraft alternately entered and exited obscuration by the planet. Open-loop and closed-loop receivers together with wide- and narrow-band digital recorders were used to capture the S- and X-band radio science data on each occasion. Between occultations, normal telemetry data capture continued.

The initial occultations occurred only over DSS 14. The spacecraft periapsis passage was allowed to move later each day until it reached the DSS 14/DSS 43 mutual view period in January 1979. Thereafter radio science was conducted at both stations until passing completely to DSS 43 later in 1979.[9]

Occultation radio science observations on the *Orbiter* were suspended for two days on 8 and 9 December to allow the stations to give their full attention to the probe entry event.

DSS 14 and DSS 43 had prime responsibility for the supporting probe entry events in real time. They were to provide telemetry and Doppler data, and analog recordings of the open-loop receiver outputs, and to establish and maintain an uplink to the large probe during the entry and descent period. After the probes had impacted the surface, they were to support the bus until it, too, entered the atmosphere and was destroyed. In conjunction with the stations of the STDN at Santiago, Chile, and Guam, they would also provide recording for the Differential Long Baseline Interferometry (DLBI) experiment.[10]

A special configuration had been implemented as part of the MDS at DSS 14 and DSS 43 for the *Pioneer Multiprobe* mission to allow simultaneous reception of five carriers, and for the processing of four carriers in real time. For backup purposes, four open-loop receivers with bandpass filters and analog recorders allowed for recording a wideband spectrum around the anticipated carrier frequencies of the four entry probes. These data could be played back in non-real time at CTA 21, and converted back up to S-band for reception and processing the Mark III-75 telemetry equipment.

DSS 11 and DSS 44 would support the bus during the entry period with command, telemetry, and Doppler data while the 64-meter stations were tracking the four probes. DSS 42 would provide equipment for generating Doppler data from the third small probe. At DSS 14 and DSS 43, the operational complexity of dealing with ten signal flow paths for the four probes and the bus gave cause for concern. Recalling that these paths were all active simultaneously and that the time available for activating them once the probes were released was about five minutes, it is not surprising that the operator crews at the stations approached the *Multiprobe* entry event with some trepidation.

The fact that the entire event was supported perfectly, and resulted in the capture of 90 percent of the possible data from all four probes during the brief period (50–60 minutes) of their descent to the surface of Venus, attests to the care in planning, thoroughness of the training, and the skill and dedication of all the DSN people involved.

The Viking Era: 1974–1978

NETWORK ENGINEERING AND IMPLEMENTATION

The DSN Mark III System Design

The enduring design of the DSN Mark III Data System originated with a presentation by Walter K. Victor, N. A. Renzetti, J. R. Hall, and Douglas J. Mudgway to the Tracking and Data Acquisition Organization and the Technical Divisions in October 1969. At that presentation, Victor first proposed a radical new approach to the DSN architectural structure.[11] It would be called the DSN Mark III-73 and would include the DSN, NOCC, GCF, and the MCCC. Due to institutional restructuring, the MCCC would be separated from the DSN in 1972, and thereafter was excluded from the scope of the Mark III Data System.

Walter K. Victor deserved to be numbered among the founders of the DSN. Bringing with him an engineering degree from the University of Texas (1942) and a brilliant mind, Victor came to JPL in 1953 to work on radio guidance systems for the Corporal and Sergeant missiles then being developed for the Army. He was soon collaborating with a colleague, Eberhardt Rechtin, in developing more efficient receivers for tracking very weak radio signals from distant missiles or space probes. Based on "phase-lock" tracking principles developed earlier by Rechtin and R. Jaffe of JPL, the "phase-lock receiver" eventually became a key element of all deep space tracking stations. Walt Victor's association with the "phase-lock receiver" became legendary throughout the DSN.

When JPL made the transition to NASA in 1958, Victor became Manager of Communications Systems Research in Rechtin's Telecommunications Division. In the years that followed, he moved to successively higher positions of authority within the tracking and data acquisition organization. During that time, he directed many of the major technical advancements that led to improvements in the performance, reliability, and productivity of the Network. He left the Tracking and Data Acquisition Office in 1973 for higher responsibility elsewhere at JPL.

Tall, autocratic, and conservative in approach, Victor would not "suffer fools gladly." As one of his engineers described his meetings with Victor, "With Walt Victor, there was no middle ground, no gray area. You were either right or wrong; you either knew, or you did not know, the answers to his always detailed questions." His penetrating mind, breadth of experience, piercing gaze, and abrupt manner brought terror to many an inadequately prepared engineer in the course of a technical presentation.

A very private man, not much given to socializing, Victor elicited great respect from his coworkers, at both JPL and NASA Headquarters, by virtue of his unequivocal commitment to excellence in every problem, task, or project he addressed. He retired from JPL in 1987.

As visualized by Victor, the DSN Mark III-73 Data System would comprise one subnet of 64-meter diameter antennas and one subnet of 26-meter diameter antennas supplemented by an additional 26-meter subnet shared with the Manned Spaceflight Network. Two launch support stations, one at Cape Canaveral (MIL 71) and one at Johannesburg, South Africa (DSS 51), would provide near-Earth tracking coverage for planetary missions, and a spacecraft-to-DSN compatibility test facility at JPL (CTA 21) would be used to verify the integrity of uplinks and downlinks before a spacecraft moved to the prelaunch phase of activity. The station in South Africa was eventually dropped from the plan for political reasons. Network control was to be exercised from a central facility at JPL (NOCC), which would have the necessary capability for the processing, display, and storage of network control data. (In its original form, the plan also included a data processing facility for mission control, but this became irrelevant when the MCCC was separated from the DSN.) The addition of a "simulation center" to the DSN would provide simulated spacecraft data streams for use in end-to-end testing of new installations at the tracking stations over real communications circuits, and training station personnel in new operational procedures.

Two important, new, and lasting basic concepts were introduced to form the basis for future design of the Mark III data system: closed-loop control of major data systems to minimize loss of data, and specific validation by the DSN of all spacecraft data delivered to the flight projects.

The first of these, "closed-loop control", implied the development of a new capability that would continuously monitor the status of the Network and the real-time data streams, compare these to preestablished standards, and generate alarms when these values exceeded specified limits. Immediate corrective action would follow. The new capability would become the DSN Monitor and Control System.

The "data validation" concept implied no alteration of, or delay to, the incoming spacecraft data by the validation technique. The validation process also included the ability to determine the status of the DSN in real time, and to make these data available to the flight projects.

The Viking Era: 1974–1978

Finally, as a basic framework on which to design and manage the future Mark III Data System, Victor identified a hierarchy of "systems and subsystems" for the Network and its (then) three facilities: the SFOF, GCF, and DSIF. (As mentioned previously, the SFOF was later deleted from the plan and the DSIF became known as the DSS.)

The definition of the DSN Systems was logically based on the functions to be performed, and became formalized as listed below:

DSN Tracking System
DSN Telemetry System
DSN Command System
DSN Monitor System
DSN Simulation System
DSN Operations Control System

A similar hierarchy was identified for each of the facilities based on its specific functions as an element of the Network.

Because of its intuitive simplicity and inherent ability to create clear manageable interfaces between them, the system hierarchy for the DSN, GCF, and DSS continues in use today throughout the Network.

In a form substantially the same as originally proposed, the Mark III-73 Data System was implemented in time to support missions in the closing years of the Mariner Era, and was soon followed by an upgrade to a Mark-III-74 model to support the major new missions of Pioneer and Helios. Then came the Viking mission to Mars.

Implementation for Viking

The process of DSN implementation to support missions in the Viking Era was essentially one of addition and modification. Except for the 64-meter stations in Australia and Spain, which had come into service in 1973 and 1974, respectively, and the NOCC, all the facilities had been in existence and had supported many previous flight projects in more or less the same form for several years.[7] Further, the network configuration required for Viking launch and early cruise (up to February 1976) was substantially that which already existed. The major increase in capability was not required for full planetary testing with the project until February 1976, well before the planetary encounter in June 1976. Therefore, the actual implementation was carried out in phases corresponding to the phase of the mission activity for which it was required.

Prior to March 1972, Network implementation plans had included the SFOF, since it was at that time a part of the DSN. Following reorganization of the Offices of Tracking and Data Acquisition and Computing and Information Systems in March 1972, the SFOF was separated from the DSN, and the plans for a Network upgrade were rearranged to encompass only the deep space tracking stations, compatibility test stations, ground communications, and the NOCC. The former SFOF then became known as the Mission Control and Computing Center (MCCC) to distinguish it from the NOCC, which resided in the same building (230) at JPL. The DSN and SFOF functions had originally been conducted as a single entity under the control of the DSN. They were separated in 1961 only to be recombined in 1963 and separated again in 1972.[7]

During 1972 and 1973, implementation activity related to Viking was relatively small, partly because the Viking mission was being replanned from the cancelled launch in 1973 to a 1975 launch, and also because DSN attention was focused principally on the Mariner Venus-Mercury mission.

By September 1973, however, implementation for the new CTA 21 had begun to move forward, and the longer range Viking implementation was finally planned. CTA 21 was urgently needed to allow the all-important Viking multiple spacecraft testing to get underway.

The implementation of CTA 21 made such good progress that a year later (August 1974) the Viking radio frequency and data system compatibility tests had been completed using two Orbiter streams and one Lander stream simultaneously between CTA 21 and the Mission Control Center. In the course of these tests, up to six data streams had been passed across the interface and processed correctly, simultaneously with the generation of two command links. This completed the implementation of CTA 21. Implementation at the STDN-MIL 71 site at Cape Kennedy and DSS 14 at Goldstone for support later in the year was in progress.

The end of 1974 saw the Block IV receiver installed at MIL 71 and the new station integrated with the NOCC. Block IV receivers for DSS 43 and DSS 63 were in production and expected to be installed before Viking launch, then projected for July 1975. Construction of facilities for the overseas high-power transmitter installations were about complete at the Madrid site while an Australian contractor was about to begin work in Canberra.

Despite the interruption for most of July to reconfigure for the Helios launch, and a further diversion to cover the *Pioneer 11* Jupiter Encounter in November 1974, the

The Viking Era: 1974–1978

installation of the Viking planetary configuration was complete at Goldstone stations DSS 14 and DSS 11 by January 1975. Testing of the new capabilities began immediately and soon indicated areas in which improvements were needed, particularly with regard to the analog and digital recording of original data records. Implementation of the Viking planetary configuration at overseas stations in Madrid and Canberra were slowed somewhat by the restrictions on configuration changes imposed during the Viking prelaunch test period.

Nevertheless, all implementation required for launch was completed by the time of the first DSN-Viking Launch Readiness Review on 9 July 1975, and the DSN was declared fully prepared to support both Viking launch and cruise operations.

DSN Mark III-75 Model

With the advent of full planetary operations on 20 July 1976, following the successful landing of *Viking Lander 1*, the entire resources of the DSN were required to provide simultaneous support for the two Orbiters and the *Viking Lander*.

In addition to Viking, the DSN was also being called upon to provide support for *Pioneers 6–11* and the *Helios 1* and *2* solar missions. The extent of the coverage involved can be adjudged from Figure 3-1, which shows all the missions in flight at the time.

The support required by all of these users of the Network encompassed telemetry, command and tracking support, error-free data transmission services from Madrid, Spain; Canberra, Australia; and Goldstone, California; as well as Network control and monitoring and production of complete data records.

The resources included three 64-meter stations, six 26-meter stations, DSN and NASCOM ground communications facilities, and the NOCC. The configuration of the Network was then called the DSN Mark III-75.

The DSN Mark III-75 was, at this time, the latest in a progressive succession of configurations or models based on the original DSN Mark III Data System design beginning with the Mark III-73. Each Model represented a new increment of DSN capability, driven by the tracking and data acquisition requirements of the future missions (see Figure 3-1). The key telemetry characteristics of each of the models through the period 1973–76 reflect the steady development of DSN capability in this period. As can be seen in the following table, additional telemetry characteristics were added to the

Mark III-73 Model and the result was the Mark III-74 Model, to which other characteristics were added to create the Mark III-75 Model.[13]

Key Telemetry Characteristics for Mark III System Models 73 Through 75

Deep Space Network Mark III-73
- Multimission capability at each station for receiving and formatting uncoded and block-coded telemetry.

- Special software for Pioneer convolutional coded telemetry.

- Centralized monitoring of Telemetry System performance by observation of project displays in the MCCC.

- Capability at each station for a single carrier, dual subcarrier, and formatting for high-speed and wideband communications.

- Recording of pre- and post-detection analog records with nonreal-time playback.

- Production of digital Original Data Record (ODR) at each DSS and playback via manual control or automatic response to project inputs.

- Real-time reporting of telemetry status at the station to the Monitor and Control Subsystem.

Deep Space Network Mark III-74
- Multimission capability for sequential decoding of long constraint convolutionally coded telemetry data up to 4,000 bits per second using the Fano algorithm, for both Pioneer and Helios.

- Real-time monitoring of Telemetry System performance at the NOCC.

Deep Space Network Mark III-75
- Capability at the 64-meter DSS to handle multiple carriers (up to four) and multiple subcarriers (up to six) with decoding, ODR, and formatting for communications circuits.

The Viking Era: 1974–1978

- Centralized monitoring of Telemetry System performance at the NOCC, and reporting via the DSN Monitor and Control System.

- Central logging (on a Network Data Log) of all data received at the DSS, with gap accounting, and automated recall of missing data from the DSS ODRs.

- Generation of a time-ordered, gap-free record of all received data called an Intermediate Data Record (IDR) for delivery to the flight project.

- Generation of Telemetry System predicts and configurations for transmission to each DSS for use in the manual control of system elements.

The ground communications capability associated with the Mark III-75 model provided for transmission of formatted telemetry data from each DSS to the NOCC and MCCC. The communication circuits were furnished by NASCOM, and consisted of one 4,800-bit-per-second high-speed line from each DSS, and one additional 28.5-kilobit-per-second wideband line from each 64-meter DSS in Spain and Australia, with 50 kilobits per second from Goldstone.

The 64-meter Stations

A functional block diagram of the DSN Mark III-75 configuration for the 64-meter stations (DSS 14 at Goldstone, DSS 43 at Canberra, and DSS 63 at Madrid) is shown in Figure 3-10.

Two Block III receivers and two Block IV receivers provided a redundant capability for up to three radio frequency carriers simultaneously; e.g., two Viking Orbiters and one Viking Lander. Block III receivers were capable of only S-band reception, while Block IV receivers could receive at either S-band or X-band. Two open-loop receivers provided for radio science occultation and solar corona experiments.

Six subcarrier demodulators were connected by an extremely flexible switching matrix to associated block decoders symbol synchronizers, and data decoders which, together with the telemetry and command processor, provided the capability to handle six simultaneous data streams, or two subcarrier data streams per spacecraft.

The block decoders were used to decode the block-coded data streams from the Viking Orbiters at 16,000 bits per second and 1,000 bits per second from the Landers. Convolutionally coded data from Pioneer or Helios were handled by the data decoders at 512 bits per second and 4,096 bits per second, respectively. Uncoded data, up to about 33 bits per second, could be handled directly by the telemetry and command processor in addition to its other command and telemetry data handling tasks.

The Data Decoder Assembly (DDA) and the Telemetry and Command Processor (TCP) delivered the processed telemetry streams to the ground communications subsystems for transmission via global NASCOM high-speed and wideband communication circuits to the GCF terminal at JPL. From that point the data could be directed to the appropriate Mission Control Center for flight project use and to the NOCC for monitoring and validation and recording purposes.

Under TCP control, two Command Modulation Assemblies and dual transmitter exciters afforded redundant paths to either a 20-kilowatt or 100-kilowatt transmitter for uplink command purposes, one uplink at a time. At Goldstone only, a 400-kilowatt transmitter could be used in place of the 100-kilowatt transmitter, if necessary.

Radiometric (Doppler and ranging) data were generated by the station tracking subsystem, which consisted of two Doppler counters and two Range Demodulation Assemblies. This arrangement, together with a Planetary Ranging Assembly, provided S- or X-band ranging simultaneously with S- or X-band Doppler. When tracking a spacecraft close to the Sun, the Planetary Ranging Assembly could be replaced by the Mu-II Ranging Machine, which embodied a different technology to give better ranging performance in the presence of solar noise.

Digital data records were made by pairs of 9-track high-density tape recorders attached to the DDA and analog records of baseband, and detected data were made by Ampex FR1400 analog tape recorders. Two high-performance Honeywell machines were available for baseband playback at the stations when necessary.

Rubidium frequency standards were the basis for all station tracking and frequency references. Interstation time synchronization to 20 microseconds was accomplished by means of an X-band Moon bounce link from the Madrid and Canberra stations to the Network master clock at Goldstone.

The Mark III-75 Data System was designed to permit the maximum flexibility in switching and interchange of assemblies in the telemetry, command, and tracking subsystems.

The Viking Era: 1974–1978

This system also provided some degree of redundancy when stations were called upon to support planetary operations for three spacecraft simultaneously. This approach enabled the stations to be rapidly reconfigured to support the different needs of the Viking, Pioneer, and Helios inflight missions.

The specific configurations needed for each mission were separately defined and coded into a Network Operations Plan for use by the station operators in manually setting up the station equipment.

The 26-meter Stations

A functional block diagram of the DSN Mark III-75 configuration for the six 26-meter stations (DSS 11 and 12 at Goldstone, DSS 42 and 44 at Canberra, and DSS 61 and 62 at Madrid) is shown in Figure 3-11.

Because DSS 42 and 61 shared the same control room and some of the same equipment as the 64-meter stations (DSS 43 and DSS 63), they were known as "conjoint" stations in the Viking Era.

Two Block III receivers provided a capability for receiving two S-band downlink carriers if necessary. Generally, they were used for only one carrier with a backup receiver, since the two subcarrier demodulators that followed could accommodate only one low-rate (33 bits per second) and one medium-rate (2,000 bits per second) data channel. Block coding at 2,000 bits per second or less was accomplished in the data decoders feeding the symbol synchronizers.

Two XDS 920 computers, known as TCPs, performed the data processing functions for two telemetry channels and two command channels. Each TCP could format and output telemetry data to a single GCF high-speed communications line for transmission to JPL. Each TCP could also deliver command data to a command modulation assembly for modulating a single S-band exciter and 20-kilowatt transmitter for transmission of commands to the spacecraft.

S-band ranging and differenced range versus integrated Doppler (DRVID), data generated by the planetary ranging assembly was available only at DSSs 12, 42, and 61. At the conjoint stations, the ranging capability was shared with the 64-meter stations, so that either, but not both stations, could provide ranging support at any time. Using their Block III Doppler extractors, all 26 stations could also generate S-band Doppler. The Tracking Data Handling (TDH) assembly was used to sample and condition the radiometric data for transmission to JPL via the GCF high-speed communication circuits.

Digital data were recorded at the output of the TCP. Baseband and detected analog records were made at the input and output of the Subcarrier Demodulator Assembly.

At the conjoint stations, a single crew was used to operate both stations from the one control room. To reduce the demands upon these crews during operations, the station monitor and control functions were exercised from a newly developed Station Monitor and Control Assembly known as the SMC IIB. The SMC provided remote control of some functions, such as receiver tuning, and, by bringing other monitoring functions to a single console, permitted the station to be operated during periods of high activity with fewer operators than would otherwise have been required.

HIGH-POWER TRANSMITTER

To meet an early Viking requirement for simultaneous commanding to two spacecraft (Orbiter/Orbiter or Orbiter/Lander) from a single antenna, the DSN planned to provide dual 10-kilowatt uplinks from a single 100-kilowatt transmitter.[14, 15] In early 1973, the dual carrier single transmitter scheme presented technical difficulties that were not resolvable at the time and a recommendation was made that separate antennas with command capability at the same longitude be used to meet the Viking requirements. However, further evaluation of the uplink performance of the Viking Lander under worst-case conditions revealed that, irrespective of the dual-carrier requirement, there remained a need for a 100-kilowatt transmit capability at the two 64-meter stations (DSS 43 and DSS 63), and in addition, a 400-kilowatt transmit capability at DSS 14 at Goldstone. These latter enhancements would "avoid unacceptable risks and/or constraints to the Viking Lander operations, particularly Mission B." Accordingly, these big new transmitters now became part of the Viking planetary configuration required at all the 64-meter stations.

Originally, the DSN stations included 10-kilowatt to 20-kilowatt S-band transmitters to provide the command and tracking data uplink to the spacecraft. However, in 1970, a 100-kilowatt high-power S-band transmitter had been installed at the new 64-meter antenna at Goldstone, and planning was initiated to provide similar transmitters at the other 64-meter antennas in Spain and Australia. Driven by the Viking requirement however, the plans were accelerated to have the DSS 63 subsystem installed and transferred to operations for Viking support by July 1975, followed by the DSS 43 subsystem in mid-October 1975.[16]

The Viking Era: 1974–1978

To save time, vendors shipped many of the components directly to the overseas sites. Other components were sent to the high-power transmitter test facility at (DSS 13) Goldstone for testing before overseas shipment.

By June 1974, elements of the DSS 63 transmitter had been assembled into a test fixture to simulate other parts of the transmitter; they were run up to full power (1 megawatt) for 24 hours. These tests assured sufficient wattage from the power supply group for either a 100-kilowatt klystron (Varian X-3060) or a 400-kilowatt klystron, which was being developed for the DSN at Varian but was not available at that time. In July, four technical personnel from Spain arrived at Goldstone for a short period of training prior to dismantling and shipping the subsystem to Spain for installation on the 64-meter antenna. After following a similar procedure, technicians shipped the DSS 43 transmitter to Goldstone in December 1974.

The construction of facilities for the new transmitter subsystem began in August 1974 at DSS 63. The facilities consisted of a building to contain the transformer/rectifier, a filter choke and control junction box, and a set of concrete pads for mounting the 1,750-horsepower (1,300-kilovoltampere), 400-cycle-per-second motor generator set. Housing was also required for the motor control center, auxiliary heat exchanger, the distilled water replenishing unit, and underground water tanks. The high-power klystron would be mounted in the tri-cone on the antenna itself and connected via waveguide couplings to the antenna microwave system. High-voltage power supplies and control equipment were to be mounted in the antenna pedestal, and a heat exchanger mounted in the alidade (non-tipping part of the antenna structure) would supply the klystron with distilled water coolant.

As the facilities became available, equipment was assembled and installed, and checkouts proceeded rapidly. Final overall and load testing was completed without incident and the 100-kilowatt high-power transmitter at DSS 63 became operational and available to support Viking in July 1975. Events at DSS 43 followed a similar pattern and the 100-kilowatt transmitter became operational at DSS 43 in October 1975.

In April 1974, the high-power transmitter at DSS 14 was shut down for approximately four weeks to install an experimental 500-kilowatt klystron that would raise the maximum transmit power to 400 kilowatts as required for Viking.

Early in 1976, the klystron tube developed a short circuit in its filament assembly and had to be returned to the manufacturer, Varian Associates, for repair. By that time, however, it had been determined that the real advantages of the 400-kilowatt capability as compared to the 100-kilowatt capability were slight, and the 500-kilowatt klystron was held as backup in the event of some unforeseen emergency.[17] Fortunately, neither of the Viking Landers for whose emergency support the backup was intended, ever had need to use it. Even as the last Viking Lander fell silent in February 1983, an analysis of the telecommunications link showed that at a transmitter power of 100 kilowatts nominal (actual power of 80 kilowatts), there was a 10- to 14-decibel margin for command with the Earth antenna pointing-angle passing through the side lobes of the lander spacecraft antenna.[18]

Mars Radar

In contrast to the 1971 and 1973 Mars radar opportunities, which were conducted at S-band frequencies, the 1975 Mars observations of the Viking landing sites were conducted at X-band. This was due in part to the availability of an experimental high-power (400-kilowatt) X-band transmitter at DSS 14, and in part to the increased emphasis on an evaluation of the Mars landing site characteristics, particularly surface roughness.

The observations were due to commence in December 1975, while the Viking spacecraft were in the cruise phase of their missions.[19, 20]

At that time,[21, 22] the experimental X-band transmitters included two, 250-kilowatt, Varian VA949J klystrons in a four-port hybrid arrangement, which would combine the two individual outputs into a single waveguide to deliver a nominal output power of 400 kilowatts. Klystron failures and other hardware problems plagued the early attempts to make successful radar observations, but in the last days of December 1975, two good observations were made at a power level of 165 kilowatts.

As the Earth-Mars range increased in the first months of 1976, intense efforts were made to increase the Mars radar transmitter power to its full 400-kilowatt rating. Varian Associates contributed significantly to this effort by responding rapidly to DSN engineers' calls for modification or refabrication of klystron components. While this work was in progress, Mars radar data gathering continued as opportunity allowed. By the end of February 1976, problems in the crow-bar protective circuits, high-voltage power supply, and klystron body current had been identified and corrected, and the radar was delivering its full rated output of 400 kilowatts at X-band. From that point on, all Mars radar tracks were conducted at full-power output.

By the end of April 1976, the strength of the radar signal returns from the planet had fallen to about one-eighteenth of their strength in December 1975, when Mars was closest to Earth. As a consequence, the data became too noisy for practical use, and the data processing technique was changed from the range-gated technique to a continuous

The Viking Era: 1974–1978

spectrum technique. This change resulted in less noise, but there was some loss in resolution compared to the former technique.

During May and June 1976, the Mars radar facility at Goldstone continued to support an intensified program of Mars observations. Also during this time, additional observations of overlapping areas of coverage were provided by the Cornell Radio Astronomy Observatory at Arecibo, Puerto Rico.

By mid-June, the data had become too noisy for analysis and the X-band radar support for Viking was concluded.

The planetary radar work continued to expand and improve in the years ahead to become a unique part of the DSN Science program. Eventually it became known as the Goldstone Solar System Radar (GSSR) and is discussed in more detail in later chapters.

The Introduction of X-band

The improvement in uplink and downlink performance that could result from operation at frequencies higher than S-band (approximately 2,000 to 3,000 MHz) had long been recognized in DSN engineering and planning. However, such improvements could not be realized without corresponding improvements in the microwave components, which together comprise the antenna/microwave/receiver subsystem. Simply changing the operating frequency alone would not be effective.

A considerable effort in research and development had been in progress during the early 1970s to develop feed-cones,[23, 24, 25] antenna surfaces, masers, and waveguide components that would operate effectively at X-band frequencies (approximately 8,000 to 9,000 MHz) in the operational Network environment.

By 1973, this work progressed to the point where the first inflight demonstration of an X-band deep space telecommunications link could be carried out using the *Mariner 10* (Mariner Venus-Mercury [MVM]) spacecraft to carry an S/X-band transponder.[26] In addition to the standard S-band uplink and downlink, the spacecraft transponder would transmit an X-band downlink carrier signal that would be coherent with the normal S-band carrier. The X-band carrier would not carry telemetry but could be switched to carry a range modulated signal. The demonstration was officially called an S/X-band experiment to distinguish it from the prime MVM mission objectives, and because the outcome of the effort was far from certain.

There remained the problem of how to adapt the DSN 64-meter antennas to receive X-band signals in addition to their existing transmit-and-receive capability on S-band. The DSN's Advanced Technology program provided the answer. Known as the "Dichroic Feed," this complex microwave device is discussed in more detail in chapter 6 In concept, this would be accomplished by adding an ellipsoid reflector over the S-band feed and a dichroic mirror over the X-band feed, with each feed-horn mounted in a separate cone structure. The dichroic mirror would pass an X-band signal directly into the X-band feed, but would direct an S-band signal on to the ellipsoid reflector. The ellipsoid reflector would then refocus the S-band signal to enter the S-band feedhorn.

The tri-cone on the 64-meter antenna at DSS 14 would be the first to be modified. When this very complex installation was completed in early 1973, performance measurements attested to the excellence of the design. No signal degradation was observed at S-band, and measured loss at X-band was less than 0.2 decibels. There was no warping or distortion of the structures, sometimes caused by thermal effects, when the S-band uplink was transmitting at full power.

The other components required for the S/X-band experiment, the new Block IV receiver, the Block IV exciter, and a new S/X-band ranging system, were available by the middle of 1973, and verified for compatibility with the onboard flight radio system.

During the early portion of the MVM mission, the S/X-band equipment experienced many problems with noise, instability, and Doppler cycle slips. Gradually, the problems were identified and corrected, and by the Mariner 10 superior conjunction during May and June 1974, the DSN was able to provide dual frequency S/X-band Doppler, range, and open-loop receiver data for radio science support. This support provided the first opportunity for spacecraft dual frequency analysis of solar corona and gravity effects, and marked the first demonstration of X-band downlink capability in the Network.

The MVM S/X experiment had demonstrated the viability of an X-band downlink and the considerable advantages of an S/X-band for radio science; this experience was to be carried a step further in the Viking mission.

Like *Mariner 10* (MVM), both Viking Orbiter spacecraft were fitted with S/X-band transponders; i.e., they would receive an S-band uplink and would transmit an S-band downlink together with a coherent X-band downlink.

The Viking Era: 1974–1978

For Viking support, both DSS 43 and DSS 63 were fitted with the dichroic plates and the Block IV receivers and exciters necessary for S- and X-band receiving capability, as had been done at DSS 14. Based on the results from the MVM mission, the Viking radio science and ranging requirements were much more stringent in terms of noise, stability, and cycle slips, than had been the case for *Mariner 10*.

Because of this new capability, the Radio Science experiments became an important part of the DSN support for Viking. Earth occultation, solar corona, and general relativity experiments were carried out successfully using the DSN S/X facilities.

In late 1975, at the end of the Viking Prime Mission, the Viking Radio Science Team reported the following:

> Since November 1975 when the X-band transponders on both Orbiters were turned on, several hundred S- and X-band Doppler and ranging passes of excellent data quality have been acquired during Mars planetary operations. These data were extremely useful in evaluating the new X-band system performance, monitoring solar flares and solar corona noise, and for the charged particle calibration of Doppler and ranging data. The data afforded the first practical demonstration of the validity of the dual frequency calibration of charged particle effects in the interplanetary medium. Also, the achievement of a 0.1 percent accuracy in the Viking general relativity experiment depended entirely upon the near simultaneity of the S-band and X-band tracking of an Orbiter and the S-band tracking of a Lander.[27]

The DSN S/X-band capability had arrived and would form the basis for all radio science observations for all flight missions from this time forward.

The 64-meter Antenna Problems

In mid-August 1976, the 64-meter antenna at DSS 14 (Goldstone) developed a potentially serious problem when it became difficult to maintain a satisfactory oil film height between the hydrostatic bearing and the azimuth bearing pads. The area of concern lay between the 128- and 140-degree azimuth positions of the antenna. This particular area of the hydrostatic bearing had been a problem since the antenna first came into service in 1966. The previous history predicted a failure condition arising about every three months, although the film height had been stable for several months at that time. The last maintenance that had been carried out about three months earlier (in June) used a shimming technique that had reached its limits and could no longer be used for further maintenance. Further delay in correcting the problem would only increase the chance of a catastrophic failure of the antenna's azimuth rotation bearing.

A new technique, which required an 18-hour downtime on seven successive days, had been proposed but never tried out, and therefore carried with it some question as to how effective the proposed "fix" might be. Occurring as it did at such a critical time in the Viking mission, just before the second Mars landing, the situation and issues involved were cause for great concern to the DSN.

Rather than proceeding with the "fix," a plan was put into effect which required daily monitoring and reporting of the oil film height against an established set of safety criteria. In this way some warning of a deteriorating condition would become visible, permitting a reevaluation of the risk involved in continuing operations. It was also decided that a firm work plan for refurbishment would be scheduled for 15 November 1976, or sooner if the bearing showed any sign of deterioration prior to that time. This plan worked very well. The oil film height remained stable through 5 November, at which time the operational usage of DSS 14 was such that the work was carried out on alternate weekdays and brought to a successful conclusion on 19 November 1976.

A further incipient problem had been known for some considerable time in the ball and socket joint on the antennas at DSS 43 and DSS 63. An inspection carried out in late 1975 revealed that the ball joint that permitted the pad to rotate as it followed the contours of the hydrostatic bearing runner was frozen because of improper lubrication. A cross-sectional diagram of the hydrostatic bearing and the ball joint assembly is given in Figure 3-12.

Prolonged attempts to force grease into the joint under high pressure were unsuccessful, so a spare unit was shipped to Canberra in January 1976. However, prior to installation on site at Canberra, this unit was found to be defective and was returned to the United States. New units were ordered for both DSS 43 and DSS 63 only, since the DSS 14 units were satisfactory. Installation was postponed to coincide with the Viking solar conjunction period in November and December, and the station downtime for the MDS work in Australia. The installation was successfully completed during that time.

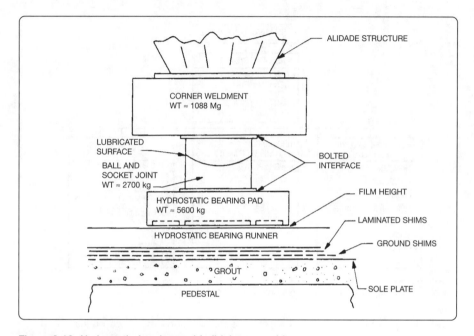

Figure 3-12. Hydrostatic bearing and ball joint assembly.

DSN Mark III Data System Project

The concepts of the DSN Mark III Data System established by Victor in 1969 were updated, refined, and developed into a Mark III Project Plan by Easterling in 1974.[28]

Easterling's fundamental objective in planning the Mark III project was to provide support to flight projects in the 1973–1978 period and establish a standard, easily managed interface with the MCCC at JPL and with other remote POCCs. A key factor in accomplishing this objective was seen to be the development of a Network Control System that would be entirely separate from and independent of the hardware and software used by the flight projects. The upgraded Network would include new technology in radiometric data generation using two station radio-interferometry techniques, increased use of automation throughout the Network, and real-time transmission of spacecraft video data from the DSSs to the MCCC using a global network of wideband communication circuits.

The plan advocated a phased design approach to hardware implementation with formal reviews at the completion of each phase, culminating in acceptance tests and a formal

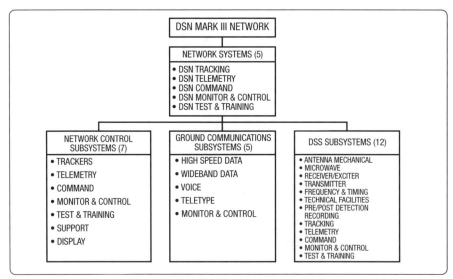

Figure 3-13. Mark III-77 Network systems and subsystems.

transfer to operations status. An efficient, cost-effective software design philosophy developed by a software technology group would employ structured, top-down programming techniques to provide for simple interfaces, well-controlled development, and easily modified programs. A phased design approach culminating in acceptance tests and transfer to operational status would be used in producing all DSN software programs. After transfer, DSN software would be subject to a rigorous program of configuration control and management, maintained by a centralized DSN Software Library.

The Network was defined to include five Network systems: Tracking, Telemetry, Command, Monitor and Control, and Test and Training. Lower tier subsystems were defined for each of the three elements of the Network: Deep Space Stations, Ground Communications, and Network Control. The Network systems and subsystems are shown in Figure 3-13.

The upgraded Network, which no longer included the SFOF, would be identified as DSN Mark III-77 and is shown in Figure 3-14.

Working within this hierarchical framework, engineers developed key characteristics for all systems and subsystems which reflected the known and anticipated requirements of the NASA Mission Set through 1981.

The Viking Era: 1974–1978

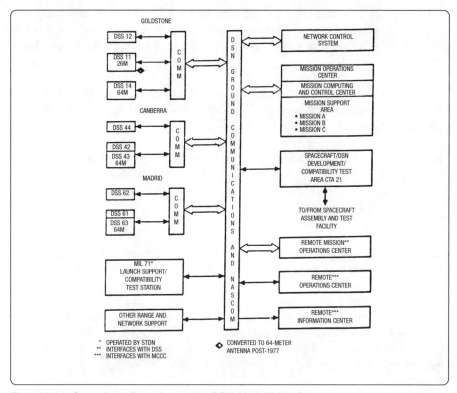

Figure 3-14. General configuration of the DSN Mark III-77 Network.

The telemetry system would take advantage of improved telecommunications capability to handle data rates up to 115 kilobits per second at Jupiter distance. It would accommodate multiple data streams and perform the block decoding, sequential decoding, and maximum-likelihood convolutional decoding functions, using standard techniques. The command range and command data rate would be extended and an improved "store and forward" command transmission process would be introduced. Radiometric data types would then include long baseline interferometer data, as well as planetary ranging and Doppler observables. The introduction of Hydrogen Maser frequency standards in place of the older Rubidium Standards would improve the accuracy and stability of all radiometric data types to improve the uncertainties in spacecraft navigation targeting parameters.

Network operations would be conducted from an NOCC using dedicated facilities to serve multiple missions at several locations simultaneously. It would also include a formalized DSN scheduling procedure based on a 7-day, 8-week, 52-week, and 5-year Network allocation plan.

Improved management of the operational interfaces between the flight projects and the DSN would result from the newly defined DSN Monitor and Control, and Test and Training capabilities.

The GCF capability for duplex high-speed data transmission would be raised from 4,800 bits per second to 7,200 bits per second, while the wideband capability would increase from 28.5 kilobits per second to 50 kilobits per second from all sites to JPL. Up to 230 kilobits per second would be available from Goldstone to JPL. The GCF would create a GCF log of all flight project data generated by the DSN and deliver it on a negotiated schedule.

The DSSs would be upgraded with S- and X-band downlink capability for tracking and telemetry and for acquiring up to five spacecraft in one antenna S-band beam width.

Finally, the plan proposed a management approach to carry out the Mark III-77 project. It included an overall Project Manager, as well as an Implementation Manager for each of the Network Data Systems, and for Quality Assurance and Reliability. The managers were to be supported by cognizant system and subsystem engineers. Guidelines were also given for scheduling, and progress and problem reporting.

It was early 1974, and the window of opportunity for implementation of these major changes in the Network was approaching. The window would open at the end of the Viking Prime Mission (last quarter of 1976) and close just before the Pioneer Venus Prime Mission started (second quarter of 1978). There was no time to lose. Implementation Task Managers were appointed for the data systems, and began to develop technical requirements derived from the key characteristics contained in the plan, and schedules based on the rapidly approaching window.

The largest of the tasks included several of the Network systems, and for that reason was called the Mark III Data Systems Implementation Task; it was established in February 1975 with Paul T. Westmoreland as Task Manager.

Paul T. Westmoreland started at JPL in February 1960 as one of the original engineers in Renzetti's Communications Engineering and Operations Section. Except for a two-year lateral transfer to the Mission Control and Computing department in 1977, he rose through the DSN to become the assistant laboratory director for tracking and data acquisition in 1997. Born and educated in California, Westmoreland received a B.Sc. in electrical engineering from USC in 1960, just prior to joining JPL, and a master's degree in the UCLA Engineering Executive Program in 1968.

Westmoreland's early work led to the introduction of digital computing techniques for operational control of the Network in the mid-1960s. Based initially on Scientific Data Systems (SDS) 910- and 920-type computers, and subsequently on Modcomp machines, and later still, on distributed computing techniques, this approach provided the operational flexibility that was the key to the future success of the Network control process. As the applications for digital technology proliferated in the Network, so did the requirements for software to drive the digital processes. In collaboration with Walt Victor, Westmoreland introduced new and innovative software standards that provided management with insight and control of the cost and status of formerly runaway software development processes.

Later in his career, he directed the DSN Systems Engineering Office, where he was ultimately responsible for the overall architectural design and growth potential of the Network. In 1997, he was appointed director of the Telecommunications and Mission Operations Directorate.

Paul Westmoreland was a pleasant though serious man for whom pride in his appearance complemented obvious pride in his achievements. His well-cut suits with fashionable ties were notable among his more casually dressed coworkers. He was an expert ballroom dancer, and he cast gold costume jewelry as a hobby. He was highly respected at JPL and at NASA Headquarters for both his technical judgment and his fiscal astuteness. His management style owed much to his early years at JPL when Rechtin and Victor held the positions that he would later hold. In a real sense, the Network's ability to support many of the major spacecraft-related projects described in this history owes much to the technical foresight and budgetary awareness of Paul Westmoreland.

He retired from JPL in 1998.

Mark III Data System (MDS) Implementation

A plan for implementing the portion of the Mark III project that involved the processing and transport of all data within the Network was issued by P. T. Westmoreland in February 1975.[29] The "Westmoreland plan" was intended to embody the principal features of project management, system and hardware design, and software design philosophy enunciated by Easterling in the Project Plan.[30]

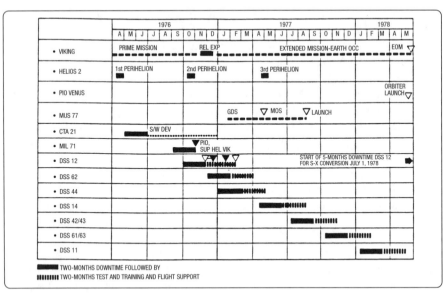

Figure 3-15. DSN Mark III Data System implementation schedule, 1977.

This major Network-wide reconfiguration was called the DSN Mark III Data System (MDS) implementation task and involved all stations of the Network in sequence, throughout 1977 until completion, on schedule and under budget, in mid-1978.

In carrying out this task, the DSN would replace all its obsolete SDS-920 computers with new MODCOMP II-25 minicomputers. Each of the nine tracking stations in turn would have its old telemetry and command processors replaced with new machines and software, and then be requalified to resume inflight support. The implementation task schedule, which was carefully planned to avoid impact with the inflight mission support activity and the launch of the new Voyager mission to Jupiter and beyond, is shown in Figure 3-15.

Because of the size of the effort involved in this task, the basic implementation was accomplished by an integration contract with Univac Corporation to provide for the fabrication, test, and integration of the Modcomp and other assemblies to form station-level configurations for onsite installation and test by traveling teams of DSN engineering personnel. Three basic station-level configurations were designated, one for the 26-meter stand-alone stations (DSS 11, 12, 44 and 62), one for the 26/64-meter conjoint stations (DSS 42/43 and DSS 61/63), and a 64-meter configuration

The Viking Era: 1974–1978

for DSS 14 and DSS 63, as well as for CTA 21 at JPL and MIL 71 at Merritt Island, Florida. Each configuration contained up to three of each of the following major assemblies:

- Telemetry Processor Assembly (TPA)-Assumed the telemetry functions of the existing DSS Telemetry and Command Processor (TCP) by replacing the older XDS-920 computers with new Modcomp minicomputer to form a separate Telemetry Subsystem. The existing Symbol Synchronizer remained in place.

- Command Processor Assembly (CPA)-Assumed the command functions of the existing DSS Telemetry and Command Processor (TCP) by replacing the older XDS-920 computers with new Modcomp minicomputer to form a separate Command Subsystem. The existing Command Modulation Assembly remained in place.

- Metric Data Assembly (MDA)-Replaced the existing DSS Tracking Data Handling Subsystem (TDH) and functioned as part of the DSS Tracking System.

- Communications Monitor and Formatter (CMF)-Functioned as part of the existing GCF to provide for input/output formatting of high-speed data and wideband data between the MCCC and the CPA, MDA, and TPA assemblies. It also carried out the function of making Original Data Records (ODR).

- Digital Instrumentation Subsystem (DIS) to Mark III Interface (DMI)-A non-minicomputer assembly that provided for the interconnection of the CMF, MDA, and TPA minicomputer assemblies, and interfaced them with the existing DIS.

- Time Format Assembly (TFA)-A non-minicomputer assembly that functioned as an extension of the existing Frequency and Timing Subsystem (FTS) to provide Greenwich Mean Time (GMT) inputs and various interrupt pulses to the TPA, CPA, MDA, and CMF.

- Data Systems Terminal (DST)-A non-minicomputer assembly that provided a central means of connecting associated Megadata terminals (computer monitor displays) and G. E. Terminets (computer input terminals) to any two of up to seven minicomputer assemblies.

Each station-level configuration was designated an "MDS" and was based on the arrangement shown in Figure 3-16.

In total, the MDS assemblies and peripherals occupied 214 DSN standard size cabinets assembled into 87 individual modification kits, tested and shipped to the stations for installation during the 2-year period of implementation.

As Viking conducted important relativity experiments with DSSs 14, 42, and 63, the MDS implementation activity started at DSSs 12, 44, and 62. As each of these stations was withdrawn in turn from Viking, Helios, and Pioneer support, the MDS reconfiguration was carried out and the station was returned to operational status following an appropriate period of testing and crew retraining.

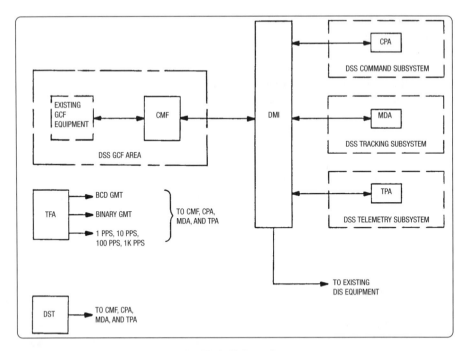

Figure 3-16. Basic organization of the Mark III Data System.

The Viking Era: 1974–1978

By mid-July 1977, DSSs 12, 14, and 62 had completed system performance testing in their new configurations and resumed full operational support. Stations 42 and 43 were released from tracking support on July 15, began a ten-week period of reconfiguration and retest, and were returned to operation in mid-October. By the end of January 1978, the work had been completed at DSSs 61 and 63 and the stations returned to operations support. The last of the six stations, DSS 11, returned to service on 26 April 1978.

The accomplishment of this major reconfiguration throughout the entire Network, in the midst of a heavy and demanding tracking support schedule that included two major launches (*Voyagers 1* and *2*), presented a formidable and exacting task. That it was carried out on schedule, without any significant problems, and without impacting the ongoing flight support was a credit to all the engineering and operations personnel involved.

DSN Mark III-77 Model

At the completion of the MDS implementation task in April 1978, the new DSN model Mark III-77 would have all the capabilities needed for support of the remaining missions in the Viking Era Mission Set (see Figure 3-1), plus future missions including the (then-planned) 1983 Galileo mission to Jupiter. These missions include *Pioneers 6–11*, the Pioneer Venus and Pioneer Multiprobe, Helios, Viking, and Voyager.

The continued growth of the DSN capabilities, in response to flight project requirements, is obvious to anyone who examines the key characteristics of the telemetry system. The additional telemetry capabilities resulting from the MDS transformation of the Mark III-75 model to the Mark III-77 model are listed below:[31]

1) High-rate X-band telemetry capability (up to 250 kilosymbols per second) at both a 34- and a 64-meter subnet in addition to S-band telemetry at all subnets.

2) Maximum likelihood decoding (Viterbi algorithm) of short-constraint-length convolutional codes at all DSSs. Deletion of block decoding after completion of the Viking mission.

3) Replacement of the TCP with a dedicated processor, the TPA, for telemetry.

4) Precise measurement of received signal level and system noise temperature.

5) Simultaneous reception of five carriers at a selected DSS for Pioneer Venus Multiprobe (to be deleted after the mission).

6) Replacement of the Data Decoder Assembly by incorporating its functions into the TPA.

7) Formatting of all decoded data for high-speed or wideband transmission to the NOCC and MCCC via the GCF.

8) Real-time arraying of signals received from two stations at the same longitude.

DSN Mark III-77 Telemetry System

A simplified block diagram of the telemetry system for the DSN Mark III-75 model is shown in Figure 3-17.

Prediction messages are initially generated at the NOCC for transmission by a high-speed data line (HSDL) to the DSSs for the purpose of establishing the proper telemetry modes and configurations. Such messages consist of predicted frequencies, data rates, signal levels, and tolerances.

At the DSS, the radio frequency spacecraft signal is collected by the antenna, amplified by the Antenna Microwave Subsystem and passed to the Receiver-Exciter subsystem. The radio frequency (RF) carrier is tracked by the receiver, and the telemetry spectrum is routed to the Subcarrier Demodulation Assembly (SDA) where the subcarrier is regenerated, and the symbol stream is demodulated. The resulting demodulated symbol stream is passed to the Symbol Synchronizer Assembly (SSA) where it is digitized. The digitized stream for convolutional encoded data is then routed to the Maximum Likelihood Convolutional Decoder (MCD) for decoding of short-constraint-length convolutional codes, or to the TPA for decoding of long-constraint-length codes, or uncoded data. The digitized symbol stream for block encoded data is sent to either the TPA or the Block Decoder Assembly (BDA) depending on the data rate. All the data are formatted by the TPA for transmission by high-speed or wideband data line to the NOCC and MCCC.

The Viking Era: 1974–1978

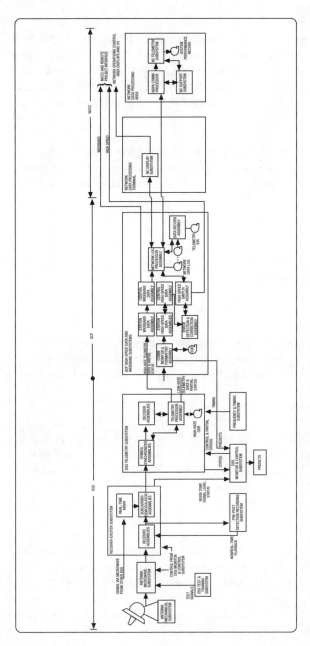

Figure 3-17. Functional block diagram of the DSN Mark III-77 Telemetry System.

At each DSS, an ODR of the decoded data is written in GCF blocks by either the TPA for high-rate data or the Communications Monitor and Formatter (CMF) for low-rate data. The data are passed to a high-speed or wideband buffer depending on the data rate, and then transmitted to the MCCC at JPL or in some cases to a Remote Mission Operations Control Center (RMOC), as in the case of Pioneer (ARC) or Helios (GSOC) and, in parallel to the NOCC for network control and monitoring purposes.

At the NOCC, a limited amount of decommutation of engineering telemetry data is performed to analyze system performance and to obtain certain spacecraft parameters useful in controlling the Network. The NOCC also receives and displays DSN Telemetry System parameters.

All the data received by the NOCC either in real time or by recall from the ODR are written to the Network Data Log by the Network Log Processor. The Data Records Assembly provides for the recall of any missing data by activating the ODR after each station pass to complete the NDL. It uses the NDL as a source to create the IDR for delivery to the flight project. The IDR is a complete, time-ordered record of all spacecraft data received at each DSS during a spacecraft tracking pass.

DSN Telemetry System performance is displayed and validated in the NOCC by a comparison of the telemetry system monitor data, which are transmitted from the DSS, with preset standards and limits for the particular configuration being used.

DSN Mark III-77 Command System

In the DSN Mark III-75 Command System model, the DSS command processing function shared the same computer (XDS-920) as the telemetry processing function. In the Mark III-77 model, the command function resided in a new dedicated computer (MODCOMP II-25) to be known as the CPA. The CPA provided higher reliability, greater processing speed, and more core memory for the command function. The increased processing speed allowed the DSN to support higher spacecraft command data rates. In the TCP XDS-920, the 8-bits-per-second command bit rate used by Helios was near the upper limit of the Mark III-75 Command System. Voyager required a command rate of 16 bits per second, and the new CPA would accommodate this and even higher bit rates anticipated for future missions.[32, 33]

Later, in 1978, using the increased data storage capability of the CPA, a new command technique known as "store and forward" was introduced. This technique allowed

The Viking Era: 1974–1978

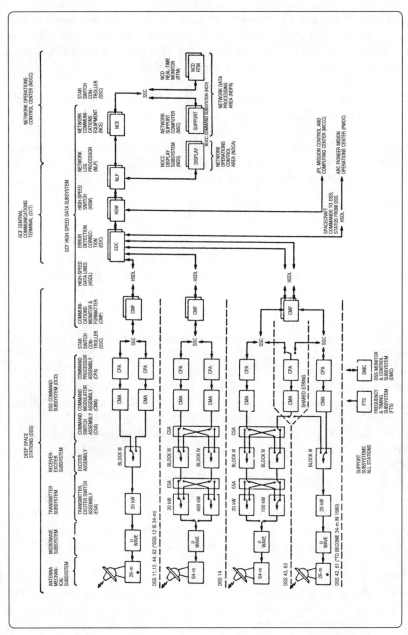

Figure 3-18. DSN Mark III Command System, 1978.

the mission operations controllers to prepare large files of spacecraft commands in advance and then forward them to the DSS at the beginning of a spacecraft tracking period. At the DSS, the files were stored on the CPA disk and placed in a queue for radiation in the specified order at the designated time.

The DSN Command System as it existed after the MDS reconfiguration and subsequent enhancements is shown in Figure 3-18.

In addition to these improvements, the overall Network command capability was further enhanced by an increase in the uplink transmitter power at DSS 12 and DSS 62 from 10 to 20 kilowatts to meet the Network standard.

DSN Mark III-77 Tracking System

The art of spacecraft navigation is based upon the three types of radiometric data generated by the DSN Tracking System. The radiometric data types are S- and X-band Doppler, S- and X-band range, and antenna tracking angles. Doppler and range in various combinations are the most powerful data types, and are used directly in the spacecraft orbit determination process. The Mark III-75 Tracking System had provided these data types for Pioneer, Helios, and Viking. To support Venus and Voyager, additional new capabilities were needed, and these were provided in the Mark III-77 model.[34]

By improving the DSS frequency stability by two orders of magnitude, the new Hydrogen maser frequency standards that replaced the older Rubidium frequency standards extended the precision range and Doppler capability of the DSN Tracking System to beyond 20 AU to meet the long-range navigation requirements for Voyager. Radiometric data could now be time tagged to the 10-nanosecond level relative to the DSN master clock.

Calibration of the DSN radiometric data for the effects of the tropospheric and ionospheric media along the transmission path between spacecraft and Earth was made possible by a new Meteorological Monitor Assembly that measured these data at the stations and transmitted them via the new GCF to the Network users.

At the DSSs, all of the old hardware and software associated with the Tracking System was replaced with new units based on the MODCOMP computers. The planetary ranging capability was improved for operation when the line of sight to the spacecraft passes close to the Sun, and a new unit called the MDA carried out

The Viking Era: 1974–1978

all metric data handling functions and controlled the interface with the upgraded GCF.

At the NOCC, improvements were made in the generation of tracking data predicts based on a long span of trajectory data supplied by the flight project. The display, real-time monitoring, and data validation functions were upgraded to supplement the DSS and GCF improvements in the Mark III-77 model.

DSN Mark III-77 Ground Communications

For the Mark III-77 model, the GCF capability that provided for transmission of formatted telemetry data from each DSS to the NOCC and MCCC was also increased. The high-speed data line rate from all stations was increased from 4,800 to 7,200 bits per second. The wideband capacity from each 64-meter DSS in Spain and Australia was raised from 28.5 to 56 kilobits per second with a capacity of 230 kilobits per second being available from Goldstone.

End-to-end encoding with new 22-bit error polynomials provided automatic error detection and correction to significantly reduce the possibility of undetected high-speed transmission errors. With this correction scheme the DSN was able to approach nearly 100 percent delivery of real-time data, and cut down on the station time and personnel effort previously required for post-track replay of missed telemetry data. These capabilities as they were set up for Viking and the later missions of the Viking Era, are shown in Figure 3-19.

In mid-1978, the GCF high-speed-data subsystem was converted from the older line switching method of routing messages that had been used for Helios, Viking, and earlier missions, to programmable message switching capability. The upgrade also included changes to the NOCC and MCCC interfaces with the GCF. However, the line switching interface for Viking was maintained until March 1979. By then, message switching was standard for all missions.[35]

DSN Mark III-77 Arraying Capability

As part of the Mark III-77 capability for Voyager, a two-station arraying capability was to be provided at each complex. The combining function was to be carried out at the subcarrier level using the subcarrier spectrum outputs from a 64-meter station and a 34-meter station to provide a single output for telemetry processing. The arraying was to accommodate subcarrier frequencies of 24 to 500 kilohertz and symbol rates from 2 to

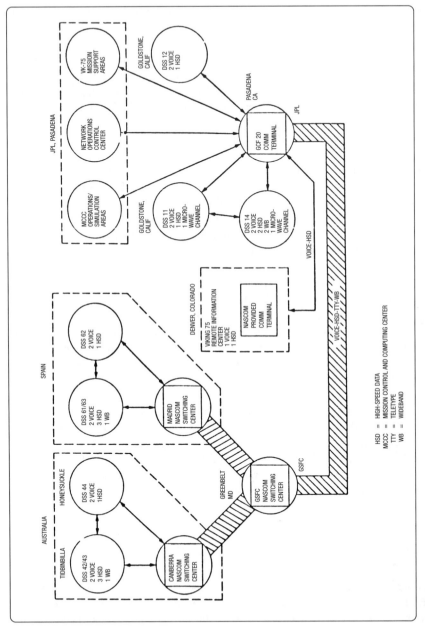

Figure 3-19. DSN-GCF-NASCOM ground communications, 1976–78.

250 kilosymbols per second. The combining process was to introduce less than 0.2 decibel degradation and to operate at a received signal-to-noise ratio as low as 0.8 decibel from a 64-meter DSS and -5.2 decibels from a 34-meter DSS. This capability would find its first operational application as the DSN entered the Voyager Era and the Voyager spacecraft approached its first encounter with Jupiter.[36]

DSN Mark III-77 Decoding Capability

To accommodate the decoding requirements of the Viking Era mission set the DSN Mark III-77 model provided for the following three types of telemetry channel decoding:[37]

- Viking-Block Codes 32:6; biorthogonal codes handled by the BDA at the 64-meter DSS at data rates up to 16,000 bits per second. At other DSS, the block decoding is performed by the TPA software at rates limited to 2,000 bits per second.

- *Pioneers 6–11* and Helios-Long-Constraint Length Convolutional Codes (32:1/2); Sequential decoding by the Fano algorithm using TPA software at output rates limited to 4,000 bits per second.

- Pioneer Venus, Voyager, and future missions-Short-Constraint-Length Convolutional Codes (7:1/2); maximum likelihood decoding by the Viterbi algorithm using the MCD at output rates up to 250,000 bits per second.

The block decoding functions for Viking were not substantially affected by the transition to the Mark III-75 model. The long-constraint-length, convolutional decoding functions for Pioneer and Helios were also essentially unaltered except that they were now carried out within the TPA by a separate processor that interfaced to the new MODCOMP II-25 computers.

However, the short-constraint-length, convolutional decoding function required by *Pioneer 78*, Voyager, and future missions was an entirely new capability. This new decoder was called a Maximum Likelihood Convolutional Decoder (MCD) and utilized the Viterbi decoding algorithm operating on short-constraint-length codes (K=7) to provide a gain of about 5 decibels over uncoded data at data rates from 10 to 250,000 bits per second. The obvious bit rate advantage of maximum likelihood decoding is due to the fact that a large part of the computations can be performed in parallel, rather than in sequence, as in sequential decoding.

The MCD was a special purpose digital computing device that received coded, quantized symbols from an SSA and output decoded data to the TPA for formatting and transmitting to the Mission Control Center at JPL. The functional requirements for the MCD and the factors used in their selection had been the subject of a great deal of study, development, test, and evaluation within the DSN for several years prior to their implementation in the MCD.

The salient features of the MCD are described below:

1) Short-constraint-length (K=7) provided simplified encoder design, adequate decoding performance, and reasonable computational load.

2) Minimum node synchronization time (grouping of incoming symbols into pairs for rate 1/2 are triplets for rate 1/3).

3) Code rates of 1/2 and 1/3 are provided to satisfy Voyager and provide growth capability for future requirements. (Note: For a given symbol rate, rate 1/3 provides about 0.3-decibel improvement over rate 1/2, but at the expense of reduced data rate.)

4) Maximum data rate of 250,000 bits per second was chosen to be about double the maximum SSA capability to allow for future increase in SSA capability.

5) Decoder bit-error rate performance represented a state-of-the-art capability and is given as a function of the ratio of energy-per-bit (E_b) to noise-spectral-density (N_o) as shown below:

Energy-per-bit/Noise-spectral Density

Bit-Error Rate	Code (7:1/2)	Code (7:1/3)
10(-3)	3.0	2.7
10(-4)	3.8	3.5
10(-5)	4.5	4.2
10(-6)	5.2	4.9

For a comprehensive discussion of the theory and application of telemetry channel encoding and decoding by Joseph Yuen, and other experts in that field, the reader is referred to the book "Deep Space Telecommunications Systems Engineering."[38]

The 26-meter Antenna S-X Conversion Project

In consonance with the Mark III-77 project guidelines, a study group was formed early in 1975 to prepare an implementation plan and budget estimates for upgrading the 26-meter antennas to larger diameter and adding an X-band downlink capability. It was called the 26-Meter S-X Conversion Project. Trade-off studies using telecommunication link design parameters for the Pioneer Venus S-band mission and the Voyager S/X-band mission showed that the optimum diameter for the new antennas would be 34 meters. Following a review at NASA Headquarters in December, it was decided that the stations to be modified would be DSS 12, DSS 44, and DSS 62 and that they would be completed on the following schedule: DSS 12 (Goldstone), 1 December 1978; DSS 44 (Canberra), 1 April 1980; and DSS 62 (Madrid), 1 October 1980.[39]

The implementation plan issued by N. A. Renzetti on 1 July 1976, called for extensive changes in the existing Antenna Mechanical Subsystem, the servo and antenna pointing control mechanisms, the Antenna Microwave Subsystem, and the Receiver-Exciter Subsystem. These changes were to be made on a negotiated schedule to minimize conflict with the ongoing flight missions.[40]

The aperture increase from 26-meters to 34-meters required extensive modification of the existing antenna structure. These changes, together with changes to the microwave elements of the antenna, are shown in Figure 3-20.

The new X-band masers for the 34-meter S/X-band stations would have the same basic amplifying portion and cryogenics as the existing operational masers on the 64-meter antennas, but would be provided with superconducting magnets for improved gain and stability. These X-band masers were also repackaged to make a more compact and easily maintained unit, and they were supported by much improved instrumentation. In particular, the microwave switching control system for the new S/X-band waveguide assembly was completely redesigned to be compatible with the centralized station control and monitoring facilities, and included a graphics display of each selectable microwave configuration.[41]

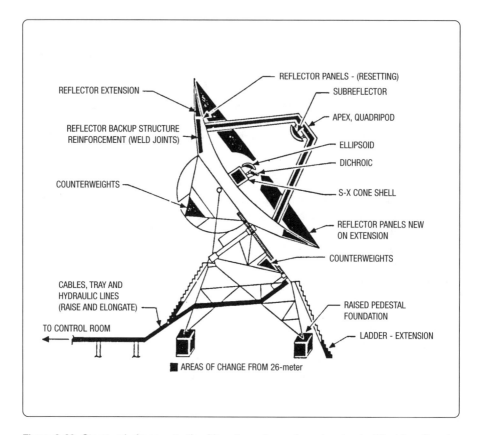

Figure 3-20. Structural changes to the 26-meter antenna for extension to 34-meter diameter.

The addition of the X-band receive capability at the new 34-meter stations also required the addition of a fourth harmonic filter to the transmitter subsystem. The purpose of the filter was to suppress the fourth harmonic output from the S-band transmitter klystron amplifier, which otherwise could cause interference with the X-band signals in the X-band maser receiving pass band.

In order to provide sufficient long-term frequency stability for outer planet navigation and potential interferometry applications, an upgrade to the existing 26-meter frequency standards was required. For this purpose, two high-grade Cesium standards were provided, which could be referenced to the hydrogen maser frequency standards at the nearby 64-meter stations.

The Viking Era: 1974–1978

The most cost-effective way of providing receivers to complement the X-band maser front end was to first down-convert the X-band maser output to S-band, and then feed the S-band signals to the existing Block III S-band receivers. To maintain uplink and downlink frequency coherence, the reference frequency for the downconverters was obtained through appropriate frequency multipliers and dividers from the transmitter-exciter frequency source.

When completed, the 34-meter S/X-band antennas were required to have the following RF performance:

S-band: Transmit: 2,110 (±10) megahertz
  Gain: 55.3 (±0.7) decibels
S-band: Receive: 2,285 (±15) megahertz
  Gain: 56.1 (+0.3/-0.9) decibels
  System Noise Temperature:
  Diplexed, 27.5 (±2.5) degrees kelvin
  Listen only, 21.5 (±2.5) degrees kelvin
X-band: Receive: 8,420 (±20) megahertz
  Gain: 66.9 (+0.3/-0.9) decibels
  System Noise Temperature:
  Diplexed, 25 (±3.0) degrees kelvin

The first station in the series, the Goldstone 26-meter antenna at DSS 12, was decommitted from flight project support in June 1978 to undergo the upgrade to 34-meter, S/X-band. The work was carried out in the period June through September 1978. System and subsystem tests were conducted at the station, in October 1978, to verify that the antennas met the design specifications. A period of extensive crew training was followed by ground data system tests and demonstration tracks on Voyager before being declared available for full operational support in November 1978. A photograph of DSS 12 after its conversion to the 34-meter S/X-band configuration is shown in Figure 3-21.

The large concrete blocks on which the antenna pedestal stands were required to provide sufficient ground clearance for the increased antenna diameter. At the 34-meter sites in Australia and Spain, large, concrete-lined trenches were formed in front of the antennas as an alternative method of providing the necessary ground clearance.

Following the completion of DSS 12, effort on the 26-meter S-X conversion project was transferred to the two overseas sites; they were completed in the Voyager Era, but

Figure 3-21. Goldstone Station, DSS 12 antenna after conversion to 34-meter, S-X-band Configuration, October 1978.

by that time the original plan had been changed. Rather than convert the 26-meter stations at Cebreros and Honeysuckle as originally planned, the conjoint stations, DSS 42 and DSS 61, were upgraded to 34-meter diameter. DSS 44 and DSS 62 faced an uncertain future and were eventually retired.

The Viking Era: 1974–1978

OVERVIEW

Dominated by Viking and driven by the requirements for tracking and data acquisition support for the other major new planetary missions of Helios, Voyager, and the later Pioneers, DSN capabilities expanded in the years of the Viking Era, 1974 through 1978. Competition between the flight projects for antenna tracking time led to a much more formalized approach to forecasting and allocation of these finite DSN resources. While the initial approach left much to be desired, it was continually refined on the basis of experience and became more effective as time went on. The introduction of a formalized architecture for the individual stations of the Network in the form of an integrated assemblage of interactive systems did much to enhance the manageability of further improvements and new implementation in the Network as a whole. The Mark III Data Systems Project that preceded the Mark III-77 model of the Network is a prime example. New technology in low-noise X-band receivers, arraying of multiple antennas, and advanced coding techniques were combined to push the limits of downlink sensitivity to new levels to match the demands for higher data rates at ever increasing distances. Advances such as these were augmented with improvements to existing resources such as the conversion of the existing 26-meter antennas to 34-meter diameter and retrofitting them with X-band receivers.

Although it was not apparent at the time, the launch of the Pioneer multiprobe spacecraft to Venus in December 1978 marked the beginning of the longest hiatus in planetary spacecraft launchings in the history of NASA. For the next ten years, until the launches of Magellan and Galileo in 1989, the DSN would be involved in supporting the deep space missions already in flight and supporting new and existing Earth-orbiting missions. It is to these events that we now turn in that period of the history of the DSN dominated by the remarkable Voyager spacecraft—the Voyager Era.

Endnotes

1. Edward C. and Linda N. Ezell, *On Mars: Exploration of the Red Planet, 1958–1978*, NASA SP: 4212 (Washington, DC: NASA 1984), p. 452.

2. Jet Propulsion Laboratory, "Deep Space Network to Flight Project Interface Design Handbook," DSN Document Number 810-5, Latest Revision, 1995.

3. D. J. Mudgway and M. R. Traxler, "Tracking and Data Acquisition Support for the Viking 1975 Mission to Mars," JPL Technical Memorandum TM 33-783, Vol. I (15 January 1977), pp. 5, 35, 123.

4. David Morrison and Jane Sanz, *Voyage to Jupiter*, NASA SP-439 (Washington, DC: NASA, 1980), p. 25.

5. J. E. Allen and H. E. Nance, "Voyager Mission Support," DSN Progress Report PR 42–49, November/December 1978 (15 February 1979), pp. 29–33.

6. N. A. Renzetti, "A History of the Deep Space Network From Inception to January 1969," JPL Technical Report TR 32-1533 (September 1971), p. 41.

7. William R. Corliss, "A History of the Deep Space Network," NASA CR-151915; (Washington, DC: NASA, 1976), pp. 92, 194.

8. R. B. Miller, "Pioneer 10 and 11 Mission Support," DSN Progress Report PR 42-33, March/April 1976 (15 June 1976), pp. 21–25.

9. T. W. Howe, "Pioneer Venus 1978 Mission Support," DSN Progress Report PR 42-51, March/April 1979 (15 June 1979), pp. 19–30.

10. R. B. Miller, "Pioneer 1978 Venus Mission Support," DSN Progress Report PR 42-31, November/December 1975 (15 February 1976), pp. 11–14.

11. W. K. Victor, N. A. Renzetti, J. R. Hall, and D. J. Mudgway, "DSN Mark III Data System Development Plan, Oct-Nov 1969," presentation given at JPL, 21 October 1969, Deep Space Network Records, Jet Propulsion Laboratory Archives.

12. R. Cargill Hall, *Lunar Impact: A History of Project Ranger*, NASA History Series; NASA SP-4210 (Washington, DC: NASA, 1977), pp. 91, 23.

13. M. L. Yeater, "DSN System Requirements: DSN Telemetry System 1977 Through 1982," DSN Document 821-2: Rev. B (15 September 1978).

14. D. A. Bathker and D. W. Brown, "Dual Carrier Preparations for Viking," JPL Technical Report TR 32-1526, Vol. XIV: January/February 1973, (15 April 1973), pp. 178–199.

15. D. A. Bathker and D. W. Brown, "Dual Carrier Preparations for Viking," JPL Technical Report TR 32-1526, Vol. XIX: November/December 1973, (15 February 1974), pp. 186–92.

16. J. R. Paluka, "100Kw X-band Transmitter for FTS," JPL Technical Report TR 32-1526, Vol. VIII: January/February 1972 (15 April 1972), pp. 94–98.

17. D. J. Mudgway and M. R. Traxler, "Tracking and Data Acquisition Support for the Viking 1975 Mission to Mars," JPL Technical Memorandum TM 33-783, Vol. II (15 March 1977), pp. 94, 125, 137, 208.

18. D. J. Mudgway, "Tracking and Data Acquisition Support for the Viking 1975 Mission to Mars; May 1980 to March 1983," JPL Publication 82-107 (15 May 1983), pp. 3–13.

19. G. S. Downs, R. R. Green, and P.E. Reichley, "A Radar Study of the Backup Martian Landing Sites," DSN Progress Report PR 42-36, September/October 1976 (15 December 1976), pp. 49–52.

20. M. A. Gregg and R. B. Kolby, "X-band Radar Development," DSN Progress Report PR 42-20, January/February 1974 (15 April 1974), pp. 44–48.

21. C. P. Wiggins, "X-band Radar Development," JPL Technical Report TR 32-1526, Vol. XII, September/October 1972 (15 December 1972), pp. 19–21.

22. R. L. Leu, "X-band Radar System," JPL Technical Report TR 32-1526, Vol. XIX, November/December 1973 (15 October 1973), pp. 77–81.

23. R. W. Hartop, "X-band Antenna Feed-Cone Assembly," JPL Technical Report TR 32-1526, Vol. XIX: November/December 1973 (15 February 1973), pp. 173–175.

24. P. D. Potter, "Improved Dichroic Reflector Design for the 64-m Antenna S-and X-band," JPL Technical Report TR 32-1526, Vol. XIX: November/December 1973 (15 February 1973), pp. 55–62.

25. D. L. Trowbridge, "X-band Travelling Wave Maser Amplifier," JPL Technical Report TR 32-1526, Vol. XVII: July/August 1973 (15 October 1973), pp. 123-30.

26. E. K. Davis, "Mariner Venus/Mercury 1973 Mission Support," JPL Technical Report TR 32-1526 Vol. XIV, January/February 1973 (15 April 1973), pp. 5–13.

27. D. J. Mudgway and M. R. Traxler, "Tracking and Data Acquisition Support for the Viking 1975 Mission to Mars," JPL Technical Memorandum TM 33-783, Vol. III (1 September 1977), p. 105.

28. M. F. Easterling, "DSN Data System Development Plan, Mark III Project," DSN Document 803-1, Rev. A (15 March 1974).

29. P. T. Westmoreland, "Technical Requirements Document for DSN Mark III Data Systems Implementation," JPL Technical Requirements Document, TRD 338-300, Rev. B (7 February 1975).

30. D. C. Preska, "DSN Standard Practices for Software Implementation," DSN Progress Report PR 42-29, July/August 1975 (15 October 1975), pp. 119–30.

31. E. C. Gatz, "DSN Telemetry System Mark III-77," DSN Progress Report PR 42-42, September/October 1977 (15 December 1977), pp. 4–12.

32. W. G. Stinnett, "Mark III-77 DSN Command System," DSN Progress Report PR 42-37, November/December 1976, (15 February 1977), pp. 4–11.

33. Thorman, H. C., "DSN Command System Mark II-78," DSN Progress Report PR 42-49, November/December 1978 (15 February 1979), pp. 11–18.

34. W. D. Chaney, "DSN Tracking System Mark III-77," DSN Progress Report PR 42-40, May/June 1977 (15 August 1977), pp. 4–13.

35. M. S. Glen, "DSN Ground Communications Facility," DSN Progress Report PR 42-36, September/October 1976 (15 December 1976), pp. 4–12.

36. E. C. Gatz, "DSN Telemetry System Mark III-79," DSN Progress Report PR 42-49, November/December 1978 (15 February 1979), pp. 4–10.

37. J. H. Wilcher, "A New Sequential Decoder for the DSN Telemetry Subsystem," DSN Progress Report PR 42-34, May/June 1976 (15 August 1976), pp. 84–87.

    M. E. Alberda, "Implementation of a Maximum Likelihood Decoder in the DSN," DSN Progress Report PR 42-37, November/December 1976 (15 February 1977), pp. 176–83.

    J. W. Layland, "Convolutional Coding Results for the MVM 1973 X-band Telemetry Experiment," DSN Progress Report PR 42-48, September/October 1978 (15 December 1978), pp. 18–21.

38. Joseph H. Yuen, "Deep Space Telecommunications Systems Engineering," JPL Publication 82-76, Jet Propulsion Laboratory, California Institute of Technology, (July 1982), pp. 220–47.

39. V. B. Lobb, "26-meter S-X Conversion Project," DSN Progress Report PR 42-39, March/April 1977 (15 June 1977), pp. 157–67.

40. N. A. Renzetti, "DSN Data System Development Plan, 26-meter S-X Conversion Project," DSN Document 803-7 (1 July 1976).

41. H. R. Buchanan, "S-X 34-meter Conversion Receiver and Microwave Performance," DSN Progress Report PR 42-50, January/February 1979 (15 April 1979), pp. 219–25.

CHAPTER 4

THE VOYAGER ERA: 1977–1986

THE VOYAGER ERA

From the overall DSN point of view, the transition from the Viking Era to the Voyager Era was not marked by a clearly defined change in activity or objectives. Rather, driven by the press of real-time mission events, the focus of DSN attention merged gradually from Viking-related matters to those of Voyager.

It began with the successful launches of two large planetary spacecraft in 1977 and 1978. For almost the next decade, the two Voyager spacecraft repeatedly astonished the world with a flood of dazzling science data and images as they transmitted their data from Jupiter, Saturn, and Uranus. At the same time, many other inflight missions vied with Voyager for the support of the tracking stations of the Deep Space Network. The appearance of a new class of mission, the Earth Orbiters, to the DSN "inventory" further complicated the situation in this period.

DSN resources were stretched to the limit, not only by the number of simultaneous missions but by the requirements for new and enhanced capabilities to support the multiplicity of missions and the ever-extending uplinks and downlinks from the two *Voyagers*.

For the purpose of giving an account of DSN history between the Voyager launches in 1977 and the first appearance of Galileo in 1986, this period may be identified as the Voyager Era, and it is to this era that we now turn our attention.

The scale of Network activity during the Voyager Era is best appreciated in the context of the Deep Space and Earth-orbiting missions which were then of concern to the DSN. The chart shown in Figure 4-1 depicts the situation.

This workload, together with engineering and implementation of upgrades to the capabilities of the Network, is the subject of the discussion that follows.

Deep Space Missions

Except for the activity associated with the receiver problem on *Voyager 2*, the Voyager launches and early cruise events in 1977 and 1978 were "nominal." DSN Operations resources were dominated more by the multiple spacecraft requirements of the Viking mission and the complexities of the Pioneer Venus missions during that period. The Mark III Data System (MDS) implementation effort, together with the start of the 26-m to 34-m antenna upgrade and the S/X-band conversion tasks, fully occupied DSN engineering and implementation personnel.

The Voyager Era: 1977–1986

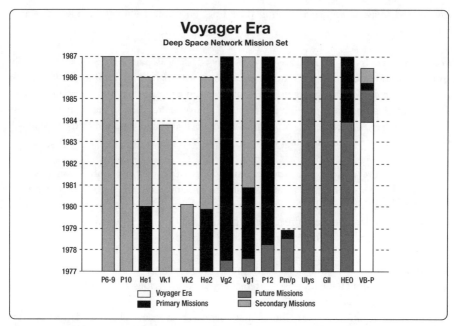

Figure 4-1. Voyager Era Deep Space mission set, 1977–86. Each of the vertical bars in this chart represents a deep space or Earth Orbiter mission which was in the DSN "inventory" during the Voyager Era. During the period shown on the vertical scale, all missions passed through various phases of their lives. Future missions were in the planning stage, primary missions were in flight, and secondary missions had already completed their primary missions and were being tracked by the DSN stations on a lower priority basis as "extended missions." The number and phase of the missions in any year represents the relative level of activity in the Network at that particular time. For example, from mid-1977 through 1980, the Network handled five, and for a short time six, primary missions simultaneously with four secondary missions and three missions, including Earth Orbiters, in the planning stages.

However, by the beginning of 1979, the focus of DSN attention was rapidly changing. The Mark III Data System implementation task had been completed, the new capabilities required for the Voyager Jupiter Encounter were in operation throughout the Network, and operations teams were trained in their use. The Viking mission had been extended through May 1978 and then further continued through February 1979. Both Pioneer Venus missions had been successfully completed in 1978, and both Voyager spacecraft were approaching Jupiter and were expected to carry out a full program of science experiments and imaging sequences during their brief encounters with the planet in March and July 1979.

For approximately two years, the Voyagers moved along separate trajectories to equally spectacular and scientifically rewarding encounters with Saturn in 1980 and 1981. From there, *Voyager 2* would be retargeted to Uranus for an encounter in 1986, leaving *Voyager 1* to continue along an inevitable path toward the outer reaches of the solar system. At Uranus, *Voyager 2*, because of its remarkable longevity, would be retargeted yet again—this time for an encounter with Neptune in 1989.

For their success, each of the Voyager encounters depended not only upon operable spacecraft, but also upon ever greater significant enhancement of the uplink and downlink DSN capabilities of the Network. At the same time, a heavy expenditure of DSN operational resources in personnel, training, and facilities was required simply to maintain a viable science data return from existing missions.

As the mission set in Figure 4-1 clearly shows, there were as many as ten inflight, deep space missions throughout the solar system, supported at various times during this period.

A closer look at the chart also shows that there were no new planetary launches after the Pioneer Venus missions in 1978. To a large extent this hiatus was due to debates within NASA about the objectives and funding for the Lunar spaceflight program using the Space Shuttle, and the robotic space exploration programs using expendable launch vehicles.[1]

Galileo was the new flight project most affected by this situation. As the vertical bar in Figure 4-1 shows, Galileo remained in "new mission" status for ten years. Between 1978, when the original plan for a 1982 launch was approved, and 1987, when the final plan for a launch in 1989 was approved, the Galileo launch vehicle was changed four times. Each change to the launch vehicle necessitated a complete redesign of the mission with corresponding changes to the requirements for tracking and data acquisition support. With each change, the DSN responded with a plan for new or modified capabilities and schedules. Change was the name of the Galileo game in those years.[2]

The Ulysses mission also remained in a state of suspension during the Voyager Era. Ulysses was a joint NASA/ESA mission to study the poles of the Sun and interplanetary space both above and below. As part of the joint effort, NASA had committed a Shuttle launch to ESA for the purpose of deploying the *Ulysses* spacecraft into Earth orbit. From there, *Ulysses* would use a solid motor Inertial Upper Stage (IUS) to inject it into a trajectory toward Jupiter. The gravitational field of Jupiter would redirect *Ulysses* out of the plane of the ecliptic and into a heliocentric orbit passing over the poles of the Sun. First planned for launch in 1983, *Ulysses* finally made it to the launch pad in October 1990 after two postponements, one in 1983 and one in 1986. The design of the mission and its requirements on the DSN were not affected by the delays in the Shuttle program.

Toward the end of the Voyager Era, a truly international cooperative mission made its appearance. The Venus-Balloon mission in mid-1985 involved the Soviet, French, and United States space agencies. Although of very short duration, it presented a complex engineering and operational challenge for the DSN. Its successful completion established a basis for future relationships between these agencies in the area of tracking and data acquisition support for deep space missions.

Earth-Orbiting Missions

The DSN began the Networks Consolidation Program, which expanded the capability of DSN tracking and data acquisition support to include support for high-Earth orbiting (HEO) missions as distinct from Deep Space (DS) missions in 1980. HEO missions were defined to be those which were beyond the capability of the newly developed NASA space network called the Tracking and Data Relay Satellite System (TDRSS). The HEO class of missions would have orbits with an apogee in excess of 12,000 km, and would be supported by the 26-m and 9-m networks added to the DSN as part of the Networks Consolidation Program. Mainly through the persistent efforts of W. R. Martin, this criterion was eventually modified to allow the DSN to provide support, in the near-Earth region below 12,000 km, for a whole new class of non-NASA missions known as "reimbursables." Under the reimbursable arrangement with NASA, foreign space agencies were able to employ the services of the DSN for part or all of their missions, subject to compliance with DSN technical and resource conditions. NASA would be reimbursed at a standard hourly rate for the antenna time provided to the foreign space agency. Within a few years, the term "HEO" fell out of common usage and became simply "Earth Orbiter (EO)," to include these latter missions.

DSN support for Earth-orbiting missions began in 1984 with ICE (International Cometary Explorer). Under a different name, ICE had been placed into a heliocentric orbit by a Delta launch vehicle in 1978. As a cooperative mission with ESA, it was to monitor solar phenomena. Later, the name was changed to ICE and its orbit was changed to encounter the Comet Giacobini-Zinner in September 1985. It was soon joined by the

Active Magnetospheric Particle Explorer (AMPTE) series of missions. Designed to study the transfer of mass from the solar wind to the magnetosphere from an Earth orbit above 100,000 km, these missions were also a cooperative effort between NASA, the United Kingdom, and the Federal Republic of Germany.

DSN support for ICE and AMPTE continued through the entire Voyager Era, supported a number of short duration reimbursable missions, and reached a peak of thirteen in 1985.

Since 1984, support for Earth-orbiting missions in the form of cooperative, reimbursable, or NASA projects has become a permanent feature of the DSN purpose in the United States space program.

The Voyager Era: 1977–1986

NETWORK OPERATIONS

Prime Missions

Voyager at Jupiter

The Voyager Project was the ultimate realization of a much earlier NASA plan to send two robotic spacecraft to visit all of the outer planets of the solar system in the later years of the 1970s. As early as 1969, even as Congress was approving funds for the first exploratory missions to Jupiter (*Pioneers 10, 11*), NASA was making grand plans for future planetary exploration.

Based on a report from the United States National Academy of Sciences, which found that "exceptionally favorable astronomical opportunities occur in the late 1970s for multiplanet missions," this "mother of all missions" came to be known as the Grand Tour. An additional report, published in 1971, concluded with a specific recommendation "that Mariner-class spacecraft be developed and used in Grand Tour missions for the exploration of the outer planets in a series of four launches in the late 1970s." At least in theory, this unique juxtaposition of the planets would allow a passing spacecraft to use the gravitational pull of one planet to alter its trajectory in such a way as to redirect it toward a flyby of the next planet. The process could be repeated, as required, to make a complete tour of all the outer planets. The technical challenges of such a mission were enormous. Amongst them were precise celestial navigation and deep space communications, both primary functions of the DSN.

In its original form, the NASA plan for the Grand Tour encompassed dual launches to Jupiter, Saturn, and Pluto in 1976 and 1977, and dual launches to Jupiter, Uranus, and Neptune in 1979. A total cost, over the decade, was about $750 million.[3] Later, however, political and budgetary constraints forced NASA to scale back the original plan to two missions to Jupiter and Saturn, with an option for an encounter with Uranus. The total cost of the new missions was to be $250 million, a more acceptable figure in the fiscal climate of the early 1970s.

Congressional approval was soon forthcoming and the official start of the Voyager mission was set for 1 July 1972. NASA designated JPL as the Lead Center and Edward C. Stone, a distinguished expert on magnetophysics from Caltech, as Project Scientist. Because the new mission was based on the proven, JPL-designed Mariner spacecraft, it was initially named MJS, for Mariner Jupiter-Saturn; the name became Voyager in 1977 but most of the early documentation retained the original name. At JPL, Raymond L. Heacock became Project Manager with Richard P. Laeser as his Mission Director, and Esker K. Davis as Tracking and Data Systems Manager, representing the DSN. Although these names changed several times as the mission progressed over the next twenty-five years, the functions always remained the same.

In both concept and execution, the Voyager Project was one of the most ambitious planetary space endeavors ever undertaken. The *Voyager 1* spacecraft was to investigate Jupiter and several of its large satellites, and Saturn and its rings and large satellite, Titan. *Voyager 2* was also to observe Jupiter and Saturn and several of their satellites after which it was to be redirected toward an encounter with Uranus in 1986. After their final encounters, both spacecraft would eventually cross the boundary of the solar system into interstellar space. Each spacecraft would carry instrumentation for conducting eleven scientific investigations in the fields of imaging science, infrared radiation, ultraviolet spectroscopy, photopolarimetry, planetary radio astronomy, magnetic fields, plasma particles, plasma waves, low energy charged particles, cosmic ray particles, and radio science. A new onboard computer system gave the Voyagers greater independence from ground controllers and more versatility in carrying out complex sequences of engineering and scientific operations than on the Mariners earlier.

On each spacecraft, uplink communications with the DSN were provided by two S-band radio receivers while downlink communications used four 25-watt transmitters, two of which operated at S-band and two at X-band. Each spacecraft carried a large, 3.7-meter diameter, high-gain antenna (HGA) in addition to a smaller, low-gain antenna intended for use as backup. Combined with the HGA, the X-band downlink was designed to deliver telemetry data up to 115.4 kilobits per second to the DSN 64-m antennas from the distance of Jupiter.

That was the plan and the basis for the Voyager project requirements for DSN tracking and data acquisition support for the first part of the overall mission—namely the launch phase, cruise phase, and with the Jupiter Encounter.[4] The Network responded to these requirements with the DSN MDS. The MDS and the events associated with the Voyager launch and cruise phases marked the closing stages of the Viking Era. Events associated with the approach of the two Voyager spacecraft to Jupiter, toward the end of 1978, soon showed that transition to the Voyager Era had already begun.

As the year 1979 opened, the space drama unfolding at JPL and featuring *Voyager 1* at Jupiter had begun to attract the attention of space scientists and observers throughout the world. Excerpts from "Voyage to Jupiter"[5] describe the mounting interest. "In mid-January, photos of Jupiter were already being praised for 'showing exceptional details of the planet's multicolored bands of clouds.'" "By early February, Jupiter loomed too large in the narrow-angle camera to be photographed in one piece." Sets of pictures called mosaics were necessary to cover the entire planet body. A spectacular movie covering ten Jupiter days and displaying the swirling vortices of the upper atmosphere would be assembled from thousands of images transmitted from the spacecraft as it approached the planet. These images, transmitted at the rate of about one image per 90 seconds over the X-band downlink at 117 kilobits per second, required 100 hours of continuous DSN coverage to complete.

The *Voyager 1* spacecraft reached its point of closest approach to Jupiter at 4:42 A.M. PST, on 5 March 1979, at a distance of 350,000 km from the planet center and 660 million km from Earth. The governor of California was present at JPL, and a special TV monitor was set up in the White House for President Carter to witness this historic event. A short time before the closest approach to Jupiter, Voyager had begun an intensive sequence of observations of Io. Much of the data, taken during the Canberra pass, was stored on the spacecraft tape recorder and played back later over the Goldstone stations so that it could immediately be transmitted to JPL.

Typical examples of the many images of Jupiter and Io taken by *Voyager 1* are shown in Figures 4-2 and 4-3.

The passage of *Voyager 1* through the Jupiter system is shown in Figure 4-4.

The entire event, from the time it crossed the orbit of Callisto inbound to the time it recrossed it outbound, took only 48 hours. In that short time, equal to two complete passes around the DSN, *Voyager* made observations on a major planet, Jupiter, and five of its satellites—Amalthea, Io, Europa, Ganymede, and Callisto. Never before had a planetary encounter yielded such a wealth of new and unique scientific data. It was truly a major milestone in the history of planetary exploration. In his essay, "Voyager: The Grand Tour of Big Science,"[6] Andrew Butrica noted, "Voyager is planetary exploration on a grand scale." The mission was only just beginning.

As *Voyager 1* receded from Jupiter, it continued to work through the carefully planned post-encounter mission sequences, and accelerated on to a new trajectory that took it to an encounter with Saturn in November 1980.

Voyager 2, four months behind *Voyager 1*, moved into the Jupiter observatory phase and took imaging sequences for another time-lapse movie and activated its UV and fields and particles instruments.

Figure 4-2. *Voyager 1* image of Jupiter, 13 February 1979, at a distance of 20 million kilometers, showing Io (left) and Europa (right) against the Jupiter cloud tops. One of the most spectacular planetary photographs ever taken was obtained on 13 February 1979, as *Voyager 1* continued to approach Jupiter. By that time, at a range of 20 million kilometers, Jupiter loomed too large to fit within a single narrow-angle imaging frame. Passing in front of the planet are the two Galilean satellites. Io, on the left, already shows brightly colored patterns on its surface, while Europa, on the right, is a bland, ice-covered world. The scale of all of these objects is huge by terrestrial standards. Io and Europa are each the size of our Moon, and the Great Red Spot is larger than Earth.

For the next several months, DSN operations activity would be dominated by these two high profile planetary missions. *Voyager 1* post-encounter activities were kept at a low level so that the majority of the support facilities could be devoted to preparation for the *Voyager 2* encounter. The Doppler tracking loop problem in the *Voyager 2* radio receiver complicated each DSN tracking pass by requiring a continuous change in the uplink frequency to compensate for the Doppler shift at the spacecraft. This change could now be predicted and included in the computer driven predicts, but it was also necessary to monitor the spacecraft telemetry to detect any drift in receiver rest frequency and modify these predicts in real time to maintain the spacecraft receiver in lock.

The Voyager Era: 1977–1986

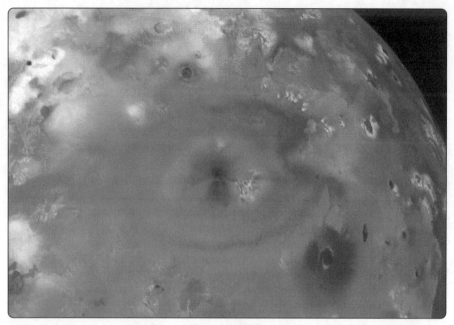

Figure 4-3. *Voyager 1* image of Io, 4 March 1979. Taken at a distance of 400,000 kilometers, the picture shows the giant volcano Pele and strangely colored deposits of surface material.

Despite these difficulties, all transmitted command sequences were accepted by the spacecraft, and uplink and downlink communications were maintained successfully.

As *Voyager 2* closed with Jupiter, the returning images showed clear evidence of the great changes that had taken place in the Jovian atmosphere since the *Voyager 1* encounter in March. In passing through the Jupiter system, *Voyager 2* would be able to fly by the same satellites as *Voyager 1*. It would "see" different faces of Callisto and Ganymede and pass much closer to Europa and Ganymede. Unfortunately, it would pass much further away from the volcanic satellite Io. Although it was not realized at the time, the data from the closer pass to Europa and the further pass from Io would be of great significance when the *Galileo* spacecraft returned to Jupiter seventeen years later. Some of the Voyager sequences were changed as a result of the *Voyager 1* experience, but in general they followed the same pattern.

The Jupiter Encounter occurred at 3:29 P.M. PDT, on 9 July 1979, amid great excitement as new data on the satellites and the planet itself poured in and were released to

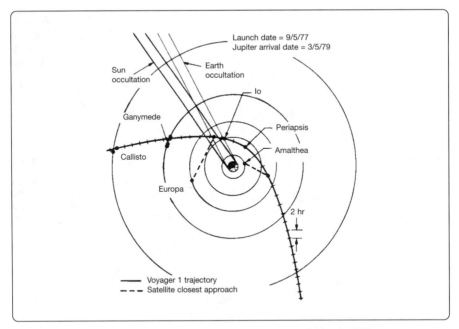

Figure 4-4. *Voyager 1* trajectory through the Jupiter system, 5 March 1979.

the public in the daily press conferences. A new satellite of Jupiter, Andastra, was discovered and a ring quite unlike those of Saturn was observed for the first time. The NASA Associate Administrator for Science, Thomas A. Mutch, was moved to remark that "such events are clearly read into the record. And I submit to you that when the history books are written a hundred years from now . . . the historians are going to cite this particular period of exploration as a turning point in our cultural, our scientific, our intellectual development."[7]

During and after the encounter, there was some concern about the effects of the Jupiter radiation environment on the spacecraft radio receiver. This caused more rapid drift in frequency than expected and actually resulted in loss of the uplink connection with the spacecraft on the day after encounter. By sending repeated commands at various frequencies, the DSN operators were finally able to find a frequency at which the spacecraft receiver would accept the commands, just in time for the commands needed to fire the thrusters for the trajectory correction maneuver that would enable the spacecraft to make the turn for Saturn.

The Voyager Era: 1977–1986

While *Voyager 2* added considerably to the volume of science delivered by *Voyager 1*, it also added to the number of unanswered questions about the "King of Planets." The answers would be a goal for Galileo, the next Jupiter mission, which lay many years in the future. At that time, however, the next exciting goal for both Voyagers lay less than two years away. It was the planet Saturn.

The demand for contiguous support for Voyager and the Pioneer Venus Orbiter in 1979 had forced the DSN to consider economies in the use of tracking station time for pretrack preparation, i.e., the configuration and calibration of the station hardware and software needed for an upcoming spacecraft tracking pass. Prior to this time, the DSN had generally been able to assign a complete station to a flight project in a critical encounter phase, to the exclusion of all other users. In this situation pretrack preparation times of as much as six to eight hours were normally used. Time needed for essential station maintenance consumed additional time. Early in 1979 this was no longer possible because of end to end Voyager and Pioneer view periods of 8 to 10 hours each at the same 64-m sites. To contend with this situation, DSN operations devised a new strategy for pretrack preparation. Pretrack and post-track activities for both spacecraft were drastically shortened and carried out during a single allocated time period.[8] "Quick turnaround" reduced the time allowed between the end of one track and the start of the next to the time required to reposition the antenna and to mount new tapes. In addition, time for routine station maintenance was based on a new formula which ensured an essential, monthly minimum number of hours for each type of antenna.

From this time on, the DSN was under constant pressure to increase the amount of tracking time available for flight project use, often at the expense of reducing or even eliminating, essential maintenance time with consequent risk to station reliability. Over the years, the basic rules devised in 1979 for allocation of pretrack and post-track preparation time have served the DSN well in striking a balance between conflicting demands for maximum antenna tracking time and maximum station reliability.

Voyager at Saturn

In terms of uplink and downlink capability, the requirements of the Voyager missions to Jupiter had been met by the DSN Mark III-77 Data System described in detail in the previous chapter. In fact, several years earlier, the Voyager requirements for telemetry, command, and radiometric data had been one of the principal drivers for that major upgrade to the Network capability and the schedule on which it was carried out (see chapter 2). When completed in April 1978, it provided the Voyager missions with an X-band, high-rate (up to 115 kilobits per second) telemetry downlink at Jupiter range from all three 64-m stations. Radiometric products, in the form of two-way Doppler and ranging data, provided for precise navigation and the S-band, 20-kilowatt, transmitters were more than adequate for command purposes at Jupiter range.

However, to support the Voyager spacecraft at Saturn, double the range of Jupiter, and possibly even Uranus, four times the Jupiter range, substantial enhancements to the Mark III-77 capability, particularly in the detection of weak signals, were required. Because the downlink signal strength diminishes as the square of the spacecraft distance from Earth, the signals reaching the DSN antennas when the spacecraft reached Saturn would be only one-fourth of those received from Jupiter, one-fourth less again when the spacecraft reached Uranus.[9]

These and other issues related to DSN operational support for Voyager were routine topics on the agenda of the weekly Voyager project meetings at JPL. Led by the project manager or the mission director, project meetings were the established JPL forum for negotiation of all requirements and interactions between the DSN and all flight project organizations. Once the actual mission began, the agenda of the regular project meetings was expanded in scope (and duration) to include status and progress reporting, as well as future requirements on the DSN. The meetings comprised representatives from each of the institutional organizations involved with that particular flight project, of which the DSN was one. Marvin R. Traxler represented the DSN to the Voyager project. The DSN appointed similar representatives to each of the flight projects to speak on DSN matters.

It should be pointed out that negotiations between the flight projects and the DSN, which involved the design and implementation of new DSN capabilities, were completed as much as five years prior to the "time of need," due to the long lead time required for the DSN approval, funding, contractual, implementation, testing, and operational training processes. In some cases, even five years proved to be insufficient time. That is the reason for the long planning periods represented by future missions in the charts of the DSN mission sets for the various eras (see Figure 4-1 for the Voyager Era). A formal set of top level documentation conveyed the project "requirements" and the DSN "commitments" to the two Program Offices at NASA Headquarters, the Office of Space Science (OSS) and Tracking and Data Acquisition (OTDA), for formal, top-level approval. Lower tier documents within the project and the DSN disseminated the necessary technical and operational detail to the implementing organizations. These steps infrequently occurred in serial fashion. Rather, driven by the always-pressing flight project schedules, work proceeded on the assumption that the necessary formalities would be eventually completed. Interactions between the flight projects and the DSN were iterative in nature; as space communications technology advanced, the technical requirements of deep space missions advanced to justify the implementation of the new technology in the Network, and vice versa.

Such was the case for Voyager. Even as the two Voyager spacecraft left their launch pads bound for Jupiter in 1977, engineers in Robertson Stevens's Telecommunications Division at JPL had turned their attention and considerable talents to the enhancements that would be needed for the existing Network capabilities to support the Voyager spacecraft, if and when they reached Saturn, Uranus, or even Neptune. Of course these enhancements would benefit all deep space missions; NASA insisted that they be multimission in nature, and so they were, although they were generally attributed to the first mission to make use of them.

Ultimately, the flight project requirements that had been approved for implementation in the Network were published in the form of a plan, jointly signed by project and DSN representatives. The referenced DSN Preparation Plan for Voyager is typical of those prepared for all flight projects which used the Network at various times.[10]

The final plan to meet the Voyager project requirements for DSN support of the two Saturn encounters included the addition of major new or improved capabilities, to the existing Network, in the following technical areas:

1. S-band and X-band antennas: Capability to receive both S-band and X-band downlinks at three 34-m stations and three 64-m stations. All 64-m antennas were optimized with X-band, low-noise masers for improved downlink performance, and with special microwave feeds for radio science experiments during encounter.

2. Downlink signal enhancement: Two-station arraying at each Complex using the 34-m and 64-m antennas and the Real-Time Combiners to improve the signal margin by about 1.0 dB (approximately 25 percent) compared to the 64-m antenna alone. This would allow the downlink telemetry rate to run as high as 44.8 kilobits per second.

3. Precision navigation: Up to six different kinds of radiometric data to enhance spacecraft navigation and radio science experiments by allowing for the removal of charged particle effects in the interplanetary media. These data types consisted of various combinations of S-band and X-band Doppler and ranging data. Besides basic improvements in the accuracy of the ranging system, the DSN Tracking System also included automatic uplink frequency tuning to compensate for the failed frequency tracking loop in the *Voyager 2* transponder.

4. Radio science augmentation: New precision powered monitors, spectrum signal indicators, open-loop receivers and multiple-channel, wideband, digital recorders together with appropriate software were installed at DSS 63 Madrid, the prime radio science station designated by Voyager for covering the occultation and Saturn rings experiments during the *Voyager 1* encounter. Later, some of this equipment was moved to the Canberra site and installed at DSS 43 to cover the *Voyager 2* encounter. This equipment measured changes to the inherent radio frequency characteristics (polarization, phase delay, spectral spreading, scintillation, etc.) of the Voyager radio signal as it grazed the Saturn atmosphere or passed through the rings. Later analysis of wideband recordings of these data provided significant new scientific data on the composition of the atmosphere and the structure of the rings.

The manner in which this work, together with other project-related activity, was carried out during the Voyager Era is described in a later section on Network engineering and implementation.

Following their highly successful encounters with Jupiter in 1979, both Voyager spacecraft commenced the Jupiter-Saturn cruise phase of the mission. The Saturn trajectories had been established as the spacecraft passed Jupiter and, in the months that followed, both spacecraft carried out routine engineering, science, test, and calibration activities. While these routine activities were in progress, the spacecraft carried out a number of special activities.

In February 1980, a delicate cruise science maneuver, which involved turning the spacecraft away from the Earth pointing direction, was performed. The maneuver went well and the spacecraft reacquired the uplink after the antenna came back to Earth point. The scan platform was calibrated in March, and the Canopus star tracker sensitivity was checked in April. Numerous navigation cycles were carried out to refine the radiometric data used for spacecraft orbit determination. A navigation cycle consists of one continuous uninterrupted pass around the Network. Radiometric data, consisting of Doppler, ranging, and a new data type called Delta Differential One-way Ranging (Delta-DOR), are generated by each of the participating stations in turn, as the spacecraft passes through each longitude.

While the real-time operations elements of the Network were supporting all of these Voyager related activities, other parts of the DSN operations organization were also busy, preparing for the upcoming Saturn Encounter.

The Voyager Era: 1977–1986

In March 1980, operations and engineering representatives from the stations in Spain and Australia arrived at Goldstone for training in the installation and operation of a new device called a Real-Time Combiner (RTC). The RTC enabled the signals from two or more separate antennas to be combined electronically to produce a single output of considerably greater strength than either of the input signals alone. When used in this manner, the antennas were said to be arrayed. The RTC had been used earlier in a two-station array at Goldstone during the *Voyager 2* Jupiter Encounter, with encouraging results. The X-band downlink signals from the DSS 14, 64-m antenna had been combined with the signals from the new DSS 12, 34-m antenna to give a telemetry signal gain of 1.1 dB relative to the 64-m antenna alone. This result, obtained under real-time operational conditions, agreed well with the theoretical, predicted value of 1.1 (±0.2) dB. There was great interest in adding a similar capability at the two other 64-m sites where the 26-m to 34-m antenna upgrades were just being completed. After completing the classroom courses and receiving operations experience, the trainees would return to their home sites to replicate the Goldstone installations.

The overseas sites, Deep Space Station 42 (DSS 42) in Australia and DSS 61 in Spain, were being requalified for operational support following the upgrade of their antennas from 26-meter to 34-meter diameter. In May, DSS 14 (Goldstone) and DSS 62 (Spain) were returned to operational status after requalification following antenna downtime for replacement of the subreflector at DSS 14 and repair of the antenna drive gear boxes at DSS 62.

All of the DSN upgrades and modifications to hardware planned for the Voyager Saturn Encounter had been completed and declared operational by mid-1980. In addition, seven new software packages needed for the antenna, communications, command, radiometric, meteorological, occultation, and planetary ranging systems were installed throughout the Network and certified for operational use. The DSN was ready for the first Voyager Saturn Encounter.[11]

Voyager 1 began its concentrated observations of Saturn on 22 August 1980, just 82 days before its closest approach to the ringed planet. At that time, the spacecraft was travelling with a heliocentric velocity of 45,675 miles per hour, at a distance of 67.6 million miles from the planet. The radio signals from the DSN antennas were taking 80 minutes to travel the distance of 901 million miles from Earth to the *Voyager* spacecraft.[12]

The DSN began a series of navigation cycles around the Network to provide precise orbit determination data on which the spacecraft Navigation Team would base the parameters for the final trajectory correction maneuver to fly close by, but not impact, Titan.

In addition, the unique geometry of the Saturn Encounter, zero declination, required highly accurate ranging measurements from two stations simultaneously to provide radiometric data from which the declination of the spacecraft orbit could be determined.

Earlier, the spacecraft had experienced some minor hardware problems in the Canopus star tracker and the scan platform supporting the cameras, but neither was expected to pose a serious problem to the planned Saturn Encounter activities.

The Canopus star tracker helped stabilize the spacecraft and keep it properly oriented by tracking Earth, the Sun, and a reference star, normally Canopus. A backup star tracker was available for use if needed. The scan platform supported the imaging cameras and other science instruments, and under certain conditions of operation had shown a slow drift which could be compensated for, if necessary.

By making use of the 28 percent gain in downlink signal power that would result from using the DSN 34-m and 64-m antennas in the arrayed mode at all three complexes, the Voyager mission controllers planned to run the spacecraft downlink at a telemetry data rate of 44.8 kilobits per second. Without arraying, the maximum data rate possible from Saturn would have been only 29.9 kilobits per second. By contrast, the data rate at the Jupiter Encounter had been 115.2 kilobits per second due to the much shorter spacecraft to Earth range of 413 million miles.

The onboard planetary radio astronomy experiment had been used to determine the rotation rate of Saturn with greater precision than was possible with Earth-based measurements, and cyclical bursts of nonthermal radio noise had been detected. Daily scans of the planet by the ultraviolet spectroscopy instrument were searching for hydrogen sources, and fields and particles instruments were constantly monitoring the interplanetary medium near Saturn. Hundreds of photographs were being taken by the imaging system to compile a full color "time-lapse" movie as the Voyager spacecraft zoomed in on the planet. One of the spectacular pictures from this period is shown in Figure 4-5. At the time this picture was taken, *Voyager 1* was 80 days from Saturn Encounter at a distance of 66 million miles. In addition to the planet itself, three satellites could be seen, and the inner and outer rings, separated by Cassini's division, were clearly visible. The imaging data was transmitted from the spacecraft over DSS 43 at a data rate of 44.8 kilobits per second.

Before the spacecraft reached Saturn, it passed behind the Sun as viewed from Earth. During this period of solar conjunction, from 3 September through 6 October, the angle defined by the Sun, Earth, and Voyager became 15 degrees or less, and the radio noise emitted by the Sun gradually degraded the radio downlink to the DSN tracking sta-

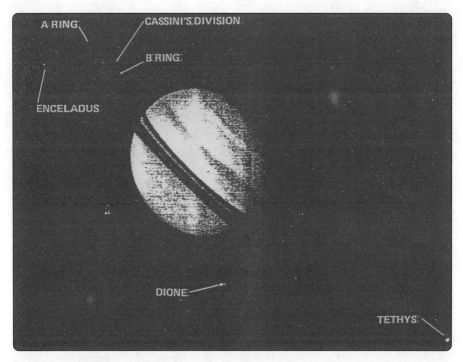

Figure 4-5. *Voyager 1* image of Saturn and its rings, August 1980. Taken on 24 August 1980, 80 days before closest approach at a distance of 66 million miles, this image shows the Saturnian satellites Enceladus, Dione, and Tethys.

tions. However, these conditions provided a unique opportunity for the radio astronomy observations of the Sun and heliosphere as the radio signals from the spacecraft passed through the solar corona. At each of the Deep Space Communications Complexes, the radio science equipment had been upgraded with new and improved hardware and software in anticipation of these imminent events.

Following the solar conjunction period, the uplink and downlink performance returned to normal as the spacecraft continued to rapidly approach the planet.

By 24 October, 19 days before encounter, the field of view of the narrow angle camera could no longer cover Saturn in one frame. It took four pictures to image the entire planet. Ten days later, the Saturn image was larger still and more mosaic pictures were needed to cover it. Attention was then focussed on more detailed examination of the

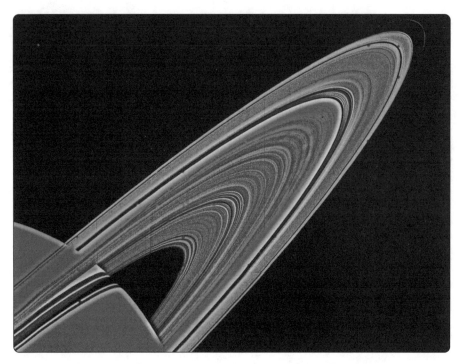

Figure 4-6. *Voyager 1* image of Saturn and its rings, November 1980, shows approximately 95 individual concentric rings.

planetary features. The extraordinarily complex structure of the Saturn ring system was shown in a mosaic of two images taken on 6 November from a distance of 4.9 million miles. (See Figure 4-6.)

On 11 November 1980, twenty-six hours before reaching the point of closest approach to the planet, the spacecraft began executing the encounter sequences which had been preprogrammed into its onboard computers months earlier. Downlink telemetry rates varied between 19.2 kbps and 44.8 kbps as the spacecraft automatically switched through various formats to return imaging, general science, and playback data in rapid succession. Taking high-resolution pictures as it went, the spacecraft would first pass close (7,000 km) by the haze-covered satellite Titan before dipping below the ring plane as it accelerated toward the point of closest approach. Eighteen hours later, on 12 November 1980, *Voyager 1* reached its closest point to Saturn, 184,000 km from the center and 124,000 km above the cloudtops of the shadowed southern hemisphere.

The Voyager Era: 1977–1986

Some 100 minutes after closest approach, the spacecraft passed behind Saturn and remained obscured from Earth for 90 minutes. During this time, and at entrance and exit from occultation, DSS 43 and DSS 63 carried out valuable radio science measurements related to an understanding of the structure and composition of the Saturn atmosphere and ionosphere. At this point the radio signal from the spacecraft was taking four and one-half hours to reach the Earth-bound antennas of the DSN.

Voyager 1 continued to observe the planet and its satellites through December 15, at which time the Saturn post-encounter phase ended and a new phase of scientific data collection began. For as long as the DSN could maintain an uplink and a downlink with the spacecraft it would continue to observe the planetary medium, participating in celestial mechanics and solar experiments with other spacecraft travelling through the solar system. It was now on a solar system escape trajectory that would take it out of the plane of the ecliptic, the plane of Earth's orbit around the Sun. Someday, *Voyager 1* and other planetary spacecraft will reach the heliopause, the outer edge of the solar system. The DSN will continue to track these spacecraft, each going in a different direction to try to determine the size of this invisible region of our solar system. It was expected that *Voyager 1* would cross the heliopause in about ten years, 1990, at a distance of 40 AU.[13]

Voyager 1 had completed its primary mission of planetary exploration. There were no other planets along its new path nor could its path be changed even if there were. Eventually, it would exit the solar system altogether, climb above the ecliptic towards the constellation Ophiacus, and chase but never catch it for all time.

The attention of the Voyager project returned to *Voyager 2*, now seven months away from its encounter with Saturn in August 1981.

During the encounter phase of the *Voyager 1* mission, the mission operations activity on *Voyager 2* had been relatively quiet, with most of the DSN tracking support being provided by the 26-m network, while *Voyager 1* dominated the 64-m and 34-m networks. Beginning in February 1981, however, the pace on *Voyager 2* picked up as the DSN raced to complete several major new capabilities needed to support the *Voyager 2* Saturn Encounter.

The equipment used for the *Voyager 1* radio science observations had to be moved from DSS 63 in Spain to DSS 43 in Australia, the prime viewing site for *Voyager 2* solar conjunction. This equipment, which included the four-channel narrow band and wide band receivers and associated digital recording assemblies, had to be integrated with new software in the Occultation Data Assembly and tested while operational procedures were developed and crew training progressed in parallel.

The receiver problem on *Voyager 2*, described earlier, had now become a "fact of life" with which the DSN had to deal. To this end, the DSN had developed new software to automatically control the Digitally Controlled Oscillator (DCO) which drives the frequency of the uplink transmitter. The DCO would be programmed to compensate for the inability of the *Voyager 2* spacecraft receiver to acquire and track the uplink frequency transmitted by the DSSs, because of its failed tracking loop. It would reside in the Metric Data Assembly (MDA).

There was, however, a complication introduced by spacecraft internal temperature change. It had been determined that various spacecraft activities would cause compartment temperature changes, which would in turn cause the center frequency of the spacecraft receiver to drift in an unpredictable manner. This would make it very difficult or impossible for the DSN to set the transmitter frequency to the exact Best Lock Frequency (BLF) needed to track the spacecraft receiver. To provide background data on the frequency offset and drift caused by spacecraft temperature variations, the DSN had been supporting a special tracking procedure known as "adaptive tracking." In an "adaptive tracking" sequence, the 34-/64-m station carries out a series of uplink frequency ramps estimated to pass through the BLF. By observing the spacecraft receiver reaction to the uplink frequency ramps, a real-time determination of the correct value of the BLF and its drift can be made. These data can then be used to program the DCO to keep the uplink frequency centered in the slowly drifting receiver pass band.

The DSN would need to use "adaptive tracking" during the *Voyager 2* Near Encounter period to ensure rapid and reliable acquisition of the spacecraft receiver for uplink command purposes. The key to this new capability was the new MDA software which drove the DCO.

This software was installed at all sites by midyear. After the operations crews were trained in its use, it was verified for encounter operations by conducting demonstration tracks with the live spacecraft.

Early in 1981, the DSS 12 antenna, upgraded to 34-m in time for the Jupiter Encounter, was decommitted from support operations for six weeks for further upgrade work to improve its radio efficiency. The two outer rows of antenna surface panels were replaced and reset, and the subreflector surface was replaced with one designed for better illumination of the primary antenna. The controller for the subreflector was also upgraded.

The Voyager Era: 1977–1986

The net result of this work was an improvement of 0.7 dB in antenna gain, demonstrated on the first operational pass when the antenna returned to service in April. The DSS 12/14 array would also benefit during encounter operations.

On 5 June 1981, the observation phase of the *Voyager* 2 Saturn Encounter began. The first activity was the movie sequence, which started over the Madrid Complex with Deep Space Stations 61 and 63 (DSS 61 and DSS 63) tracking the spacecraft, and was concluded on 7 June over the Canberra Complex with tracking support by DSS 42 and DSS 43. During the movie phase, the arrayed 34-/64-m configuration was used at all complexes to enhance the received imaging telemetry data. The performance of the arrayed stations was well within the predicted tolerances and resulted in excellent picture data quality.

While the spacecraft team conducted a sequence of scientific observations somewhat similar to those conducted by *Voyager 1*, the DSN completed the remaining hardware and software items needed for encounter, and enhanced the proficiency of the DSS crews with Operational Verification Tests.

Beginning with the passes over DSS 61 and DSS 63 on 13 August, and continuing around the Network for several days, high rate imaging data were obtained by the DSN. The 34-m and 64-m stations were arrayed for this support and the real-time data were received at 44.8 kbps with playback data at 29.8 kbps. These images, which were taken under better lighting conditions and at a better approach angle than had been possible on *Voyager 1*, were later used by the project to compile a Saturn rings movie.

A final pre-encounter trajectory correction maneuver on 18 August was supported over DSS 12 and DSS 14. During the maneuver sequence, the spacecraft antenna was placed off Earth point for over an hour resulting in loss of the downlink for that time. DSS 14 reacquired at the proper time and the telemetry data indicated that the trajectory correction had been performed correctly.

In preparation for the extensive radio science activities which would occur over DSS 43 during this encounter, a final radio science Operational Readiness Test was conducted with the Canberra Complex on 19 and 20 August. All of the encounter equipment was operational, and the actual operational sequence was used for the test. With the successful completion of this test, the DSN was declared ready to support the second Voyager encounter of Saturn.[14]

The Near Encounter Mission phase started on 25 August 1981. Recording of celestial mechanics data in the form of closed-loop Doppler and ranging data had begun earlier and was continued through the encounter period. High-rate imaging data of the closest approach sequences from the narrow-angle camera was obtained at 44.8 kbps in the arrayed mode. Image reception in the arrayed configuration was excellent and no images were lost due to DSN operations. With most of its Saturn observations completed, *Voyager 2* passed within 63,000 miles (101,000 km) of the Saturn cloudtops, at 9.50 P.M. (Earth Received Time), 25 August 1981. Still to come were observations of the dark side of the planet and southern hemisphere, the underside of the rings, and several satellites.

Shortly after 10 P.M. PDT, the imaging sequence ended and the spacecraft began to disappear behind the disk of Saturn. For the next 90 minutes, it would remain occulted by the planet. As it entered occultation, and again as it exited, the DSN radio science System at DSS 43 recorded open-loop and closed-loop receiver data. The data would yield information regarding the Saturn atmosphere and ionosphere and the microwave-scattering properties of the rings.

When the spacecraft exited occultation and the DSS 43 downlink was reestablished, telemetry data indicated that the scan platform had not carried out its programmed pointing sequences while it was behind the planet. Only black sky image frames were being received. Playback of the tape recorder data from the spacecraft indicated that the scan platform had functioned properly while the spacecraft was occulted and the fault had occurred just prior to egress. Investigation of the problem by the project resulted in restoration of the scan platform capability some days later and the Saturn imaging sequences resumed. Throughout all of this action, the DSN continued to provide Network support and accommodate numerous schedule and sequence changes as the situation required.

A trajectory correction maneuver on 29 September refined the *Voyager 2* flight path to Uranus with a swing-by assist toward Neptune.

Preliminary science results from *Voyager 2* revealed that subtle changes had taken place in the Saturn atmosphere since the *Voyager 1* visit nine months earlier. *Voyager 2* saw more detail in the atmosphere and much more detail in the rings, which could be numbered in the thousands. *Voyager 2*'s trajectory took it on a wide arc through Saturn's magnetic field, exploring different regions and adding to the data obtained by *Pioneer 11* and *Voyager 1*. On its passage through the Saturn system, *Voyager 2* passed closer to some of the satellites, further away from others, than did *Voyager 1*, and it returned a magnificent set of images of the planet and all its major satellites.

DSN support of the *Voyager 2* Near Encounter activities at Saturn was accomplished without any significant problems. Radio science played a major part of the encounter operations at DSS 43. During the closest approach period, DSS 43 generated 10 medium-band and 40 wide-band Digital Original Data Records (DODRs) for radio science. These were used to produce 484 radio science intermediate data records (IDRs). All DSN Near Encounter operations were conducted in the arrayed mode. The quality of telemetry imaging data was excellent and no images were lost.

Even with the scan platform problem, it was considered that the Voyager Saturn mission objectives had been met.

The DSN contributed to this success in two most significant ways. First, the successful operational use of "adaptive tracking" mode enabled the DSN to accommodate the uplink problems created by the failed spacecraft receiver tracking circuit, and second, the excellent performance of the Network in using the "arrayed" configuration to enhance the downlink performance to the point where 44.8 kbps real-time telemetry data at Saturn range became a reality for the first time ever.

By the end of August 1981, *Voyager 2* had completed nearly half of its three-billion-mile journey to Uranus, measured from the launch in August 1977 to Uranus Encounter in January 1986. Then four years old, the spacecraft was in good condition except for the radio receiver problem discussed above and some difficulty with the scan platform pointing sequences which stuck, shortly after closest approach to Saturn on 25 August. Since then, the anomaly had been thoroughly analyzed and understood, and the scan platform had been maneuvered successfully several times. Prospects for a successful Uranus Encounter appeared to be good.

Nevertheless, at the final press conference for the Saturn 2 Encounter, several of the speakers, including Ed Stone, the Imaging Team Leader, and Bruce Murray, the JPL Director, reminded those present of the long hiatus in deep space missions that lay ahead. It would be five years before the DSN would see the launch of another deep space mission, and that would be Galileo—or so they thought.

Venus Balloon/Pathfinder

In the period between 1983 and 1985, as international relations between the Soviet Union and the United States began to improve, the first signs of scientific collaboration in space-related endeavors appeared in the form of a cooperative project involving scientists of the Soviet, French, and American space agencies. It would be called called Venus Balloon/Pathfinder. As Bruce Murray observed, "This technical partnership between the United States, Europe, and Russia came about despite the absence of any formal relations between NASA and the Soviet Union. The original U.S.-USSR bilateral space agreements of 1972 (which facilitated, among other endeavors, the Apollo-Soyuz handshake in space in 1975) expired in 1982. Renewal became a casualty of U.S. hostility to the USSR, triggered by the Soviets' suppression of the Solidarity movement in Poland and their invasion of Afghanistan."[15]

On 11 and 15 June 1985, the Soviet *VEGA 1* and *VEGA 2* spacecraft released two instrumented balloons into the atmosphere of Venus. The VEGA spacecraft continued past the planet on their way to a rendezvous with Comet Halley in March 1986. Drifting with the Venus winds at an altitude of about 54 km, the balloons travelled one-third of the way around the planet during their 46-hour lifetimes. Sensors carried by the balloons made periodic measurements of atmospheric pressure and temperature, vertical wind velocity, cloud particle density, ambient light level, and frequency of lightning. The data were transmitted to Earth and received at the DSN 64-m antennas and at several large antennas in the USSR. Approximately 95 percent of the telemetry data were successfully decoded at the DSN complexes and in the Soviet Union, and were provided to the international science community for analysis.[16] These data would supplement current knowledge of the Venus atmosphere obtained by earlier Soviet Venera spacecraft and the NASA Pioneer Venus Probes.

Ground-based tracking support for the Venus balloon experiment involved an international network of about a dozen radio astronomy antennas organized by the French space agency, CNES, a more limited internal Soviet Network and the three 64-m antennas of NASA's Deep Space Network.[17]

Consequent upon the negotiation of appropriate diplomatic and technical agreements in Moscow, Paris, and Washington, the DSN managers determined the trajectories of both the *VEGA 1* and *2* bus spacecraft during the Venus flyby phase to recover telemetry from the balloon signal and, as part of the international network, to acquire VLBI data from each balloon/bus pair while the two were within the same antenna beamwidth.

The Voyager Era: 1977–1986

To meet these commitments, the DSN would employ, as far as possible, existing capabilities normally used for planetary spacecraft navigation, radio science, and radio astronomy. However, there were some special requirements connected with the Soviet spacecraft's L-band downlink and telemetry system.

To deal with the downlink, JPL engineers designed, built, and installed L-band feedhorns and low noise amplifiers at all three 64-m antennas. An L-band to S-band frequency-up converter provided an S-band signal spectrum to the radio science and receiver-exciter subsystems for subsequent extraction and recording of telemetry spectra, one-way Doppler and the essential DDOR and very long baseline interferometry (VLBI) data. The telemetry system was not so straightforward. The peculiarities of the Venus Balloon telemetry system required the development of special software to extract the telemetry data burst from an open-loop recording of the L-band spectrum transmitted by the balloon. Owing to the somewhat unusual nature of this mission, the normal JPL software development resources were not available to the DSN which had, therefore, to seek help elsewhere. The necessary help and expertise needed to produce the software came from Spain in the form of a software development team led by Jose M. Urech and the engineering staff of the Madrid complex.[18]

Jose M. Urech was director of the Madrid Deep Space Communications Complex at Robledo from 1981 until his retirement in 1999. Prior to becoming Director, he had served for fifteen years as servo engineer and station analyst for NASA's 26-m station at Cerebros near Robledo. At retirement, he had been associated with NASA tracking stations as a foreign national for thirty-three years.

Dr. Urech had roots in Madrid, Spain, and was educated in both Spain and France prior to earning a doctorate in engineering from the Polytechnic University of Spain in 1969.

Quiet-spoken, courteous, and low-key in manner, with a partiality for good food and wine, he engendered confidence in all who dealt with him, at whatever level, in NASA business. His social graces, technical ability, and leadership skills were of great help in integrating his team of Spanish engineers into the American-oriented methodology of the space program, and in minimizing the inevitable effect of cultural differences in resolving issues that occasionally arose between the two.

In addition to carrying out his management responsibilities, Jose Urech made time to pursue his technical interests. In 1969 and 1970, he first developed and demonstrated the concept of combining the output of two antennas to improve telemetry reception. This experiment, the first of its kind in the DSN, arrayed two antennas, 20 km apart, to enhance the downlink signal from *Pioneer 8*. In 1985, Jose Urech led a team of engineers from the Madrid station in the development of the special software needed to process the telemetry signals from the Venus Balloon mission. He also contributed to that mission by coordinating the various technical efforts of participants from JPL/NASA, CNES (France), and IKI (USSR).

Throughout his career, Urech was highly regarded by JPL as a valuable consultant and additional resource that could be, and frequently was, called upon to address technical questions related to the performance and productivity of the tracking stations.

He retired from NASA/INTA in 1999 to pursue his interests in active outdoor activities, music, and science, and to assist his wife with her business ventures.

The success of the Venus Balloon experiment depended on two new, very precise orbit determination techniques—Delta differential one-way ranging (DDOR), which was used for the Voyager planetary spacecraft navigation by the DSN, and the VLBI, which was used by radio astronomers for determining the position of extra-galactic radio sources.

The DSN would use DDOR and one-way Doppler techniques to determine the main spacecraft (bus) trajectory, for a two-week period, around the time of Venus flyby. Balloon position and velocity would be obtained from VLBI measurements between the main spacecraft and the balloon, taken by an international network of ground-based radio astronomy antennas, which included the three 64-m antennas of the DSN, over a period of approximately two days near each Venus Encounter, when both bus and balloon were in the same DSN antenna beamwidth.

The DDOR navigation technique employed by the DSN for very precise determination of spacecraft orbits required a measurement on a nearby extra-galactic radio source (EGRS). To find EGRS suitable for VEGA orbit determination, the DSN first had to make measurements of the L-band correlated flux density of 44 potential EGRS taken from the existing JPL radio source catalog of 2.3 GHz and 8.4 GHz sources.[19]

During the actual encounter period, the DSN performance in all three areas of support was satisfactory. Balloon telemetry data obtained by the DSN were provided to the science teams on computer-compatible magnetic tape in the form of original recorded spectra and in the form of demodulated and decoded data streams. Most of the decoded telemetry data was provided from MDSCC where Spanish engineers succeeded in adaptively adjusting their data processing software to compensate for the totally unexpected wind conditions on Venus that created Doppler rates up to fifty times greater than the predicted values on which the software designs were based.

Only seven of the ninety-two balloon telemetry transmissions were missed, mainly due to downlink signal variations caused by the balloon itself.[20]

During the fifteen-day period of Venus Encounter observations, the DSN succeeded in obtaining good DDOR data on 85 percent of the attempts. These data formed the basis for the VEGA flyby orbit determination. The DSN also obtained good VLBI data on sixty-seven of the sixty-nine balloon transmissions. These data, in conjunction with the VEGA orbit data, were used to estimate balloon position and velocity.

The end of the Venus Balloon experiment did not, however, spell the end of DSN involvement with the two VEGA spacecraft. Nine months later, in early 1986, the DSN would again be providing DDOR data from the VEGA spacecraft in support of Pathfinder operations for the Giotto mission to Comet Halley.

DSN support for these missions took place as the DSN moved into the Galileo Era described in the next chapter.

Voyager at Uranus

The Downlink Problem

As the excitement associated with the successful Saturn encounters in 1980 and 1981 subsided and the two Voyager spacecraft began to move along their new trajectories, *Voyager 1* toward the edge of the solar system and *Voyager 2* toward Uranus, the Voyager Project and the DSN began negotiations for even more enhanced tracking and data acquisition support than that provided for Saturn. At a distance of 20 AU from Earth, double the range of Saturn, with the downlink signal correspondingly reduced to 25 percent of its strength, the Voyager requirements for Uranus posed a further challenge to the ingenuity, innovation, and expertise of JPL's Telecommunication Division, which provided engineering support to the DSN. Basically, the problem was that of compensating for the loss of downlink due to increased spacecraft range (at Uranus) with increased signal gathering capability on the ground.

In the early 1980s, the DSN had adequate technical capability to support the planetary missions then in the mission set, at distances of 5 to 10 times the Earth-Sun distance, 5 to 10 Astronomical Units (AU). It was not until the successful Saturn Encounter in

August 1981 that a Uranus Encounter at 20 AU in 1986 became a real possibility. The DSN antennas did not then have the additional 4 to 6 dB of downlink capability that would be required to support the desired imaging data rate from *Voyager 2* at the distance of Uranus.[21]

The NASA planetary program was entering the hiatus period of the 1980s and, other than *Voyager 2* at Uranus and possibly Neptune, there were no future missions in view that would justify additional, large, new antennas. However, earlier DSN studies had drawn attention to the significant improvement in downlink performance that could be obtained from an array of existing antennas.[22] Furthermore, it was (theoretically) possible, by the use of data compression techniques in the spacecraft, to increase the efficiency of the telemetry data stream itself. By combining the improvements derived from DSN arraying with the improvements resulting from spacecraft data compression, the downlink capability required to obtain the desired science data return at Uranus could be realized.

Voyager 2 at Uranus in 1986 could not be regarded as merely a five-year extension of the 1981 Voyager Jupiter-Saturn mission, for it was indeed a new mission from a DSN point of view. It had been renamed the Voyager Uranus Interstellar Mission.

To this end, a comprehensive study of DSN options for employing arraying techniques to enhance DSN downlink performance for Voyager encounters of Uranus and possibly Neptune was undertaken in 1982. The study was not limited to DSN antennas and therefore included a survey of all known large antenna facilities, including those of Australia, England, Germany, Japan, Italy, and Russia. The "Interagency Array Study" Team was led by J. W. Layland and published its recommendations in April 1983.[23]

In considering arrangements for interagency arraying, the study recommended that permanent ties should be sought with other space agencies such as Japan's Institute for Space and Aeronautical Science (ISAS), but that for shorter term goals, such as support of Voyager at Uranus, a radio astronomy observatory seemed more appropriate. In this context, the array of the Canberra Deep Space Communications Complex (CDSCC) with the 64-m antenna of the Australian Commonwealth Scientific and Industrial Research Organisation (CSIRO) radio astronomy observatory at Parkes, New South Wales, was recommended as most viable DSN configuration for support of the Voyager at Uranus. A photograph of the 64-m antenna of the CSIRO radio astronomy observatory at Parkes, New South Wales, Australia, is shown in Figure 4-7.

The Voyager Era: 1977–1986

Figure 4-7. Parkes 64-m radio astronomy antenna.

The downlink performance of all the DSN arrayed complexes and the enhancement that would be provided by the addition of the Parkes antenna to the CDSCC array is shown in Figure 4-8, which is taken from this study.

In addressing the specific needs of Voyager at Uranus, the study had started with the Voyager project requirement for an imaging science data return of 330 full frame images per day, with about 30 to 50 images per day in the few months preceding encounter. Coupled with this requirement was a possible improvement in the efficiency of the spacecraft telemetry data transmission system resulting from the use of new coding and data compression techniques in the spacecraft Flight Data Systems. The new spacecraft capability would be implemented by inflight reprogramming of the two redundant Flight Data System processors to function as a dual parallel data processor. This enabled the spacecraft to transmit the science telemetry at lower data rates than would otherwise have been possible.

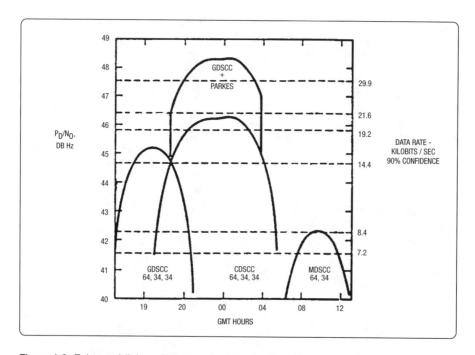

Figure 4-8. Enhanced link performance for *Voyager 2* at Uranus. The figure predicts how the telemetry data downlink between the Voyager spacecraft at Uranus and the DSN antennas on Earth would vary as Earth rotated. The vertical scales depict the maximum sustainable rates at which science data could be transmitted over the downlink (right) with their equivalent signal-to-noise ratios (left). The rotation of Earth is represented on the horizontal scale in terms of GMT. The lower curves represent the situation for arrayed antennas at each of the DSN sites in Goldstone, California; Canberra, Australia; and Madrid, Spain. The upper-center curve shows the greatly enhanced performance that would result from coupling the DSN arrayed antennas at Canberra with the Australian radio astronomy antenna at Parkes.

An assessment of the imaging science data return resulting from this new spacecraft capability and the link performance shown in Figure 4-19 concluded that the combined capability of all three DSN complexes, supplemented at CDSCC with the Parkes Antenna, would be able to return 320 full frame images per day under optimum conditions.

Although there were some qualifications to that estimate relating to spacecraft tape recorder strategy and telecommunications link uncertainties, these were understood and accepted. The requirement was very close to being satisfied in a feasible and cost-effective way,

The Voyager Era: 1977–1986

and planning to support Voyager at Uranus with each Complex fully arrayed with one 64-m and two 34-m antennas (only one 34-m at Madrid) supplemented with the 64-m antenna at Parkes, went forward.

This would be the first time that interagency support had been provided for a major planetary encounter, and as such the Parkes-Canberra Telemetry Array (PCTA) would be a "pathfinder" for further applications of this technology. And much sooner than was foreseen at the time, interagency arraying would become established as an alternate resource for DSN support of many NASA and non-NASA deep space missions.

It would be a busy time for Director Tom Reid and his staff engineers as implementation of the PCTA and preparations for the Voyager Uranus Encounter played out simultaneously at the Canberra Complex. Nevertheless, it was a situation Reid had experienced on several occasions in the past. In fact, most of the major expansion of the tracking facilities at the Canberra Deep Space Communication Complex (CDSCC), in Tidbinbilla near Canberra, took place during Tom Reid's directorship of the Complex. In that eighteen-year period, the 64-m antenna was built and later increased in size to 70 m; the 26-m antenna was enhanced to 34 m; a new, high-efficiency, 34-m antenna was built; and X-band uplinks and downlinks were added. Their successful integration into the Network and subsequent record of outstanding service to NASA spaceflight programs owed much to his cooperation.

A native of Scotland and an electrical engineering graduate of the University of Glasgow, Reid served in both the Royal Navy and the Australian Navy before taking charge of telemetry services for the Australian Weapons Research Establishment's rocket test range at Woomera in 1957. His association with NASA began in 1963 with radar support for the Gemini program, at the Red Lake site near Woomera. In 1964 he took charge of the NASA tracking station at Orroral Valley, moved to Honeysuckle Creek in 1967, and in 1970 he was appointed to succeed R. A. Leslie as Director of the Tidbinbilla Station.

His crisp management style and penchant for clear lines of authority, particularly in his relations with JPL and NASA personnel, made a visit to "his" Complex a memorable experience for many Americans. He ran the station in a disciplined, formally organized way that attracted and retained the best technical staff available. As a direct result of their teamwork and his leadership, the CDSCC played a critical role in all of NASA deep space missions in the years 1970 to 1988.

At the time of his retirement in 1988, his wife, Margaret, represented the Australian Capital Territory as a senator in the Australian Parliament.

The design and operational use of the arraying system would be based on the successful demonstration of arraying techniques in a DSN operational environment that had been observed during the Jupiter and Saturn encounters, and the additional new DSN capabilities that were becoming available as part of the Mark IVA implementation effort.

Unlike the DSN antennas, which were colocated at each Complex, the Parkes antenna was some 280 kilometers CDSCC. In addition, it required completely new receiving and telemetry equipment to make it compatible with the Voyager downlink. The design and implementation of the PCTA was a joint effort of JPL, CSIRO, and the European Space Agency (ESA).[24] ESA became involved because of an existing arrangement with CSIRO to support an ESA mission to Comet Halley called Giotto, in 1987, shortly after the Voyager Uranus Encounter.

In summary, CSIRO was to provide a new feedhorn and antenna with upgraded surface for X-band operation, ESA was to provide the X-band maser and down converter, and JPL was to provide an interfacing microwave assembly, 300-MHz receiver, formatting and recording equipment and a baseband data interface to the 280-km intersite microwave link supplied by the Australian Department of Science.

Together, these facilities provided the critical elements of a real-time combining system similar to that used at Goldstone for the Saturn Encounter with DSS 14 and DSS 12. However, in this case a special, long-baseline, baseband combiner would be required at CDSCC to compensate for the much longer delay on the baseband signal path between Parkes and CDSCC. At CDSCC, baseband signals from DSS 14, DSS 42, and DSS 45 would be combined first, before being combined with the Parkes baseband signal in the long-baseline combiner. In addition, the new Mark IVA monitor and control facilities provided greatly improved conditions for correctly operating these very complex configurations in a critical real-time environment. A functional block diagram of the Parkes-CDSCC telemetry array configuration is shown in Figure 4-9.

Because of the critical role of CDSCC and Parkes in the Uranus Encounter strategy, both sites were provided with pairs of Mark IVA digital recorders to back-up the real-time system with baseband recordings, at both sites.

Arrangements at the Goldstone and Madrid complexes were similar, but simpler. At Goldstone, the baseband signals from DSS 14 and the two 34-m antennas, DSS 12 and DSS 15, were combined in the new Mark IVA Baseband Assembly (BBA) which, in addition to the baseband combining function, carried out the functions of subcarrier demodulation and symbol synchronization. These two latter functions had been carried

The Voyager Era: 1977–1986

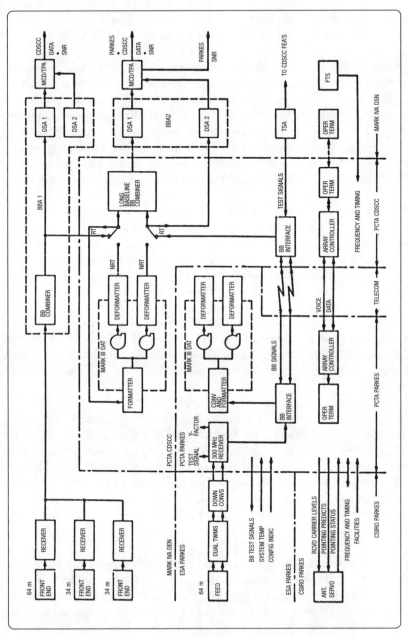

Figure 4-9. Functional block diagram of the Parkes-CDSCC telemetry array for Voyager Uranus Encounter, January 1986.

out by separate assemblies in the former Mark III system. At Madrid the array consisted of just the 64-m antenna and the 34-m standard antenna at DSS 61.

In addition to its downlink capability, a further item of great concern in connection with the Uranus Encounter was the pointing capability of the DSN antennas.[25] To avoid degrading the downlink signal from the spacecraft significantly, it was determined that the DSN antennas must be pointed to less than 0.16 of the antenna half-power beamwidth. This corresponded to less than six one-thousandths (6/1,000) of a degree or 6 millidegrees (mdeg) for the 64-m antennas, and 11 millidegrees for the 34-m antennas. The new Mark IVA antenna pointing system contained new computers and new equipment, all of which required new calibrations and procedures for making the calibrations.

There were two ways of pointing the DSN antennas. In "blind pointing," an antenna pointing prediction program derived from the spacecraft ephemeris, the antenna computers are fed data for necessary corrections before sending pointing signals to the servos to drive the antenna. This method did not rely on any downlink signal from the spacecraft, and was used during radio astronomy occultation observations and spacecraft acquisitions. In "conical scanning" (conscan) the antenna was moved in a conical scanning motion about the approximate direction of the spacecraft. Error signals, derived from a comparison of the downlink signal strength at opposite parts of the scan pattern, were used to drive the antenna to a "null" position. The "conscan" mode thus enabled the antenna to sense the apparent direction of the spacecraft radio signal. This method was useful in the absence of accurate predictions, or for searching for a "lost spacecraft" signal.

During the pre-encounter period, a process for initial pointing calibration of the Mark IVA antenna system was developed at the Madrid Deep Space Communications Complex (MDSCC), using radio stars as distant target points of known position. After a series of radio star observations, the data were reduced to pointing angle offsets and used to produce a systematic pointing error model. These data then provided the corrections needed by the antenna predicts program for "blind pointing." Typical accuracies obtained from these observations were less than 4 mdeg for the 64-m antennas and less than 6 mdeg for the 34-m antennas.

However, radio stars near the 23-degree south declination of the *Voyager 2* spacecraft were not available. For that declination, "conscan" offset data from live Voyager tracking passes were used. Comparison of the apparent direction of the spacecraft signal given by conscan, with the direction given by the corrected predict program, was the conscan offset. To the extent that the conscan properly sensed the apparent direction of the spacecraft signal, the

The Voyager Era: 1977–1986

conscan offsets determined the total system pointing errors. Using this method on a large number of tracks at Goldstone, it was determined that the conscan defined axis and the actual antenna beam axis were coincident within 1 to 2 millidegrees. Subsequent tests at Goldstone verified that the downlink signal level was not significantly affected by the method used to point the receiving antenna. With this basic information in hand, the CDSCC carried out an extensive program of blind pointing exercises in preparation for support of the radio science events connected with the Uranus encounter.

These investigations substantially improved DSN confidence in the ability of its antennas to support the outer planet missions without incurring significant downlink degradation due to pointing errors, and laid to rest the initial concerns about antenna pointing errors relative to the imminent Voyager Uranus Encounter.

The regular weekly Voyager project meetings provided a forum for discussion and resolution of the issues and concerns described above. Based on these discussions, Voyager and DSN representatives gradually developed a mutually acceptable plan for DSN support of what was by then called the Voyager Uranus Interstellar Mission.[26] Marvin R. Traxler continued to represent the DSN in these negotiations, while the project was represented by George P. Textor, the newly appointed Mission Director for the Voyager project.

The DSN plan for Uranus met the project requirements for downlink telemetry at a maximum data rate of 29.9 kilobits per second, improved the quality of radiometric data for spacecraft navigation purposes, and further improved and expanded the radio science data gathering capability of the Network. Antenna arraying techniques, based on past experience with the Jupiter and Saturn encounters, and including the new, 34-meter high efficiency antennas and the Parkes radio astronomy antenna in Australia, were used to enhance the downlink signal power received by the DSN stations, while the Mark IVA-1985 model of the Network provided the new capabilities required to provide telemetry, command, radiometric, and radio science services for *Voyager 2* at the Uranus encounter.

Both of these initiatives represented major increments in Network capability and were subsequently used to great advantage not only by *Voyager 1* and *2*, but by all of the other ongoing missions. The engineering and implementation particulars of these concurrent streams of activity and the sensitive task of bringing them into service to meet the Voyager need date (mid-1985), without disruption to the routine operational support of the ongoing missions, are discussed later this chapter.

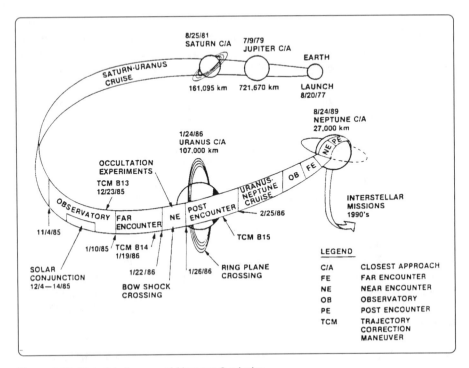

Figure 4-10. Pictorial diagram of *Voyager 2* mission.

Uranus Encounter

On 4 November 1986, 81 days before *Voyager 2* was due to flash by the outer planet Uranus, the Voyager flight team began continuous, extended observations of the Uranian system at better resolution than possible from Earth. The one-ton spacecraft was then travelling at nearly 15 km per second (relative to Uranus), and radio signals travelling at the speed of light were taking 2 hours and 25 minutes to reach the DSN antennas on Earth, 2.88 billion km away. Uranus lay 103 million km ahead. This was the start of the observatory phase during which the spacecraft and DSN would be fine-tuned to prepare for the close encounter observations. A pictorial diagram of the *Voyager 2* mission including all its planetary encounters is given in Figure 4-10.

Two weeks earlier, the Voyager Flight Team had conducted a Near Encounter Test (NET) to validate the readiness of all elements of the project, including the DSN Flight Team and Science Team, to commence the Uranus Encounter operations depicted in Figure

4-22. DSN participation in the NET demonstrated that while the DSN had the basic capability to support the Near- Encounter phase, additional operational proficiency would be needed to reach the level deemed appropriate for those critical operations. This was in large part attributable to the schedule slips that had occurred in the Mark IVA implementation project. Hampered by budget changes and software problems, completion of the implementation tasks had consumed the time that was originally planned for operational training and proficiency activity.

On 5 November 1986, the DSN reviewed its status and responded to the deficiencies that had shown up in the NET. Arraying, radio science, and operator training were the principal areas of concern. The review found that, with the help of specialist engineering personnel supplementing the station operations crews during critical periods, the DSN was ready to support the observatory phase. Further, the observatory phase would provide sufficient additional operational experience to enable the stations to deal with the Far- and Near-Encounter phases. With this proviso and several hardware and software liens against various elements of the new Mark IVA system, the DSN prepared to meet what was possibly its greatest challenge yet, the *Voyager 2* encounter of Uranus.

The spacecraft made its closest approach to the planet on 24 January 1986. Despite the earlier anxiety about its scan platform, onboard computers, and one remaining active radio receiver, the Voyager spacecraft performed perfectly. A final trajectory correction maneuver scheduled for 19 January was canceled since the flight path was deemed satisfactory without further refinement. After travelling an arc of nearly 5 billion km to Uranus, the Navigation Team estimated that the spacecraft passed within 20 km of the aim point. This astonishing navigational accuracy had been achieved by the use of spacecraft optical navigation to complement the Doppler, ranging and Delta differential one-way ranging (DDOR) data types provided by the DSN. Navigation accuracy is critical to the success of crucial science observations that depend not only on knowledge of the spacecraft position but also on proper, accurate pointing of the instruments aboard the steerable scan platform. The spacecraft executed these complex sequences without incident. The science data return quickly grew to avalanche proportions as each of the science instruments carried out its preprogrammed observations. Previously unknown facts about the planet, its atmosphere, rings, magnetosphere, winds, and satellites were soon being presented to the "happily bewildered" scientists. Typical of the many beautiful imaging science results is the striking picture of Miranda, one of the inner moons of Uranus, shown in Figure 4-11.

Figure 4-11. *Voyager 2* image of the Uranus satellite Miranda.

At a distance from Earth of two billion km (1.212 billion mi) and traveling at 65,000 km (40,000 mi) per hour, *Voyager 2* passed within 31,000 km (19,000 mi) of Miranda's surface. Image motion compensation prevented smearing of the image during the long exposure times necessitated by the low lighting conditions at Uranus. The clarity of this image amply demonstrates the capability of the Voyager imaging system combined with the benefits of image motion compensation and the excellent performance of the DSN downlink communications channel.

The new Mark IVA systems all worked well throughout the encounter period. At all sites, array performance, including the Parkes element at Canberra, was close to predicted values. The complex and lengthy radio science sequences at CDSCC were executed without significant loss of data. The new Mark IVA Monitor and Control System created some difficulty at first, while the operations crews adjusted to the concepts of centralized control in a critical operational environment.

The Voyager Era: 1977–1986

For the DSN, the Voyager Uranus Encounter in January 1986 had been a challenge in deep space communications and operations complexity. To meet the Voyager requirements for science data return, the DSN had designed, implemented, and put into operation a complex system of arraying all the DSN antennas at each longitude. At Canberra, the array had been supplemented with a non-DSN 64-m antenna, 280-km distance, belonging to a foreign agency. On the project side, a risky but successful inflight reprogramming of onboard computers had increased the efficiency of the telemetry data stream to complement the DSN improvements in the downlink performance. Special uplink tuning sequences were required to compensate for the Voyager spacecraft receiver problem.

At the same time, the DSN had completed a Network-wide upgrade of the entire data system to increase its overall capabilities and decrease operations costs by introducing centralized control to station operations.

Together, the new technology, new methodology and limited time, created a difficult position for the DSN in the weeks prior to encounter. Responding to this situation, DSN operations and engineering personnel in Pasadena, Goldstone, Canberra, and Madrid were able to meet the challenge and the science returned from the Voyager Uranus Encounter attests to their success.

In the Voyager Uranus Encounter, three major advances in DSN capability, represented by the Mark IVA Data System, antenna arraying, and centralized operations control, had been successfully demonstrated under the most critical operational conditions. These capabilities would carry the DSN forward into the Galileo Era, where once again the Voyager mission would stretch the DSN downlink capability to its limits—the next time at Neptune.

EXTENDED MISSIONS

Helios

The two Helios spacecraft successfully completed their prime missions in mid-1976 and, with both spacecraft in good operating condition, continued to orbit the Sun and return prime scientific data during repeated perihelion and superior conjunction periods for many years. The DSN continued to support Helios as an extended mission by providing tracking support as resources became available, mainly on the 26-m Network. This support was supplemented by the German tracking station, DSS 68 at Weilheim. Under these conditions, the supportable downlink data rate was 64 bps.

In March 1980, the downlink transmitter on the *Helios 2* spacecraft failed, and despite numerous attempts by the DSN to restore communications with the spacecraft, no useful data could be recovered. On 8 January 1981, a command was sent to turn the *Helios 2* spacecraft transmitter off to prevent it from becoming an uncontrolled source of radio interference.

Helios 1, however, continued to operate normally and, supported primarily by a network consisting of the two DSN 26-m stations, DSS 11 (Goldstone) and DSS 44 (Honeysuckle), and the German station DSS 68 (Weilheim), it participated in important collaborative experiments with Voyager, Pioneer, and ISEE-3 to study scientific phenomena in the inner solar system.

When time became available on the 64-m subnetwork in December 1981, *Helios 1* was able to obtain continuous 64-m coverage for a six-day period coincident with the solar conjunction period following the fourteenth perihelion passage. During this time, the DSN radio science equipment was used to make measurements of Faraday rotation from a Sun-Earth-Probe (SEP) angle of 4.1 degrees on entry, to an SEP on exit of 3.8 degrees. To carry out this type of experiment, the 64-m stations had been equipped with rotatable, microwave, linearly polarized feeds and closed loop polarimeters to automatically and precisely measure the orientation of the linearly polarized downlink signal. Measurement of the Faraday rotation caused by passage of a radio signal through the solar corona provides important scientific information about electron density and the solar magnetic field.

Helios 1 continued to perform functions of this kind for several more years into the Voyager Era, supported mainly by the 34-m antennas at low data rates (8–256 bps). As the power from the solar generators weakened, the spacecraft could be operated only for a few months in each orbit. Eventually, both spacecraft receivers ceased to operate, and without an uplink command capability to send correcting signals, the high-gain antenna gradually drifted off Earth "point." The downlink from *Helios 1* was last seen at DSS 43 Canberra on 10 February 1986, nearly 12 years after its launch in 1974. It was predicted that, with diminished power available, the spacecraft would automatically shut down its transmitters soon after.[27]

VIKING AT MARS

In June 1978, NASA gave approval to continue the Viking Extended Mission for a further Martian year (1.88 Earth years). This period of the Viking mission was called the Viking Continuation Mission (VCM). Its purpose was to replace Orbiter imaging data that had been lost or degraded during Sun occultation periods, to gather radio science data including near simultaneous Lander Orbiter ranging measurements, and to continue to make occultation and gravity fields observations. Both of the Viking Landers were placed in an automatic mode of operation to reduce the work load needed to manage Lander operations.

At the beginning of the VCM, management of the Viking project was transferred from the Langley Research Center to the Jet Propulsion Laboratory. The original flight team was reduced from about 800 personnel to 150, and DSN support for Viking mission operations was cut from 168 to 40 hours per week. The operational workload was further reduced by placing both Landers in an automatic mode of operation to eliminate the need for daily attention by ground controllers. These economies in resources allowed DSN personnel to redirect their efforts to support the more critical mission phases of the Voyager and Pioneer Venus flight projects.

The depletion of expendable attitude control gas had started to take its toll on the Viking Orbiters. Although *Viking Orbiter 1* (VO-1) was still performing normally, *Viking Orbiter 2* (VO-2) continued to be plagued by intermittent gas leaks in the yaw-axis control jets. On 25 July 1978, the transmitter on VO-2 ceased operating and the spacecraft was presumed to have finally lost attitude control capability. By the end of the year, both of the S-band transmitters on VL-2 had also failed, leaving the Lander dependent on the relay link via VO-1 for returning its data to Earth. Nevertheless, both Landers continued to collect meteorology and imaging data from the Mars surface and return it to Earth either directly or via the relay through VO-1. Throughout the year, VO-1 remained in good condition and returned excellent radio science, imaging, and general relativity data.

DSN support for the two Landers and the one remaining Viking Orbiter was curtailed even further in 1979 to make resources available for the Voyager Jupiter encounters. VO-1 was placed in a housekeeping mode in March and transmitted only spacecraft engineering data. The science systems were deactivated and new operating and protection instructions were stored on the spacecraft to make it as self-sufficient as possible. Daily contact was not required, but the spacecraft was interrogated every two weeks to check on its status.

In late 1979, VL-1 became a separate mission called the Viking Lander Monitor Mission, operated autonomously by command sequences stored in the Lander onboard computer. At the same time, VO-1 was designated as the Orbiter Completion Mission and DSN support for it was increased.

Resuming its normal operating mode, VO-1 began returning high-resolution photographic coverage of areas of the Martian surface that had not previously been obtained or adequately covered in the earlier phases of the Viking mission. Included in this coverage were possible landing sites for future Landers. (The Mars Pathfinder mission in 1997 used these images.) Frequent occultation events provided opportunities for radio science observations of Mars atmospheric fluctuations, correlation of ionospheric plasma temperature with solar activity, and improved knowledge of Mars topographic features and gravity anomalies. Both VL-1 and VL-2 remained in the automatic mode, performing repetitive observations on 37-day cycles. VL-1 ranging data obtained directly from the surface of Mars was collected to aid in improving the existing models of the ephemeris of the planet. Meteorological data and imaging data were also returned periodically, as well as data from the inorganic analysis of the final sample of Mars soil.

In July 1980, the Office of Public Information at JPL issued the following press release: "After more than four years exploring Mars, NASA's *Viking Orbiter 1* has almost reached the end of its mission. The *Orbiter* has used most of its attitude-control gas, which keeps its solar panels pointed to the Sun and the antenna aimed at Earth. When the gas was exhausted, about 23 July, controllers at JPL sent commands to turn off *Viking Orbiter 1* to end its long and productive mission.

"Meanwhile, *Viking Lander-1* is programmed to operate unattended on Mars into 1990, perhaps even into 1994 . . . . As long as it survives, *Viking Lander-1* will continue to collect photos and weather data from Mars and, on command from Earth, transmit them on approximately a weekly basis."

The Voyager Era: 1977–1986

And so it was that at 20:15 GMT on 7 August 1980, a command from DSS 61 in Spain turned off the S-band and X-band transmitters on VO-1. This would prevent them from becoming an uncontrolled source of radio interference. *Viking Orbiter 1* would continue to silently orbit Mars for many decades until its decaying orbit would lead it to destruction in the upper reaches of the Mars atmosphere.

The Viking Orbiter Project Scientist compiled the following statistics of general interest for the Viking Orbiter missions.

Viking Orbiter Statistics of General Interest

Item	VO-1	VO-2
Days from launch to end of mission	1,814	1,050
Orbits of Mars	1,488	706
Pictures recorded in orbit	36,622	16,041
Data bits played back from the tape recorders	358×10^9	161×10^9
Distance of tape travel across recorder heads (km)	2,955	1,397
Total number of commands sent by DSN	269,500	
Tracking passes provided by DSN	7,380	
Hours of tracking time provided by DSN	56,500	

The *Viking Orbiter 1* mission to Mars ended after five years of continuous DSN operational support. The demise of the last Orbiter also deprived *Viking Lander 2* of its Earth relay capability and its mission, too, was ended.

The last of the Vikings, *Lander 1*, which was located in Chryse Planetia, had been operating in an automatic mode since March 1979 transmitting stored engineering, imaging, and meteorological data on command from one of the DSSs. During the first extended mission, one of the two spacecraft radio receivers failed, leaving the other receiver permanently connected to the Lander high-gain antenna. Since that time, the Lander had been operating quite successfully but was completely dependent for its uplink and downlink communications on a long-term antenna pointing program, previously stored on the spacecraft, and was valid through 1994.

Managed and controlled by fewer than thirty people at JPL, this was to be the Viking Monitor Mission and hopes were high for a low-cost, long-lived science observation post on Mars. Radio science, meteorology, and comparative time-lapse imaging were to be the principal mission objectives. These goals were achieved, over the next two years, using whatever time became available in the busy DSN tracking schedules.

Despite the steady flow of good science data in 1982, all was not well with *Lander 1*. Early in the year, three of the four nickel-cadmium storage batteries showed signs of losing their energy storage capacity. For several months, the normal recharging sequences failed. Finally, in an attempt to extend the life of the batteries, advice was sought from experts in this field all over the country. These reviews led to recommendations for a new battery charging strategy which would be incorporated into new *Lander* sequences and uplinked to the *Lander* on 19 November 1982 from DSS 43.

After the uplink was transmitted to the *Lander*, the station searched for the expected downlink, but could find no sign of an S-band carrier signal. Shortly thereafter, the spacecraft controllers realized with horror that the new battery-charging sequence had been inadvertently written to the memory locations occupied by the *Lander* high-gain antenna pointing program.

It followed that the high-gain antenna was no longer pointing to Earth and that until the situation could be corrected by giving the high-gain antenna the correct Earth pointing instructions, communication with the spacecraft was no longer possible. In the months that followed, strenuous efforts were made to get commands into *Lander 1* to retrieve control of the high-gain antenna but to no avail. Telecommunications link analyses showed that with the 100-kW DSN transmitters sending commands there was sufficient signal strength at the *Lander* to reach the receiver even through a side lobe of the high-gain antenna. Other failure modes were proposed, tried, and failed to produce a downlink.

Finally the Mission Director decided that no further effort was warranted and, in March 1983, the Viking Lander Monitor Mission was terminated.

During the course of the Viking mission, a substantial contribution to the field of radio science was made by investigators using the Orbiters, or Landers, or both. Based on these results, scientific papers were published in the areas of general theory of relativity, solar-wind scintillation, dynamics shape and gravity of Mars, gravity waves, solar gravity, solar-wind electron density, and improvement in the Mars ephemeris.

Observations from the VL-1 meteorology sensors provided major insight into many Mars atmospheric processes such as frontal systems, annual climate variations, and dust storms.

The Voyager Era: 1977–1986

The VL-1 cameras were programmed to view each Martian scene periodically with the same camera parameters, including pointing. In this way, comparative imaging could be used to observe for dust, condensation, or erosion over a period of time. Until the mission was terminated, several imaging frames showing these processes were obtained.

The *Viking Lander 1* was eventually renamed the "Thomas A. Mutch Memorial Station." Dr. Thomas A. Mutch was the Viking Project Scientist who disappeared while on a climbing trip in the Himalayas in 1980.

The world press noted the passing of the last Lander with many tributes and accolades. For nearly twenty years, the Viking presence on Mars had provided a topic for a continuous stream of articles and reports of fact and speculation. In May 1983, the New York *City Tribune* published an article by international science writer J. Antonio Huneeus, which recognized the Viking endeavor quite succinctly: "Together the two Landers took more than 4,000 photos of the Martian surface, while the Orbiters took over 50,000 photos. They mapped 97 percent of the Mars' surface with a resolution of 25 meters. More than three million Martian weather reports were received between the two Landers and a number of other experiments were conducted.

"However, what received most publicity at the time were the experiments designed to search for 'evidence of living organisms and organic molecules on the Martian surface.' Again, it was the first, and so far only, time that such an experiment had been conducted on another planet. Although some scientists had high hopes that the experiment would reveal important principles about extraterrestrial life processes, the results were inconclusive.

"In NASA's own words, 'no clear, unambiguous evidence of life was detected in the soil near the landing sites.' But the space agency adds that, 'the question of life on Mars, according to Viking biologists, remains an open one.' "

And that is where it remained for the next twenty years until the new Mars initiative opened in 1997, with the spectacular appearance of Mars Pathfinder on the world's space scene. It, too, was renamed to memorialize one of its strongest advocates. The new station became the "Carl Sagan Memorial Station."[28]

Pioneers 6, 7, 8, 9

The *Pioneer 6, 7,* and *8* spacecraft remained in an orbit around the Sun similar to that of Earth, varying between 0.8 and 1.2 AU. All three spacecraft remained in good operating condition as they continued to be supported as extended missions, through the Voyager Era. They provided field and particle data individually and in conjunction with other solar wind measuring spacecraft when conditions of alignment were favorable.

The last successful DSN acquisition of *Pioneer 9* took place on 18 May 1983. Until that time, the spacecraft had been tracked periodically as schedules permitted, but on the last scheduled track in October, a downlink could not be detected. Over a span of several years, commands were sent to configure the spacecraft into all possible combinations of receivers, decoders, transmitters, and antennas in an effort to reestablish the downlink. The SETI multispectral analyzer at Stanford was used to try to detect a weak signal, but with no success. In all, eighteen separate attempts were made to revive the spacecraft to no avail. The Pioneer Project Office declared the end of the *Pioneer 9* mission on 5 March 1987.

Pioneers 10, 11

The *Pioneer 10* spacecraft, launched in March 1972, reached Jupiter in December 1973 and continued on a path 2.9 degrees above the ecliptic that would eventually take it out of the solar system. Throughout the Voyager Era, it continued to maintain an uplink and a downlink with the DSN while returning a continuous stream of particle and field science data, and participating in several collaborative experiments with other spacecraft in different parts of the solar system.

The *Pioneer 11* spacecraft was launched in April 1973 and reached Jupiter in December 1974. Using the strong gravitational field of Jupiter to redirect its trajectory, *Pioneer 11* reached Saturn in September 1979. Throughout the long cruise time between Jupiter and Saturn, the spacecraft remained in excellent condition with few system problems. During this period, the DSN provided routine tracking and data acquisition support while the mission was controlled from the Pioneer Mission Operations Control Center at Ames Research Center (ARC) in Mountain View, California. The heliocentric geometry of the *Pioneer 10* and *Pioneer 11* trajectories are shown in Figure 4-12.

After its Saturn Encounter, *Pioneer 11* escaped the solar system travelling a direction opposite that of *Pioneer 10*, which was already on a solar escape trajectory. It was expected that *Pioneer 11* could be tracked by the DSN through 1987.

The DSN prepared to support the *Pioneer 11* Saturn Encounter with new S-band masers in all three 64-m stations. Additional downlink performance would be obtained for a few days, during the closest approach, by operating the 64-m stations in the receive-

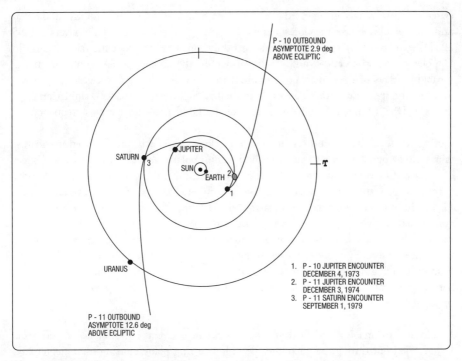

Figure 4-12. *Pioneer 10* and *11* heliocentric trajectories, September 1979.

only mode to gain an extra 0.7 dB. Because of the extensive uplink command activity required, the uplink would be provided by the 26-m stations consisting of DSS 44 and DSS 62 and the recently completed 34-m antenna at DSS 12.

To further enhance the downlink capability during the 14-day encounter period, an experimental arraying technique would be used at Goldstone to combine the signals from the 34-m antenna at DSS 12 with those from the 64-m antenna at DSS 14 in a real-time combiner. An improvement in signal to noise ratio of 0.4 dB to 0.5 dB was predicted at antenna elevations above 13 degrees. This would be verified by demonstration tracks using live spacecraft data at 1,024 b/s (coded) over DSS 12 and DSS 14 prior to encounter.

Because the closest approach distance would be less than two Saturn radii from the center of the planet, the rate of change of the S-band Doppler frequency at encounter would reach as high as 70 Hz per second. This would be too great for the DSN receivers to track unaided. This fact, coupled with the low downlink signal level resulting from the extreme Saturn range, necessitated the use of preprogrammed receiver tuning at the stations that would observe the closest approach event of the encounter. The receivers at DSS 12, DSS 14, and DSS 63 would be automatically tuned to ramp out the rapid change in Doppler as the spacecraft swung by the planet. In the absence of an automatic tuning capability, relays of operators were employed to manually tune the receivers at DSS 62. To assist the spacecraft receiver in tracking the uplink frequency changes, similar arrangements at DSS 12 and DSS 62 would control the DSN transmitter exciter frequency.

During the seven-day period on either side of closest approach, the DSN efforts to sustain a 1,024 bps data rate to support the imaging polarimeter met with various degrees of success. On occasions, problems with the Goldstone array, with telemetry decoder performance, Canberra weather, and some radio interference from a Russian satellite forced the project to reduce the data rate to 512 bps. The net result was a return of approximately twenty pictures better than Earth-based resolution instead of the forty pictures that would have been possible had the 1,024 bps data rate been sustained for the entire period. Even so, the science data return was still very extensive, leading the Pioneer Imaging Team to estimate that the scientific data return from the *Pioneer 11* Saturn Encounter exceeded that from both the *Pioneer 10* and *Pioneer 11* encounters of Jupiter.

In addition to imaging and field and particle science, several radio science experiments were carried out during the encounter period. The ring plane crossing on 1 September, the Saturn occultation shortly afterwards, and superior conjunction beginning on 11 September afforded opportunities for radio science observations.

Following its Saturn encounter on 1 September 1979, *Pioneer 11*, too, began to move along an endless path above and out of the ecliptic in the opposite direction to *Pioneer 10* and at a considerably steeper angle. All systems aboard the spacecraft were performing normally. Like *Pioneer 10*, it would remain a fully operating spacecraft, collecting fields and particle data and participating in collaborative experiments, as its signal slowly weakened with increasing range.

Managed from the ARC Mission Control Center in Sunnyvale, California, both spacecraft would remain a presence on the DSN tracking support schedules throughout the entire Voyager Era.

Performance estimates, based on measured performance in 1980, indicated that the 64-m stations using the 18.5-kelvin low-noise S-band receivers would be able to support the *Pioneer 10* spacecraft at a data rate of 16 bps through 1989, 4 years and 13 AU greater than previous estimates. By 1990, the distance from Earth would be 7 billion km, the DSN radio transmissions would be taking over 13 hours to travel to *Pioneer 10* and back, and the signal level received at the 64-m DSN antennas would be 172 dB below 1 milliwatt (-172 dBm). At the same time, the *Pioneer 11* distance would be 4.5 billion km and its data rate would be 32 bps. The life of both spacecraft would ultimately be determined by the life expectancy of their Radioisotope Thermoelectric Generator (RTGs) power sources, rather than the downlink telecommunication limits of the DSN antennas.[30]

Pioneer 12 (**Venus Orbiter**)

Pioneer 12, the orbiting spacecraft of the two Pioneer Venus missions of 1978 (the other was a Venus atmospheric entry multiprobe) continued to operate normally through the Voyager Era, collecting a wealth of new information about Venus. In 1980, the first detailed radar maps of Venus were compiled using data returned from *Pioneer 12*. As the shape of its 24-hour orbit gradually changed, *Pioneer 12* observed the bow shock and the ionospheric tail of Venus in 1983 and 1985. Making use of its ultraviolet spectrometer, *Pioneer 12* observed Comets Encke in 1984, Giacobini-Zinner in 1985, and Halley in 1987, as well as several other comets.

Although the spacecraft used an S-band uplink and downlink for regular communications with the DSN, it also carried a 750-milliwatt, X-band transmitter intended specifically for carrying out radio science experiments during occultation periods. The DSN supported numerous radio science experiments with *Pioneer 12* using this capability. Celestial mechanics experiments determined irregularities in the Venus gravitational field and dual frequency (S- and X-band) occultation experiments provided new data about the Venus atmosphere.

Still active in its prime mission phase, *Pioneer 12* was supported regularly on the 64-m network through the entire Voyager Era.

HIGHLY-ELLIPTICAL EARTH ORBITERS

INTERNATIONAL COMETARY EXPLORER (ICE)

The ICE spacecraft started life as the *International Sun Earth Explorer* (ISEE) and spent the first five years of its life in heliocentric orbit observing solar wind effects in the region between Earth and the Sun. During this period, tracking support was provided by the 26-m stations of the Ground Spaceflight Tracking and Data Network (GSTDN). In 1983, ISEE-3 was retargeted to intercept Comet Giacobini-Zinner on 11 September 1985, and renamed the International Cometary Explorer (ICE). The Spacecraft to Earth distance at comet encounter would be 0.47 AU. On its new trajectory, ICE soon exceeded the tracking capability of the 26-m antennas and, in January 1984, became the first of the highly elliptical Earth-orbiting satellites to depend on DSN 64-m antennas for tracking and data acquisition support.

Designed for use in a near-Earth orbit of about 0.01 AU, the ISEE-3 downlink would be reduced in strength by a factor of 2,500 when the spacecraft, as ICE, encountered the comet. Not only would the DSN 64-m antennas be required, but significant, innovative enhancements to the existing DSN receiving and telemetry data processing capabilities would also be necessary to support a telemetry data rate of 1,024 bps from ICE at comet encounter.

The ICE spacecraft carried two identical, S-band, 5-watt transmitters that could be used interchangeably to carry telemetry or ranging data. One transmitter operated at 2,270 MHz, the other at 2,217 MHz. To overcome the problem of downlink signal margin, the DSN proposed to transmit telemetry on both downlinks simultaneously and to combine the separate signals at the DSSs before feeding the combined signal to the demodulation, synchronizing, and decoding processors. In addition to the two downlinks from a single 64-m station, single downlinks from nearby 34-m stations could be added to the combiner, to further enhance the downlink prior to data processing.

Earlier studies had evaluated the feasibility of this approach and estimated that, under optimum conditions, this arrangement could provide an improvement of 3 to 4 dB compared to the use of a single 64-m antenna and a single downlink.

Implementation of the scheme, however, proved to be very difficult. Simultaneous reception of the two downlink carriers on a single 64-m antenna required the addition of new wide-bandwidth masers, which initially experienced problems with gain and bandwidth stability. Special multiport resistive combiners were built and installed at each of the DSCCs.

Several hundred hours of link performance on the ICE data were accumulated and evaluated in mid-1985 as the encounter time approached. The marginal signal power available from the spacecraft at this range made it essential to account for every 0.1 dB of telemetry performance that could be extracted from the combined downlinks.

The large radio astronomy antenna at Arecibo, Puerto Rico, was also scheduled to support the ICE for a short time during the crossing of the comet tail. For this purpose, new receiving and telemetry equipment for the 2,270 Mhz downlink was provided by GSTDN.

Additional support near the Australian longitude was to be provided by the Japanese 64-m antenna at Usuda, Japan. Under a joint NASA-ISAS agreement, the station would be implemented with the necessary low-noise maser amplifier and telemetry processing equipment by the DSN. The high aperture efficiency of the Usuda antenna, together with the low-noise maser configuration, resulted in a downlink performance almost equivalent to the DSN 64-m dual channel configuration.[32] A block diagram of the DSN configuration for support of the ICE encounter with comet Giacobini-Zinner is shown in Figure 4-13.

Configuration testing, resolution of problems, and intensive crew training continued at all sites right up to the actual encounter time. ICE passed through the tail of Comet Giacobini-Zinner on 11 September 1985. The three sites having view of the event, Madrid, Goldstone, and Arecibo, successfully acquired the spacecraft and reported good downlink performance at or above the predicted level. Telemetry data was sent via high-speed data line to the ICE Principal Investigators at GSFC from all three sites simultaneously. The Madrid Complex sent arrayed data from the 64-m antenna and one 34-m antenna, while the Goldstone Complex sent data from its 64-m antenna and two 34-m antennas. The Arecibo site sent data from one downlink during the overlap of the Madrid and Goldstone view periods. The multiple data streams allowed the mission controllers at GSFC to choose the best data stream for delivery to the investigators for scientific analysis.

The return of telemetry data at 1,024 bps, from this historic first encounter with a comet, was made possible to a large degree by the application of DSN analysis, engineering, and operations expertise to refinement of the ICE downlink.

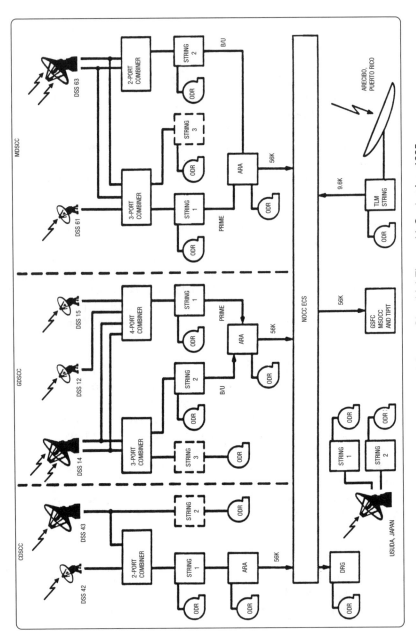

Figure 4-13. DSN configuration for ICE encounter of Comet Giacobini-Zinner, 11 September 1985.

For the rest of the Voyager Era, the DSN would track ICE as schedules permitted while it continued to return science data, generally at a lower data rate, as it pursued its highly elliptical Earth orbit around the Sun.[31]

Active Magnetospheric Particle Tracer Explorer (AMPTE)

The AMPTE spacecraft consisted of three separate satellites: Charge Composition Explorer (CCE) provided by the U.S., Ion Release Module (IRM) provided by the Federal Republic of Germany, and the United Kingdom Satellite (UKS) provided by the U.K. From highly elliptical Earth orbits, the three satellites would engage in an international, cooperative study of the interaction of the solar wind with Earth's magnetosphere. The prime missions would last about one year for the IRM/UKS, and four years for the CCE. Project management was exercised by Goddard Space Flight Center (GSFC), for NASA, in cooperation with DFVLR.

Following launch from Cape Kennedy by a single Delta 3924 launch vehicle in August 1984, all three satellites were placed in highly elliptical Earth orbits by onboard propulsion units. The final orbit for CCE was 1,148 km by 7.78 Earth radii (Er), while that of the IRM and UKS was 407 km by 17.7 Er.

All S-band uplinks and downlinks for telemetry, command, and ranging communications provided by the spacecraft were compatible with the capabilities of both ESA and DSN ground tracking stations.

The AMPTE would be the first Highly-elliptical Earth Orbiters (HEO) to be supported by the DSN following completion of the Network consolidation project. Beginning in August 1984, the three 26-m antennas recently transferred from GSFC (DSSs 16, 46, 66), would supplement the DSN 34-m standard antennas to provide coverage for the CCE. The 34-m standard antennas would support all three spacecraft. The three 64-m antennas would be required to support critical downlinks only. The ESA tracking stations provided additional support for the IRM/UKS satellites.

Following a critical period from August 1984 through April 1985, when daily 8-hour passes were required, tracking coverage was reduced to two-hour tracks, twice daily, for CCE spacecraft tape recorder data dumps and radiometric data support.

Downlink telemetry data rates for the CCE varied from 3.3 kbps (real time) to 105.6 kbps (playback). Uplink command data rate was 125 bps. Radiometric data consisted of standard DSN ranging data only. Telemetry data was provided to the project at GSFC in the form of digital Original Data Records (ODRs). Telemetry data from the 26-m stations were provided in the form of analog recordings.

The AMPTE missions introduced the DSN to the new HEO mission types which were to become a very significant part of the total DSN effort in ensuing years. Although the IRM/UKS satellites ceased to be of concern to the DSN in mid-1985, the CCE continued to require support as an extended mission through 1989.

REIMBURSABLE MISSIONS

By 1984, the reputation, maturity, and unique capabilities of the Deep Space Network, as well as the international value of the dollar, had made the DSN attractive to many non-NASA and foreign space agencies as a viable alternative to their own facilities for tracking and data acquisition support of Earth-orbiting missions. In this environment, NASA had begun to explore the possibilities of providing DSN support for non-NASA programs on a reimbursable basis, in addition to that which it already provided to non-NASA missions on a cooperative basis. Representatives of the NASA Office of Tracking and Data Acquisition (OTDA), supported by the technical staff from the DSN, began to establish working relationships with several other national space centers and foreign space agencies to explore the possible use of DSN resources for non-NASA, Earth-orbiting missions, on a reimbursable basis.

As a result of these actions, requirements for support of non-NASA missions began to appear, along with those for NASA missions, in the standard DSN documentation of the period.[33] Suddenly, the DSN was involved, not only with a new class of Earth-orbiting missions, known generically as "reimbursables," but also with many new foreign space agencies, some of which are listed below.

1. Europe: European Space Agency (ESA)

2. Italy: Agencia Spaciale Italiana (ASI)

3. Germany: Deutsche Forschungs-und Versuchsanstalt fuer Luft und Raumfahrt (DFVLR)

4. Germany: Bundesministerium fuer Forschung und Technologie (BMFT)

5. Japan: National Space Development Agency (NASDA)

6. Japan: Institute of Space and Aeronautical Science (ISAS)

7. Spain: Instituto Nacional de Technica Aeroespacial (INTA)

8. France: Centre Nationale d'Etudes Speciale (CNES)

9. Russia: (Former Soviet Space Agency) Astro Space Center (ASC)

Prior to the formalization of the reimbursable process in 1991, interactions between those taking part in the interagency negotiations were rather informal. Joint technical Working Group Meetings initiated by the DSN usually alternated between the various foreign agencies and JPL, and were used to plan support for upcoming missions, disseminate information of mutual interest regarding DSN current status and future plans, and assess potential DSN support for future reimbursable missions. Science considerations for these missions were not a topic at these meetings. Where technical and operational support by the DSN was the subject of discussion, Robert M. Hornstein, or another associate, represented NASA Headquarters, while R. J. Amorose, or one of his managers, spoke for the DSN.

It was the joint Working Group Meetings that afforded a medium for the definition, discussion, and approval of the detailed technical and operational interfaces between the DSN and the non-NASA user agency. Later, these interfaces would be published in the appropriate interface control documents. Quite frequently these activities were taking place while the formal approvals were still in progress. It, too, was an iterative process which, since NASA looked to the DSN for an estimate of the costs involved, necessitated a thorough understanding of the prospective user's requirements before the desired support could be formally approved. These meetings were a necessary precursor to the formal request to NASA for reimbursable support.

Generally, the DSN appointed a Tracking and Data Systems Manager (TDSM) to convene the meetings, represent it in the negotiations, and be responsible for all aspects of DSN support once the mission began. The TDSM for the reimbursable missions, of which there were several in progress at any one time, was also accountable for the DSN antenna time and other resources which were expended in support of the mission, and for reconciling with the original estimates on which the support had been based.

In a directive issued in 1991,[34] the Director of Ground Networks, Robert M. Hornstein, set out a "NASA policy covering the use and reimbursement of the NASA ground networks by non-NASA users." This document applied to the Deep Space Network, in addition to the two other NASA ground networks, the Spaceflight Tracking and Data Network (STDN), and the Wallops Tracking and Data Facility. It identified the available, standard services, which included tracking, radiometric data for orbit determination, acquisition of telemetry data, transmission of commands, radio science data recording, prelaunch support planning, and documentation and delivery of the data to a NASA Communications Network (NASCOM) gateway. Other special services, unique to particular missions, were also identified. It was, in short, a "full service" policy.

The Voyager Era: 1977–1986

Requests for service were to be provided "36 months before the requested service date for NASA's assessment and determination if the service could be provided." Users were required to "submit a letter of intent, together with an earnest money deposit to initiate planning activities leading to the provision of services." Work could not begin until a reimbursable arrangement describing, "the support to be provided, the schedule for the support activities, the cost estimate, the progress schedule and any necessary terms and conditions for the support," had been completed.

The costs for reimbursable services were based on "the number of actual support hours provided by each antenna multiplied by an appropriate rate schedule," which was revised by NASA each fiscal year. To this were added miscellaneous costs for Headquarters and DSN technical staff, travel and administration, software modification, DSN navigation support, documentation, testing, and support from other elements of the ground network.

A NASA Management Instruction (NMI), issued in 1991, conveyed the essence of this policy to all NASA Field Centers, including JPL. The NMI included specific conditions and procedures by which non-NASA organizations could request DSN support. Within the DSN, these instructions were translated to the working level for the guidance of the technical and administrative staff that would implement reimbursable mission operations in the Network.[35]

The costs and number of reimbursable missions in the DSN Mission Set varied considerably over the years following their first appearance in 1985. The hourly cost for a 26-m or 34-m antenna from 1985 through 1990 was $920 per hour. By 1992 this had risen to $1,100 per hour. The hourly rate for the 64-m antennas was double those rates. In 1993, this rate, which was then adjusted annually to reflect the current operating costs of the Network rose to $2,825 per hour for the 26-m and 34-m antennas, and $5,650 per hour for the 70-m antennas.

Reimbursable missions made their first appearance in the Network toward the end of the Voyager Era, with four television broadcast satellites, TV-SAT (German), TDF-1 (French), and BS-2B and MS-T5 (Japanese). DSN support for each of them was of relatively short duration (days to weeks) and involved only the 26-m and 34-m antennas with existing capabilities. From this modest beginning in 1985, the number of "reimbursables" in the total DSN mission set would steadily increase to become a major effort, as the DSN moved into the Galileo Era.

NETWORK ENGINEERING AND IMPLEMENTATION

NEW 34-METER HIGH-EFFICIENCY ANTENNAS

With the success of the DSN 26-m to 34-m antenna conversion program fresh in mind, the earliest design for the NCP/MK IV program was predicated on upgrading the 26-m GSTDN antennas which had been moved to the DSN sites as part of the Consolidation program, to 34-m. These antennas were, however, X-Y mount design and not at all suitable for DSN, deep space tracking applications. Further reconsideration and cost versus performance studies led to a decision to build a set of completely new antennas, based on a design study on larger antennas versus multiple antenna arrays, which had been made for the DSN by Ford Aerospace in the mid-1970s. The new antennas were to be built as 34-m array elements having receive-only capability. There would be no transmitter and the common aperture feed would be optimized for X-band at the expense of S-band performance. DSS 15 and DSS 45 were funded in FY 1981 to meet the immediate need of Voyager at Uranus, and funding for DSS 65 was deferred until FY 1984. By that time, the future Magellan requirements for an X-band uplink were in part responsible for the completion of DSS 65.[36]

The high-efficiency antennas, so-called because the aperture efficiency at X-band was relatively high (approximately 70 percent), were intended to supplement the 34-m standard antennas and the 64-m antennas in an arrayed ensemble of antennas. They differed from the standard 34-m antennas mainly in having an azimuth-elevation (Az-El) wheel and track mount rather than an Hour-Angle-Declination (HA-DEC) mount. They had a single, common feedhorn for both S-band and X-band and no transmitter. A Block II maser, transferred from the 64-m stations, provided for X-band reception and a cooled Field Effect Transistor (FET) was used for S-band reception.[37]

The antenna structure was designed to maximize stiffness, minimize distortion, and provide accessibility for maintenance. It was mounted on a four wheel azimuth turntable consisting of a square platform with self-aligning wheels, rotating on a circular, machined track. The Az-El wheel and track mounting structure can be clearly seen in this photo (Figure 4-14) of the first HEF antenna to be completed, DSS 15 at Goldstone.

The main reflector surface was shaped by the contiguous positioning of individual panels to present a homologous paraboloid for improved microwave efficiency. The quadripod and apex structure was designed to limit aperture blockage to five percent. The alu-

Figure 4-14. 34-meter high-efficiency (HEF) antenna: DSS 15, Goldstone.

minum subreflector was constructed from six individual panels overlaid with aluminum tape to form a continuous conducting surface. The subreflector three-axis positioning mechanism served to minimize microwave performance loss due to gravity deformation of the antenna primary structure, quadripod, and primary reflector when the antenna was operating at various elevation positions.

The antenna drive system was integrated with the Mark IVA configuration in the SPC via interconnecting microprocessor computers. Antenna position control signals, trajectory prediction data, systematic error correction tables, transformation of coordinates, and monitor data were handled by drive and control equipment located on the antenna, and by the pointing computers in the SPC. Microwave subsystem configuration control was also exercised from the SPC. No operators were required at the antenna site.

When completed, the high-efficiency antenna cost about $4,800,000 (1986 U.S. dollars) each.[38]

Typical microwave performance parameters are given below:

Parameters	X-band (8,400 GHz)	S-band (2,300 GHz)
Antenna Gain (dB)	68	56
System Noise Temp. (kelvin)	20	40

The parameters given in the table above are intended for illustration only. Under operational conditions these values changed significantly, depending on the microwave configuration in use, the specific low-noise front-end amplifier, the antenna elevation, weather, etc. These effects were important factors to be taken into account in designing uplinks and downlinks, and were given, explicitly for that purpose, in the "Deep Space Network to Flight Project Interface Design Handbook."[39]

The first HEF antenna at Goldstone, DSS 15, was brought into service with the Mark IVA and SPC 10 in October 1984, followed by DSS 45 at SPC 40, Canberra, in December 1984. The HEF antenna at the Madrid Complex, DSS 65, became operational with SPC 60 in April 1987 with the additional capability of a 20 kW X-band transmitter to support the Galileo and Magellan missions. The added X-band uplink transmit capability was retrofitted to DSS 45 and DSS 15 in 1987 and 1988.

The 64-meter to 70-meter Antenna Extension

Nowhere in the Deep Space Network was the ardent pursuit of refinement in the uplink and downlink more apparent than in the 70-meter antennas. They were the single most obvious structure to draw and hold the attention of the thousands of people who visited the complexes in Australia, Spain, and California each year. Technical and nontechnical visitors alike were fascinated by the huge, gleaming white structures that readily conveyed the sense of deep space communications links with spacecraft in distant parts of the solar system. These were not fragile laboratory instruments continuously nurtured by white-suited laboratory technicians. These instruments were rugged engineering marvels that were designed to run continuously, exposed to all kinds of environmental conditions, from the searing heat of a Mojave Desert summer, to the driving winds and freezing rain of a Canberra winter, while maintaining a mechanical and electrical performance equivalent to the most sophisticated of laboratory instruments. Their size was deceiving. Seen alone against the natural contours of the surrounding country, the size of the antennas did not appear to be their most striking feature. It was the intricate complexity of the structure that impressed the viewer. But seen against a human figure at the base or on the antenna, or a truck or building, the visitor was confounded by the size of the structure. Taller than a 20-story building and over one acre in surface area, the antenna itself was impressive even when stationary. When it was in silent motion, slewing from the horizon position of a setting spacecraft to acquire the downlink from another spacecraft, rising like a planet from the opposite horizon, the sense of awe was complete.

Impressive though the external appearance of the 70-meter antenna was, the exquisite design and engineering that allows it to function as a major element in the uplink and downlink between Earth and the distant spacecraft could not be seen from outside. Indeed most of it could not be seen at all, for it lay embedded in the technical details of the electrical and mechanical performance specifications that were associated with an operational 70-m antenna of the Deep Space Network.[40]

It was no accident that this "jewel in the crown" of the DSN looked and performed as well as it did, for it represented the culmination of a process of continuous development, refinement, and operational experience extending over twenty years since the first 64-meter antenna became operational at Goldstone, in 1966. From that time until 1987, the network of three 64-m antennas served the DSN well in supporting the mission sets of the Mariner, and Viking, and Voyager Eras described in this and earlier chapters.

By 1982, however, interest in enhancing the capability of the 64-m antennas had reached the point where the chief engineer of the Tracking and Data Acquisition Office, Robertson Stevens, would recommend "that we embark on a program to design and implement the full technical performance potential of the 64-meter antennas." He went on to give performance and cost predictions for several options.

For Robertson Stevens, JPL was his first, and only, employer. Hired straight after college graduation (1949) with a master's in electrical engineering from the University of California at Berkeley, Robertson (Bob) Stevens remained with the DSN throughout his working life. With a family background associated with Naval aviation, Bob Stevens's education path moved through San Diego, Sacramento, and the Naval Academy, gathering qualifications before reaching Berkeley.

Stevens's career began with antennas and communications systems for the Corporal and Sergeant missiles, during JPL's Army period. By 1960, the start of the Ranger period, Stevens had become head of the Communications Elements Research Section in Rechtin's Telecommunications Division. While Stevens concerned himself with developing the elements of communications systems; antennas and amplifiers, masers and microwaves, transmitters and timing systems, Walt Victor's Communications Systems Research Section focussed on research at the prototype systems level. In due course, Stevens followed Victor to become manager of the Telecommunications Division. Most of the significant advances in communications technology that eventually found their way into the operational Network were made in that division, under his direction.

A great believer in physical fitness, Bob Stevens rode his bicycle five miles to and from JPL each day across the busy Pasadena city streets. On the job, he always seemed to be around to encourage or help his engineers when the going was rough, sleepless like everyone else during all-night tests at Goldstone, or calm and confident in the Network Control Center at JPL, wearing his good-luck red vest during a launch or critical planetary encounter. Together with Mahlon Easterling, who introduced the DSN to block-coding as an enhancement for the Surveyor downlink, this author and Bob Stevens appeared regularly with his flute or guitar to play carols at midday Christmas parties around the "Lab" in the freewheeling days when that was permitted. Tousled dark hair, a ruddy complexion, a ready smile, and a hearty laugh were typical of the man. Stevens's casual, slightly ruffled appearance and somewhat shy manner belied the intensity of the attention that he brought to bear on matters that concerned him. He had a wonderful ability to resolve issues between upper management, who often did not understand the technicality of the issues before them, and engineers, who did not appreciate the constraints those technicalities an issue. Later, as chief engineer for the Telecommunications and Data Acquisition Office, he applied his many talents to the role of "troubleshooter" for the office.

He retired from JPL in 1991. As Stevens saw it, the "full potential of the 64-m antennas" would only be realized by the "Ultimate" option described by the performance and cost parameters shown below. And that, he said, was his recommendation. It was a powerful statement, and on that basis the project went forward.

Options for Upgrading the 64-Meter Antenna

Upgrade Option	X-band Perf. Increase	S-band Perf. Increase	Cost per Antenna (FY 1982)	X-band Value in $M/dB[4]
Ultimate[1]	1.9 dB	1.4 dB	$6,000,000	3.2
Intermediate[2]	1.4 dB	1.4 dB	$3,300,000	2.4
Gravity[3] Compn.	0.3 dB	0.0	$330,000	1.1

1. Ultimate: Extend 64-m to 70-m; shaped precision reflectors; gravity compensation.
2. Intermediate: Extend 64-m to 70-m; shaped reflectors; gravity compensation.
3. Gravity compensation: Braces and Y-axis correction to 64-m or 70-m tipping structure.
4. For comparison, the arrayed addition of one new 34-m antenna to a standard 64-m antenna provides approximately 1.3 dB performance increase for a cost of $6.0M, as installed and equipped for X-band reception only. The equivalent value is $4.6M/dB. However, the added 34-m antenna would provide a capability for simultaneous multiuser support which is not given by the single, upgraded 64-m antenna option.[41]

The 64-meter Antenna Rehabilitation and Performance Upgrade Project was established in September 1982 with D. H. McClure as project manager. The guidelines to the project manager were to "increase the technical performance of the 64-m antennas in terms of gain, efficiency, and reduced (radio) noise," and to "carry out the rehabilitation of the 64-m subnet."

The following account of the execution and successful accomplishment of this task is an abridged version of an unpublished report on the project by D. H. McClure and F. D. McLaughlin.[42]

Overall program management was under the cognizance of the NASA Office of Space Operations. Responsibility for project management was assigned to the Office of Telecommunications and Data Acquisition at JPL. Implementation responsibility was assigned by that office to the manager for the project, D. H. McClure. The personnel required to carry out the project consisted primarily of JPL employees with support from contractor personnel for some engineering, administrative, and documentation functions. Major technical areas were monitored by cognizant specialist engineers assigned to the project.

The project consisted of two main initiatives, rehabilitation and upgrade. The rehabilitation task included repair of the hydrostatic bearing and replacement of the radial bearing, at the Goldstone 64-meter antenna, and investigation and determination of the cause of the tilt on the 64-m antenna at Canberra. The performance upgrade task involved increasing the existing 64-m antenna diameter to 70-m, improving the stiffness of the structure by modifying the structural braces, and improving the subreflector focus capability by adding automatic Y-axis focussing. This latter task was eventually carried out as part of the Mark IVA task.

In summary, the project costs were $3.8 million for the rehabilitation task and $4.1 million for the upgrade task.

The rehabilitation task was completed at Goldstone by June 1984 and the DSS 43 tilt investigation was completed in early 1985.

The upgrade task was completed and the (now) 70-m antennas returned to mission operations service with 70-m performance as follows: DSS 63, July 1987; DSS 43, October 1987; DSS 14, May 1988.

Rehabilitation of DSS 14 Pedestal

The "hydrostatic bearing" assembly, which forms part of the azimuth rotation system, was a key feature of the 64-m antenna design. In this design, the 13.2-million-kilogram rotating structure was supported by a film of high pressure oil approximately 0.25 millimeter thick. The oil film formed the load-bearing medium between each of three "pads" that carried the entire rotating structure, and a large, horizontal circular steel "runner" bearing on which the shoes rotated in azimuth. The runner bearing had a finely machined upper surface to support the pads and was itself supported by a massive circular concrete pedestal. Between the upper surface of the concrete and the lower surface of the runner lay a thick layer of grout which was supposed to provide a stable, impervious, interface between the steel bearing and the concrete pedestal. The dimensions of the components of the hydrostatic bearing were impressive. Each pad was approximately 1 meter wide, 1.5 meter long, and 0.5 meter deep. The runner bearing was 24.4 meters

The Voyager Era: 1977–1986

Figure 4-15. Rehabilitation of the DSS 14 hydrostatic bearing and pedestal concrete, June 1983–June 1984.

in outside diameter, 1.12 meters wide, and 12.7 cm thick. The walls of the pedestal were more than one meter thick and were topped by a massive "haunch" 2.1 meters high and 1.8 meters thick. The haunch provided the foundation for the grout and the runner. A cross-section showing the hydrostatic bearing components, and the portion of the concrete pedestal that was removed and replaced during the year-long rehabilitation task at Goldstone, is shown in Figure 4-15.

Throughout the life of the 64-m antennas, deterioration of the grout and the concrete had resulted in variations in the clearances between the pad and runner surfaces as the pads moved around the runner during azimuth rotation of the antenna. On numerous occasions, the film height reached minimum value and the antenna was automatically halted, bringing mission operations to a stop. The antenna was then out of service until the faulty section of the runner could be repaired by jacking it up and inserting long tapered shims between the grout and the runner, or even replacing the grout itself.

During the twenty-month period between January 1981 and August 1982, 12,500 hours of effort were expended in maintaining the grout. By this time, the antenna was operating with film heights at or near the alarm level (0.005 inch).

During the *Voyager 2* Encounter with Saturn in August 1981, the unevenness in the runner surface reached the safety limit. It was only by posting an engineer on the antenna to personally monitor the film height that DSS 14 was able to complete its final tracking pass with *Voyager*. It was a very close call. After that, the antenna was out of service for several weeks while temporary repairs were made. Had this problem occurred a few weeks earlier, the Saturn Encounter would have been significantly affected. Reliable operation of the antenna was essential to support not only the ongoing flight projects but the also the *Voyager 2* Encounter with Uranus in January 1986. It was clear that drastic remedial measures were called for.

Over the years since the hydrostatic bearing problem first appeared, a great deal of study and analysis of the problem had been carried out in an effort to determine the fundamental reasons for the failure of the load-bearing grout. This work led to the discovery in 1983 that the principal cause of the problem was not the grout but the concrete in the pedestal region beneath the grout. Briefly, it was determined that the pedestal concrete had deteriorated due to a chemical reaction between the aggregate and the cement. The aggregate reacted with the alkali in the cement to form silica gel, which in turn absorbed moisture, expanded, and caused microcracks in the concrete. The net result was a decrease in the stiffness of the concrete to the point where it was no longer able to provide the proper support for the hydrostatic bearing. Based on this and previous engineering analyses, the decision was made to remove all the concrete in the haunch at DSS 14 and replace it with concrete of a new type of mix which would not be subject to the former aggregate problem.

In June 1983, the antenna was removed from service and repair of the pedestal and refurbishing of the hydrostatic bearing had begun. The six-million-pound rotating structure was raised and placed on temporary supporting columns to allow the hydrostatic bearing and the runner to be removed for rework. All the concrete in the pedestal haunch was then removed one section at a time by a four-person crew with jackhammer and drills. Because of the enormous amount of reinforcing steel embedded in the concrete, it took the crew about five days to remove a 40-foot-long section. As each section was removed, the new concrete was poured and allowed to cure. When the new "haunch" was completed, the radial bearing was replaced and the hydrostatic bearing components installed and aligned. Final tests on the new type of concrete showed that the original specifications for stiffness had been met or exceeded. Various phases of this work as it progressed are shown in Figure 4-15.

The DSS 14 antenna was returned to service almost exactly one year after it was taken down for the pedestal rework. Raising the six-million-pound antenna, placing it on a temporary support structure, removing the hydrostatic bearing and runner and removing and replacing 450 cubic yards of high-strength concrete had been a monumental task. The task was completed on schedule, within budget, and without a lost-time accident. As a result of this work, the DSS 14 pedestal was expected to provide excellent support for the hydrostatic bearing for many years.[43]

DSS Azimuth Radial Bearing

While the work described above was in progress, another key element of the antenna's azimuth rotation system was also being repaired. This was the radial bearing.

The radial bearing consisted of a 30-foot-diameter steel ring surrounding a concrete collar at the top of the pedestal. A vertical wearstrip was attached to the runner to provide a track for three wheel assemblies equally spaced and attached to the rotating alidade structure. The wheel assemblies, bearing on the vertical wearstrip and runner, maintained the correct vertical axis of rotation for the entire antenna.

In the twenty years after the antenna was built, the grout behind the radial bearing runner and wearstrip had deteriorated to the point where the risk of failure in the entire radial bearing assembly was unacceptably high. Failure of this bearing would result in many months of downtime for repair. Accordingly, the decision was made to take advantage of the current downtime to repair the radial bearing as well as the hydrostatic bearing. A new runner and wearstrip were fabricated and installed. It took several weeks to adjust and align the runner and wearstrip to be concentric with the center of the antenna foundation. New grout was placed and cured for two weeks for maximum strength. The wheel assemblies were removed and shipped to a contractor for refurbishing. Despite some problems with correct realignment of the wheel assemblies during the reinstallation process, the work was completed successfully, on time, to support the first rotation of the antenna on the new pedestal.

DSS 43 Pedestal

The final item in the rehabilitation task concerned the pedestal tilt at DSS 43 in Canberra. This had been discovered in 1973 and later attributed to nonuniform loading on the pedestal. When it was decided to expand the 64-m antennas to 70-meter diameter, a special analysis was made with new data to determine the cause of the tilt and the potential impact of the increased weight on the foundation. The ensuing report concluded that the tilt was due to the presence of softer supporting soil and bedrock at the south side of the pedestal, compared to that which supports the pedestal on the north side. The report further concluded that the soil had compressed to its maximum amount, and that the proposed additional weight would not increase the tilt enough to be cause for concern.

Antenna Performance Upgrade

The concept of enhancing the performance of the 64-m antennas had been studied for many years. There were a number of options and alternatives available, each with its own advantages in terms of performance, cost, and viability. Chief amongst the supporters of this initiative was Robertson Stevens, a significant partner in the development and implementation of the original 64-m antenna at Goldstone in the mid-1960s. His report on the evolution of large antennas in the DSN traces the course of these studies from 1958 to the 1982 recommendation to proceed with the 70-m extension program.[44]

By 1983, with a solution to the problem of supporting *Voyager 2* at Uranus (20 AU) in 1986, the DSN addressed the next question of how to support the mission at even greater distance, namely at Neptune (30 AU) in 1989. Using a metric called an aperture unit (Ap.U.), equivalent to the effective aperture of a DSN 64-m antenna with a system noise temperature of 25 kelvin at X-band, Stevens showed that for the Neptune Encounter, 10.92 Ap.U. would be required. The Uranus configuration had provided only 5.25 Ap.U., 4.45 Ap.U. from an array of DSN antennas, and 0.8 Ap.U. from the Parkes antenna. Where would the shortfall in downlink performance come from?

Drawing on a background of previous engineering studies, Stevens proposed to increase the DSN contribution from 4.45 Ap.U. to 6.45 Ap.U. by upgrading the existing 64-m antennas for a gain of 1.65 Ap.U., and by adding a new high-efficiency antenna at Madrid for an additional 0.35 Ap.U. The balance of 4.47 Ap.U. would be contributed by non-NASA facilities consisting of the Parkes, Australia, 64-m radio astronomy antenna, the very large array antennas of the U.S. National Radio Astronomy Observatory in New Mexico, and the 64-m antenna of the Japanese Space Agency in Usuda, Japan. This approach was considered viable and cost-effective, and with the Voyager Neptune encounter as the driving motive, planning for upgrading the three existing 64-meter antennas to 70-meter diameter, on that basis, went forward.

The Voyager Era: 1977–1986

Performance analysis studies had predicted that in terms of gain improvement relative to the 64-m antennas, 0.6 dB would come from the microwave feed system, 0.5 dB from the precision surface of the subreflector and main reflector, and 0.8 dB from the increased diameter, for a total of 1.9 dB or about 55 percent. Overall performance at S-band would be about 0.5 dB less due to the fact that the reflector surface refinement had little effect at the lower frequency.

Notwithstanding its importance to the Voyager Neptune Encounter, upgrading the 64-m antenna had many other technical advantages. The additional antenna gain applied to the uplink as well as the downlink, affording a considerable enhancement of command capability under adverse conditions. In a new navigation technique called very long baseline interferometry (VLBI), which involved two 70-m antennas alternately tracking a spacecraft and a natural radio source such as a quasar, the increased sensitivity of the 70-m antenna allowed a much wider choice in the selection of radio sources that were suitable for this purpose. In arraying with smaller antennas, the signal from the larger antenna could be used to aid the carrier tracking of the signal from the smaller antenna at or near signal threshold. Furthermore, while the cost of extending the 64-m antennas to 70-m diameter was about the same as building a new 34-m antenna, the cost per aperture unit was only about 60 percent as much. Finally, estimates of the maintenance and operations cost for an additional 34-m antenna amounted to $200,000 per year, while the costs for maintenance and operation of the existing 64-m antennas were not expected to increase significantly by their enlargement to 70-m diameter.

When the project began, it was estimated that modification of the 64-m antennas to 70-m would require about 12 months for each antenna, or three years if they were modified in sequence. However, the demand for the antennas to support ongoing missions was such that the time period for the entire job had to be reduced to fit the 18-month period following the Galileo launch scheduled for mid-1986, and the Galileo Probe release scheduled for the end of 1987. This 18-month window was the only period in the foreseeable future when the antennas could be taken out of service without significant impact to in-flight missions. The challenge became one of finding a way to complete the implementation in half the time it would take under normal circumstances.

Normally, an implementation task of this type would be done by a single contractor who would be responsible for the design, fabrication, erection, and testing. The lead time for this approach would run into many months. Instead, to save time, the design was done by JPL engineers with assistance in detail design from TIW Systems of Sunnyvale, California. Except for the Australian antenna, JPL acted as the prime contractor for the major items on a build-to-print basis. In Australia, the fabrication of the structural elements and the quadripod and the erection work were contracted locally. In addition to this time saving strategy, it was planned to return each 70-m antenna to service with a gain performance at least equal to the 64-m antenna, and to perform the final fine-tuning to reach predicted 70-m performance later as the opportunity provided.

For various reasons associated with the support of ongoing missions, particularly Voyager and the upcoming Galileo launch, urgency of the work at Goldstone, the seasonal weather conditions at each site, and availability of subcontractors and equipment, it was decided to carry out the antenna expansion work in this order: Spain, Australia, and Goldstone.

The way the plan eventually worked out is shown in the following table.

Summary of 64-m to 70-m Antenna Expansion Start and Finish Dates

Location	Start Date	Return to Operations with 70-m Performance
Madrid	1 Aug 1986	29 July 1987
Canberra	1 Feb 1987	21 Oct 1987
Goldstone	1 Aug 1987	1 June 1988

The loss of the Shuttle *Challenger* in January 1986 caused a delay in the launch of Galileo, and as a result the eighteen-month downtime constraint was removed. This left the Voyager Neptune Encounter in August 1989 as the next scheduled mission constraint. However, the project elected to hold to its June 1988 completion date for the last antenna at the expense of some overlapping downtime between the Madrid and Canberra antennas.

Microwave Design Considerations

Working within the constraints of antenna diameter (70 m), existing feed location, and existing 64-m antenna rib shape, DSN engineers designed a microwave optics system that would optimize each of these constraining parameters. An asymmetrically shaped subreflector of approximately 25 feet in diameter would optimally illuminate the new, high-precision, shaped surface of the 70-m main reflector. Microwave energy, focussed by the subreflector, would enter a single hybrid-mode horn, to feed the existing low-noise X-band maser. The choice of microwave feed was based on an existing, proven design, and offered a performance and cost advantage over either the S/X-band feedhorn designed for the 34-m high-efficiency antennas or the dual hybrid-mode horn which had been used on the 64-m antennas.

The single hybrid-mode horn, mounted on the X-band Receive Only (XRO) cone would provide the ultimate in X-band receive performance. S-band receive and transmit capability would be provided by the existing dichroic plate and reflex feed arrangement on the S-band Polarization Diversity (SPD) cone, while the third cone carried an X-band/K-band Receive (XKR) feed, used principally for radio and radar astronomy purposes.

Measurements of RF efficiency and system temperature were made at each 64-meter antenna before it was taken out of service and again before it was returned to service after upgrade to 70 meters. That design objectives were met is plainly obvious from the results given in the table below.

Summary of Measured Antenna Gain Before and After Upgrade

Gain (dBi)	DSS 14 X-band	DSS 43 X-band	DSS 63 X-band	All Ants. S-band
64-m	72.33	71.94	71.91	61.40
70-m	74.17	74.10	74.28	63.34
Improvement	+1.84	+2.16	+2.37	+1.94

The reasons for the differences between stations would be understood and corrected as time progressed. But, as the project ended, the gain improvement of 1.9 dB had been realized and all three antennas were back in service in time to meet the immediate requirements of the *Voyager 2* Neptune Encounter in August 1989.[45]

Kicker Braces

Before the 70-m expansion work on the two overseas 64-m antennas could begin, it was necessary to install additional structural bracing or "kicker braces" on each antenna. This additional bracing was required to mitigate the effects of gravitational deflection, with consequent radio performance loss, as the antenna was moved in elevation. The modification that involved replacement of two short, existing knee braces with very substantial steel members, had proven very effective on the Goldstone 64-m antenna.

The modification to the Australian antenna was started in October 1984 and completed six weeks later. S-band performance measurements on the antenna in January 1984 showed a gain versus elevation curve which was similar in shape but 0.4 dB lower than the corresponding Goldstone data. The reason for the discrepancy was not clear. When the Spanish antenna was modified some months later, its gain versus elevation curve also differed from those of both DSS 43 and DSS 63. It was decided to defer any further investigation of these discrepancies until after the 70-meter extension work had been completed.

Subreflector

The shaped subreflector for the 70-m antenna was almost 8 meters in diameter, about two meters deep, and 1.6 cm thick. It was constructed of six, cast aluminum panels, welded together to form two large pieces, and then bolted together for machining. At the time, it was probably the largest cast aluminum structure that had ever been built. It required state-of-the-art casting, welding, and machining technology to achieve the desired asymmetrical surface shape, with an accuracy of 2.54 millimeters (mm). The three subreflectors were cast and machined by specialty contractors in Spain. However, the Spanish contractors soon ran into technical problems with the casting and machining processes, and the resulting delays to the work were compounded by transportation, economic, language, and cultural difficulties. Early estimates of subreflector fabrication time changed from 27 weeks to 40 weeks and directly threatened the completion schedule for the Australian antenna.

A contingency plan involving fabrication of a reinforced plastic subreflector of the same weight, diameter, and mounting configuration as the aluminum subreflector, was put into effect. The plastic unit, which had a surface of revolution rather than a shaped surface, was covered with a fine, expanded aluminum mesh. Although its performance would not be as good as that of the full aluminum subreflector, it would suffice to allow work to be completed to the 64-m performance level at DSS 43. The plastic subreflector was fabricated in California in two halves, shipped to the antenna site near Canberra and installed without incident.

Until the aluminum subreflector for DSS 63 arrived, a concrete weight was installed in its place so that servo tests and counterweight adjustments could proceed on schedule. The subreflector eventually arrived in time for the RF testing to take place as scheduled.

Surface Panels

The surface of the 70-m antenna was over 4,185 square meters in area and was constructed of 1,272 individual aluminum panels of 17 different sizes. Each panel could be adjusted individually to conform to the desired antenna surface shape which, when complete, was accurate to 0.6 mm (root-mean-square) over the entire surface. The panels were made by a special new facility in Italy, using a metal-bonding process to produce the high-precision panels to an rms tolerance of 0.1 millimeter. Apart from some currency problems in 1987 caused by devaluation of the U.S. dollar, the work proceeded smoothly, and all three sets of panels were delivered to the sites on time.

Madrid Site Implementation

With the necessary scaffolding and a 500-ton crane in place, disassembly of the DSS 63 antenna began on 1 August 1986. Spanish contractors had already fabricated and delivered the quadripod and all the new structural elements needed to replace the old ribs and extend the diameter to 70 m. Reinforcing for the antenna backup structure and elevation wheel was to be welded in place. Additional counterweights would be needed to balance the added weight of the new subreflector, quadripod, and primary reflector. Three months into the job, the contractor advised JPL that the downtime would have to be extended by eight weeks due to various accumulated delays in the work. Efforts to recover the lost time proved futile. Since the antenna conversions were planned to be done in serial fashion, this delay would seriously affect completion of the Canberra and Goldstone antennas. It was decided to start the DSS 43 downtime on 1 February 1987, as originally planned, even though this would mean that two 64-m antennas would be out of service simultaneously until DSS 63 was completed. At the same time, the Canberra and Goldstone schedules were extended from six to eight months based on the Madrid experience. This decision was soon vindicated when the *Challenger* accident in January 1987 resulted in a three-year delay to Galileo, leaving the Voyager Neptune Encounter in August 1989 as the main driver for project completion.

The structure was complete and the 1,272 panels installed in mid-December 1986. Alignment was delayed due to some of the worst winter weather in Europe in 100 years. Using the concrete dummy weight for the subreflector, the counterweight was balanced and the antenna servo systems tested and calibrated.

The aluminum subreflector finally arrived in March, and was installed on the quadripod so that RF gain measurements could be made. The first set of results indicated a serious deficiency in antenna performance. All 1,272 panels were readjusted with even worse results. This was a grave situation which had to be understood before the alignment work on the Canberra antenna could begin. An intense effort, called in DSN parlance a "Tiger Team," set about analyzing the problem. The "kicker braces" were tightened and another panel alignment was carried out using a new measurement technique called holography. These actions were successful in assuring that the 70-m antenna exceeded the performance of the 64-m by 2.1 dB.

On this basis, the antenna was returned to service on 27 July 1987, with a plan to complete the final, more detailed measurements at a later date.

Canberra Site Implementation

Working under a somewhat different funding arrangement from the other two sites, it was Australian rather than American contractors who provided the new rib trusses, reinforcing steel, quadripod, counterweight, and services for erection of the Canberra antenna. With all material and equipment in place, the downtime started on 1 February 1987, as scheduled. Work proceeded smoothly despite some loss of time due to misalignment of the truss modules and a problem with mounting some of the surface panels.

Because the reason for the anomalous performance on the Madrid antenna had not been determined by the time the contractors were ready to start panel alignment on the Canberra antenna, it was decided to set that antenna using holography for the final alignment settings. Although this time consuming effort added three weeks to the completion date, DSS 43 was returned to service with 70-m performance on 21 October 1987. Although the measured antenna performance was somewhat less than that measured at Madrid (attributed to the use of the plastic subreflector), it met the 70-m performance specifications and remained in operational service for over two years before finally being replaced by the aluminum version in December 1989.

Goldstone Site Implementation

The implementation of the 64-m to 70-m antenna extension at DSS 14 benefitted immeasurably from the DSN experience at the other two sites. The U.S. contractors had the structural material on site, ready for installation at the start of the downtime set for 1 October 1987. There were no problems in aligning the trusses at Goldstone, and the installation of quadripod, rib modules, and panels was quickly accomplished as planned. The aluminum subreflector from the Spanish contractor finally arrived on-site in February 1988. Since the "Tiger Team" had not resolved the Madrid antenna performance anomaly by that time, it was decided to set the panels as had been done at Canberra, with a rough setting first, followed by a holograph-aided procedure.

As a result of the experience gained in Spain and Australia, the final panel settings and the gain and efficiency measurements were conducted without problems, enabling the DSS 14 antenna to be returned to service with 70-m performance on 29 May 1988, two days ahead of schedule. Various aspects of the antenna work described above are shown in Figure 4-16.

The Voyager Era: 1977–1986

Figure 4-16. Work in progress on the 60-m-to-70-m antenna extension at DSS 14, Goldstone, December 1987.

So ended the 64-m Antenna Rehabilitation and Performance Upgrade Project. When it was finished, the DSN had a complete subnetwork of 70-m antennas whose performance met or exceeded the design values under all conditions and whose serviceability had been vastly improved as a result of many years of continuous operational experience. It met the mission support requirements of the *Voyager 2* encounter of Neptune in 1989 and would enhance the downlink and uplink for all flight projects thereafter. Nevertheless, this antenna would be the final chapter in the DSN relationship with large antennas. Studies had shown that larger antennas were associated with diminishing returns in cost per/dB of performance gain.

The path to future enhancement of the downlink lay in the array of smaller, higher efficiency antennas, and the pursuit of other avenues of telecommunications technology, as future events would show.

NETWORKS CONSOLIDATION

In the past, the DSN managed by JPL provided tracking and data acquisition support for NASA deep space planetary missions. Managed by Goddard Space Flight Center (GSFC), a similar service was provided for NASA Earth-orbiting missions by a worldwide network of tracking stations known as the Spaceflight Tracking and Data Network (STDN). GSFC had for some time been developing a replacement for the STDN which would consist of three geostationary satellites with a capability for tracking and relaying, to a single ground station, data from Earth-orbiting satellites with a periapsis of less than 15,000 km. This new facility, called the Tracking and Data Relay Satellite System (TDRSS), but known in general terms as the Space Network, became operational in 1984. The few remaining satellites with Highly Elliptical Earth Orbits (HEEO), which lay outside the TDRSS zone of coverage, were serviced by a portion of the STDN called the ground network or GSTDN. Since the HEEO missions to be supported by GSTDN would have operational parameters that were in many respects similar to deep space missions supported by the DSN, the question naturally arose as to whether there was any combination of GSTDN and DSN that could result in a more economical way of supporting those missions that could not be supported by the TDRSS.

To address this question, NASA Headquarters commissioned a study to be carried out jointly by representatives from its Office of Space Tracking and Data Systems (OSTDS), JPL, GSFC and the overseas Complexes. Known as the Networks Consolidation Working Group, this body first met in May 1979 and completed its study in October 1979.[46]

In this study, a consolidated network comprised of various elements from the present DSN and the GSTDN showed significant cost savings and greater capability than the sum of the separate networks. This option was planned around the then-existing plan for the next generation of DSN improvements called the Mark IVA Model, augmented with GSTDN facilities and resources to support HEEO missions. A top-level overview of the DSN and GSTDN Network Facilities that would eventually comprise the Consolidated Network is shown in Figure 4-17.

In January 1980, a Networks Consolidation Program (NCP) office was established, in the Tracking and Data Acquisition (TDA) Office at JPL, as a consequence of NASA approval of the recommendations of the Networks Planning Working Group. Simultaneously, program offices were established at GSFC and in OSTDS at NASA Headquarters. JPL and GSFC management and technical counterparts were identified, and monthly working sessions were initiated. Based on the work of these groups, a management plan, transition plan, mission support plan, and a contingency plan were developed

The Voyager Era: 1977–1986

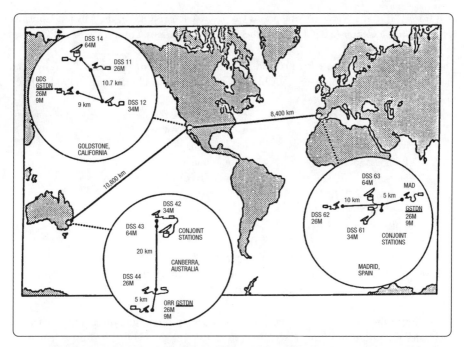

Figure 4-17. Elements of the DSN and GSTDN, 1982.

and published in 1981.[47] Budgetary considerations, changes in requirements, the potential mission set, antenna cost trade-offs, staffing, and implementation schedules became matters of great concern as the requirements and constraints of the Network Consolidation program were integrated with those of the Mark IVA program.

The major elements of the GSTDN and DSN which were to be consolidated were the tracking antennas and their associated support facilities. The current location of these elements at Goldstone, Canberra, and Madrid is shown in Figure 4-13. In the existing Mark III DSN, and also in the GSTDN, each antenna was associated with its own support facilities and control room. (At the overseas complexes, the two stations share the same operations and control building.) In the Mark IV DSN, the fundamental elements of each consolidated DSCC would consist of multiple antennas operated and controlled from a single Signal Processing Center (SPC). As far as economically possible, all antennas at each complex were to be colocated with the 64-m antenna to permit unattended operations. Data output from each of the three complexes is routed from the SPC via the GCF to the Central Communications Terminal (CCT) at JPL. The data are distrib-

243

uted from JPL to the Network Operations Control Center (NOCC), the Mission Control and Computing Center (MCCC), and other Project Operations Control Centers (POCC).

The 1981 Mission Support Plan considered the mission set for the years during which the Network Consolidation-Mark IVA Implementation Task would take place (1982–1985) in the following way:

Deep Space Missions	Earth Orbiter Missions
Active Missions	Active Missions
Viking Lander	International Comet
Pioneer 10, 11, 12	Explorer (ICE)
Voyager 1, 2	
Planned Missions	Planned Missions
Galileo Planned Missions	Active Magnetosphere Particle Tracer (AMPTE)

This set of missions would form the basis for the schedule and sequence in which the huge implementation task would be carried out.

To minimize the impact to ongoing missions and to protect the critical mission readiness dates for future missions, a set of "ground rules" was established. Implementation for both the consolidation effort and the Mark IVA effort would have to be planned around these constraints. The "ground rules" included coverage for the Shuttle and Earth satellites, and set the conditions under which a 64-m antenna or a 34-m antenna could be taken out of service for modification. This meant that both projects had to be accomplished with no interruption of service to the flight projects. To satisfy these constraints, three separate and parallel streams of activity were planned:

First, an interim configuration would be implemented around the 34-m subnetwork between March 1983 and February 1984 to support the two HEEO missions, AMPTE and ICE, beginning in March 1984. During this period, DSSs 12, 42, and 61 would be configured in sequence.

In parallel, a consolidation configuration would be implemented between March 1983 and January 1985. In this period the actual consolidation aspects of the task would be accomplished.

The Voyager Era: 1977–1986

Finally, while this work was in progress throughout the Network, the stream of activity directed to completing the final Mark IVA configuration by mid-1985, in time to support the *Voyager 2* encounter of Uranus in the following January, would be moving ahead.

It began in June 1983 at Goldstone. All the existing Mark III equipment was installed in the DSS 12 Control Room to create an interim configuration. Ongoing mission operations were conducted from there. Similar rearrangements of the equipment at DSS 42 and DSS 61 were made to create interim configurations at the overseas sites. This provided one full subnetwork of antennas in the interim configuration to support ongoing mission operations round the Network.

Under the Consolidation plan, the first antenna to be moved would be the 26-m antenna from the GSTDN station at Honeysuckle Creek near Canberra (DSS 44). Starting in March 1983, it would be dismantled and re-erected near the 64-m antenna on the DSN Complex site at Tidbinbilla where it would be identified as DSS 46. GSFC would provide the necessary electronics equipment to support a single uplink and downlink. The data and communications circuits from DSS 46 would be routed through NASCOM direct to the POCC at GSFC. The relocation was complete by January and System Performance Tests verified the integrity of the station in July, and readiness to support AMPTE in August 1984.

The 26-m antenna, DSS 16 at Goldstone, did not need to be moved. Microwave, and later fiber-optic, links would be used to connect it to the new Signal Processing Center on the Goldstone Complex. The existing equipment would be reduced to a single uplink and downlink with communications direct with the POCC at GSFC.

In January 1984, the second relocation of a 26-m antenna would begin. It was to be moved from the GSTDN site at Fresnadillas, Spain, to a site near the 64-m antenna on the DSN Complex at Robledo and identified as DSS 66. GSFC provided the electronics to support a single uplink and downlink, and communications circuits were routed via NASCOM directly to the POCC at GSFC. The relocation was completed in October, and System Performance Tests verified the integrity of the station in January 1985.

With this move completed successfully, the tasks related to the Networks Consolidation phase were now accomplished.[48]

MARK IVA IMPLEMENTATION

To accommodate the new Signal Processing Center (SPC10) at Goldstone, the existing control room at DSS 14 had to be enlarged and modified. This work began as soon as the interim configuration at DSS 12 had begun supporting mission operations. While SPC 10 was under construction, DSS 14 was taken offline for repair work on the hydrostatic bearing and repair work on the concrete pedestal. (Note: The existing control rooms at the two conjoint stations were already large enough for their SPC and no significant modifications were needed.) A few weeks later, construction work began on DSS 15, a new, 34-m high-efficiency (HEF) antenna, at a site close to both the DSS 14 antenna and the new SPC. Later that year work began on the construction of a similar antenna—DSS 45 at the Canberra Complex.

The old DSS 11 Control Room at the Goldstone Pioneer site was reactivated as a minimum hardware test bed and called the Network Test Facility (NTF). It was used as a staging area to assemble the equipment, get it turned on, and get the connections with the Local Area Network (LAN) working. This was the first time the DSN had used a LAN, and there was some concern about moving from the familiar (but now obsolete) star-switch controller system and going into a new, unproven, LAN-based system.

A key factor in the Implementation Plan required each set of complex equipment to be assembled and tested by the people who were going to use it. Teams from the Madrid and Canberra complexes arrived at Goldstone for this purpose. They were the ones who were going to install it and they had to understand it and learn how to use it. They were not reticent in expressing their opinion of the so-called "user-friendly" design. The Australian Team thought "user-surly" would be more appropriate.

By October 1984, DSS 15 had been completed and equipment installed, the Mark IVA design had been validated, and SPC 10 was completed. System Performance tests with DSSs 14, 15, and 16, and SPC 10 were started.

At the Canberra Complex, DSS 45 was completed, the Mark-IVA equipment installed in SPC 40, and System performance tests with SPC 40, DSS 43, and DSS 45 were completed by January 1985. A photograph of the Canberra Deep Space Communications Complex, taken in early 1985, is shown in Figure 4-18. The Signal Processing Center (SPC 10), 64-m antenna (DSS 43), 34-m standard antenna (DSS 42), recently completed 34-m HEF antenna (DSS 45), and the re-located 26-m X-Y antenna (DSS 46) are clearly seen.

The Voyager Era: 1977–1986

Figure 4-18. Canberra Deep Space Communications Complex, January 1985.

A 34-m high-efficiency antenna at the Madrid Complex was not planned for this time, and with the relocation of DSS 66 at the end of 1984, implementation of the Mark IVA equipment into SPC 60 could begin immediately. System Performance Tests with SPC 60 and DSS 63 were completed in May 1985.

All that remained to complete the Mark IVA configuration throughout the Network was to upgrade DSSs 12, 42, and 61, all still in the interim configuration, to the Mark IVA status. With these three stations added to those already integrated with the new SPCs, the Mark IVA upgrade task would be completed.

The final configuration of the Mark IV Model DSN, which includes the antennas and capabilities provided by the consolidation program, is shown in Figure 4-19.

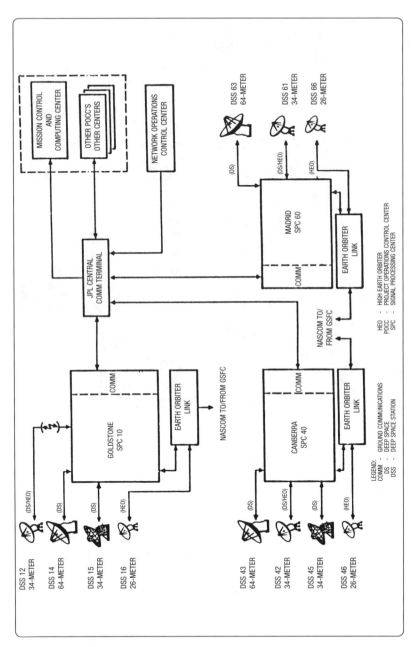

Figure 4-19. The DSN Mark IVA-85 configuration.

THE VOYAGER ERA: 1977–1986

There were four other elements of the DSN, which although not directly involved in consolidation tasks were deeply involved in the Mark III to Mark IV transition. These were the Ground Communications Facility (GCF), the Network Operations Control Center (NOCC), and the two test facilities, CTA 21 at JPL and MIL 71 at KSC. The transition plan for the GCF allowed the GCF to simultaneously support the existing Mark III and the gradually evolving Mark IVA configurations. As the Mark IVA upgrades were completed at the complexes the GCF communications channels were switched over to the Mark IVA GCF, and the old Mark III equipment was deactivated.

A similar approach was used for the NOCC, due to the need to support a hybrid arrangement of Mark III/Mark IVA, while the interim configuration was current in the Network. Cut over to the Mark IVA NOCC was effected as each Complex came into service in the Mark IVA configuration.

Both CTA 21 and MIL 71 contained single string Mark III equipment. At CTA 21 this was adequate for spacecraft compatibility testing, and hardware or software development associated with the interim configuration. In early 1985, when the validation and testing of the Mark IVA equipment at Goldstone was complete, one of the new Mark IVA strings from SPC 10 replaced the obsolete Mark III equipment. At MIL 71, the Mark III equipment would be used to support the AMPTE launch. Later, it would be replaced by Mark IVA equipment.

Inevitably, there were delays to the original schedule and changes to the original plan. Throughout its entire life, the Mark IVA project was adversely affected by attrition of the original budget.

A typical 1982 budget profile against which the progress of the Mark IVA and Networks Consolidation programs would be measured is illustrated in the table below:

Typical Budget Profile (1982)

Fiscal Year	1981	1982	1983	1984	1985	Total
Networks Consolidation ($M)	6.5	21.5	20.4	16.1	5.3	69.8
Mark IVA ($M)	10.0	19.5	22.1	21.9	14.9	88.4
Total ($M)	16.5	41.0	42.5	38.0	20.2	158.2

Under constant pressure from the 1980 Reagan Administration, the NASA budget was cut several times during the course of the NCP/Mark IVA program. These reductions inevitably impacted the funding available for the work in progress, resulting in programmatic changes, deferment of some tasks, and schedule delays. The DSN could only deal with this continuing condition by deleting planned capabilities from the original design.

Despite these setbacks, the program was completed in time to support the Voyager Uranus Encounter in January 1986. There were many liens outstanding against the system and little time left for operations verification testing. It was mainly due to the heroic efforts and dedication of the operations and engineering people at the three complexes that the Mark IVA worked so well at the time it was needed.

It was the perception of those who were in the best position to know that the ultimate cost of the Mark IVA project, including the Consolidation elements, was 10 to 20 percent less than the overall cost estimated in 1982. This, however, resulted from mandated budget cuts met by deleting capabilities, rather than from DSN management or engineering processes.

Although there were some questionable features of the internal design, particularly in the subsystem operation, the overall goal of centralized control was met and amply demonstrated in the years that followed. The main purpose of the Mark IV Network upgrade to centralized control was to reduce operations costs by reducing the number of operators required to operate each antenna. That this goal was successfully achieved is shown graphically in Figure 4-20.

In retrospect, the conversion of the DSN Mark III Data System to the Mark IVA Model was considered, at the time, to be the most complex implementation project undertaken in the DSN since its inception, surpassing even the MDS effort of five years earlier. Initiated in 1980, the objective of the Mark IVA project was to reduce operations costs, improve reliability and maintainability, and to increase telemetry reception capability for the Voyager Uranus encounter and all subsequent planetary missions.

That these immediate goals had been achieved soon became evident in the successful Uranus Encounter and the events that soon followed. The validation of the long-term goals became apparent as the DSN moved through the remaining decade and a half of the century. Additions to the Mark IVA were made as the increasing mission set required, but the original concepts on which this major undertaking were based gave no cause for reconsideration. It could well be said the Mark IVA-85 model was indeed the forerunner of the "modern" DSN.[49]

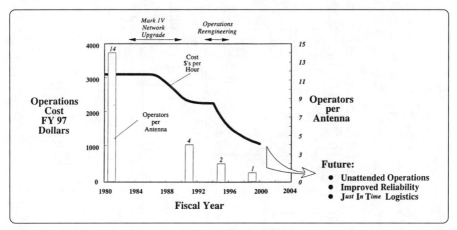

Figure 4-20. DSN antenna operating costs, 1980–2000. The data shows that the Mark IVA Network upgrade reduced antenna operations costs from over $3,000 per hour in 1980, when 14 operators per antenna were required, to $2,400 per hour in 1990, when only 4 operators were required. Further economies in operations costs lay in the future.

Signal Processing Centers

The three Deep Space Communications Complexes (DSCCs) that composed the Mark IVA DSN were located at Goldstone, California; Tidbinbilla near Canberra, Australia; and Robledo, near Madrid, Spain. There was one Signal Processing Center (SPC) and four antennas at each of the DSCCs. (The fourth antenna, a 34-m high-efficiency type was added to the Madrid DSCC in 1987.) High-speed and wideband communications circuits linked each complex to the Network Operations Control Center (NOCC) at JPL. The necessary communications terminal and data transmission facilities at each Complex and at JPL were included in the Mark IVA DSN, shown in completed form in Figure 4-19 on page 263.

Each antenna within a Complex had some locally mounted equipment—antenna drive and control equipment, low-noise amplifiers, receiver front ends, and transmitters. The antenna pointing equipment, microwave instrumentation, transmitter control, receiver, and metric data were located at the SPC. The SPC also contained the equipment which could be assigned to any of the designated antenna "links," such as the telemetry, command, radiometric, radio science, and interferometry processing equipment. Nonassignable equipment such as that required for frequency and timing, test and training, and global communications was also located in the SPC. Equipment associated with the 26-meter

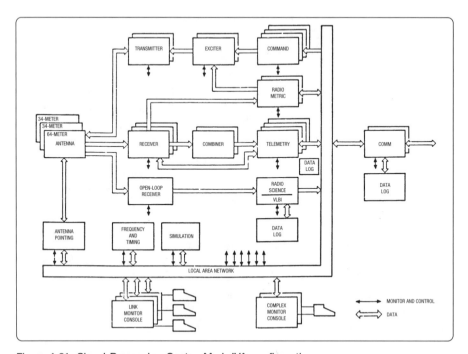

Figure 4-21. Signal Processing Center, Mark IVA configuration.

Earth-Orbiter link was not controlled from the SPC until a few years later when it, too, was integrated with the SPC.

In the Mark IVA Model, the Signal Processing Center at each DSCC was configured to support the operation of each antenna individually, or to array any combination of the antennas. A simplified block diagram of an SPC is shown in Figure 4-21.

Within each SPC, the equipment could be grouped into three "links" by operators working from the Link Monitor Console. Each "link" comprised the equipment necessary to support one spacecraft with a downlink and uplink as required by the particular mission sequence. This included the antenna and equipment associated with the tracking, receiver, telemetry, command, radiometric, radio science, monitoring and communications functions. Each "link" was then assigned to a particular spacecraft/mission by an operator working from the Complex Monitor Console according to an established schedule. Other subsystems provided test support, frequency and timing, maintenance, etc.

The Voyager Era: 1977–1986

Interactions between the two monitor consoles and the hardware and software comprising the "links" were carried by an extensive Local Area Network.

Compared to the former Mark III configuration, this arrangement resulted in a very significant reduction in the number of personnel required to conduct continuous mission operations at each Complex. Improvements in operability, reliability, maintainability, and the ability to reconfigure rapidly to work around anomalous situations, were soon apparent. The objectives of the Mark IVA project had certainly been achieved.

The control and data processing equipment for the 26-m Earth Orbiter tracking station was not integrated into the SPC. With the exception of the telemetry equipment which was updated, the equipment remained as it was when the stations were transferred from the GSTDN to the DSN in 1983. These stations could support only one uplink/downlink at a time, but a cross-support connection with the SPC enabled a spacecraft link through the 26-m antenna to be processed by one of the "links" at the SPC. Likewise, a spacecraft link through the 34-m standard antenna at the SPC could be connected to the Earth Orbiter station for data processing. Either of these operational arrangements was known as the "cross-support" mode.

Mark IVA Network Operations Control Center

The Network Operations Control Center (NOCC) at JPL, was also upgraded as part of the Mark IVA project. The NOCC Mark IVA configuration is shown in Figure 4-22.

Data from the SPCs was transported by the GCF to the Central Communications Terminal. There, the data was switched to the JPL Mission Control Center (for JPL missions) or to a remote Project Operations Control Center (for non-JPL Missions). The data was also routed in parallel to the NOCC, where data monitoring and network control functions were performed. The data monitoring functions were performed in the real-time monitor processors which were retained from the Mark III model but upgraded with Mark IVA compatible software.

Mission support data products such as DSS tracking schedules, mission sequence of events, prediction data for antenna pointing, and setting uplink and downlink parameters, were also generated in the NOCC and provided to the SPCs as required for mission support.

The Network was controlled from the NOCC by station controllers who were provided with suitable displays to enable them to carry out these functions. During the Mark IVA period the physical arrangements in the NOCC were greatly improved by substituting convex shaped monitor consoles for the original circular consoles. These

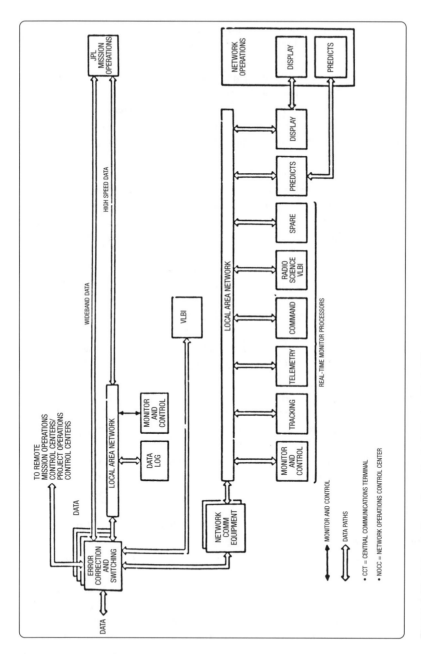

Figure 4-22. Network Operations Control Center, Mark IVA.

accommodated larger monitor screens (14-inch) arranged in two tiers of five, provided better viewing conditions and increased the work capacity and physical comfort of the individual controllers.

Mark IVA Ground Communications Facility

The communications links between the SPCs and the Central Communications Terminal (CCT) at JPL were also upgraded as part of the Mark IVA project.[50] The GCF Mark IVA model permitted a more cost-effective utilization of available circuits without losing the flexibility to protect critical data. At each SPC, new communications processors were now capable of multiplexing data from all of the antennas on the Complex (data links) on to a single 224 kbps circuit for transmission to JPL. In addition, critical data could be transmitted on a single 56 kbps circuit with error detection and correction from the Complex to JPL. Complete, error-free and gap-free records of all data transmitted from the SPC would be generated automatically by the Data Records Subsystem residing in the Central Communications Terminal at JPL. This Intermediate Data Record tape file would be the end product, deliverable to each flight project as a permanent record of the data captured from each spacecraft during each tracking station pass. Future plans called for a modular expansion of the transmission data rate capability to 1,544 kbps, enabling error correction on all data transmitted from the complexes.

At Goldstone, new microwave links were required to connect DSS 12 to the new SPC 10 now located 21 km away. Networks Consolidation added the former STDN station, DSS 16, to the DSN and necessitating the construction of a new fiber-optic link, 8 miles in length, to bring its baseband data to the Goldstone GCF communications terminal for transmission to JPL.

Consistent with the future introduction of packetized telemetry, plans were made to expand GCF capability to accept variable size data blocks which would be sized to match the characteristics of the data stream being transmitted. This capability would further contribute to Network efficiency by minimizing the amount of data conditioning needed for ground transmission by the older Mark III system.

DSN MARK IVA SYSTEM UPGRADES

Mark IVA Telemetry

At the Deep Space Communication Complexes (DSCCs), the DSN Telemetry System was changed in two major ways as a result of the Mark IVA implementation project.[51]

First, the addition of baseband combining equipment allowed for the arraying of up to seven antennas at each Complex. Arraying the 64-m antenna with two 34-m antennas in this way would provide a gain of 1.84 dB relative to the gain of a single 64-m antenna. The enhanced downlink performance derived from arraying was driven by the Voyager project requirement for support during the Uranus Encounter where the DSN antenna array at CDSCC was supplemented with the 64-m antenna at the Parkes Radio Astronomy Observatory.

Second, the telemetry system was configured to support up to three missions simultaneously, with one or two of them being highly elliptical Earth-orbiting (HEO) missions.

In addition to these major changes in capability, the Mark IVA telemetry system provided for demodulation of Manchester-coded data modulated directly on the carrier with coded data rates up to 211.2 kilosymbols per second (ksps). This capability would be used by many of the HEO missions. For planetary missions, maximum likelihood decoding of short-constraint length convolutional codes and sequential decoding of long-constraint length codes was provided. Precise measurement of received signal level and system noise temperature, was available to all DSN users.

At each Complex, new features oriented towards improved DSN operations included centralized control and real-time status reporting by the Mark IVA Monitor and Control System. Original data records produced by each of the four telemetry groups could be played back to the NOCC when necessary by local manual control or by automatic response to GCF instructions from the Central Communications Terminal at JPL.

The principal features of the Mark IVA telemetry arrangement at a DSCC are shown in Figure 4-23.

At each Complex there was one 64-m antenna and two 34-m antennas. The 64-m antenna and 34-m standard antenna could receive an S-band and an X-band carrier simultaneously. The 34-m high-efficiency antenna could receive one X-band carrier only. The high system noise temperature at S-band made it unsuitable for deep space reception at S-band. The telemetry data processing equipment was arranged in four groups, each of which contained two data channels. Each channel could process one data stream from either an HEO or deep space spacecraft depending on its capability. Hardware in Groups 1 and 2 closely resembled the Mark III model. Hardware in Groups 3 and 4 contained the new Baseband Assembly (BBA), which was a new design incorporating the baseband combiner, subcarrier demodulator, and symbol synchronizer functions. A functional block diagram of the new BBA is shown in Figure 4-24.

The Voyager Era: 1977–1986

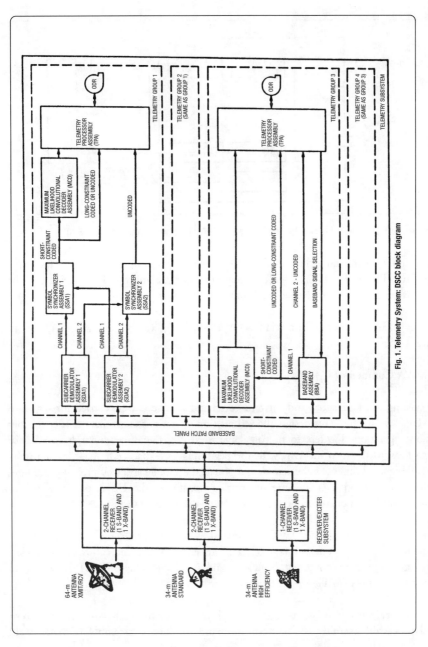

Figure 4-23. Principal features of the DSCC Mark IVA Telemetry Subsystem.

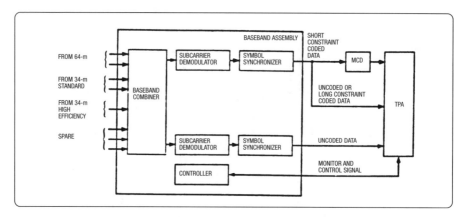

Figure 4-24. Functional block diagram for Mark IVA Baseband Assembly (BBA).

In addition to performing the functions required to format the decoded data for transmission to JPL via the GCF, the Telemetry Processor Assembly (TPA) carried out the functions required by the centralized Monitor and Control system.

The key characteristics of the Mark IVA 1985 Telemetry System are tabulated in Figure 4-25.

Mark IVA Command

The existing Mark III store and forward command functions were carried forward into the Mark IVA system. This design allowed mission operations controllers at JPL or other Mission Operations Centers to prepare large files of spacecraft commands in advance and to forward them to the DSCC at the beginning of a spacecraft track for transmission at designated times. The system design provided elaborate and well-proven procedures to protect command data file integrity during transfer to the DSN station as well as bit-by-bit confirmation of proper radiation from the DSN transmitter.

In the Mark IVA model, the DSCC Command System utilized elements of the upgraded antenna, microwave, transmitter, receiver-exciter and monitor, and control subsystems.[52] These capabilities were configured, controlled, and driven by the new Mark IVA Command Subsystem as shown in Figure 4-26.

Four identical strings of command processing equipment, each containing a command processor assembly (CPA) and command modulator assembly (CMA), could

Parameter	Antenna		
	64-meter	34-meter Standard	34-meter High-Efficiency
Frequency Range, MHz			
S-band	2270–2300	2200–2300	2200–2300
X-band	8400–8440	8400–8500	8400–8500
Gain, dBi			
S-band	$61.7^{+0.3}_{-0.4}$	$56.1^{+0.3}_{-0.7}$	$55.8^{+0.0}_{-0.5}$
X-band	$72.1^{+0.6}_{-0.6}$	$66.2^{+0.6}_{-0.6}$	$67.3^{+0.5}_{-0.8}$
System Noise Temperature, K Zenith			
S-band with maser			
Diplex	18.5 ± 3	27.5 ± 2.5	
Listen-only	14.5 ± 2	21.5 ± 2.5	
S-band with FET			
Diplex		130 ± 10	
Listen only			115 ± 10
X-band with maser	20 ± 3	25.0 ± 2.5	18.5 ± 2

Figure 4-25a. DSN Mark IVA Telemetry System key characteristics. RF reception characteristics.

Functions	Channel 1	Channel 2
Baseband Combining	N/A	N/A
Subcarrier Demodulation	100 Hz to 1 MHz, squarewave or sine wave	100 Hz to 1 MHz, squarewave or sine wave
Symbol Synchronization	6 s/s to 268.8 ks/s	6 s/s to 268.8 ks/s
Data Format	NRZ-L, NRZ-M, Bio-L	NRZ-L, NRZ-M, Bio-L
Sequential Decoding	$K=24$ or 32; $R=1/2$; frame length selectable; 16 s/s to 20 ks/s	N/A
Maximum Likelihood Convolutional Decoding	$K=7$; $R=1/2$ or $1/3$ 10 b/s to 134.4 kb/s	N/A
Uncoded	6 b/s to 268.8 kb/s*	6 b/s to 268.8 kb/s^2

* Record only with non-real-time playback above 250 kbps.

Figure 4-25b. DSN Mark IVA Telemetry System key characteristics. DSCC Telemetry Subsystem Channel (Groups 1, 2).

The Voyager Era: 1977–1986

Functions	Channel 1	Channel 2
Baseband Combining	Up to seven basebands	N/A
Subcarrier Demodulation	10 kHz to 2 MHz, squarewave or sine wave	10 kHz to 2 MHz, squarewave or sine wave
Symbol Synchronization	4 s/s to 4 Ms/s	4 s/s to 4 Ms/s
Data Format	NRZ-L, NRZ-M, Bio-L	NRZ-L, NRZ-M, Bio-L
Sequential Decoding	$K=24$ or 32; $R=1/2$; frame length selectable; 16 s/s to 20 ks/s	N/A
Maximum Likelihood Convolutional Decoding	$K=7$; $R=1/2$ or $1/3$ 10 bps to 250 kbps*	N/A
Uncoded	4 bps to 500 kbps*	4 bps to 500 kbps*

* Record only with non-real-time playback above 250 kbps.

Figure 4-25c. DSN Mark IVA Telemetry System key characteristics. DSCC Telemetry Subsystem Channel (Groups 3, 4).

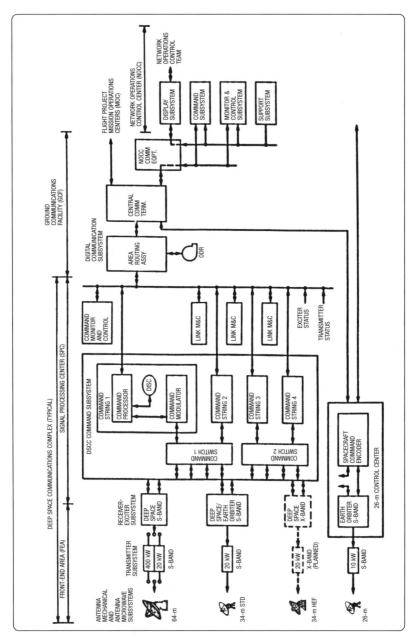

Figure 4-26. DSN Command System Mark IVA, 1985.

be connected via a switch assembly to the exciters of any of the S-band transmitters associated with the 64-m or 34-m antennas as shown in Figure 4-26. At the proper time for radiation, validated command data files previously stored on disk by the CPA would transfer to the CMA for subcarrier modulation before being passed to the appropriate exciter for generation of the command-modulated S-band carrier for radiation by the selected transmitter. Suitable interfaces with the monitor and control system provided for real-time status and configuration monitoring. Although the DSN command system was limited by the available transmitters to S-band uplink transmissions only, the need to provide an uplink on X-band for future missions was recognized, and development work to that end was already in progress in the engineering sections.

Mark IVA Tracking

In 1980 and 1981, the DSN Tracking System Mark III-79 model had been extensively modified to improve its performance, operability, and adaptability to support anomalous spacecraft conditions (notably, the *Voyager 2* uplink tuning problem).[53] The addition of a new microprocessor-based exciter frequency controller or Digitally Controller Oscillator (DCO) in place of the former programmed oscillator (POCA) to the receiver-exciter subsystem, provided a large repertoire of precision, time-dependent frequency ramps, controlled via the Metric Data Assembly or a computer terminal. Under the control of the Metric Data Assembly, the DCO would generate the special uplink tuning necessary to accomplish the special uplink tuning required to support the *Voyager 2* spacecraft, as well as standard tuning for all other spacecraft uplink operations.

The Meteorological Monitor Assembly transmitted non-real-time meteorological data to the NOCC for use in calibrating radio science and VLBI observations.

The quality of ranging data generated by the Planetary Ranging Assembly, as the angular separation between a spacecraft and the Sun decreases, was improved by the addition of selectable higher frequency components to the range code and better filters to decrease waveform distortion.

In the NOCC, older software used to generate frequency and pointing predicts for the stations, based on a Probe Ephemeris Tape supplied by the flight projects. It was redeveloped to make better use of the capabilities of the Sigma V computers in the NOCC Support Subsystem (NSS).

In the Mark IVA implementation, further improvements were made to most of the elements of the Tracking System, and all of them were integrated with the new DSCC Monitor and Control Subsystem (DMC) to provide centralized control for all subsystems and collection and distribution of the monitor data required by each subsystem.

At the NOCC, a new, Digital Equipment Corporation VAX 11/780 processor was added to perform two, key Tracking System functions. The first was an orbit determination function, similar to that carried out by the JPL flight projects, but intended specifically for DSN support of the new cooperative and reimbursable missions. The other was the predicts generation and distribution function previously carried out in the Sigma V-based NSS. Using radiometric data generated by the tracking stations, the NOCC Navigation Subsystem (NAV) could generate orbit data files and state vectors for use by the NSS, or for dissemination to flight project navigation teams, or on a reimbursable basis to external users such as the European Space Operations Center located in Darmstadt, Germany; the German Space Operations Center at Oberpfaffenhofen, Germany; and the Japanese Tracking and Control Center located at Tsukuba, Japan.

More than any of the other DSN systems, the Tracking System reflected the impact of changes to existing hardware and software and procedures introduced by the Mark IVA implementation. The quality and quantity of the S-band and X-band radiometric data being delivered to the flight projects, especially Voyager, in the immediate post Mark IVA period was not satisfactory. In March 1985, a special team was formed to monitor performance and recommend changes to improve radiometric data delivery prior to the Voyager Uranus Encounter in January 1986. Many changes to improve the operability of the Mark IVA Tracking System were proposed and implemented over the following months. By January 1986, the usable data return had improved to an acceptable level to meet the *Voyager 2* encounter navigation requirements. The excellent quality of the radiometric data provided by the DSN was considered to have been a strong contributor to the remarkable targeting accuracy of the satellite Miranda, as *Voyager 2* transmitted the Uranian system.

Over the next few years, several major new capabilities would be added to the DSN Tracking System. These would include an X-band uplink, Media Calibration Subsystem, the Sequential Ranging Assembly, and a real-time interface between the radiometric and navigation delivery systems.

The Voyager Era: 1977–1986

AN ABRUPT TRANSITION

Unlike the gradual transition between the Viking Era and the Voyager Era, the transition from the Voyager Era to the Galileo Era was marked by an abrupt and tragic event—the Space Shuttle *Challenger* disaster in January 1986. At that time, both the Ulysses and Galileo missions had been included in the DSN mission set for the Voyager Era for several years (see Figure 4-1). Both spacecraft were to be launched by the Space Shuttle using a high-energy, liquid-hydrogen fueled, Centaur G upper-stage combination. They would be the first spacecraft to be launched from the Shuttle in this manner.

Since both spacecraft were bound for Jupiter, both launches were planned to take advantage of the May 1986 Jupiter launch opportunity. The Ulysses launch window opened on 15 May, and the Galileo launch window opened on May 20. Agreements between the two flight projects had been made for the resolution of conflicts for launch priority (in case of launch pad delays) or overlap of view periods at the DSN tracking stations.

The DSN planned to support both missions with the recently completed Mark IVA network configuration, 64-m antennas, 34-m standard antennas, 34-m high-efficiency antennas and the newly consolidated 26-m antennas.

In January 1986, both spacecraft were well advanced in their prelaunch preparations at Kennedy Space Center. At each of the DSN complexes, however, attention was focused on the critical Voyager Uranus Encounter operations then in progress. At the same time, a team from the DSN Operations Office at JPL was en route to each of the complexes to discuss support for the Galileo and Ulysses launches later in the year. They had reached Madrid when the news came. Two days later the Voyager spacecraft made its closest approach to Uranus.

Although the weeks that followed were a period of great uncertainty and confusion in NASA and the Galileo and Ulysses projects, DSN was not immediately affected. Voyager and other inflight missions continued to receive DSN attention. Then, on 12 February 1986, both flight projects were informed that the planetary missions would not be launched in May 1986, and they were directed to work towards the next Jupiter launch opportunity in June 1987. The change for the missions affected the DSN much more significantly than was realized at the time, as subsequent events would show.

For various reasons, the launches did not take place. The reasons for the further deferment of the launches, their effect on the DSN, and the eventual accomplishment of Ulysses and Galileo missions are discussed in chapter 5, "The Galileo Era."

Endnotes

1. Bruce C. Murray, *Planetary Report* XVII (Number 3, May/June 1997): 14.

2. Bruce C. Murray, *Journey into Space* (New York: W. W. Norton, 1989).

3. David Morrison and Jane Sanz, *Voyage to Jupiter*, NASA SP-439 (Washington, DC: NASA, 1980).

4. E. K. Davis, "Deep Space Network Preparation Plan for Mariner Jupiter Saturn 1977," Document 618-706, Jet Propulsion Laboratory, Mariner Jupiter Saturn Project Office (15 December 1976).

5. David Morrison and Jane Sanz, *Voyage to Jupiter*, NASA SP-439 (Washington, DC: NASA, 1980), pp. 58, 60.

6. Andrew J. Butrica, "From Engineering Science to Big Science," chapter 11, NASA SP-4219 (Washington, DC: NASA, 1998), pp. 251–56.

7. David Morrison and Jane Sanz, *Voyage to Jupiter*, NASA SP-439 (Washington, DC: NASA, 1980), p. 108.

8. J. E. Allen and H. E. Nance, "Voyager Mission Support," TDA Progress Report PR 42-53, July-August 1979 (15 October 1979), pp. 4–5.

9. N. A. Fanelli and H. Nance, "Voyager Mission Support," TDA Progress Report PR 42-54 September/October 1979 (15 December 1979), pp. 19–23.

10. M. R. Traxler, "Deep Space Network Preparation Plan for Voyager (Saturn)," Document 618-706, Jet Propulsion Laboratory, Voyager Project Office; Rev. A (25 March 1980).

11. N. A. Fanelli and H. Nance, "Voyager Mission Support," TDA Progress Report PR 42-61 November/December 1980 (15 February 1980), pp. 22–26.

12. Jet Propulsion Laboratory, Voyager Project Office, "Voyager Bulletin; Mission Status Report No. 52" (27 August 1980).

13. Jet Propulsion Laboratory, Voyager Project Office, "Voyager Bulletin; Mission Status Report No. 57" (7 November 1980).

14. N. A. Fanelli and H. E. Nance, "Voyager 1 Mission Support (I), Voyager 2 Mission Support (II)," TDA Progress Report PR 42-66, September/October 1981 (15 December 1981), pp. 24–29.

15. See endnote 2.

16. C. T. Stelzried, "The Venus Balloon Project," TDA Progress Report PR 42-85, January-March 1986 (15 April 1986), pp. 191–98.

17. R. A. Preston and J. H. Wilcher, "The Venus Balloon Project," TDA Progress Report PR 42-80, October-December 1984 (15 February 1985).

18. J. M. Urech, A. Chamarro, J. L. Morales, and M.A. Urech, "The Venus Balloon Project Telemetry Processing," TDA Progress Report PR 42-85, January-March 1986 (15 April 1986), pp. 199–211.

19. K. M. Liewer, "Selection of Radio Sources for Venus Balloon-Pathfinder DDOR Navigation at 1.7 GHz," TDA Progress Report PR 42-87, July-September 1986 (15 November 1986), pp. 279–84.

20. See endnote 16.

21. R. Stevens, "Implementation of Large Antennas for Deep Space Mission Support," TDA Progress Report PR 42-76, November/December 1983 (15 February 1984), pp. 161–69.

22. C. T. Stelzried and A. L. Berman, "Antenna Arraying Performance for Deep Space Telecommunications Systems," TDA Progress Report PR 42-72, October-December 1982 (15 February 1983), pp. 83–88.

23. J. W. Layland et al., "Interagency Arraying Study Report," TDA Progress Report PR 42-74, April-June 1983 (15 August 1983), pp. 117–48.

24. D. W. Brown et al., "Parkes-CDSCC Telemetry Array; Equipment Design," TDA Progress Report PR 42-85, January-March 1986 (15 May 1986), pp. 85–110.

25. R. Stevens, R. L. Riggs, and B. Wood, "Pointing Calibration of the Mark IVA DSN Antennas (for) Voyager 2 Uranus Encounter Operations Support," TDA Progress Report PR 42-87, July–September 1986 (15 November 1986), pp. 206–39.

26. Jet Propulsion Laboratory, Voyager Project Office, "Voyager Bulletin; Mission Status Reports Nos. 72-81." See reports beginning 4 November 1985.

27. A. L. Berman and C. T. Stelzried, "Helios Mission Support," TDA Progress Report PR 42-67 November/December 1981 (15 February 1982), pp. 18–24.

28. W. E. Larkin, "Tracking and Data Acquisition Support for the Viking 1975 Mission to Mars; June 1978 to April 1980," Jet Propulsion Laboratory Publication 82-18 (15 April 1982). D. J. Mudgway, "Tracking and Data Acquisition Support for the Viking 1975 Mission to Mars; May 1980 to March 1983," Jet Propulsion Laboratory Publication 82-107 (15 May 1983).

29. R. E. Nevarez, "Pioneer 10 and 11 Mission Support," TDA Progress Report PR 42-61, November/December 1980 (15 February 1981), pp. 17–21. D. Lozier, "Pioneer Spacecraft," personal communication, 1 August 1997.

30. G. M. Rockwell, "Pioneer 11 Saturn Encounter Support," TDA Progress Report PR 42-53, July/August 1979 (15 October 1979), pp. 10–20. R. B. Miller, "Pioneer 11 Saturn Encounter Mission Support," Tracking and Data Acquisition (TDA) Progress Report PR 42-52, May/June 1979 (15 August 1979), pp. 4–7. R. B. Miller, "Pioneer 11 Saturn Encounter Mission Support," TDA Progress Report PR 42-54, September/October 1979 (15 December 1979), pp. 24–27.

31. J. W. Layland, "ICE Telemetry Performance," TDA Progress Report PR 42-84, October–December 1985 (15 February 1986), pp. 23–213. N. A. Fanelli and D. Morris, "ICE Encounter Operations," TDA Progress Report PR 42-84, October–December 1985 (15 February 1986), pp. 176–85.

32. J. P. Goodwin, "Usuda Deep Space Center Support for ICE," TDA Progress Report PR 42-84, October–December 1985 (15 February 1986), pp. 186–96.

33. Jet Propulsion Laboratory, "Deep Space Network Mission Support Requirements," JPL Document D-787, Rev. O (July 1987).

34. National Aeronautics and Space Administration, Office of Space Operations, "Reimbursable Policy for the Use of Ground Network Facilities," Ground Networks Division, Internal Document (19 April 1991).

35. National Aeronautics and Space Administration, Office of Space Communications, "Obtaining Use of Office of Space Communications (OSC) Capabilities for Space, Suborbital and Aeronautical Missions," NASA Management Instruction, NMI 8430.1C (31 December 1991).

36. R. J. Wallace, personal interview with D. J. Mudgway (13 August 1997).

37. P. L. Parsons, "Antenna Microwave Subsystem (Mark IVA)" TDA Progress Report PR 42-79, July-September 1984 (15 November 1984), pp. 165–69.

38. M. F. Pompa, "The New 34-Meter Antenna," TDA Progress Report PR 42-85, January-March 1986 (15 May 1986), pp. 127–38.

39. Jet Propulsion Laboratory, "Deep Space Network to Flight Project Interface Design Handbook," DSN Document 810-5, Module TCI-30 (1997).

40. D. H. McClure and F. D. McLaughlan, "64 Meter to 70 Meter Antenna Extension," TDA Progress Report PR 42-79, July-September 1984 (15 November 1984), pp. 160–64.

41. R. Stevens, "Report on Maintenance, Rehabilitation, and Upgrade of the DSN 64-meter Antennas," TDA Office Internal Memo (22 March 1982). B. Wood, "Report on the Mechanical Maintenance of the 70-Meter Antennas," DSN Document 890-257, Jet Propulsion Laboratory, p. 5; Document D-10442 (31 January 1993).

42. D. M. McClure and F. D. McLaughlin, "64-m Antenna Rehabilitation and Performance Upgrade Project Report," private communication (13 August 1997).

43. D. M. McClure, "Repair of DSS-14 Pedestal Concrete," TDA Progress Report PR 42-81, January-March 1985 (15 May 1985), pp. 136–48. H. McGinnes, "Rehabilitation of the 64-m Antenna Radial Bearing," TDA Progress Report PR 42-65, July-August 1981 (15 October 1981), pp. 151–61.

44. See endnote 21.

45. See endnote 42.

46. M. L. Yeater and D. T. Herrman, "Networks Consolidation Program," TDA Progress Report PR 42-59, July/August 1980 (15 December 1980), pp. 83–84. M. L. Yeater, and D. T. Herrman, "Networks Consolidation Program," TDA Progress Report PR 42-65, July/August 1981 (15 October 1981), pp. 19–24.

47. M. L. Yeater et al., Jet Propulsion Laboratory, Deep Space Network, "Networks Consolidation Program Plans and Mark IVA Implementation Plans," DSN Document 803-008, Vol. I, Management Plan (9 October 1980); Vol. II, Transition Plan (15 June 1981); Vol. III, Mission Support Plan (15 December 1981); Vol. IV, Mark IVA Implementation Plan (15 May 1984); Vol. V, NCP Contingency Study (15 August 1981); Vol. VI, Mark IVA Contingency Plan (15 June 1983).

48. D. D. Gordon, "Mark IVA 26-Meter Subnet," TDA Progress Report PR 42-79, July-September 1984 (15 November 1984), pp. 152–59.

49. R. J. Wallace and R. W. Burt, "Deep Space Network Mark IVA Description 1986," TDA Progress Report PR 42-86, April-June 1986 (15 August 1986), pp. 255–69.

50. R. A. Crowe, "The GCF Mark IV Implementation and Beyond," TDA Progress Report PR 42-82, April-June 1985 (15 August 1985). M. R. Traxler, Jet Propulsion Laboratory, "Deep Space Network Preparation Plan for Voyager (Uranus) Interstellar Mission," DSN Document 870-19 (1 April 1984).

51. D. L. Ross, "Mark IVA DSCC Telemetry System Description," TDA Progress Report PR 42-83, July-September 1985 (15 November 1985), pp. 92–100.

52. H. C. Thorman, "DSN Command System Mark IV-85," TDA Progress Report PR 42-84, October-December 1985 (15 February 1986), pp. 135–42.

53. J. A. Wackley, "The DSN Tracking System," TDA Progress Report PR 42-63, March-April 1981 (15 June 1981), pp. 8–14. J. A. Wackley, "The Deep Space Network Tracking System, Mark IVA, 1986," TDA Progress Report PR 42-85, January-March 1986 (15 May 1986), pp. 139–46.

CHAPTER 5

THE GALILEO ERA: 1986–1996

THE GALILEO ERA
A Defining Moment

Unlike the transition from the Viking Era to the Voyager Era, the transition to the Galileo Era was clearly defined by the aftermath of a tragic event of great significance to NASA in general and to the Deep Space Network (DSN) in particular. On 28 January 1986, 73 seconds after a perfect lift-off from the Cape Kennedy launch pad, the Space Shuttle *Challenger* exploded in full view of a worldwide television audience. The investigation that followed would forever change the course of DSN support for the two major deep space missions of that time, Ulysses and Galileo.

At the time of this disaster, both the *Galileo* and *Ulysses* spacecraft were at Cape Kennedy being prepared for separate launches in May 1986, from the *Challenger*. After deployment in Earth orbit by the Shuttle, both spacecraft would be injected into their individual mission orbits towards Jupiter by NASA's high-energy liquid hydrogen-fuelled Centaur booster rockets.

Both *Ulysses* and *Galileo* had been carried by the DSN as future missions for several years, and DSN plans to support the May launches were well advanced. After reaching Jupiter, Ulysses was to swing up and out of the plane of the ecliptic to enter a solar orbit passing over the poles of the Sun. En route to Jupiter, *Galileo* would release an atmospheric Probe targeted to enter the Jupiter atmosphere at the same time as the spacecraft arrived. On arrival, *Galileo* would be inserted into an orbit around Jupiter from which it would first return the telemetry data from the atmospheric Probe before making imaging and other science observations from numerous close encounters with the Jovian satellites.

Bruce Murray[1] describes what happened next: "A presidential commission to investigate *Challenger*'s demise was immediately formed. Chairman William Rogers soon concluded that 'the decision to launch was flawed.' By then, a sobered NASA had already acknowledged that Galileo could not be launched in May 1986. Galileo and Ulysses would have to slip at least thirteen months more, to June 1987, when Earth and Jupiter next would be in favorable locations" for a Centaur-based planetary launch. Meanwhile, based mainly on crew safety considerations, objections to the use of Centaur as a Shuttle-borne upper stage were increasing. In June 1986, the NASA Administrator canceled the Centaur as a Shuttle payload. Without the Centaur, the 1987 missions were not possible. The only option for an upper stage injection vehicle left to *Galileo* and *Ulysses* was the much lower energy solid motor booster developed by the Air Force, called the Inertial Upper Stage (IUS).

The Galileo Era: 1986–1996

Ultimately, both *Galileo* and *Ulysses* were rescheduled for launching by the Shuttle in October 1989, using the solid motor IUS as an upper stage injection vehicle. Except for dates, the Ulysses mission was not altered by this change; however, the Galileo mission was significantly changed in ways that materially affected the DSN. It became a new mission. It would be referred to as the Venus-Earth-Earth-Gravity-Assist (VEEGA) Galileo mission, and it was planned for launch in October 1989, with Jupiter arrival in December 1995. In April 1991, just over a year into the mission, an intractable problem with the spacecraft high-gain antenna led to a complete redesign of the mission once again. The timely DSN response to this potentially disastrous situation was a major factor in the ultimate recovery of the Galileo mission objectives.

Mission Set

These momentous events would cause the Galileo project to follow a path that would affect DSN engineering, implementation, and operations in a major way for the next ten years. For historical purposes, therefore, it is convenient to designate the period 1986–1996 as the Galileo Era. The DSN mission set for the Galileo Era is shown in Figure 5-1.

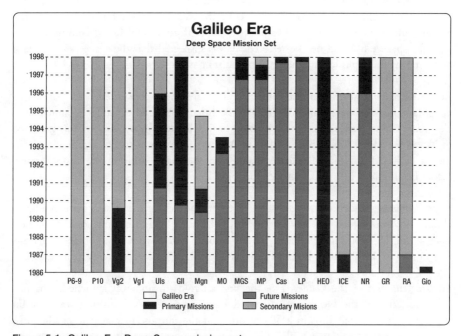

Figure 5-1. Galileo Era Deep Space mission set.

The following paragraph gives an overview of the Galileo Era mission set and missions that it included.

OVERVIEW

Following successful flybys of Venus in 1985, the *Vega 1* and *Vega 2* spacecraft continued on to close encounters with Comet Halley in March 1986. Renamed Pathfinder, they relied on DSN tracking support to refine the orbit of Comet Halley for use by Giotto and several other comet-bound missions in 1986. DSN support for all the foreign missions to Comet Halley in 1986 was provided on the basis of cooperative or reimbursable agreements.

Using radar techniques to penetrate the dense Venusian atmosphere, the Magellan mission to Venus in 1989 observed and mapped the surface features of the entire planet. At the end of its long extended mission, it gathered information on aerobraking techniques which would be of great value in the design of the future Mars Global Surveyor mission.

Mars Observer was launched in September 1992 and reached the vicinity of Mars in August 1993. Three days before it was to be inserted into orbit for a Mars mapping mission, the downlink was lost. Despite persistent efforts by the DSN, communication with the spacecraft could not be restored. It was suspected that the spacecraft had been destroyed by an explosion in the retro-propulsion system.

Reflecting a renewal of scientific interest and a substantial constituency for Mars, two new Mars missions were launched in 1996. Both launches used expendable launch vehicles. *Mars Global Surveyor*, a similar mission to the ill-fated *Mars Observer*, was launched in November 1996. On arrival at Mars in August 1997, it was to use aerobraking techniques derived from the Magellan data to trim the initial orbit to the correct shape, before beginning a full global surface mapping mission.

Mars Pathfinder, launched in December 1996 was the first of the "faster, better, cheaper" class of missions, which was to characterize the Cassini Era and the missions of the new millennium. Using a unique inflated "bouncing ball" system, its landing on the Mars surface in August 1997 was to be the first since the Viking Landers', twenty years earlier.

Towards the end of the Galileo Era, the first asteroid mission was launched. The primary objective of the Near Earth Asteroid Rendezvous (NEAR) mission was to orbit the near-Earth asteroid Eros in a 50-km orbit for the purpose of determining the gross physical properties, mineralogical composition, and morphology of the asteroid surface.

The Galileo Era: 1986–1996

Launched in February 1996, it would arrive at Eros in January 1999 after an Earth swingby in January 1998.

Throughout the Galileo Era, the DSN continued to support *Voyagers 1* and *2*, *International Cometary Explorer* (ICE), and *Pioneer 10* as extended missions.

By 1990, the number of Earth-orbiting missions for which the DSN was given responsibility had increased to the point where they were managed in the DSN scheduling system as a unique class, termed high-Earth orbiter (HEO) missions. Mostly, these were inflight missions which were more suited to the ground-based network of the DSN than to the space-based network of the Tracking and Data Relay Satellite System (TDRSS). At about the same time, the number of non-NASA spacecraft for which the DSN provided tracking and data acquisition services on a reimbursable basis to NASA began to increase significantly. These also were defined uniquely as "reimbursables" and were managed within the DSN, independent of the deep space missions. From this point on, DSN resources were assigned according to predetermined priorities to the deep space, HEO, and reimbursable missions.

We turn now to examine in more detail the remarkable events that took place in the Galileo Era and the DSN involvement in all of them. The DSN transition to the Galileo Era began with its continuing support for the foreign missions to Comet Halley.

NETWORK OPERATIONS

Deep Space Prime Missions

The Comet Halley Missions

In "Journey into Space," former Jet Propulsion Lab Director Bruce Murray describes his intense efforts to interest NASA in a U.S. mission to Comet Halley in late 1985 and 1986. It would, he pointed out, be seventy-six years, the year 2061, before another opportunity arose. Murray started lobbying early, in 1980, right after President Reagan had taken office and media interest in JPL was still sparked by the stunning visual products of Voyager's encounter with Saturn.

"Where is the United States headed in space?" Murray was asked.

"Halley," Murray replied. What Murray had in mind was a Halley Intercept Mission (HIM), a NASA spacecraft that would intercept the comet's path and photograph and make scientific observations of the structure and composition of the Halley nucleus. Murray pushed hard for the support of Reagan's new NASA administrator James Beggs, presenting HIM as a "unique, affordable opportunity for NASA and the United States." At NASA Headquarters, the arguments went back and forth for months. Central to the debates were the costs and scientific merit in light of other foreign missions already planned, and the rising demands of the Shuttle program for limited NASA funds. It was all to no avail. In September 1981, NASA Headquarters officially directed JPL to cease all further activity associated with the HIM. JPL would have to settle for the United States mission then flying near Halley, ICE, or those observing from afar, Pioneers, and for DSN support of the foreign missions.

In discussing the DSN role in the Comet Halley missions in 1986, Charles Stelzried, Leonard Effron, and Jordan Ellis wrote, "In man's continual quest to explore and understand the nature and origin of the solar system, 1985 through 1986 may well be noted as the 'Year of the Comet.' The first in-situ measurements of a comet occurred on 11 September 1985, with the passage of the International Cometary Explorer (ICE) through the tail of Comet Giacobini-Zinner, 7800 kilometers down-stream of the nucleus. This unprecedented encounter was followed six months later by the spectacular rendezvous of a fleet of five spacecraft in the vicinity of Comet Halley during the four-day long period March 9-13, 1986. The Halley armada included Japan's *Sakigake* (MS-T5) and *Suisei* (Planet-A), the Soviet Union's *Vega-1* and *Vega-2*, and ESA's *Giotto*. Twelve days later, ICE also passed within Halley's sphere of influence. In addition to the aforemen-

The Galileo Era: 1986–1996

tioned six spacecraft, two other interplanetary explorers turned their instruments in the direction of Comet Halley in 1986. The long-lived *Pioneer 7* and the Venus-orbiting *Pioneer 12* contributed to the library of space-based measurements made during the Halley exploration period . . . . It was also an unprecedented period in the arena of international cooperation between national space agencies and the worldwide scientific community."[2]

Together with two other national space agencies, the European Space Agency (ESA) and the National Space Development Agency (NASDA) of Japan, NASA (through the DSN) was a major contributor to the Comet Halley mission initiatives. The DSN provided telemetry, command, and navigation support for *Giotto* in the near-Earth phase and for the comet encounter. The DSN also provided telemetry, command, and navigation support for the near-Earth phase of the Japanese missions, as well as for the short period around the actual comet encounter. Following the completion of their Venus Pathfinder missions described in the previous chapter, the Soviet *Vega-1* and *Vega-2* spacecraft, in order to improve their encounter orbit determination, made use of the Very Long Baseline Inteferometry (VLBI) techniques that the DSN had used for the Giotto mission. The Comet Halley observations made by ICE, *Pioneer 7*, and *Pioneer 12* were carried out as part of the continuing DSN support for these missions.

Giotto

DSN support for the Giotto mission to Comet Halley in 1986 was provided on the basis of a reimbursable agreement between the European Space Agency (ESA) and NASA. The *Giotto* spacecraft was a spin-stabilized vehicle that used S-band and X-band downlinks and an S-band uplink to communicate with the DSN antennas. The DSN committed to providing telemetry, command, and navigation support for the near-Earth phase and for the Comet Encounter. The uplinks and downlinks were fully compatible with DSN telecommunication standards, and there were no special or additional implementation requirements imposed on the DSN by the Giotto mission. It was recognized that the late cruise and encounter periods in February and March 1986 would coincide with the Uranus Encounter period of *Voyager 2* and that schedule conflicts for DSN antenna time were to be expected.

Giotto was launched aboard an Ariane-1 launch vehicle from Kourou, French Guiana, on 2 July 1985, and it was injected into a heliocentric orbit by the spacecraft's onboard propulsion system the following day. DSN support began with DSS 46 at Canberra, Australia, as the initial acquisition station. For the next seven days, continuous support was provided by the 26-m subnet and the telemetry and radio metric data was routed to the European Space Operation Center (ESOC) at Darmstadt, Germany, via JPL and the NASA Communications (NASCOM) switching center in Madrid, Spain. After that, DSN support was reduced to one pass per day on a 34-m antenna, DSS 12 or DSS 42. Over a two-week period in mid-September 1985, the DSN participated with ESA in the first of several Navigation Campaigns to validate the radio metric data collected by the ESA tracking stations. Radio metric data (two-way Doppler and ranging) from the station pairs DSS 12/42 and DSS 61/42 was provided in the form of Orbit Data Files for comparison with the ESA data taken during the same passes.

Preparations for encounter with Halley began in October with an extensive series of rehearsals using *Giotto* as a data source and the actual sequences that would be used during encounter. As the test progressed, anomalies were identified and corrected, and the procedures and techniques were refined. Some changes were made to the data flow paths for delivering the Madrid and Canberra data to ESOC. All *Giotto* data from these sites would be routed to NASCOM at the Goddard Space Flight Center in Greenbelt, Maryland, rather than to JPL. NASCOM would transmit the data directly to ESOC via satellite link. Because this configuration reduced the transmission path by 6,000 miles, it was expected to make a significant reduction in the number of transmission errors at the high encounter data rate of 46 kbps. The rehearsals and Navigation Campaign passes continued right up to the start of the encounter.

The DSN responded to a *Giotto* spacecraft emergency in late January by making DSS 12 available for support. However, the downlink from the distressed spacecraft was too weak for reliable telemetry data acquisition, and the project requested support from the Goldstone 64-m antenna. DSS 14 was released from *Voyager 2* support of Uranus encounter by negotiation between the Voyager and Giotto Project managers. Although this was its first support of Giotto, the station was able to acquire the downlink signal and flow the telemetry engineering data to ESOC within 37 minutes for analysis. After it was determined that the spacecraft antenna was mispointed, the ESA station in Carnarvon, Australia, was used to transmit a corrective command when the spacecraft came into view at that longitude.

Three days of continuous support by the 64-m subnetwork to cover Halley's Comet encounter events started on 11 March 1986 and was accomplished without incident. Two seconds before the time of closest approach, the downlink signal was lost due to mispointing of the spacecraft high-gain antenna. This was attributed to nutation of the spacecraft spin axis induced by comet dust particles impacting the spacecraft. Although nutation dampers on the spacecraft slowly compensated for this effect, 34 minutes of close-encounter, real-time, telemetry data was lost. Some of this was later recovered from the digital recordings made at the tracking stations.

The Galileo Era: 1986–1996

Nevertheless, the project reported that the overall encounter results were spectacular and that pictures had been obtained to within about 1,500 kilometers of the nucleus before the downlink was lost.[3]

Analysis by ESOC, using radio metric data from the DSN and the ESA Deep Space Tracking System (DSTS), augmented with the *Vega-1* and *Vega-2* Halley observations (described in the previous chapter) indicated a miss distance of 610 kilometers with an uncertainty of 40 kilometers.

An independent estimate of the spacecraft arrival time and position at Halley was derived at JPL from DSN radio metric data obtained during the final Navigation Campaign (just prior to encounter). A predicted comet miss distance of 610 kilometers was obtained with an uncertainty of 138 kilometers. The times of closest approach agreed to within two seconds.

It was suggested that differences in the comet ephemerides and spacecraft attitude information used by DSN and ESA could easily account for the small disagreement in the miss distance solutions.[4]

Phobos

After a hiatus of some ten years following the Mariner and Viking missions to Mars, scientific interest in exploration of that planet resumed in 1988 with the Soviet Union's Phobos 88 missions. The USSR Space Research Institute (abbreviated IKI in Russian) in Moscow was the mission planning organization. More than a dozen countries were involved in putting together the scientific payload. It would be the first international deep space initiative of any scope since the successful Halley flyby missions, Vega and Giotto, in March 1986.

The mission was to be devoted to studying the planet Mars and its environment, particularly its natural satellite Phobos, which was thought to be an asteroid captured by Mars. The study of Phobos was the prime scientific objective of the mission and was to be conducted by means of two Landers on the Phobos surface. One of them, a long-term automated Lander (abbreviated DAS in Russian), would carry celestial mechanics experiments, spectrometers, seismometers, and cameras. The other lander, called a "Hopping Lander" or "Frog," was fitted with a jumping mechanism to allow it to move from place to place, analyzing the chemical composition and physical and mechanical properties of the surface at each location.

The Landers were to be deployed to the surface of Phobos from a "bus" spacecraft whose orbit had been carefully adjusted to be synchronous with that of Phobos. The bus would remain in synchronous orbit for about a month, making topographical studies of Phobos before initiating the approach and Lander descent sequences. The bus was to descend to within 50 meters of the surface before releasing the Landers to make individual, controlled descents to the surface. Once on the surface, the Landers would begin their automated science missions, the DAS from a single fixed location, the Frog from various locations determined by its 20-meter jumps. The bus would return to its circular orbit, from which it would observe Mars, the Sun, and the interplanetary environment for a few months. The total duration of the mission was to be fifteen months.

Communications with the spacecraft were planned for every five days during cruise and every three days in the orbital phase. In the interval, data were to be recorded on board the spacecraft. The link with the DAS was to be established three times per day, once during each orbit of Phobos, and last 20 to 40 minutes. The Frog would transmit its data directly to the "bus" for recording and later retransmission to Earth.

The main receiving station was at Eupatoria in the Soviet Union. From there, data was transmitted to IKI in Moscow for processing and distribution to the experimenters via the French Space Agency, Centre National d'Etudes Spatiales (CNES), in Toulouse, France.

DSN involvement in Phobos followed naturally from its involvement with the Vega and Halley missions two years previously. Of particular interest to the DSN was the opportunity that the Phobos DAS Lander provided for refinement of the ephemerides of Mars and Phobos, and for improving the accuracy of the existing radio-to-planetary frame-tie measurements.[5]

Ranging measurements between Earth and the DAS Lander were to be used to determine very precisely the orbit of Phobos around Mars. They would also improve current knowledge of the orbit of Mars, which had been most accurately determined prior to this by the ranging measurements of the two Viking Landers from 1976 to 1978. Together, these measurements should allow the Mars orbit to be determined to an accuracy of about 5 nanoradians, relative to the orbit of the Earth around the Sun. This was much better than the then-current level of knowledge of the "planetary reference frame."

In addition to the Mars ephemeris improvement, the Phobos mission would also provide a rare opportunity for delta VLBI measurements, which could be used to tie together the "planetary" and "radio" reference frames used for interplanetary navigation.

The Galileo Era: 1986–1996

Each Lander had a "VLBI broadcast mode," consisting of two coherent tones spaced 14.71425 MHz apart on the downlink frequency of 1.7 GHz (C-band). In this mode, delta-VLBI measurements between a Lander and a nearby compact radio source in the sky would locate the Lander, and therefore Phobos, in the "radio" frame. The ranging measurements described above would be used to tie the positions of Phobos and Mars together, so that the delta-VLBI data effectively placed Mars in the "radio frame," thereby completing the "planetary to radio" frame tie. It was on the strength of this argument, and its importance to future JPL navigation technology, that the DSN went forward with plans to support the Phobos missions.

Apart from the problem of finding suitable compact radio sources for the delta-VLBI observations, no significant additional effort was required to meet the Phobos requirements for celestial mechanics. However, in two other areas, a very considerable effort was required.

The Soviet-designed Phobos Landers were complex spacecraft capable of receiving commands, transmitting telemetry, providing two-way range and Doppler and supporting delta VLBI measurements. Spacecraft-to-Earth communications were carried by a 1.7-GHz (L-band) uplink and a 5.01-GHz (C-band) downlink.

During each Phobos-Earth view period, the Lander telemetry data was transmitted in 2,048-bit frames encoded in one of three different modes. One of these modes (convolutional code, $K = 6$, $r = 1/2$) could be decoded using the standard DSN MCD capability. To accommodate the other two modes, special software was developed at Madrid as a logical follow-on to the Venus-Balloon experience. This was then integrated into a Phobos telemetry data processor.[6]

Since neither the uplink nor the downlink frequencies were compatible with the DSN standard operating frequency bands, it was necessary to design and implement a special dual-frequency feed to handle simultaneous two-way communications with the Landers. The dual-frequency feed was composed of an L-band feedhorn enclosing a coaxially mounted C-band surface-wave antenna.[7] This feed geometry provided greater than 38 dB of isolation between the C-band uplink and the L-band downlink, and it could safely handle 15 kW of uplink C-band transmitter power. The concept for this design was derived from the L-band feedhorn that had been implemented at all 64-m antennas for the earlier Vega and Halley missions.[8] The way in which the horn was positioned in the antenna to correctly illuminate the subreflector is shown in Figure 5-2.

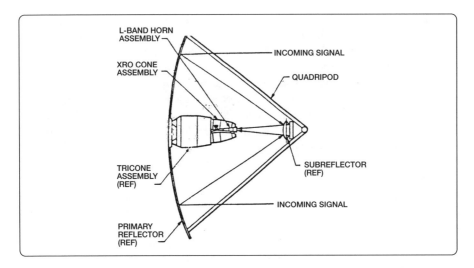

Figure 5-2. Phobos L-band feedhorn mounted on the 64-m antenna.

Following the pattern of previous Soviet space projects, the Phobos mission comprised two nearly identical spacecraft. Each spacecraft was launched on a Proton launch vehicle in July 1988 from Baikonour in the Soviet Union. *Phobos-1* was disabled by a ground control error before reaching Mars orbit. *Phobos-2* completed the Earth-Mars transfer phase and was successfully inserted into an elliptical Mars orbit in January 1989.

Using Soviet and French onboard imaging data supplemented with two-way Doppler and VLBI data provided by the DSN, the complex process of changing the elliptical spacecraft orbit to be circular and "synchronous" with the orbit of Phobos began. On 21 March, the spacecraft entered a nearly circular orbit with a period of 7.66 hours, which kept it close (about 275 km) to Phobos. On 27 March, without prior warning, all communications with the *Phobos-2* spacecraft were lost. Extended efforts to reestablish contact with the spacecraft were unsuccessful.

Despite the premature termination of the mission, all was not lost. A useful body of Phobos frame-tie data had been obtained before the spacecraft was lost. The data consisted of two passes of VLBI observations of the *Phobos-2* spacecraft and angularly nearby quasars. The data was recorded with the DSN wide-channel bandwidth VLBI systems during an hour of observing passes on Goldstone-Madrid and Goldstone-Canberra baselines on 17 February and 25 March. Sparse though this data was, its subsequent reduction and analysis yielded satisfying results for the "frame-tie" experiment. Prior to 1989, the planetary and radio source reference frames had been aligned with an accuracy of approximately 250 nanoradians. The analysis of the Phobos data reduced this uncertainty to 20 to 40 nanoradians, depending on direction.[9]

Background for Magellan

By the time the Galileo Era began, Venus was one of the most visited planets in the solar system. Since the early 1960s, no less than twenty spacecraft from the United States and the Soviet Union had carried out missions to Venus. Despite all that attention, there existed only a sketchy, general knowledge of the planet's surface because the dense, constant cloud cover precluded surface observations by conventional optical means. In the late 1970s and early 1980s, the Pioneer Venus (U.S.) and Venera (USSR) spacecraft had used radar to image its surface. Earth-based radar, too, had been used to penetrate the clouds and provided valuable surface data, but this coverage was limited by the fact that Venus always presented the same hemisphere to Earth view. Although these endeavors had provided answers to many intriguing scientific questions and disclosed most of Venus's large-scale geography, they raised many more questions that remained unanswered.

To pursue these questions further, NASA planned to launch a new radar imaging spacecraft called Venus Orbiting Imaging Radar (VOIR) from the Space Shuttle in 1984. The program was subsequently modified to defer the launch to 1988, and the mission name became Venus Radar Mapper (VRM). The *Challenger* accident in 1986 forced a further deferment of the launch to April 1989. The new Mission was called Magellan.[10]

The plan was for Magellan to arrive at Venus in August 1990, spend eight months in orbit about the planet, and map most of surface at a resolution nearly ten times better than the resolution of existing surface maps. On arrival, the spacecraft's rocket propulsion system would be fired to inject the spacecraft into an elliptical orbit around Venus.

After establishing the correct orbital conditions and checkout of the radar systems, the science acquisition phase of the mission would begin. This phase would last 243 Earth days or one Venus day, the time required for Venus to make one complete rotation under the elliptical orbit of the spacecraft. During the part of each orbit when the spacecraft was near the planet surface, the spacecraft would be oriented so that the high-gain antenna pointed slightly to one side of the ground track, and the radar was turned on. The surface return data would be recorded on the spacecraft tape recorders until the radar was turned off at the end of the mapping sequence about 30 minutes later. Altimetry and radiometer data were also to be taken during this period.

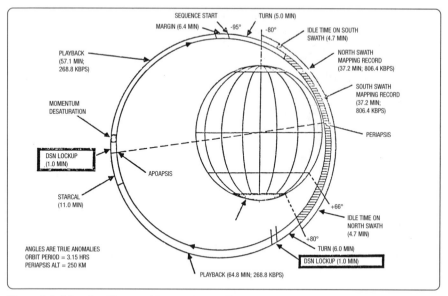

Figure 5-3. Magellan Venus orbit mapping profile.

As the spacecraft moved further away from the surface toward apoapsis, it would be reoriented to point the HGA toward Earth and the stored radar and spacecraft engineering data would be transmitted to the DSN. Playback, at a downlink data rate of 268.8 kilobits per second, would take place in two periods of about one hour each, with an interruption of 14 minutes between them for a gyro star calibration. Data-taking on the succeeding revolution would start about 10 minutes after completion of playback.

These data-taking and playback cycles were to be repeated on each orbit, at regular intervals of three hours and nine minutes. Figure 5-3 shows these extremely busy orbital activities and highlights the small margin of time between them.

On each successive orbit, the radar would scan a new swath of surface as the planet rotated under the spacecraft orbit. Based on a very precise knowledge of the position of the spacecraft determined from two-way and three-way Doppler data, the Synthetic Aperture Radar (SAR), altimetry, and radiometry data would be processed into images and maps for scientific study.

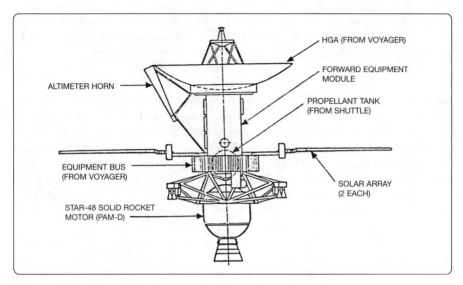

Figure 5-4. Diagram of *Magellan* spacecraft in cruise configuration.

Magellan Mission to Venus

The *Magellan* spacecraft was powered by two Sun-tracking solar panels and stabilized in three axes using gyros and a star sensor for attitude reference. The spacecraft carried a solid rocket motor for Venus orbit insertion, as well as a small hydrazine system for trajectory correction maneuvers and certain attitude control functions. Low-gain and medium-gain antennas and a 3.7-m high-gain antenna (HGA) rigidly attached to the spacecraft provided S-band and X-band uplinks and downlinks with the DSN. The HGA also functioned as the SAR mapping and radiometer antenna during orbital operations. A separate horn antenna, for the radar altimeter, was fixed to the side of the main body structure. A diagram of the *Magellan* spacecraft is shown in Figure 5-4.

The capabilities of the DSN Mark IVA systems in 1989 were adequate to meet the tracking and data acquisition requirements of the Magellan mission. However, the requirements for 70-m antenna time could not be fully met due to conflicting requirements from other high-priority missions such as *Galileo*, *Ulysses*, and *Voyager 2*. In such cases where the 70-m antenna conflict could not be resolved through scheduling, *Magellan* was downloaded to an arrayed configuration of two 34-m antennas at the same location. Further economy in antenna usage was effected by making a small modification to one of the two receivers at all 70-m and 34-m standard antennas, which allowed both of the *Magellan* subcarriers to be received through a single X-band receiver channel. Since *Pioneer 12* was also in orbit about Venus, the second receiver channel at the same antenna could then be used to pass the S-band *Pioneer 12* signals.

The downlink coverage of important events associated with the launch and initial acquisition sequence were the subject of an immense amount of planning and coordination by the DSN in the year or two preceding launch.[10] DSN responsibility, in this so-called "near-Earth" phase of the mission, was profoundly changed by the NASA decision to use the Space Shuttle rather than the former expendable launch vehicles for all future planetary launches. The original "near-Earth" planning for Shuttle launches was related to *Galileo*, since at that time, *Galileo* was to be the first of the planetary spacecraft to use the Space Shuttle as a launch vehicle.[11] As it turned out, *Magellan* became the first planetary spacecraft to use a Shuttle launch, and the successful tracking coverage, data flow, and generation of predicted data products during the "near-Earth" phase of the first Shuttle-based launch would validate the thoroughness of DSN planning. The interagency agreements, interfaces, and communication configurations developed for *Magellan* were used on all subsequent planetary launches, beginning a few months later with *Galileo*.

The Magellan mission to Venus began on 4 May 1989, eleven years after the launch of *Pioneer Venus*, the last planetary spacecraft to use an expendable launch vehicle. *Magellan* thus became the first planetary spacecraft to be launched from a reusable space vehicle, and it was for JPL the first planetary mission since *Voyager 1* in September 1977.

After deployment from the Space Shuttle *Atlantis* cargo bay, the IUS/Spacecraft combination coasted to the correct point for injection. At that point, the two-stage IUS injected the spacecraft into an Earth-Venus transfer orbit with an arrival date of 10 August 1990. The spacecraft then separated from the spent IUS, and initial acquisition of the *Magellan* downlink was accomplished without incident by the first viewing, using DSN 26-m and 34-m antennas at Canberra.

Mapping began in September 1989 and continued for three cycles of 243 days each. During this period, *Magellan* mapped over 98 percent of the Venus surface. The startling detail produced by the *Magellan* SAR data is obvious in the image of an impact crater in the Atalanta Region of Venus, reproduced in Figure 5-5.

In January 1992, partway into Magellan's third cycle, one of two telemetry modulators on the X-band downlink from spacecraft transmitter A failed, leaving the spacecraft unable to transmit the high-rate mapping data. The S-band downlink and telemetry performance were

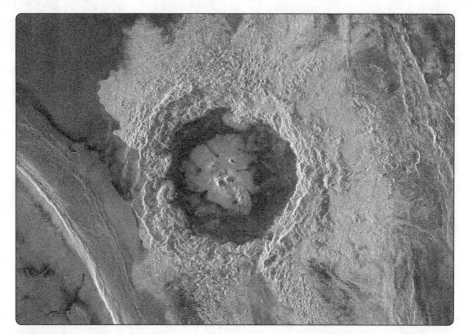

Figure 5-5. *Magellan* SAR image of an impact crater in the Atalanta Region of Venus.

not affected. In an effort to continue the mapping mission, the project switched back to transmitter B. However, transmitter B, which had been used earlier in the mission, had a history of generating spurious signals and noise which severely degraded its normal use for transmitting the mapping data. Various schemes to continue the transmission of the mapping data at lower data rates were tried with varying degrees of success. Finally, mapping was ended on 14 September 1992, and the radar was turned off on the fourth cycle. Transmitter B continued to provide low-rate engineering data on the X-band downlink, and the S- and X-band Doppler data were used to obtain a high-resolution global gravity field map.

Between cycles 4 and 5, the existing elliptical orbit necessary for radar mapping was modified to near circular for gravity studies by aerobraking and propulsive maneuvers. This was the first time an interplanetary spacecraft had used a planet's atmosphere to significantly change a spacecraft orbit. The then-new technique was used very effectively until almost ten years later, when a new generation of lightweight spacecraft began to arrive once more at Mars. Repeated aerobraking maneuvers rapidly reduced the orbital period from over 3 hours to 94 minutes, and its apoapsis altitude from 8,469 kilometers to 542 kilometers. With the near-circular orbit, *Magellan* could collect high-resolution gravity data over the poles, an area for which there was no previous data. A gravity field map covering 95 percent of the globe resulted from the fifth and sixth cycles. In addition to the gravity field measurements, during several occultations that occurred during this period, *Magellan* performed bistatic radar experiments to measure the specular reflection of the radio signal from the surface, and radio science experiments to determine the characteristics of the atmosphere, ionosphere, and interplanetary medium. The DSN was an integral part of all of these activities.

Due to funding limitations that precluded continuing the mission further, *Magellan* was intentionally destroyed by an aerobraking maneuver that caused the spacecraft to enter the atmosphere on orbit number 15,032, near the end of the sixth cycle. All subsystems remained functional to the end, albeit with some loss of redundancy. The last confirmed downlink contact with *Magellan* was made by DSS 65, Madrid, at 10:04:35 UTC on 12 October 1994. This was almost five and a half years after the first downlink contact with the DSN following its launch in May 1989.

In retrospect, the Magellan mission was an enormous success, returning more data than all other planetary missions combined. Despite difficulties with the X-band downlinks toward the end of the mission, the mapping data return from the mission far exceeded the primary mission objectives. Over 98 percent of the planet was imaged at resolutions as small as 75 meters, and a high-resolution gravity map covering more than 95 percent of the planet surface was generated.[12]

Voyager at Neptune

The year 1989 was a most notable year for space exploration in general and the DSN in particular. The notable events of 1989 began in January with the arrival of the Soviet spacecraft *Phobos-2* at Mars. In February, the third and final satellite in the new Tracking and Data Relay Satellite System (TDRSS) was placed in geosynchronous Earth orbit. The Magellan launch in late April marked the first new planetary spacecraft launch in the 12-year period since the 1 launch in 1977. It was also the first planetary spacecraft to be launched on the Space Shuttle. In June, the Voyager Neptune Encounter period began. The spacecraft made its closest approach to the Neptune surface in August, and the prime mission continued through September. The second Space Shuttle launch of a planetary spacecraft took place in October with the *Galileo* spacecraft bound for Jupiter. The year ended with the long-awaited launch of the Hubble Space Telescope in December. The DSN was involved to a greater or lesser extent in all of these events. Here, we discuss the Voyager Neptune Encounter, in which the DSN was involved in a major way.

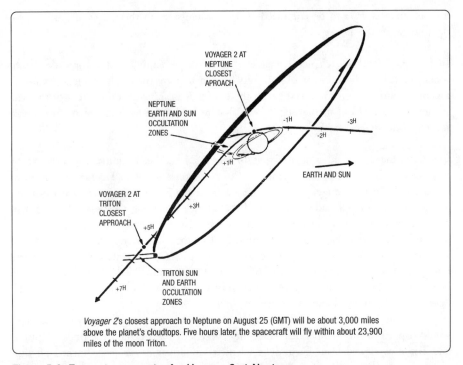

Figure 5-6. Encounter geometry for *Voyager 2* at Neptune.

The Neptune Encounter would be *Voyager 2*'s closest encounter with any of the planetary bodies in its twelve-year journey through the outer solar system. The close flyby had been designed to maximize science observations of the planet's magnetic field, charged particles, and deep atmosphere, as well as to bend the flight path close to Triton, while limiting the risk of radiation damage to the spacecraft. Figure 5-6 shows the Neptune Encounter geometry.

While the previous Uranus Encounter took place at a spacecraft-to-Earth range of about 20 AU, the distance to Earth would be about 30 AU for the Neptune Encounter. Although the spacecraft would be 1.6 billion kilometers (1 billion miles) further from the antennas of the DSN, it was necessary to keep the downlink telemetry rates about the same as were used at Uranus in order to return the vast amount of science data that was expected from this very unique event. How was this to be accomplished?

Two techniques were to be employed—data compression on the spacecraft, and antenna arraying in the DSN.

In the spacecraft, the telemetry data was modified by editing and compression techniques to reduce the number of data bits needed to transmit a full-size image without loss of resolution. This improvement in the efficiency of information content of the data stream was carried out in the Flight Data System before the telemetry signal was transmitted from the spacecraft.

On Earth, the signal-gathering sensitivity of the DSN had been significantly increased by the enlargement of the 64-m antennas to 70-m antennas in 1987. Further improvement could be obtained by arraying 34-m antennas with the 70-m antenna at each Complex. The Uranus encounter had demonstrated the efficacy of adding the 64-m Parkes antenna to the CDSCC array in order to enhance the DSN capability even further at that longitude. The Neptune Encounter required all of that and more. Plans were made to supplement the 70-m/34-m Goldstone array with the 27 (25-m) antennas of the National Radio Astronomy Observatory's (NRAO) Very Large Array (VLA) facility in Socorro, New Mexico. In addition, the 70-m antenna of the Institute for Space and Astronautical Science (ISAS) at the Usuda Observatory in Japan was to collect radio science data in sync with the DSN stations and the Parkes Radio Astronomy antenna in Australia. This would happen during the periods when the spacecraft passed behind the rings, Neptune, and Triton as viewed from Earth. The telemetry modulation was to be turned off during these complex occultation sequences to provide maximum power in the S-band and X-band downlinks to enhance the quality of the radio science data.

By August 1988, the Goldstone-VLA link had been installed and progressively checked at the maximum general science and engineering data rates of 14.4 kbps and 21.6 kbps that would be required during Encounter operations. System performance tests with the new in-flight software verified the design parameters, and gave confidence that the correct downlink signal-to-noise ratios, error rates, and signal margins were being achieved. Much remained to be done before the DSN was fully ready for encounter, but a review of the Ground Data System in February of that year found that progress to that point was satisfactory.

At the beginning of 1989, *Voyager 2* was 4.36 billion km (2.71 billion miles) from Earth—299 million km (186 million miles) from Neptune. The Radio Science Team was completing an eight-week period of solar occultation observations, and both the DSN and Voyager Operations Teams were engaged in intensive testing and training exercises to ensure maximum readiness for the encounter events. Bright cloud features had begun to appear on the Neptune images and, with the exception of the well-understood receiver problems, the spacecraft was in good health.

In April, a large dark spot rotating round the planet in 17 to 18 hours was detected in the growing detail of the Neptune images by the Imaging Team.

The readiness of all elements of the DSN to support the encounter was reviewed again, in depth, on 1 June. The review included the Canberra, Madrid, and Goldstone Complexes, the NOCC, GCF/NASCOM, Very Large Array, Parkes Radio Telescope, and the Usuda Deep Space Station. It was determined that with minor exceptions, all the new capabilities were in place and had been demonstrated, including the new interagency capabilities and interfaces. The review also found that appropriate arrangements were in place for dealing with personnel or equipment contingencies and protecting the sites against radio frequency interference, and that the operations crews had demonstrated an acceptable level of operational competence. The DSN was declared ready.

As defined by the Mission Plan, the Neptune Encounter period began on 5 June 1989, 81 days before closest approach. The first 62 days were called the "observatory phase" and consisted of continuous observations of the Neptune system combined with numerous pre-encounter checkouts and calibrations of the science instruments. By the time the "far encounter" phase began on 6 August, the spacecraft had reached the point where the image of the planet and the rings required at least two narrow-angle camera frames to cover them. Detailed ring and satellite imaging began, and infrared observations of Neptune were made. As *Voyager 2* approached the planet, imaging scientists were able to track the features in the clouds to determine wind speeds.

The "Near Encounter" period, from 24 to 29 August, contained the highest value Neptune science data. It included a close swing over Neptune's north polar region, a flyby of the satellite Triton, a view of distant Nereid, numerous magnetic field measurements, and searches for other possible rings and satellites. Voyager passed 4,900 kilometers (3,000 miles) above the cloudtops of Neptune and, five hours later, passed within 40,000 kilometers (25,000 miles) of Neptune's largest moon, Triton.

The five-day "near encounter" period was covered by all three of the DSN complexes at various times. The complicated arrayed antenna configurations at VLA-Goldstone (VGTA) and Parkes-Canberra (PCTA) both operated perfectly to return the high-volume engineering and science data at the desired downlink data rates of 14.4 and 21.6 kilobits per second.

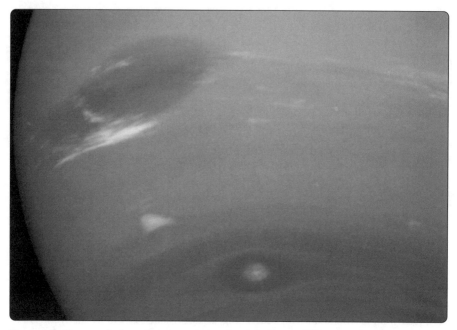

Figure 5-7. *Voyager 2* views three most prominent features of Neptune. Although Neptune is the smallest of our solar system's four gaseous planets, its volume is nearly 58 times that of Earth. Three of the most prominent features in Neptune's atmosphere are captured in this photograph reconstructed from two images taken by *Voyager 2*. At the north (top) is the Great Dark Spot. To the south of the Great Dark Spot is the bright feature "Scooter," so-named because it rotates around the globe more rapidly than the other features. Still further south is the feature called "Dark Spot 2," which has a bright core.

The remarkable images of Neptune and Triton reproduced in Figures 5-7, 5-8, and 5-9 are typical of the period. They were made possible by (among other things) the precision of the Voyager navigation and the high quality of the telemetry downlink. Both of these depended on the DSN for their realization.

Voyager 2 discovered six new moons and a number of rings at Neptune. It found an amazing assortment of icy terrains and a form of ice volcanism on the surface of Triton. Radio science occultation data from the DSN, Parkes, and Usuda antennas were used to characterize the atmospheres and ionospheres of Neptune and Triton and to investigate the structure of the rings system. Celestial mechanics determined the gravity field of the planet and the

The Galileo Era: 1986–1996

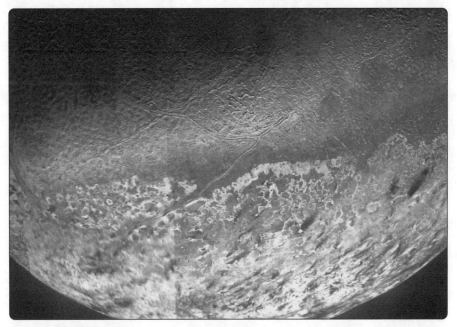

Figure 5-8. *Voyager 2* views south polar cap of Triton. Triton, Neptune's largest moon, is about two-thirds the size of Earth's Moon and orbits Neptune at about the same distance as the Moon orbits Earth. Nearly two dozen individual images were combined to produce this comprehensive view of the Neptune-facing hemisphere of Triton. The large south polar cap at the bottom of the image is highly reflective and slightly pinkish in color.

masses of its satellites. *National Geographic* described the Neptune Encounter events as "Voyager's Last Picture Show"[13] and the Voyager mission as the "Voyage of the Century."[14]

Once past Neptune, *Voyager 2* continued to observe the Neptunian system on a continuous basis for five more weeks until early October 1989. Observations of the dark side of Neptune and calibrations were carried out during the post-encounter period. Although this was *Voyager 2*'s last planetary encounter, the spacecraft continued to return data on magnetic fields and particles and to make other science observations in space as it joined *Voyager 1* and *Pioneers 10* and *11*, moving toward the edge of the solar system and beyond, into interstellar space. It would now be known as the Voyager Interstellar Mission and would remain a presence in the DSN "inventory" of deep space missions for many years to come.

Figure 5-9. *Voyager 2* views Neptune and Triton. This dramatic view of the crescents of Neptune and its largest moon Triton was acquired by *Voyager 2* about three days after its closest approach to Neptune. The spacecraft was then 4.86 million kilometers (3.01 million miles) from Neptune, and 5.22 million kilometers (3.24 million miles) from Triton. The smallest details discernible are approximately 90 kilometers (60 miles) in size.

The Neptune Encounter was over, and the DSN was changing its focus once again to the beginning of a new planetary mission to follow where *Voyager 2* had gone ten years earlier. It was the long-delayed Space Shuttle launch of the Galileo mission to Jupiter.

Background for Galileo

Carried forward by the dramatic successes of the Viking Mars Landers and the Pioneer and Voyager flybys of Jupiter in the mid-1970s, JPL proposed an ambitious mission to Jupiter for the 1980s. Known as Jupiter Orbiter Probe (JOP), the spacecraft combined a powerful orbiter with an atmospheric entry probe and would be launched directly to Jupiter by a Titan-Centaur expendable launch vehicle. However, late in 1975, NASA decreed that all new planetary missions would use the Space Shuttle, and at the time of its approval by Congress in 1977, JOP was to be the first mission to do so. In January

The Galileo Era: 1986–1996

1978, JOP became officially known as the Galileo Mission to Jupiter and was scheduled for launch in January 1982 on the thirtieth Shuttle flight. For a direct flight to Jupiter, the *Galileo* spacecraft would need the high-energy, liquid hydrogen-fuelled Centaur as an upper stage. The spacecraft, with attached Centaur, would be deployed into Earth orbit by the Shuttle. At the proper time, the Centaur would be fired to inject *Galileo* into a direct trajectory towards Jupiter. That was the plan, but it never materialized.

Over the next six years, the Galileo project became reluctantly involved in a series of NASA changes to the Shuttle launch schedule and performance, as well as changes to the availability of a suitable upper stage. Each of the changes, none of them due to the *Galileo* spacecraft itself, necessitated a change to the mission design, with consequent impact on the supporting services required, including the DSN. Despite the de-moralizing effects of so many changes on the technical design teams at JPL, the tenacity of Galileo Project Manager John Casani held the critical elements of the project together. In that period alone, the Galileo mission was redesigned three times before reaching the final version destined for a May 1986 launch. This version would use the Centaur upper stage to achieve a direct ascent trajectory to Jupiter, with arrival two years later in August 1988.[15]

That plan, too, never materialized. The *Galileo* spacecraft had been shipped from JPL to Cape Kennedy in December 1985, while preparations for its planned Shuttle-Centaur launch the following July were in progress. In January 1986, the Space Shuttle *Challenger* exploded shortly after launch and, in the aftermath of the ensuing enquiry, Centaur was cancelled as a Shuttle payload. For the fourth time, the Galileo mission had to be redesigned.

What emerged from this effort was to be called the Galileo Venus-Earth-Earth-Gravity-Assist (VEEGA) mission. VEEGA would be launched in October 1989 with a lower-risk, but also lower-energy upper stage called an Inertial Upper Stage (IUS). Since the energy available from the IUS was insufficient to take *Galileo* directly to Jupiter from Earth orbit, VEEGA would take advantage of the gravitational fields of Venus and Earth to make up the deficit. After leaving Earth orbit, *Galileo* would fall inward towards the Sun to intercept Venus in February 1990. The Venus flyby was designed to accelerate *Galileo* back toward a close Earth flyby in December 1990. A similar maneuver would use Earth's gravitational field to accelerate *Galileo* into a wide orbit round the Sun for a second and final gravitational assist from Earth two years later, in December 1992. Then, *Galileo* would finally be moved in the right direction and with sufficient energy to reach Jupiter in December 1995, six years after leaving Earth. The convoluted VEEGA trajectory is shown in Figure 5-10.

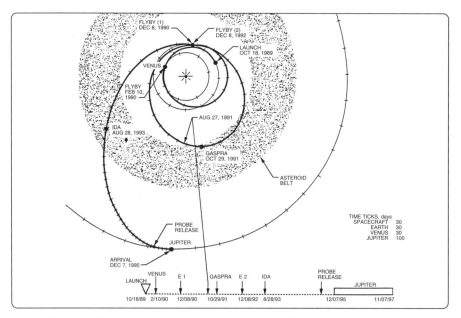

Figure 5-10. The Galileo Venus-Earth-Earth-Gravity-Assist (VEEGA) Trajectory.

The *Galileo* spacecraft was to consist of an Orbiter and an atmospheric Probe arranged as a composite unit as shown in Figure 5-11.

The Probe would enter the Jupiter atmosphere for the purpose of determining the composition and structure of the atmosphere as it descended through the various cloud layers. The Orbiter was to be inserted into Jupiter orbit for the purpose of making observations of the Jupiter system, including its satellites, during the 22-month prime mission.

Though there were grand plans for the original Galileo mission to Jupiter, the actual mission, due to unforeseen circumstances, was carried out in quite a different way, as we shall see. From the time of its launch in October 1989 until the planned HGA deployment in April 1991, the *Galileo* spacecraft relied on S-band (2.3 GHz) uplinks and downlinks for its telecommunications support by the DSN. These links were used for telemetry, command, navigation, and radio science. After the HGA deployment, the mission was to have used the X-band (8.4 GHz) capabilities of both the spacecraft and the DSN to provide the additional telecommunications link performance margin required to meet all of the mission objectives at maximum Earth to Jupiter range.

The Galileo Era: 1986–1996

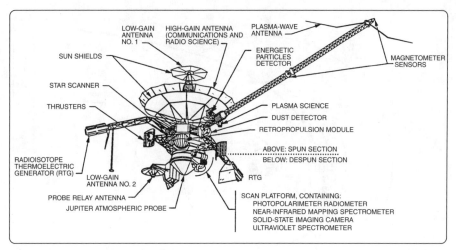

Figure 5-11. The *Galileo* spacecraft cruise configuration. In the composite spacecraft diagram, the high-gain antenna (HGA) is shown in its unfurled position. At launch, the antenna is stowed in a "furled" position, rather like a closed umbrella. At a suitable point in the mission, when gravitational forces and solar conditions are appropriate, a latching mechanism is released to allow the HGA to slowly unfurl into the parabolic shape required for it to perform as an antenna. This antenna is a key element in the spacecraft's X-band downlink communication channel to the DSN 70-m antennas on Earth. Without it, an X-band downlink is not possible. Without an X-band downlink, the High Rate Telemetry Data Channel, on which the entire mission depended for its prime science data return, is not possible.

However, when the antenna was unlatched in April 1991, the unthinkable happened. The antenna failed to unfurl properly and, despite intensive efforts over the following two years to free what appeared to be a stuck element of the unlatching mechanism, the HGA remained unusable. The X-band downlink was useless.

In February 1993, a decision was made to redesign the mission using the available low-gain antennas (LGAs) to support the existing S-band uplinks and downlinks. DSN and spacecraft engineers hastened to initiate design and implementation plans to provide the new capability that would be needed to conduct orbital operations in 1996 and 1997 on the S-band LGAs.

Five months before arriving at Jupiter, the Galileo Probe was released from the spacecraft to continue its trajectory directly to the point of entry into the Jupiter atmosphere. A few days after releasing the Probe, the Orbiter carried out a deflection maneuver

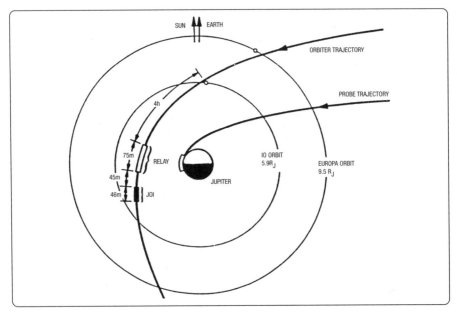

Figure 5-12. The Galileo Orbiter and Probe arrival at Jupiter, 7 December 1995.

to place it on a trajectory that would overfly the Probe during its entry and descent. When the Probe reached that point, it entered the Jovian atmosphere. After surviving the deceleration forces and deploying a parachute, it descended through the atmosphere, making scientific observations as it passed through the cloud layers. This data was transmitted on an L-band radio link to the Orbiter, which was in view of the Probe as it descended through the cloud layers. The Orbiter relayed a small sample of this data directly to Earth as it was received from the Probe during its descent. To ensure that the critical Probe data was preserved, the Orbiter also recorded the full set of Probe data for later retransmission to Earth. After the Probe relay sequence had been completed, the Orbiter used its retropropulsion engine to insert itself into permanent Jupiter orbit. A diagram of the complex arrival sequence is shown in Figure 5-12.

With the successful recovery of the full set of Probe data, the first set of mission objectives had been accomplished. The remainder of the Galileo mission was directed toward the accomplishment of the second set of mission objectives, namely, the study of the Jupiter system including its satellites.

The Galileo Era: 1986–1996

Jupiter orbital operations began in December 1995, immediately following Jupiter Orbit Injection (JOI). For approximately the next two years, the Orbiter would make scientific observations of the Jupiter system, including imaging from eleven close encounters with the Jupiter satellites Ganymede, Callisto, and Europa. The modifications and improvements that had been made to the S-band downlink as a consequence of the failed X-band antenna were fully validated. The new science data returned by the spacecraft exceeded the most optimistic expectations of the scientists.

The Galileo prime mission officially ended in December 1997, with the spacecraft in fully operable condition. By that time, a proposal to extend the mission for an additional two years to further study Europa had been approved. DSN capabilities required to support the extended mission were somewhat less than those required for the prime mission, and on this basis, the Galileo Europa mission went forward into the Cassini Era.

Galileo Mission to Jupiter

This was to be the second deep space mission to use the Space Shuttle as the launch vehicle (Magellan was the first). The Shuttle-Centaur launch in May 1986 would have placed Galileo on a direct ascent trajectory to Jupiter, with arrival two and a half years later in December 1988. The change from the former expendable type of launch vehicle to the Space Shuttle necessitated major changes to the procedures and data flow configurations used by the DSN to provide the project with telemetry and command links with the spacecraft during the launch and Earth-orbit period. This was largely due to the events associated with the Shuttle onorbit phase and the direct launch involvement of additional NASA Centers, including Johnson Space Center in Texas and Lewis Research Center in Ohio.[16]

The DSN planned to support the 1986 Galileo mission with the Mark IVA configuration described in the previous chapter. The Galileo downlinks would be on S-band and X-band, and all complexes could support the downlink telemetry rates from 40 bits per second to 134 kilobits per second. The normal uplink would be on S-band, with 20-kilowatt transmitters at the 34-m stations and 125-kilowatt transmitters at the 64-m stations. However, the DSN planned to demonstrate the then-new X-band uplink technology for command, Doppler, and ranging purposes by means of 20-kilowatt X-band transmitters located at the high-efficiency (HEF) antennas only. The command data rate for S- and X-band would be 32 bits per second. The Network planned to support the launch and cruise phases of the Galileo VEEGA mission with the configuration shown in Figure 5-13.

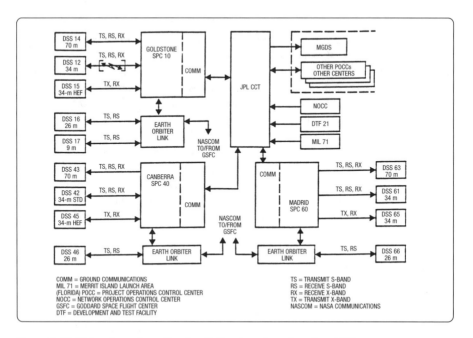

Figure 5-13. Network configuration for *Galileo* launch and cruise, 1989.

Network loading studies had identified many conflicts for Network support in 1988 and 1989. These conflicts were due to the combined requirements of Galileo at Jupiter, Voyager at Neptune, and Ulysses in solar orbit, together with requests for extended 64-m support from ICE and *Pioneers 10* and *11*. It was anticipated that these would be resolved as the mission progressed. In addition to long-term conflicts, the ESA spacecraft, *Ulysses*, was also at the pad preparing for a Shuttle-Centaur launch. Resolution of the *Ulysses* and *Galileo* competing requirements for launch priority had been documented.

In January 1986, the Network stations were completing tests to verify the operational configurations of hardware and software that would be required for the forthcoming Galileo mission. The new and complicated inter-Center interfaces needed to deliver spacecraft telemetry to the project at JPL through the Shuttle phase were being validated. The spacecraft radio system had been checked for full compatibility with the DSN frequencies, receivers, and transmitters. The DSN was ready to proceed to the final prelaunch phase.

The Galileo Era: 1986–1996

Following the *Challenger* accident, the DSN suspended further effort on Galileo-related tasks except for technical support to the mission design team. The project desperately searched for an alternate way of delivering the heavy *Galileo* spacecraft to Jupiter without the high-energy Centaur launch vehicle. The search ended in August 1986 with the discovery of the 1989 VEEGA trajectory, which was adopted as the new mission baseline design.

While the new trajectory satisfied the limited injection capabilities of the IUS upper stage, it carried with it several drawbacks of great significance to the DSN. One was the flight time of six years; another was that the spacecraft would arrive at Jupiter at maximum Earth-Jupiter range with a southerly declination of -23 degrees rather than minimum range with a northerly declination of +18 degrees, as would have been the case for the 1986 launch. The maximum range translated into minimum downlink signal strength. In addition, the southerly declination meant that the most favorable location for tracking Galileo at Jupiter would be the single site at Canberra, Australia, in the southern hemisphere, rather than from two sites, Goldstone and Madrid, in Earth's northern hemisphere.

With a 1989 launch, an arrival in 1995, and a two-year orbital objective, it would be 1997 when the mission ended. The DSN was being asked to commit to a capability almost ten years in advance. This was an unprecedented degree of commitment for NASA Headquarters to approve, considering that it had not been a simple matter to get approval for even much shorter-term agreements.

Quite apart from that, by 1995, the spacecraft RTG power system would no longer be adequate to run the spacecraft X-band transmitters at full power. They would have to run at low power. This fact, combined with the adverse effect of increased range, meant that the downlink signal power available to the DSN receivers would be about one-fifth of the original design value. This was nowhere near enough to capture the huge amount of high-rate science data called for in the original mission design.

Having come this far with the mission, the designers were not inclined to reduce their requirements, and there was no way to find a solution before launch, now less than two years away. It would be up to the DSN to find a solution to the problem, commit to implementing it, and have it in full operational readiness when *Galileo* arrived at Jupiter on 7 December 1995.

The DSN plan to address this situation is illustrated in Figure 5-14 which shows the effect of Galileo arrival at a southerly declination on the available downlink telemetry data rates.

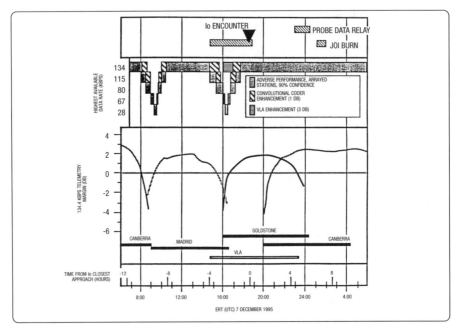

Figure 5-14. Galileo 1989 VEEGA mission: predicted X-band downlink performance Jupiter arrival, 7 December 1995. The horizontal scale in the figure shows the UTC time on 7 December 1995, (this is also called Earth Received Time or ERT) at which the downlink telemetry data rates shown in the upper vertical scale would be available at each of the Deep Space Communication Complexes. The lower vertical scale depicts the available downlink power margin relative to the maximum data rate of 134 kilobits per second.

As the figure shows, the Probe data relay of paramount scientific importance would take place in view of the southern hemisphere stations, where the view periods would be longer.

The fully arrayed 70-m and 34-m antennas at Goldstone and Canberra could support the high-rate telemetry for all of this period. However, the less vital but still important Io encounter events would occur in the downlink telemetry "gap" between the setting Madrid stations and the rising Goldstone stations. It was imperative that this gap be filled or significantly reduced in order to capture the Io Encounter data.

In June 1987, a combined Galileo/DSN team of telecommunications engineers presented John Casani, then Galileo Project Manager, with a proposal for improving downlink telemetry performance by 1 dB. The plan was to add a new convolutional decoder of

The Galileo Era: 1986–1996

type (15, 1/4) (see chapter 7 for an explanation of convolutional coding) to the existing (7, 1/2) convolutional coder on the spacecraft Telemetry Modulation Unit (TMU).[17] A suitable switch would allow the TMU to be commanded to select either one. An improvement of 1 dB in downlink telemetry performance would narrow the data gap during the Io Encounter as shown in Figure 5-4. The increase in science data return would be of great value to the Galileo project. However, because the decoders in the DSN Mark IVA telemetry system were designed to handle the existing 7, 1/2 code, an entirely new, very large decoder would be needed for the 15, 1/4 code. The DSN undertook to develop such a machine and have it operational in the Network by 1995. The costs were estimated at about $1M for Galileo and about $3–4M for the DSN.[18]

Furthermore, the DSN proposed that the downlink "gap" could be closed even further by combining the *Galileo* high-rate telemetry obtained by the DSN with a similar stream of recorded data from the Very Large Array (VLA) antennas of the National Radio Astronomy Observatory (NRAO) in Soccoro, New Mexico. Also, the VLA would provide a valuable backup in the event that weather or a technical failure disabled the Goldstone 70-m antenna at the critical time of Io Encounter. The possible addition of the VLA to the Galileo Ground Data System had been under study for some time by a special team led by James W. Layland, a telecommunications specialist in the TDA Planning Office. The team had identified a number of technical problems for which solutions were not available prior to launch.[19]

Encouraged by the successful demonstration of VLA support for the Voyager Neptune Encounter, which would take place in August 1989, the DSN was confident that a solution to the problems related to VLA support for Galileo data would be found before Galileo reached Jupiter, and planning proceeded on that basis.

Convinced by the arguments of the telecommunications engineers, Casani agreed to add the new convolutional coder to the spacecraft telemetry system. This was completed shortly before launch, just in time for the DSN engineers to verify the decoding algorithms on which the design of the new DSN decoder would be based. This complex device was to be developed by JPL's Communications Systems Research Section, where the engineer responsible for the task, Joseph I. Statman, dubbed it the Big Viterbi Decoder (BVD). The goal was to deliver a demonstration model at Goldstone in May 1991, and fully operational versions at Goldstone, Madrid, and Canberra by May 1995.

By November 1988, a DSN support plan for the 1989 mission had been approved by NASA. Galileo would be supported by the 34-m standard antennas, the 34-m high-efficiency antennas and the 70-m antennas which had been upgraded from 64-m diameter in 1987–88.

In order to capture the high-rate (134 kilobits per second) telemetry data at maximum Jupiter range, the DSN planned to use its full arraying capability of 70-m and 34-m antennas supplemented for critical Io Encounter period by the VLA. The VLA data was to be recorded and delivered to JPL for non-real-time combining with the recordings of the DSN arrayed data.

The S-band uplinks would be provided by 400-kW transmitters at the 70-m antennas and by 20-kW transmitters at the 34-m antennas. In addition, 20-kW X-band transmitters were installed at each of the three 34-m HEF antennas and would be used to demonstrate the viability of X-band uplink technology in a critical operational environment.

For spacecraft navigation purposes, the DSN would generate the standard Mark IVA radio metric data types, together with the necessary media calibration data. In addition, to meet the Galileo navigation requirement for angular measurement accuracy of 50 nanoradians (1 sigma), the DSN Very Long Baseline Interferometry (VLBI) system would process the special tones provided on the spacecraft S-band and X-band downlink radio carriers.

Radio science data generated by the Mark IVA Radio Science System during solar conjunctions and Jupiter occultations was to be supplemented with data for two other unique experiments.[20] The Faraday Rotation Experiment required the DSN to record the right hand and left hand polarization signals at the output of the S-band orthomode feed on the 70-m antennas. The data would later be used by the Radio Science team to determine the rotation of the plane of polarization (Faraday rotation) of the linear polarized S-band downlink as it passed through the magnetic field in Jupiter's ionosphere. Investigators would use this data to study the magnetic field of Jupiter at lower altitudes than was previously possible.

The objective of the other experiment was to detect the presence of gravitational waves in the interplanetary medium. The success of this experiment was critically dependent on the sensitivity and stability of the two-way Doppler frequency measuring system. Two-way X-band Doppler was now available at the 34-m HEF antennas and the necessary frequency and phase stability had been designed into the microwave and receiver-exciter subsystems for these antennas. Although the fractional frequency stability of the X-band RF carrier signals remained to be verified after launch, expectations were high that the desired level, at least 5 parts in 10^{15} (5×10^{-15}), would be achieved, and that the hitherto undetected gravitational waves would be observed during the solar oppositions in April 1994 and June 1995.

NASCOM was to provide wideband digital communications circuits with a maximum capacity of 224 kbps between all Complexes and the Mission Control and Computing Center (MCCC) at JPL. This would be sufficient to transport the 134 kbps telemetry plus ancillary data in real time from the sites to JPL. To enable the German Space Operations Center (GSOC) to support Galileo cruise operations, provisions were also made for low-rate data, voice, and teletype services from the GSOC at Weilheim, Germany, to the MCCC at JPL.

In the weeks preceding launch, the scene at Cape Kennedy was colored by the activities of a group of environmentalists protesting the use of radioisotope thermoelectric generators (RTGs) as a prime power source for *Galileo*, despite the assurances of NASA that the risk of endangering the environment or the population was minimal. NASA asserted its confidence in the detailed risk analysis studies that had been made for this and several earlier planetary spacecraft carrying similar RTG power sources, and the launch went ahead as planned.

Because this was a combined STS/IUS launch, a high degree of coordination and technical interfacing with other NASA Centers was involved. In the last few weeks before launch, the DSN participated in integrated simulation tests with the Kennedy Space Center (KSC), Johnson Space Center (JSC), and the Air Force Consolidated Space and Test Center (CSTC). In these tests, state vectors were received from CSTC and processed by the DSN Navigation Team to produce predictions for eventual use by the initial acquisition station in making first contact with the spacecraft. Data was correctly transferred across the various interfaces and delivered in a timely manner to the correct destinations. Following a final "Operational Readiness Test," the DSN was declared ready for launch on 17 September.

STS-34, carrying the *Galileo* spacecraft bound for Jupiter, was finally launched from Cape Kennedy on 18 October 1989. The launch was perfect and the subsequent deployment of the IUS and *Galileo* spacecraft from the Shuttle cargo bay on the fifth Earth revolution was completely normal. Following deployment, all the critical events leading to initial acquisition of the spacecraft downlink, first by the 26-m antenna at Goldstone and shortly after by the 34-m antenna at DSS 12, played out as exactly as planned. The success of the initial acquisition sequence testified to the excellence of the inter-Center coordination described above. *Galileo* was on its way to Jupiter via a close encounter with Venus in February 1990 and Earth flybys in December 1990 and December 1992.

Over the next few weeks, as the Earth-to-spacecraft range continued to increase, the downlink signal level reached the point where the 70-m antennas were needed to provide the telemetry data required for spacecraft management and control. This created severe conflicts for other DSN users whose needs were, of necessity, of lesser priority than those of *Galileo*. The situation was exacerbated in mid-December when the 70-m antenna at Madrid became immobilized due to a failed elevation bearing. The DSN responded immediately, and a team of JPL engineers arrived at DSS 63 two days later to start repair work. Under intense pressure because of the rapidly approaching Venus encounter, and working under extremely difficult winter conditions, the team was able to replace the massive bearing and return the antenna to operations on 25 January, fifteen days before Venus Encounter.

The *Galileo* closest approach to Venus occurred on 9 February 1990, during the mutual view of DSS 43 and DSS 63. Playback of the first images of Venus began over DSS 14 on 13 February. Three hours into the *Galileo* pass, the Goldstone antenna had to be shut down because of exceptionally high winds at Goldstone. Despite the resulting loss of data, reports from the Galileo science team commented favorably on the high quality of the downlink data delivered by the DSN. The Galileo Navigation Team used the Doppler and ranging data generated by the DSN to confirm that the gravity assist from the flyby resulted in the expected equivalent velocity increase of 2.2 km per second.

Within a few days of Venus Encounter, the sustainable downlink telemetry data rate on even the 70-m antennas had fallen to 40 bps, and by March it had fallen still further to 10 bps. This was due to an unfavorable pointing angle of the spacecraft low-gain antennas used for transmitting the downlink signal to the DSN. During the first year of its orbit, the spacecraft was oriented with the spin axis pointed towards the Sun. In this attitude, the Sun shields would protect the spacecraft and the furled X-band high-gain antenna from damaging solar thermal radiation, and the spacecraft would depend on its low-gain antennas for uplink and downlink. When the spacecraft orbit passed beyond the Earth orbit (1 AU), solar heating would no longer be a problem. The spacecraft would be oriented toward Earth, the big high-gain antenna would be deployed and the X-band uplink and downlinks could be used. Until then (April 1991), the S-band low-gain antennas would have to suffice.

Following repair of the DSS 63 elevation bearing, it was decided to inspect and carry out a bearing maintenance program on all three 70-m antennas to assure their integrity for the rest of the Galileo mission. This work, which required three weeks of downtime at each antenna, was started in May 1990 and completed successfully at DSS 43 in

The Galileo Era: 1986–1996

January 1991. Fortunately, during most of this time, the spacecraft was approaching the first Earth flyby and the need for 70-m support was considerably reduced.

As the spacecraft began to approach Earth in August, the DSN intensified its effort to deliver high-quality Delta Differential One-way Ranging (DDOR) for use by the spacecraft navigation team in refining the spacecraft targeting parameters for the first Earth Encounter.

The 26-m and 34-m stations at Canberra, Madrid, and Goldstone supported the operationally intense four-hour period of Earth-1 Encounter activity of 8 December 1990 without incident. During this period, very high downlink signal levels were able to sustain telemetry data rates of 134 kilobits per second to provide the project with 100 percent of the desired science data. Closest approach to Earth occurred at an altitude of 960 kilometers over the Caribbean Sea and out of view of the DSN, between Madrid set and Goldstone rise. The effort expended by the DSN on the DDOR campaign prior to Encounter was well justified when the Galileo Navigation Team estimated that the improved targeting errors resulting from the use of the DDOR data translated into a spacecraft fuel savings of approximately 6 kilograms, an important factor for consideration in the eventual design of the Jupiter satellite tour, still five years in the future.

The geometry of the first Earth flyby is shown in Figure 5-15. A dramatic picture of Antarctica taken by the *Galileo* spacecraft camera several hours after the Earth flyby at a distance of 200,000 km (124,000 miles), is seen in Figure 5-16. The downlink data from which this picture was constructed was received at the Goldstone 34-m antenna DSS 12 at 134 kbps a few hours after closest approach on 8 December 1990.

In the immediate post-Encounter period, the project took advantage of the prevailing downlink performance to return all of the Earth Encounter science data at high data rates. However, as the Earth spacecraft range increased, the downlink performance margins rapidly decreased, and by the end of January 1991, the telemetry data rate was back to about 40 bps. By then, all three 70-m antennas were available due to completion of elevation bearing work.

One of the final events in the Venus to Earth-1 portion of the Galileo mission was to be the unfurling of the high-gain antenna (HGA) in April 1991. At that time, the danger of solar heating was past and the spacecraft could be safely oriented towards Earth to point the big HGA directly at the DSN antennas. The X-band uplink and downlink would become the prime telecommunication channels. The command to

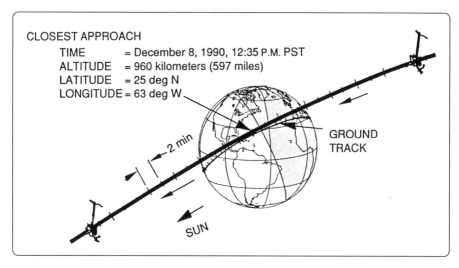

Figure 5-15. Geometry of the *Galileo* first Earth flyby.

unfurl the HGA had already been transmitted to the spacecraft and verified for completeness. It lay resident in the spacecraft computers waiting for the correct time to execute, and that time had finally arrived. A new phase of the Galileo mission was about to begin.

However, when the command to unfurl the HGA was activated, the redundant deployment motors ran for eight minutes before automatically switching off. A normal deployment should have taken three minutes. Several hundred trials had been successfully completed beforehand, but the spacecraft spin changed by less than was expected with a full deployment, and the Sun sensors indicated some partial obscurement. Something was drastically wrong with the HGA.

The rest of the spacecraft appeared to be completely normal, and the mission continued on S-band while the HGA "anomaly" was investigated. The conclusion was that three of the eighteen ribs in the HGA structure had remained stuck in the stowed or furled position, and the "anomaly" now became the "problem" of how to recover the HGA. It took the HGA Recovery Team two years to solve this problem as they subjected the antenna structure to repeated heating and cooling cycles and "hammering" shocks from the antenna drive motors.

Figure 5-16. *Galileo* image of Earth: Antarctica.

Galileo requirements for DSN operations coverage of the lengthy heating and cooling cycles were greatly increased, and conflict resolution for DSN antenna time became a major effort. The immediate impact of the HGA problem on DSN implementation plans was loss of the Big Viterbi Decoder demonstration, planned for May and June when the X-band downlink should have been available. The X-band uplink demonstration at DSS 15 during the solar opposition in March as part of the Gravitational Wave Experiment was also a casualty.

Meanwhile, the spacecraft continued outward bound towards the apoapsis of its two-year orbit around the Sun (see Figure 5-10). On the fringe of the asteroid belt in October 1991, it would make a close pass by the Asteroid Gaspra before returning to Earth for its final gravity-assist flyby in December 1992. By mid-1991, Optical Navigation (OPNAV) data, because of its potential for extreme angular precision, had become a key data type for the Galileo Navigation Team. In this scheme, the target image, the asteroid Gaspra, is photographed against a known star background from which the precise angular position of the target can be determined. Using the DDOR data to determine the spacecraft position enables the spacecraft to be targeted very precisely to the desired position for a close flyby. Because the downlink performance was now in the 10 bps to 40 bps range, it was not possible to recover the images in real time. The strategy that was employed involved recording the images on the spacecraft tape recorder and then transmitting the playback data to the DSN at whatever data rate the downlink could sustain. Recovery of the necessary images was, therefore, very time-consuming for 70-m antennas. The DSN expended nearly 400 hours of 70-m antenna time in recovering the OPNAV data for the Gaspra Encounter alone.

Closest approach to Gaspra occurred on 29 October 1991 as the spacecraft passed within 1,600 km of the asteroid, a very precise maneuver indeed, and an impressive demonstration of the power of OPNAV. Playback of all the Gaspra science at 10-40 bps occupied many months following the actual encounter and was not completed until November of the following year, 1992, just prior to the second Earth Encounter.

The second Earth Encounter took place on 8 December 1992, with near-perfect targeting results. The altitude at closest approach was only 303 km on this occasion and occurred over the south Atlantic Ocean, again out of view of any DSN tracking station. Because the gap in DSN coverage would be about two hours in duration, an elaborate arrangement of smaller antennas in Santiago, Chile; Perth, Australia; and Okinawa, Japan, was set up to cover the gap. They would accomplish this by recording the downlink Doppler signal and returning it to JPL via the NASA Tracking and Data Relay Satellite System (TDRSS). The purpose of this effort was to search for the occurrence, at closest approach, of an orbit anomaly, possibly associated with fundamental gravitational or general relativity theory. Although high-quality data was provided from all of the sites, no orbit anomaly was found.

The targeting precision for the second Earth flyby was near-perfect. *Galileo* passed within one kilometer of its intended path, one-tenth of a second ahead of its predicted time. This, the last of the three gravity assists, added 3.7 km per second to the spacecraft's speed in its solar orbit. *Galileo* was finally on an elliptical path to intersect the orbit of Jupiter, 780 million km from the Sun, on 7 December 1995. As the spacecraft departed the environs of Earth on 16 December 1992, it captured the unique picture of Earth and the Moon shown in Figure 5-17.

Except for the HGA problem, and that was a very big exception, the success of the Galileo mission to this point had been extraordinary. But the primary mission could not begin until the spacecraft reached Jupiter, and that lay three years ahead. With the downlink getting weaker every day, and no sign of progress in recovering the

The Galileo Era: 1986–1996

Figure 5-17. Earth and Moon from *Galileo*, December 1992.

HGA, the question of what could be done to save the mission was receiving a great deal of attention. Because of the long lead time involved in implementing new capability in the DSN, it was now necessary to make a decision as to whether to accept the loss of the HGA as inevitable and redesign the mission for S-band only, or to continue with the original X-band design in the hope that the HGA would eventually be recovered.

In announcing his decision in February 1993, Galileo Project Manager William J. O'Neill declared:

We are now proceeding to implement the Galileo mission using the LGA (S-band). We are absolutely confident of achieving at least 70 percent of our primary objectives including 100 percent of the Probe mission and several thousand of the highest resolution images of Jupiter ever planned. All the instruments will make their most important observations, and we will monitor the Jupiter magnetosphere virtually continuously, albeit at low data rate. The success of Galileo without the HGA at Jupiter will be a technological triumph. Developing the extensive, new spacecraft flight software and ground software and hardware, including state[-]of[-]the[-]art enhancements in the Deep Space Network, is a big challenge to complete in just three years. The project and the DSN are fully up to this challenge. Galileo will indeed fulfill its promise and be a magnificent mission at Jupiter.[21]

The new course was set, and the DSN moved forward upon it.

While attempts to free the antenna were in progress, Galileo and DSN engineers explored the possibility of continuing the mission using only the existing S-band capability and the low-gain antennas (LGAs) on the spacecraft. Earlier studies had shown that improvements in the coding and efficiency of the telemetry downlink, combined with a substantial enhancement in DSN receiving capability at S-band and suitable data processing at JPL, could indeed produce a viable S-band mission.[22] It was estimated that as much as 70 percent of the original mission objectives could be achieved.

The DSN began design and development work almost immediately, and by February 1993, when the decision to commit the Galileo mission to S-band was made, the DSN had already made substantial progress toward meeting the new requirements. The DSN continued to support the ongoing mission at low data rates consistent with the existing S-band capability, while moving toward the new configuration that would be required for orbital operations in December 1995, as the new hardware and software for the new DSCC Galileo Telemetry (DGT) subsystem became available.[23] The development of the DGT is discussed later in this chapter.

Two key elements in the design of the S-band mission were related to new spacecraft flight software. These would be transmitted to the spacecraft to reprogram its computers in two phases. The Phase 1 set of flight software would compensate for the fact that, without the HGA, the all-important Probe entry data could not be returned to Earth in real time. It would have to be recorded on the spacecraft tape recorder for later playback at the low data rates now determined by the LGA mission design. The spacecraft's redundant command and data system memory would be used to provide data storage backup for the tape recorder in case the critical Probe data got "lost" during its delivery to the project.

Once that data was safely received by the DSN, the Phase 2 flight software could be sent to the spacecraft for reprogramming its onboard computers. The Phase 2 software would provide programs to condense the huge amount of science data collected by the Orbiter during its two-year orbital mission, without losing any of the scientifically important information. That data could then be transmitted to Earth at a data rate consistent with the capabilities of the S-band downlink.

Without these enhancements, the data transmission rate via the LGA at Jupiter would be limited to 8–16 bps. When coupled with the DGT and the fully arrayed configuration of the DSCCs, including the Parkes Radio Astronomy antenna at Canberra, the data rate could be increased by as much as a factor of 10, up to 160 bps.

Although the original X-band HGA mission design had been based on a downlink data rate of 134,000 bps, most of the science investigations had been adapted to the lower data rates, and, with the exception of the Jupiter atmospheric circulation observations, few of the science experiments planned for the original mission had to be abandoned.

Through most of 1993, the Galileo S-band downlink was able to sustain a telemetry data rate of 40 bps on the 70-m antennas, but as the Earth-to-spacecraft range continued to increase, the data rate fell to 10 bps by the end of the year. The most notable event in 1993 was the close flyby of another asteroid named Ida. Because the optical images being used by the Navigation Team to target the asteroid required 30 hours of 70-m antenna time to retrieve at the low data rate, few images were provided. Nevertheless, they were sufficient to satisfy the targeting specifications, and good imaging of the asteroid was obtained during the August 1993 encounter sequence. Only five of the Ida images could be played back in 1993. The remainder were played back over a period of four months in the following year when the data rate returned to 40 bps. One of the images contained the interesting view of Ida with its own tiny satellite, shown in Figure 5-18. The image was taken 14 minutes before closest approach on 28 August 1993, at a range of 10,870 km.

In general, the Galileo mission moved through 1994 in the cruise mode with prevailing downlink conditions similar to those of the previous year. In July, the *Galileo* spacecraft was in a unique position to observe the Shoemaker-Levy Comet impacts on Jupiter. Five of the eleven science instruments onboard the spacecraft were able to make observations,

Figure 5-18. Asteroid Ida and its moon.

which were stored on the spacecraft tape recorder for later playback at 10 bps as the necessary antenna time became available during the rest of the year.

Toward the end of 1994, the first prototype models of several of the components of the DGT became available for engineering demonstration tests with live spacecraft data.[24] The new Block V Receiver (BVR) was used in downlink acquisition and tracking tests with a live signal from the spacecraft. This was a special software-based receiver designed to track the suppressed carrier downlink the spacecraft would transmit when it activated the Phase 2 flight software, in the orbital operations phase, in March 1996. Performance tests of the Full Spectrum Recorder (FSR) showed excellent agreement with theoretical predictions for signal-to-noise-ratio, harmonic contribution, and tracking loop capability. Galileo data that had been acquired simultaneously at DSS 14 and DSS 15 and recorded earlier in the year were successfully combined using a prototype of the Full Spectrum Combiner (FSC). The mean combining gain achieved with the experimental data was within 0.02 dB of the gain expected from perfect combining. Later demonstrations of intercontinental combining, using *Galileo* data from DSS 43 and DSS 14, gave equally impressive results.

The Galileo Era: 1986–1996

Verification of the operational performance of the DGT continued in 1995 as the installations progressed at each of the DSCCs. By September 1995, the BVR had been fully implemented and demonstrated at all 70-m stations, and the *Galileo* spacecraft was switched to the suppressed carrier downlink mode of operation that, together with the BVR at the 70-m stations, would be standard for the rest of the mission.

The *Galileo* S-band downlink was able to sustain a data rate of 8 to 16 bps throughout 1995, as the most critical events of the entire mission approached. As a precaution against the occurrence of problems with the 70-m antennas during these periods, each of the antennas was taken out of service for a few weeks for hydrostatic bearing and subreflector maintenance. DSN and Galileo mission operations personnel rehearsed the time-critical sequences and verified the data flow paths and configurations. In February, the new Phase 1 flight software had been uplinked to the spacecraft. The long command sequences were received correctly by the spacecraft, were loaded into the computers, and began operating in March. All was ready as the time for Probe release approached.

The Probe release sequence began on 5 July 1995. After a system status check, the Probe transferred to its own internal power supply and the "umbilical" cable connecting Probe to Orbiter was disconnected. The spacecraft turned to the correct attitude for pointing the Probe towards Jupiter, and, in response to a "GO" command, the Probe was released at 10:07 P.M. PDT, 12 July, to travel the rest of the way on its own. Nothing more would be heard from the Probe until it established its uplink relay to the Orbiter as it descended through the Jupiter atmosphere. That would be the last 60 minutes of its life.

Three weeks later, the Orbiter executed a deflection maneuver to change its trajectory slightly so that it would not enter the Jupiter atmosphere on arrival, but would arrive at the correct point for Jupiter Orbit Insertion (JOI).

Both the Orbiter and the Probe arrived at Jupiter during the overlapping view periods of DSS 43 and DSS 14 on 7 December 1995. In the final hours of its approach to Jupiter, the Orbiter was able to capture some images of distant Io as it crossed the orbit of the satellite. Limited by the capabilities of the S-band mission, the Io images were a far cry from the finely detailed pictures that the original mission design would have produced. Nevertheless, they were adequate to show the changes that had taken place since the Voyager flyby 17 years before.

The Orbiter made its closest approach to Jupiter at 2:46 P.M. PST, and the Probe started transmitting its science data to the Orbiter thirteen minutes later. Although the Probe data was not being sent to Earth in real time, telemetry from the Orbiter verified that the L-band signal from the Probe was being received and recorded by the Orbiter, much to the relief of everybody concerned. By 4:14 P.M., recording of the Probe data was complete and the spacecraft was prepared for the next major event, insertion into Jupiter orbit. This was accomplished with great precision by the 400-Newton retropropulsion engine, built by Messerschmicht-Bolkow-Blohm (MBB) and provided by Germany as a partner in the Galileo project. The first orbit of Jupiter began. Preliminary indications were that 57 minutes of Probe data had been received and recorded by the Orbiter prior to loss of Probe signal—a most remarkable feat of mission design and spacecraft technology. Actual tape playback of the Probe science data would begin at the end of January 1996 and was expected to take about two months to complete at 10 to 16 bps. The Phase 1 software had proved its worth.

From the DSN perspective, support for all the events of the critical Probe entry and JOI sequence had been very nominal. The BVR receivers had continued to track the downlink signal throughout the retroengine burn, and all desired telemetry and radio metric data products had been delivered. There were no anomalies or problems to be reported.

In each of the three years after *Galileo* entered its final Earth to Jupiter cruise phase in 1992, the DSN had supported a major scientific event, two asteroid encounters and the Comet Shoemaker-Levy observations. A solar conjunction and a solar opposition had provided good data for radio science observations. Gravitational wave observations, in conjunction with Ulysses, had also been supported. An emergency control center for Galileo had been established at Goldstone and could have supported all operations had the Mission Control Center at JPL been unavailable for any reason.

Many resource scheduling problems that appeared intractable early in 1993 were eventually resolved by DSN scheduling to fully meet *Galileo* requirements for 70-m tracking time in each of the three years. The 34-m Network was used only for test purposes during this time. The BVR had reached maturity and was now a standard operational component of the almost-completed DGT. Over the period of January 1993 to December 1995, the DSN maintained an average downlink telemetry data capture rate of 98.1 percent. In the same time, it correctly transmitted over 220,000 uplink commands to the *Galileo* spacecraft.

With the completion of the Probe and JOI events just described, the mission entered the final phase: the orbital operations phase. During the next two years, the spacecraft would make several close encounters with three of the Jupiter satellites, using the gravity assist technique to move from one orbit to another in sequence. The first encounter would be

The Galileo Era: 1986–1996

with Ganymede in June 1996, following an adjustment of the spacecraft orbit to avoid repeated exposure to the Jupiter radiation environment at each periapsis passage.

By the beginning of 1996, all the components of a single-antenna version of the DGT were onsite and being tested.[25] After a successful Ground Data System operations demonstration in May, using simulated Phase 2 flight software, this version was put online for live *Galileo* support. A short time later, inflight loading of the real Phase 2 software began. It took over sixty-five continuous hours to transmit the program to the spacecraft, the longest command sequence ever transmitted by the DSN Command System. There were no transmission errors. After verifying that both computers on the spacecraft had been correctly loaded and the data properly processed by the ground data system, the spacecraft was commanded into the new mode of operation. It immediately began transmitting the new downlink packetized telemetry at 32 bits per second for processing by the new DGT at the DSN stations.

The DGT single-antenna version was used to support the first Ganymede Encounter in June 1996. At this phase of the mission, the single-antenna DGT, in conjunction with the new downlink telemetry format, could support 40 bits per second over DSS 14 and DSS 63, and 80 bits per second over DSS 43, already a substantial improvement over the former performance of the downlink. The science data could be played back in days rather than weeks, enabling the first Ganymede data to be presented at a press conference on 10 July, just two weeks after closest approach.

A second Ganymede Encounter in September was supported by the single-antenna DGT configuration in a similar way, with equally satisfactory results. In September and October, the DSN had been engaged in end-to-end ground data system testing of the DGT in its full operational arraying mode. At SPC 40 in Canberra, the Galileo downlink was processed through the multiple-antenna version of the DGT, with signals from DSS 14 in Goldstone being combined, in real time, with signals from DSS 43, DSS 42 at Canberra, and DSS 49 at Parkes. After processing at SPC 40, the resulting enhanced telemetry stream was sent to JPL via the GCF for delivery to the project. During the tests, the spacecraft was run at all data rates up to 160 bps, the highest bit rate available. These rates required full arraying capability to recover the data but were demonstrated successfully. The demonstrations revealed a number of configuration and procedural problems that were rapidly corrected. There were no anomalies related to the design of the DGT.

Now confident in its capability to support encounter operations in the fully arrayed mode, the DSN used the DGT for the Callisto Encounter in November with very sat-

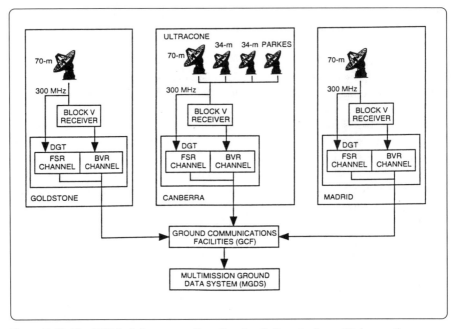

Figure 5-19. The DSN in full array configuration for Galileo Jupiter orbital operations.

isfactory results. Although the introduction of arrayed operations into the *Galileo* downlink sequences resulted in increased operational complexity, it provided almost tenfold increase in the downlink data rate that would otherwise have been possible.

Based on the operational performance of the DGT in supporting the Callisto Encounter and on confidence in the DSN ability to deal with the increased operational complexity of fully arrayed operations, the Project Manager elected to schedule every *Galileo* pass from then to the end of the mission in that mode. A functional block diagram of the DSN in the full array configuration is shown in Figure 5-19.

The Europa Encounter in December 1996 and the six remaining satellite encounters of the Galileo prime mission in 1997 were all conducted in the array mode. As it worked through these events, DSN experience in handling the complex operational processes involved in real-time arraying, rapidly improved to the point where intercontinental operations became routine. The DSN had reached a new level of competence in a very complex operational environment. A typical data rate profile for the *Galileo* downlink in 1997 routine array operations is shown in Figure 5-20.

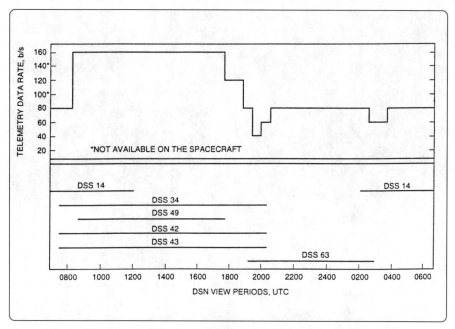

Figure 5-20. Full DSN array operations for *Galileo*: typical downlink data rate profile.

As it moved towards the conclusion of the Prime Mission in 1997, *Galileo* provided its sponsors with a cornucopia of imaging and other scientific riches. Typical of the imaging results is the striking photo of the surface of Europa shown in Figure 5-21. This image was recorded by the spacecraft during the Europa Encounter on 20 February 1997. It was downlinked as part of the total Europa science data playback sequence at 40 to 80 bps during subsequent DSN array passes over the following five weeks.

The quality of this image and the many others like it that followed is a tribute to the innovative ideas of the DSN and to the Galileo engineers who were involved in the effort to recover a viable downlink for the mission after the loss of one of its key elements, the HGA.

Background for Ulysses

In 1977, NASA and ESA agreed upon a joint mission to study the surface of the Sun and other solar phenomena. It would be called the International Solar Polar Mission (ISPM) and would comprise two spacecraft travelling on separate trajectories passing over

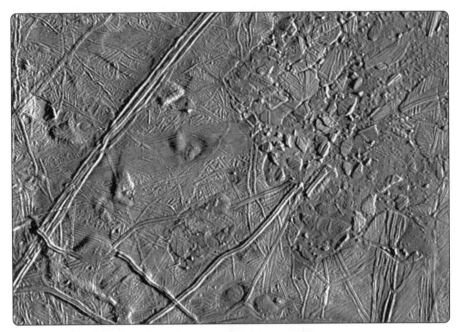

Figure 5-21. *Galileo* image of icebergs on the surface of Europa.

opposite poles of the Sun. One spacecraft would be built by ESA, the other by NASA, who in addition would provide the launch vehicle. Both spacecraft would carry a complement of experiments drawn from science communities in the United States and Europe.

ISPM was a natural successor to the Helios mission of a decade earlier. Helios had explored the solar environment from an equatorial orbit around the Sun. ISPM was to make similar observations from a polar orbit over the Sun. To create an orbit passing over the poles of the Sun, however, was much more difficult than creating an orbit lying entirely in the plane of the ecliptic. To accomplish this, mission designers planned to make use of a high-velocity gravity assist from Jupiter to change the plane and direction of the spacecraft trajectory to the extent necessary to achieve the desired north and south polar passages.

The solar polar trajectory that was eventually initiated by the single spacecraft launch of 1990 is shown in Figure 5-22. In the original plan, there were to have been two spacecraft orbiting the Sun in opposite directions.

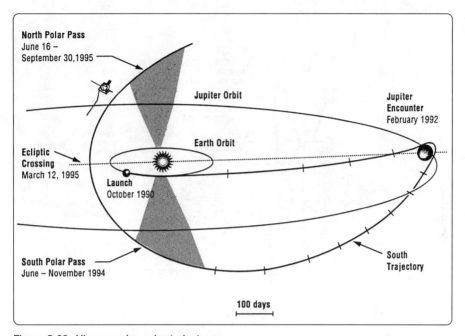

Figure 5-22. *Ulysses* solar polar trajectory.

In keeping with the NASA policy at the time of the original agreement, ISPM would use the Space Shuttle as the primary launch vehicle. A single launch was planned for 1983. The two spacecraft would be mounted in tandem on a single injection vehicle. After burnout and separation from the upper stage, each spacecraft would be injected into its individual trajectory towards Jupiter.

In 1980, because of NASA difficulties with development of the Space Shuttle, the launch was delayed to 1985. A year later, due to NASA problems with funding, the NASA spacecraft was canceled, and the launch of the remaining ESA spacecraft was further delayed to 1986. At the time, the circumstances surrounding the launch delays and cancellation of the NASA spacecraft raised a storm of protest at ESA, the aftermath of which persisted for many years in the European science community. Eventually, the ESA spacecraft was completed and placed in storage, and the ISPM mission was appropriately renamed *Ulysses*, after the legendary Greek hero who, after his return from the Trojan wars, set out to explore "the uninhabited world beyond the Sun."

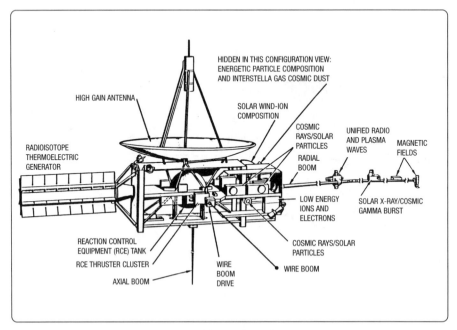

Figure 5-23. The *Ulysses* spacecraft inflight configuration.

In 1987, in the aftermath of *Challenger*, NASA and ESA jointly agreed that *Ulysses* would be launched in October 1990, following the *Magellan* and *Galileo* launches in 1989.[26]

The *Ulysses* spacecraft had a mass of 370 kg and was radiation-resistant and spin-stabilized. The flight configuration is shown in Figure 5-23. Prominent features included the 1.65-m high-gain antenna (HGA), the Radio Isotope Thermoelectric Generator (RTG), and the 5.5-m radial boom, which provided an electromagnetically clean environment for certain experiments. A 72-m dipole wire boom and a 7.5-m axial boom served as antennas for the Unified Radio and Plasma-wave experiment.

The main platform of aluminum honeycomb provided a mounting surface for all the electronic units and for the Reaction Control Equipment and its hydrazine fuel tank. Science instruments other than those mounted on the radial boom were also mounted in the body of the spacecraft.

The spacecraft carried redundant transponders, each of which included S-band receivers and 5-watt transmitters, and redundant 20-watt X-band transmitters. Located on the spin axis of the spacecraft, the HGA permanently pointed toward Earth to provide S-band uplink and downlink and X-band downlink communications with the DSN. Two low-gain antennas (LGAs) provided hemispherical coverage in the forward- and rearward-looking directions for S-band only, and omnidirectional uplink and downlink communications when the spacecraft was in the early phases of the mission.

Two fully redundant tape recorders supplemented the telemetry system to ensure continuous collection of scientific data throughout the mission. When contact with the DSN was not available, data was recorded on the spacecraft for later playback when DSN coverage became available. A variety of selectable downlink data rates between 128 bits per second and 1,024 bits per second were provided for recorded or real-time data.[26, 27]

The *Ulysses* spacecraft carried nine scientific experiments, all related to the investigation of a wide range of fields and particles phenomena of the Sun and its environment. In addition, two radio science experiments making use of the spacecraft and DSN uplinks and downlinks were to be carried out. The solar corona experiment would study the density, velocity, and turbulence spectrum of the coronal plasma near the Sun. The Gravitational Wave Experiment (GWE) would search for low-frequency gravitational waves. The GWE experiments would be correlated with similar observations made by *Galileo* in order to confirm the detection, and determine the directional characteristics, of any gravitational wave that might be detected.

Ulysses's requirements on the DSN for tracking and data acquisition support were comparatively modest in comparison to those of *Galileo* and other JPL flight projects.[28] The Mark IVA 70-m antenna upgrades and the two 34-m subnets had been completed by the time of the *Ulysses* launch, so the DSN was able to meet most of Ulysses's requirements with one 34-m subnet while *Galileo* and *Magellan* (in extended mission status) were accommodated on the 70-m and remaining 34-m antennas. For a few weeks during solar conjunctions and oppositions, radio science experiments, including the Gravity Wave Experiment, were conducted on the 70-m antennas at the rate of one or two passes a day.

The DSN configuration for *Ulysses* support was similar to that shown for *Galileo* in Figure 5-13. Within the SPCs, the Mark IVA telemetry, tracking, command, radio science, and monitor and control systems were fully capable of handling all the *Ulysses* uplink and downlink requirements. Telemetry data and S-and X-band radio metric data were delivered to the project at JPL via the Mark IVA GCF. Navigation requirements for VLBI data were handled by the NOCC Navigation Subsystem. There were no requirements for arraying.

Ulysses Solar Mission

Placed into a 300-kilometer Earth orbit by a perfect Space Shuttle launch on 6 October 1990, *Ulysses* was powered into a trajectory towards Jupiter by a combined Interim Upper Stage (IUS)/Payload Assist Module (PAM) injection vehicle. About 30 minutes after the PAM burnout, the spacecraft S-band downlink was acquired by the 26-m and 34-m antennas at Canberra, and the Earth-Jupiter phase of the Ulysses mission began. The first of several large trajectory correction maneuvers (TCMs) to target the spacecraft to the Jupiter aim point was carried out over a four-day period in mid-October.

In early November, shortly after deployment of the Axial Boom, the spacecraft X-band antenna beam unexpectedly developed an oscillatory motion called nutation. Although the resulting variations in downlink signal threatened future use of the X-band downlink, the nutation effect gradually disappeared as the Axial Boom became shadowed from solar thermal effects by the spacecraft, whose attitude slowly changed to keep the HGA pointed toward Earth. Radio science observations made during the first solar opposition in December 1990 showed no deleterious effects from HGA nutation. A small TCM carried out in July 1991 brought the trajectory within 250 kilometers of the desired Jupiter aim point.

Moving very rapidly along the corrected trajectory, and with uplink and downlink operating normally, *Ulysses* reached the Jupiter aim point on 8 February 1992. DSN radio metric data confirmed that the closest approach to Jupiter was 6.31 Jupiter radii (Rj), as expected. Note: The *Galileo* spacecraft made its closest approach to Jupiter at 4.0 Rj in 1995 with special protection to withstand the Jupiter radiation environment.

Ulysses began moving in a new orbital plane, and, for the first time, a planetary spacecraft was travelling out of the plane of the ecliptic. The trajectory took it on a big loop back toward the Sun, passing under the south polar region of the Sun, back across the equator, and over the north polar region. It was these two regions, around 80 degrees south and north latitude, that would be of greatest scientific interest.

Throughout 1992 and 1993, all spacecraft, science, and routine antenna pointing operations were performed without incident. DSN support on the 34-m antennas was routine, with 70-m passes being scheduled for special events.

On 26 June 1994, at 70 degrees south latitude, the spacecraft entered what scientists consider to be the south polar region of the Sun. For the next four months, the spacecraft collected science data on the complex forces at work in that region. Where possible, DSN schedules were adjusted to accommodate the demand for increased tracking time.

In August, the spacecraft moved into the Sun-Earth region, where the spacecraft's Axial Boom was again illuminated by the Sun. It was known from earlier experiences during the mission that this illumination would cause uneven heating of the Axial Boom, which, in turn, caused a slight wobble in the attitude of the spacecraft. The wobble was controlled by an onboard control system that maneuvered the spacecraft to keep the HGA pointed to Earth. Ground controllers in constant contact with the spacecraft carried out this technique and allowed the onboard system to detect and correct unwanted motion. The European Space Agency's tracking facility at Kourou, French Guiana, was modified and brought online to help the DSN antennas in Canberra provide 24-hour coverage of the spacecraft for this purpose.

The spacecraft reached its most extreme latitude of 80.2 degrees south in mid-September, and Ulysses mission scientists began reporting important discoveries in the structure and composition of the Sun's solar wind and magnetic field at a science conference in Noordwijk, Netherlands.

In February 1995, as the spacecraft approached a position in back of the Sun as viewed from Earth (opposition), the spacecraft S-band transmitter was turned on to complement the X-band transmitter in conducting the solar corona radio science experiment. During the "opposition period," the DSN Radio Science System would record changes to the S-band and X-band spectra induced by passage through the solar corona. The Ulysses Radio Science Team would use these data to measure the electron content of the Sun's corona.

Ulysses recrossed the Sun's equator (ecliptic) at a distance of 1.3 AU on 12 March 1995. With all systems in good operating condition, it headed toward the north polar region. When it reached 70 degrees north latitude on 19 June 1995, the second and final phase of the primary mission began. Over the Sun's north polar region, *Ulysses* continued to observe such solar phenomena as the speed of the solar wind, the rate of loss of material from the Sun, the outward pressure of the solar wind, its impact on the shape of the heliosphere and coronal structure, and the entrance or repulsion of high-energy cosmic rays at these high latitudes. This period of high science activity was supported by the Canberra, Madrid, and Goldstone stations at a rate of 16 hours per day. The observations continued through September, until the spacecraft left the high-latitude region of the north solar pole.

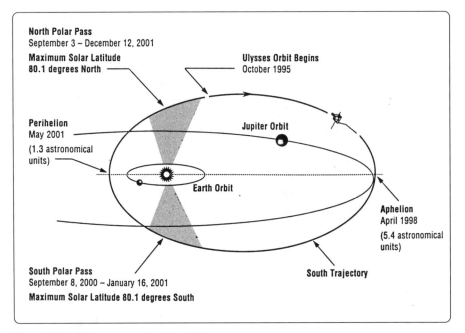

Figure 5-24. *Ulysses* second polar orbit.

On 29 September 1995, five years after it began, the end of the *Ulysses* primary mission was reached. *Ulysses* then began the long journey back out to the orbit of Jupiter, making fields and particles observations as it progressed. On reaching the giant planet's distance of 5.4 AU on 17 April 1998, the spacecraft would once again head back on a second high-latitude trajectory to reach the vicinity of the Sun in September 2000. A diagram of the second *Ulysses* polar orbit of the Sun is shown in Figure 5-24.

The DSN continued to provide 34-m antenna support for the "routine phase" of the Ulysses extended mission, at a level of about 10 hours per day through 1996 and 1997. With the spacecraft in good condition and sustaining requirements minimal, the DSN expected to support Ulysses through a second solar passage in the years 2000 and 2001.

Background for *Mars Observer*

Mars Observer was to be the next step in NASA's long involvement with exploration of the Red Planet, which had begun thirty years before with the *Mariner 4* spacecraft and the first crude pictures of the planet's cratered surface. The last spacecraft to visit Mars

The Galileo Era: 1986–1996

had been the twin Viking Orbiters and Landers in 1976, on a mission to search for signs of life from the distant past. The primary objectives of the *Mars Observer* mission were to identify and map the chemical and mineral composition of the surface, measure the topography of the surface landforms, define the gravitational field, and search for a planetary magnetic field. It was also intended to study the carbon dioxide, water, and dust in the Martian atmosphere. The mission was to last approximately one Mars year, equivalent to about two Earth years. The spacecraft carried seven science instruments designed to accomplish these objectives.

Mars Observer also carried a radio-receiver package provided by the French space agency, Centre National d'Etudes Spatiales (CNES). It was intended that the data from several penetrators and experiment packages, which were to be placed on the surface by the Russian *Mars 94* spacecraft toward the end of the mission, would be relayed to Earth by this equipment.

The mission was lost three days before it was due to enter Mars orbit to begin its primary mission. When the downlink fell silent at approximately 6 P.M. PDT on 21 August 1993, the spacecraft had traveled almost eleven months into the cruise phase of the mission with no problems and had taken its first and only photograph of Mars.

For months after the loss of the downlink, ground controllers tried to reestablish communication with the spacecraft but were unsuccessful. NASA quickly established an independent review board to determine the most likely cause for the failure and began to lay plans for a new mission to return to Mars.

Mars Observer Mission

The *Mars Observer* spacecraft was the first JPL "all X-band spacecraft." It represented the final transition of X-band technology to full operational status in the DSN. Ironically, it also carried with it an experiment to demonstrate the next generation of telecommunications technology, the use of Ka-band for space communications. Equipped with an X-band uplink and two X-band downlinks, the spacecraft used a high-gain antenna and a low-gain antenna for transmitting and two low-gain antennas for receiving the X-band uplink.

The new DSN 34-m HEF antennas with X-band uplink and downlink capability were selected to provide most of the daily communications with the spacecraft. While *Mars Observer* benefited from the general upgrades to the Mark IVA, several specific improvements were made to support the mission. These improvements included the Standard Format Data Unit (SFDU) headers for all data types except command; the Reed-Solomon decoders at each Complex; and the X-band acquisition aid antenna, mounted on the 26-m antenna at Canberra, to ensure the initial acquisition of the X-band downlink at first view after launch.

By agreement with the Multimission Operations Systems Office (MOSO) and the Mars Observer project, all DSN data, with the exception of radio metric data, was to be routed to the project data base via the new Advanced Multimission Operations System (AMMOS). Radio metric data would be routed to the DSN Multimission Navigation Team for preprocessing before delivery to the project.[29]

The *Mars Observer* launch marked a change in NASA launch policy back to expendable launch vehicles for interplanetary launches, after ten years of Space Shuttle-based launch policy. For the DSN, this resulted in some simplification of prelaunch interfaces and data flow paths compared to those required by Shuttle-based launches, but it also meant that new interfaces and agreements had to be negotiated, and data flow paths verified, before launch.

Mars Observer was placed into Earth orbit on 25 September 1992, by a Titan III first-stage vehicle launched from the Cape Canaveral Air Force Station, Florida. A Transfer Orbit Stage (TOS) booster later injected the spacecraft onto a direct ascent trajectory to Mars. Initial acquisition of the X-band downlink via the spacecraft low-gain antenna was rapidly accomplished by the Canberra 26-m and 34-m antennas, and the spacecraft settled into routine cruise activity, supported by the DSN 34-m subnet. After the first 30 days of continuous tracking, the DSN planned to support the cruise portion of the mission at a rate of one or two passes per day until the end of July, when full 34-m support on the HEF antennas would again be required for orbital operations. Science instrument calibration and checkout was completed, and the instruments began returning data.

In January 1993, the spacecraft switched to the high-gain antenna with no problems, and a downlink telecommunications experiment using a low-power Ka-band transmitter on the spacecraft was carried out. The Ka-band Link Experiment (KABLE) was designed by the DSN to test the capabilities of Ka-band (9-millimeter wavelength) radiation for use in future downlink telecommunications. The DSN research and development station, DSS 13 at Goldstone, was provided with the necessary antenna and receiving equipment to perform the tests.[30] Excellent results were obtained, and the plan was to repeat the experiment periodically throughout the cruise period.

The Galileo Era: 1986–1996

During the cruise period, over a span of three weeks, *Mars Observer* participated in a gravitational wave search with *Ulysses* and *Galileo*. The X-band Doppler data was delivered to the Radio Science Team for later analysis.

Reports from the Flight Team indicated that all spacecraft subsystems and instrument payload were working well. The first picture of Mars was taken on 26 July, one month before the spacecraft was due to arrive at the point of Mars Orbit Insertion (MOI). As the flight teams began preparing for MOI and the start of orbital operations, the DSN increased its tracking support to a full three passes per day on the 34-HEF subnet.

MOI was programmed to start on 24 August with a 29-minute burn of the spacecraft propulsion system. On Saturday, 21 August, as part of the pre-programmed orbit insertion sequence, the spacecraft downlink was turned off to protect the transmitter tubes from possible shock effects as the propellant fuel tanks were pressurized. The spacecraft was programmed to turn the downlink back on after pressurization had been completed.

The expected downlink never appeared. A declared spacecraft emergency condition gave *Mars Observer* first priority in use of DSN antennas for downlink recovery efforts. The spacecraft failed to respond to repeated uplink commands to turn the transmitter on, to switch to the low-gain wide-beam antenna, switch to a backup timer, and various other actions related to possible failure scenarios.

All three DSN complexes responded to the emergency with continuous commanding sequences and closely-monitored receiver and signal detection searches. As the days passed, each possible failure scenario was explored, but to no avail. At the end of September, an attempt was made to activate the radio beacon on the CNES radio package in the hope that its weak (1-watt) downlink might be detected to gave some indication of the location and condition of the spacecraft. The Jodrell Bank radio telescope in England was enlisted to aid the DSN 70-m antennas in the search. Again, no signal was detected after several days of searching.

Soon after the loss of downlink occurred, an independent review board was established by NASA to determine the most likely cause of the failure. This anomaly investigation was augmented by a special technical review board set up by JPL. The findings of both boards were generally consistent and pointed to a rupture in the pressurization system. If this were the case, *Mars Observer* would have tumbled out of control until the pressurant and fuel were completely exhausted. Then, electrically damaged beyond recovery, the spacecraft would have overshot Mars to become a mass of space debris in a perpetual elliptical orbit around the Sun.

Deep Space Extended Missions

The Pioneer Missions

The Pioneer program was officially terminated by NASA Headquarters on 31 March 1997. For thirty-two years, since the launch of *Pioneer 6* in 1965, the family of Pioneer spacecraft had maintained a presence as prime or extended missions on the DSN tracking schedules.

Pioneers 6 through *9* were launched from 1965 through 1968 into heliocentric orbits to make particles and fields measurements inward and outward from these orbits. In the later years, they were also used to observe the ultimate longevity and failure modes of aging flight components such as solar panels, Sun sensors, data handling systems, and radio transmitters and receivers. *Pioneer 9* was last tracked by the DSN in May 1983. The DSN continued to maintain downlinks with *Pioneers 6, 7,* and *8* until the Pioneer program was terminated in 1997. Although *Pioneers 7* and *8* had developed problems by then, *Pioneer 6* was still operating normally. An unofficial attempt to reestablish contact with *Pioneer 6* was made from DSS 63 in July 1997. Responding to a sequence of commands to switch to its backup transmitter, the spacecraft activated a coaxial switch and turned on the backup Traveling Wave Tube (TWT) amplifier for the first time in 30 years. Although the signal received by the DSN was weak (-165), and the signal to noise ratio was poor (3 dB), a small amount of Plasma Analyzer data was received. The DSN continued to monitor the spacecraft from time to time for the engineering interest it provided on long-duration deep space missions.

Pioneers 10 and *11* were launched in 1972 and 1973 to explore Jupiter and Saturn. The planetary encounters were used to place them on trajectories that would take them toward the edge of the heliosphere and eventually out of the solar system. *Pioneer 11* reached a distance of 44.1 AU before its downlink was lost on 30 September 1995. It was surmised that Earth had moved out of the spacecraft high-gain antenna beam after the spacecraft could no longer be oriented to keep it pointed in the right direction. The downlink from *Pioneer 10* was still active when the program was terminated in 1997. As with *Pioneer 6*, DSN interest in the longevity and performance of the downlink prompted an attempt to contact the spacecraft from DSS 14 on 31 July 1997. The spacecraft downlink was detected at a signal level of -178 dbm, and good two-way S-band Doppler was obtained. At that time the spacecraft was estimated to be at a distance of 68.65 AU.

The Galileo Era: 1986–1996

For most of their long lives, the *Pioneer 10* and *11* spacecraft were able to benefit from the periodic enhancements to DSN downlink capability which were driven not by the Pioneers but by the requirements of other spacecraft. In that sense, they never became obsolete and never reached the predicted limit of the DSN capability to maintain an uplink and a downlink with them.

Pioneers 12 and *13* were Venus Orbiter and Venus Multiprobe missions, respectively. Launched about three months apart, they both reached Venus in December 1978. *Pioneer 13* was destroyed on atmospheric entry, but *Pioneer 12* remained active in Venus orbit for 14 years. Over the years, the orbit gradually decayed to the point where it could no longer be raised to prevent aerodynamic damage to the spacecraft. The mission ended when no downlink was detected by the DSN after periapsis passage on 9 October 1992.

These pioneering spacecraft paid enormous scientific dividends over the more than thirty years of their diverse explorations of the solar system. Relatively inexpensive by planetary spacecraft standards, the Pioneer spacecraft established a basis for further exploration by more complex, more costly spacecraft.[31] Furthermore, they amply demonstrated the value of simple, highly reliable design principles in producing long-lived spacecraft downlinks for deep space communications.

International Cometary Explorer (ICE)

Following its successful passage through the nucleus of the Comet Giacobini-Zinner in September 1985, the ICE spacecraft continued to move along its highly elliptical orbit around the Sun as the DSN moved from the Voyager Era into the Galileo Era. Its scientific objective was to gain knowledge of how the Sun controls Earth's near-space environment.

Throughout the Galileo Era, ICE remained on the DSN tracking schedules as an extended deep space mission, supported on the 34-m subnet for a few passes per week as tracking time became available. Although the spacecraft transmitted no telemetry, the downlink carrier was recorded by the DSP during each short duration pass, and the data was delivered to the Radio Science Investigator for analysis.

Voyager Interstellar Mission

The Voyager Interstellar Mission (VIM) began in October 1989, right after the *Voyager 2* Encounter of Neptune, and included both Voyager spacecraft. The objectives of this prolonged mission were to sample the interplanetary medium, conduct ultraviolet stellar astronomy, and search for the heliopause, the region where the Sun's magnetic influence wanes and interstellar space begins. The spacecraft were expected to operate for another twenty-five to thirty years, after which the supply of electrical power generated by the RTGs would drop too low to operate the science instruments.

Voyager 1 was exiting the solar system at the rate of 525 million km per year on a heading 35 degrees above the ecliptic plane. By late 1992, it was more than 50 AU from Earth. At the same time, *Voyager 2*, diving below the plane of the ecliptic at a 48 degree angle, had reached a position more than 38 AU from Earth. All seven science instruments needed for the VIM were operating well; others designed for the earlier planetary mission had been turned off in 1990 to conserve power.

After the Neptune Encounter, the level of DSN support for Voyager was reduced to periodic contact on the 34-m antennas consistent with the new requirements of the VIM and the availability of antenna tracking time.

With the exception of the Plasma instrument on *Voyager 1*, all instruments on both spacecraft continued to collect science data and return it to the DSN. The condition of the radio subsystems on both spacecraft remained unchanged after the initial failures in 1978, and the workarounds developed to cope with them had since been improved and become normal operating procedure.

In 1997, the Voyager project developed a plan to extend the Voyager Interstellar Mission (VIM) for the next twenty-five to thirty years.[32] It was estimated that by about the year 2020, the spacecraft would no longer be able to generate sufficient electrical power for continued operation of the science instruments. Until then, both spacecraft had the potential to maintain their present capability for acquiring fields, particles, and wave science data in the interplanetary and possibly the interstellar regions.

A completely new sequencing strategy was required for spacecraft operations to provide for acquiring the desired science data, dealing with spacecraft anomalies, and maintaining each spacecraft's high-gain antenna continuously pointed toward Earth.

Not only were the VIM sequence designers constrained by the limited available onboard computer memory (1,700 words total), but the duration of the VIM introduced many considerations of special interest to the DSN.

The baseline sequence was stored in the spacecraft computer memory; it contained, in addition to a continuously executing set of spacecraft operating instructions, the HGA pointing information required to maintain the uplink and downlink with the DSN

The Galileo Era: 1986–1996

through the year 2020. All of the spacecraft events that required 70-m coverage for best downlink telemetry data return were planned to occur over the Goldstone Complex for *Voyager 1* and over the Canberra Complex for *Voyager 2*. The timing of spacecraft events had to be adjusted by the onboard sequence to keep them synchronized with station view periods. It was also necessary to avoid designated station quiet periods—Thanksgiving, Christmas, and New Year's Day—and to allow for the steady increase in one-way light time of approximately one half-hour per year of flight. For a short mission, these were trivial matters, but in terms of a potential thirty-year mission, they became very significant.

In September 1997, both spacecraft had been operating in deep space for twenty years. *Voyager 1* was 68 AU from the Sun; *Voyager 2* was 53 AU from the Sun. Traveling at the speed of light, the radio signal from the spacecraft took nine hours to reach the DSN antennas on Earth. The DSN receiving capability had improved over these years to the point where the new 34-m antennas could support a downlink data rate of 160 bps at these great ranges. Return of the recorded plasma-wave data was possible at 7.5 kbps from *Voyager 2*, and 1.5 kbps from *Voyager 1* on the 70-m antennas using the high power X-band transmitters on the spacecraft. Most of the Voyager support was therefore carried on the 34-m subnet. In a typical week in 1997, the DSN scheduled 60 to 80 hours of antenna time for each Voyager spacecraft, only 5 percent of which was provided by a 70-m antenna.

Earlier in their missions, *Voyagers 1* and *2* had revealed the scientific beauty of Jupiter, Saturn, Uranus, and Neptune. In doing so, they challenged the DSN to match the spacecraft's ability to capture the data with a capability to receive it. As both spacecraft slowly stretched the existing communications links toward the boundary of the solar system, the DSN faced the same challenge. The Voyager Interstellar Mission afforded the DSN its first opportunity to demonstrate uplink and downlink communications "to the edge"—and beyond.

Earth Orbiter Missions

General

In 1986, the DSN carried nine Earth Orbiter missions on its tracking schedules. Two of these missions were of the reimbursable type, one French and the other German in origin. All the Earth Orbiters were supported on the single 26-m subnet, which had become a DSN responsibility under the Networks Consolidation program of 1983–85. By 1997, the number of Earth Orbiter missions on the DSN schedules had increased to twenty-three, seven of which were the reimbursable type of Japanese, German, or French origin. All missions were divided among the 9-m, 26-m, and 34-m networks as appropriate.

The growth of DSN involvement with Earth orbiting missions is shown in the table below. The number of deep space missions being tracked by the DSN in the same time period is included for comparison.

Growth of Earth Orbiter Missions Supported by the DSN, 1986–97

Missions	1986	1988	1990	1992	1994	1995	1996	1997
NASA Earth Orbiter	7	5	4	4	9	19	16	23
Non-NASA Earth Orbiter	2	5	7	3	1	5	3	4
Deep Space	12	11	13	13	12	10	9	10

Under the terms of a reimbursable agreement, the space agency owning the spacecraft was charged an hourly fee by NASA for the use of the DSN antennas. In the early agreements, the costs for antenna time, supporting services, technical management, and travel were detailed separately. Later, these costs were combined into a standard fee based on the antenna type.

The reimbursable rate for the 26-m or 34-m antennas was established in 1993 at $2,825 (U.S.) per hour. By 1995, the rate had been reduced to $2,080 and further reduced to $1968 by 1997, due to economies in manpower required for station operations functions. The reduction in human resources required for DSN operations reflected the greatly expanded capabilities of improvements in the Mark IVA DSN Monitor and Control System. Corresponding hourly cost (in U.S. dollars) for the 70-m antennas was $5,650 in 1993, reduced to $4,160 by 1995 and to $3,936 by 1997.

The NASA strategic plan of 1995 called for the "transfer of operational activities, as feasible, to other federal agencies or commercial operators." In response, the Telecommunications and Mission Operations Directorate (TMOD) conducted a study to estimate the economic viability of using a commercial service as an alternative to NASA-provided tracking and data acquisition services for low-Earth orbiter (LEO) and high-Earth orbiter (HEO) missions. The results of the study were presented to the NASA Space Operations Office (Code O) in October 1995 and included a recommendation that the DSN 26-m subnet be closed by the end of fiscal year 1997. The antennas were to be decommissioned, and the equipment not needed by JPL was to be declared "surplus property" and disposed of appropriately. Procurement action for commercial services equivalent to those provided by the 26-m subnet was initiated with a target transfer date of September 1997.

Before these recommendations were implemented, however, a new proposal for a more cost-effective solution to the disposition of the 26-m network was evaluated by TMOD. This resulted in a final recommendation that the 26-m subnet be automated and utilized to support low- and high-Earth orbiting missions through the year 2001. By that time, support for low-Earth orbiters was to be provided by commercial services and support for the high-Earth orbiters would continue to be provided by the fully automated 26-m subnet.

With the authority of the new Space Operations Management Organization (SOMO), directives to this effect were issued to the Mission Services Managers at JPL and other NASA Field Centers in March 1997 by the Space Operations Data Services Manager. Within JPL this directive provided TMOD with the formal instructions it needed to proceed with automation of the DSN 26-m subnet and the eventual transfer of low-Earth orbiting missions to a commercial service. Planning toward these objectives began to move forward.

Emergency Mission Support

The introduction of the Tracking and Data Relay Satellite System (TDRSS) from 1983 to 1985 as an alternative to ground-based stations for tracking satellites in low-Earth orbits required that provision be made for emergency support in the event of failure of the TDRSS satellites, or failure of the ground station at White Sands, New Mexico. In 1983, a DSN study had shown that for a nominal cost, the 26-m subnetwork of antennas being transferred to the DSN under the Networks Consolidation program could be utilized to provide the necessary support for a limited set of emergency situations.[33] After approval of the concept by the NASA Office of Tracking and Data Acquisition (OSTDS), the necessary interfaces and operational procedures were developed to allow timely notification of an emergency situation by the Flight Projects Directorate at GSFC, and a prompt response by the DSN Operations organization at JPL. The procedures provided for the exchange of information needed by the DSN stations to point the antennas, acquire the uplinks and downlinks, transmit commands, and record and deliver the desired telemetry and ranging data.

The original list of eight Earth-orbiting missions approved for emergency DSN support under this agreement included the Hubble Space Telescope, the Space Shuttle, and the

TDRSS satellites. Through the next decade, Hubble, Shuttle, and TDRSS remained on the DSN schedules as "emergency support only" missions, while changes and additions to the original list brought the total to twelve by 1997.

Except for the coverage provided by the Goldstone 9-m and 26-m antennas for emergency Shuttle landings at Edwards Air Force Base at Mohave, California, the need for emergency support under the terms of this agreement occurred only once or twice per year in the twelve years after its inception. Shuttle support was phased out of the 26-m schedules in 1997.

Beginning in 1995, international agreements for providing emergency support for low-Earth orbiting spacecraft were negotiated among NASA (U.S.), NASDA (Japan), CNES (France), and DLR (Germany). Under these agreements, the various space agencies agreed to provide emergency support for foreign spacecraft on the antennas of each other's tracking networks, as appropriate to the particular situation.[34] Emergency support was defined to be a period of unscheduled tracking, telemetry, and/or command support necessary for spacecraft survival in the event of a spacecraft emergency. The support would be provided by existing resources on a best-efforts basis, would be of short duration, and would involve no exchange of funds or liability. Only S-band uplinks and downlinks would be used for spacecraft communications, and the support would be limited to satellites launched for peaceful purposes only. A key element of the agreements was a listing of the mission sets for which emergency support might be requested. The DSN antennas designated for this type of service included DSS 16 and DSS 27 at Goldstone, DSS 46 at Canberra, and DSS 66 at Madrid.

Lower level agreements in the form of Interface Control Documents (ICDs) were developed to manage the technical interfaces, communication configurations, protocols, operational procedures, and delivery of data products. The reciprocal exchange of telecommunications services between the agencies was viewed by the parties to the agreement as mutually beneficial. At the time of this writing, there have been no calls for support of this kind.

The Galileo Era: 1986–1996

NETWORK ENGINEERING AND IMPLEMENTATION

The 70-meter Antennas

Reporting on the maintenance, rehabilitation, and upgrade of the DSN 64-meter antennas in March 1982,[35] the DSN chief engineer, Robertson Stevens, focused attention on the history and status of the hydrostatic azimuth bearings and azimuth radial bearings, particularly those at DSS 14 and DSS 43. He also recommended a plan for continuing repair or replacement of components and the continuing upgrade of maintenance and operating procedures. The report also argued convincingly that it was indeed practical to enlarge the antennas to 70-meter diameter and to extend the useful operating frequency for these antennas.

Over the next several years, while the Network continued to support all flight missions described in the previous chapter, these recommendations were implemented. This work included the three 70-m upgrades, the last of which was completed at DSS 14 in May 1988.

By 1989, the Goldstone antenna had been in continuous operation for 23 years, and the overseas antennas had been in operation for 16 years. The specified design life for parts subject to wear was 10 years. The antennas had been specified and designed more than 27 years earlier. They had performed remarkably well, but with a resurgence in the number of planetary missions planned for the Galileo Era, even more was to be demanded of them.

To this end, a DSN Antenna Rehabilitation Team[36] was established in March 1989 to identify the resources needed to maintain the new 70-m antennas past the year 2000. The team recommended five high-priority tasks it believed would, if promptly implemented, extend the useful operating life of the antennas for the next twenty-five years.

The tasks were to develop a trend analysis program to provide early detection of incipient failures; to provide a means of lifting the antenna in the event that a pad or ball joint needed to be replaced, as had occurred in Spain and Australia in 1976; to repair or replace aging structural and mechanical elements of the antenna before further deterioration took place; to rehabilitate the elevation and azimuth gearboxes, since the gear drive assemblies ranked next to the hydrostatic bearing in the number of problems experienced over the years; and to replace the subreflector positioner and controls because these had been a longstanding source of trouble.

The hydrostatic bearings at Spain and Australia were believed to be in good condition and were not expected to require any special attention in the foreseeable future. However, there was less confidence in the future of the hydrostatic bearing at Goldstone. It was hoped that existing procedures to inhibit rust formation would be successful and that major rework would not be necessary. Only time would tell.

It was the opinion of the team that the basic design was sound and that, under normal operating conditions, the structure was not subjected to high stress levels. At the time, cyclic stresses were considered to be infrequent and of a low level, so fatigue or catastrophic failure was thought to be unlikely.

Elevation Bearing Failure

About the time that the 70-m Rehabilitation Team was publishing these findings (November 1989), the *Galileo* spacecraft was reaching the point where it became dependent on continuous support from the 70-m antennas to sustain the downlink telemetry data rate of 1,200 bps needed for its rapidly approaching Encounter with Venus on 9 February 1990. Closest approach would occur during the DSS 43 overlap with DSS 63, and full 70-m coverage was essential to the success of this first gravitational-assist event in the *Galileo* mission to Jupiter. On 13 December 1989, in the midst of this intense 70-m activity, the right-side inboard elevation bearing on the DSS 63 antenna failed, and the antenna was immobilized.

The failure occurred about 9:00 A.M. (local time) while the antenna was being used to investigate some anomalous Doppler problems at low elevation angles. It appeared that one or more of the rollers in the roller bearing race that supported the right inboard elevation bearing had cracked. The bearing sustained further damage when the antenna was driven back to the zenith position for safety. The extent of the damage can be appreciated by the photograph of the damaged bearing shown in Figure 5-25.

Obviously, a very serious antenna bearing problem now existed. The situation was exacerbated by pressure from *Galileo*, the approaching holidays, and arduous working conditions due to inclement winter weather in Spain.

The DSN immediately dispatched several antenna engineers to Madrid to investigate the problem. Dale Wells, one of the engineers involved in the mechanical maintenance of all three large antennas since they were first built, remained on site to direct the repair work, while other technical staff at JPL coordinated the shipment of spare bearings and other parts.

The unusual nature of the repair work required special tooling, much of which was fabricated on site. The right side bearing, carrying two million pounds of antenna weight, was lifted with hydraulic jacks, and removal of the damaged bearings began. This proved

Figure 5-25. Cracked bearing rollers and inner race, DSS 63, December 1989.

to be extremely difficult and was eventually accomplished by flame-cutting some of the retaining bolts and sections of the damaged race. Once the old bearing parts were removed, installation of the new bearings proceeded rapidly, and the antenna was tested and returned to service on 25 January, fifteen days before the *Galileo* encounter with Venus. The broken bearing was carefully reassembled and returned to JPL for evaluation.

An Elevation Bearing Failure Team was formed in January to investigate the DSS 63 failure and to oversee any remedial action that might be needed at the other 70-m antennas. The team first decided to return the broken bearing to SKF for a bearing failure analysis. There, it was determined that the failure had been caused by fatigue of the inner bearing rollers, which, over the years, had caused metal particles to flake off and damage the surfaces of other rollers in the bearing. This progressive action ultimately led to complete lockup of one roller, and eventually to failure of the entire bearing race.

Following the DSS 63 return to service, the elevation bearings on all the 70-m antennas (including those at DSS 63) were inspected in sequence for signs of wear. Some signs of wear were found, but none that warranted a change of the bearing races at that time. It was decided to continue with regular analysis of bearing for detection of metal particles. This work, requiring only two days of antenna downtime at each site, was carried out during April and May of 1990.

The team also directed that the loads on each of the inboard and outboard elevation bearings on all three antennas be equalized to correct the additional, unequal loading that resulted from the 70-m upgrade. This work was sufficiently important to require 21 days of antenna downtime, beginning immediately at DSS 63. It was fortuitous that the *Galileo* spacecraft was approaching its first Earth fly-by around this time, and consequently its need for 70-m support was minimal. The work was completed at DSS 14 in January 1991. In some cases, before the load equalization adjustment, the inequality of load sharing between the inner and outer bearings was as much as 80 percent to 20 percent. With the work completed, this was reduced to, typically, 55 percent to 45 percent or better.

Finally, the Team recommended that the original roller bearings, which were manufactured with 3/4-inch axial holes to facilitate the heat treatment process, should be replaced with solid roller bearings. Orders for new bearings were placed, and the work of replacement began in May 1991 at DSS 63. Using their experience and tooling from the previous bearing replacement effort, the engineers completed the work in just 17 days. To economize on antenna downtime, replacement of the elevation bearings at DSS 43 and DSS 14 was carried out in 1993 and 1994, in conjunction with the upgrade of the elevation gearboxes.

Gearbox Rehabilitation

The primary motive power to drive the huge 70-m antennas in the azimuth and elevation directions is provided by hydraulic motors activated by computer-controlled servo systems. The high-speed output shafts of the hydraulic motors are connected to the various low-speed gears (pinions) and quadrants. These physically rotate the antenna by using reducer gearboxes—four for the elevation drive and four for the azimuth drive.

Ever since the antennas were first built, persistent problems with the gearboxes had been constant cause for concern and led to expensive and antenna-time consuming repair efforts. Until about 1984, most of the gearbox problems were related to lubrication matters, but in 1985, a significant increase in the wear patterns on all antenna drive components, particularly the gearboxes, became evident. This was attributed to the introduction of the 1985 Mark-IVA servo modifications described in the previous chapter. The new servos drove the antennas at maximum velocity between the "stow" and "track" positions, and vice versa, causing additional wear and even inducing oscillations in the overall system. In addition, the increasing DSN tracking schedule in that period called for more rapid

The Galileo Era: 1986–1996

changes between multiple spacecraft; also, VLBI, clock synchronization, and Star Catalogue passes required frequent and rapid slewing of the antennas, with fast starts and stops.

By 1986, the situation had become serious enough on the DSS 14 antenna to warrant an examination and evaluation of wear on the elevation bull gear and pinion by a professional consultant from the University of California at Berkeley. On the basis of this evaluation, the elevation pinion gears at DSS 14 and DSS 43 were replaced with specially hardened gears in 1987 as part of the 70-m upgrade. Gear replacement was not required at that time at DSS 63. In April 1988, an expert consultant specializing in gear train systems was sought to examine and report on the condition of the gearboxes themselves. The two specific recommendations called for replacing the drive train gearbox components with case-hardened and precision-ground gears, and adding magnetic particle detectors to the lubrication systems to remove wear debris. DSN engineers responded with a contract to the Philadelphia Gear Company in 1990 for a completely new set of components to upgrade or replace the existing 24 gearboxes on all three 70-m antennas. The new design incorporated the experts' opinion and the DSN engineers' experience with the original design, which had provided approximately thirty years of continuous operational service.

Installation of the new gearboxes began with DSS 43 in March 1993 and proceeded around the Network as downtime could be scheduled, to DSS 14 in September 1993 and DSS 63 in June 1994.

Subreflector Drive Problems

The considerations that led the 70-m Antenna Rehabilitation Team in 1989 to recommend complete replacement of the subreflector suspension and drive systems resulted from the inadequate design of the 70-m components. Constrained in many ways by the design of the former 64-m components, they simply were not robust enough to handle the much larger and heavier subreflector used on the 70-m antennas. Frequent failures and problems occurred, and several modifications and temporary work-arounds were necessary to keep the antennas in operation. Because the estimated cost of the work was 3.6 million dollars, budget constraints at the time prevented the recommendations from being implemented immediately, and the matter was deferred.

However, before anything more could be done in this regard, nature itself intervened. On Sunday 28 June 1992, just as the DSS 14 antenna was being rotated to the horizon in preparation for a *Pioneer 10* spacecraft track, the Goldstone area was hit by two severe earthquakes within a few hours of each other.

In the shaking that followed, the 24,000-pound subreflector broke loose from the three-axis drive shaft and impacted the quadripod suspension structure. The cast aluminum subreflector was severely damaged, and its precision-contoured surface was cracked and holed. All antennas at Goldstone were halted while damage was assessed. As it turned out, only DSS 14 had sustained permanent damage, and repair started immediately. At first it was thought that the subreflector would have to be replaced, and action was started to ship the fiberglass subreflector that had been used several years earlier at DSS 43. However, it was eventually decided that the subreflector and its drive system should be repaired in place, over 71 meters above the ground, with the antenna at zenith. Over the next three weeks, all the damaged parts of the subreflector and its three-axis drive system were either repaired or replaced, and the antenna was returned to full operation on 22 July 1992. After necessary mechanical and optical realignment adjustments were made, it was estimated that the loss of RF gain attributable to damage inflicted by the earthquake (and subsequently repaired) was no greater than 0.06 dB at X-band. This small performance loss was within acceptable limits, and the antenna was approved for full operational use.

On 1 July 1992, right after the earthquake occurred, an Engineering Analysis and Corrective Actions Team was formed to analyze the damage, assist with the return to service actions, and recommend corrective action to improve the protection of the 70-m antennas from future earthquake damage. In 1995, as a result of the team's findings, new earthquake-resistant subreflector positioners were designed, built, and installed on all 70-m antennas. The 70-m antenna mechanical specifications were changed to better match an earthquake-prone environment, and future antenna designs reflected the need for seismic protection. Seismic monitoring procedures were established for the Goldstone antenna using accelerometer sensors located at strategic points in the base and apex areas of the antenna.

There have been no further earthquake incidents in the DSN as of this writing.

Toward the end of the Galileo Era, the funds available for antenna maintenance were significantly affected by overall reductions in the DSN budget. Out of this situation, yet another review board emerged to report on the physical condition of the 70-m and 34-m antennas. Based on these findings, the board was to recommend cost-effective measures that would assure their continued reliable operation in the forthcoming period of diminished resources.

The board was chaired by Richard P. Mathison, then chief engineer, and included several experts from the National Radio Astronomy Observatory (NRAO) with experience in large antennas, and some with knowledge of telescope (Keck) mounting structures.

The Galileo Era: 1986–1996

In late 1996, the board inspected the antennas at each of the three sites and closely examined specific onsite maintenance problems with the engineers involved.

The board concluded that the condition of the antennas and the level of maintenance at all sites was uniformly good. It expressed concern for incipient failures in several of the steel and concrete elements of the 70-m structures and recommended a continuing program of monitoring and inspection to establish trends in performance deterioration. To reduce antenna maintenance costs, the board suggested that the DSN should clearly establish an absolute minimum level of maintenance required for reliable operation for all 70-m antennas. This would enable proper allocation of funding between the requirements for minimum maintenance, preventative and corrective maintenance, and the replacement and procurement of spare parts.

In 1997, it was generally believed that the 70-m antennas were in sound mechanical condition, that their potential structural and mechanical weaknesses were well understood and under control, and that with the recommendations of the Mathison Review Board in effect, they would continue to provide reliable uplink and downlink service well into the new century.

INTERAGENCY ARRAYING

Parkes-Canberra Telemetry Array

As early as 1981, advanced systems planners in the DSN had begun to evaluate the possibility of enhancing the downlink performance of the DSN 64-m antennas for special, short-duration mission events of high science value, such as the future Voyager Encounters with Uranus and possibly Neptune, by arraying other large, non-DSN antennas with those of the DSN. These ideas followed naturally from the DSN's long history of arraying with its own 64-m and 34-m antennas, summarized in Figure 5-26.

Because of its past associations with the DSN and reasonably close proximity to Canberra, the 64-m antenna of the Radio Astronomy Observatory in Parkes, New South Wales, Australia, was the first non-DSN antenna to be considered for this purpose. This facility, owned and operated by the Australian Government's Commonwealth Scientific and Industrial Research Organisation (CSIRO), formed the basis for much of the early design work on the DSN's own 64-m antennas twenty years earlier.

After the necessary high-level agreements between CSIRO and NASA were completed, a technical design for a Parkes-CDSCC Telemetry Array (PCTA) was initiated.[37] At that

Mission	Date[a]	Location of antennas	Antenna types	Combiners[c]
Pioneer 8	1970	Spain	26-m + 26-m	Passive BB
Mariner Venus-Mercury (Mercury)	Sept. '74	U.S.	64-m + 2 (26-m)	R&D BB
Voyager 1 (Jupiter)	Mar. '79	U.S.	64-m + 34-m	RTC
Voyager 2 (Jupiter)	Jul. '79	U.S.	64-m + 34-m	RTC
Pioneer 11 (Saturn)	Sep. '79	U.S.	64-m + 34-m	RTC
Voyager 1 (Saturn)	Aug. '80	ALL DSCCs	64-m + 34-m	RTC
Voyager 2 (Saturn)	Aug. '81	ALL DSCCs	64-m + 34-m	RTC
International Cometary Explorer (Giacobini-Zinner)	Sep. '85	ALL DSCCs Spain/U.S. Australia/Usuda	64-m + 34-m[b] 64-m + 64-m 64-m + 64-m	Passive BB SSRC R&D SSRC R&D
Voyager 2 (Uranus)	Jan. '86	Spain U.S. Australia	64-m + 34-m 64-m + 2 (34-m) 64-m + 2 (34-m) +Parkes 64-m	RTC/BBA RTC/BBA RTC/BBA + LBC + SSRC R&D
Voyager 2 (Neptune)	Aug. '89	Spain U.S. Australia	70-m + 2 (34-m) 70-m + 2 (34-m) + VLA 27 (25-m) 70-m + 2 (34-m) +Parkes 64-m	RTC/BBA RTC/BBA + VLBC + SSRC RTC/BBA + LBC + SSRC

[a] Listing of month and year indicates encounter period.
[b] The Spain and U.S. 64-m antennas also combined dual channels with passive BB.
[c] BB = Baseband (symbol-modulated subcarrier).
LBC = Long Baseline Combiner (baseband at ~300 km).
RTC = Real-Time Combiner, first version (baseband at ~30 km).
RTC/BBA = Operational RTC—part of BBA (Baseband Assembly).
SSRC = Symbol-Stream Recording and Combining (non-real time).
VLBC = Very Long Baseline Combiner (baseband at ~1,000 km).

Figure 5-26. Twenty years of telemetry arraying in the DSN.

time, the PCTA was designed specifically to meet the downlink requirements of the January 1986 Voyager Encounter with Uranus. It was estimated that the addition of the 64-m Parkes antenna to the CDSCC configuration of one 64-m antenna arrayed with two 34-m antennas could increase the downlink capability for the Voyager/Uranus Encounter by as much as 50 percent.

The Galileo Era: 1986–1996

Prior to NASA's becoming interested in seeking the support of CSIRO for the use of the Parkes antenna for Voyager, the European Space Agency (ESA) had already completed similar agreements for Parkes support of the Giotto mission to Comet Halley. The antenna had been in service for twenty years, and a major upgrade to its facilities, computers, and servo and pointing capability had just been completed. Under this agreement, CSIRO would further upgrade the antenna for operation at X-band. Fortunately for NASA, the ESA X-band configuration for Giotto was compatible with the configuration required for Voyager, and arrangements were made with CSIRO for the two agencies to share antenna equipment and tracking time from late 1985 through the encounters with Uranus and Comet Halley in January and March 1986.

The low-noise, X-band receiving equipment that was to be shared consisted of a feedhorn provided by CSIRO, a microwave assembly provided by JPL, and two masers built in the U.S. to the standard JPL design with funding provided by ESA. The front-end equipment package was integrated with French designed down converters by ESA at Darmstadt, West Germany. A video-grade microwave link, of Japanese manufacture, was supplied by Telecom Australia, to transfer the Parkes data stream to the Canberra Complex, some 350 km distant, for combining with the DSN data stream.

At first, DSN engineers, with their long history of successful in-house experience, viewed the international flavor of the PCTA front-end package with some reservation. However, its performance under actual operational conditions left no room for skepticism, and it performed well for many years afterward, with a minimum of problems.

The functional block diagram of the Parkes-CDSCC Telemetry Array shown in Figure 5-27 identifies the various areas of responsibility among the DSN, ESA/Parkes, Telecom Australia, CSIRO/Parkes, and the PCTA at Parkes and CDSCC.

As might be expected, the transmission time delay on the 350-km microwave link from Parkes to Canberra introduced additional complexity to the combining process. The delay was found to be quite stable and equal to 1.2 milliseconds. The effect of this and other more subtle dynamic delays was provided for by special circuits in the design of the Long Baseline Combiner.

Most of 1983 was occupied with negotiating the unique NASA-ESA-CSIRO agreements and developing interface design control specifications to ensure that all the elements of the PCTA would work together when assembled end-to-end. In 1984, fabrication of the individual elements, including implementation of the intersite microwave link, was in progress. Testing and verification for compatibility with each other and with the Voyager

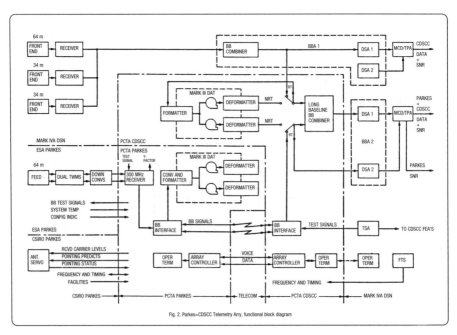

Fig. 2. Parkes−CDSCC Telemetry Arry, functional block diagram

Figure 5-27. Parkes-CDSCC telemetry array. In this diagram, the amplified X-band downlink signals from the antennas at Parkes and CDSCC were first changed down to the 300-MHz band to feed the receivers which delivered baseband (subcarrier plus data) signals at their outputs. The baseband outputs were treated differently at the two locations. At Parkes, the baseband signals were digitally recorded at the Mark III Data Acquisition Terminals (DATs) while simultaneously being transferred to CDSCC via the Telcom Australia microwave link. At CDSCC, the baseband signals from each of the three DSN antennas were first combined in the Baseband Combiner. The combined composite signal then followed two parallel paths. In one path, subcarrier demodulation and frame synchronization functions were performed by a Demodulation and Synchronization Assembly (DSA). After decoding and formatting, the data stream was returned in real time to JPL. On the other path, the baseband signal fed both a DAT (similar to the Parkes set-up) and a Long Baseline Baseband Combiner. It was in this latter combiner that the Parkes and composite CDSCC signals were finally brought together for baseband combining to provide the enhancement expected from the overall PCTA system. This operation was followed by data handling processes identical to those just described. A study of the diagram will show how the DATs allowed these same operations to be performed in non-real time whenever the need arose.

The Galileo Era: 1986–1996

and Giotto spacecraft began in 1985. The Parkes antenna began supporting Voyager and Giotto passes in September 1985, following a final set of system performance tests to validate the end-to-end system and a few more weeks of operational verification testing to refine operations procedures. By the time the Voyager spacecraft began its observatory phase in November 1985, the PCTA was running in full operational status and providing the predicted downlink enhancement that was essential for the successful Voyager Encounter of Uranus on 4 January 1986.

At the range of Jupiter, 5 AU from the Sun, the maximum downlink data rate from the Voyager spacecraft had been 115.2 kbps. By the time the spacecraft reached Saturn, at a range of 10 AU, the maximum data rate had fallen to the range of 44.8 kbps to 14.4 kbps. When the distance doubled again to 20 AU at Uranus, the data rate would have been about one-fourth of that value, without the PCTA. The enhancement of the downlink provided by the PCTA allowed the mission controllers to recover telemetry virtually error-free from the spacecraft at data rates of 21.2 kbps to 14.4 kbps.

The demonstrated operational success of the PCTA during the Voyager Uranus Encounter confirmed the DSN's confidence in interagency arraying as a viable adjunct to the DSN capability for planetary missions. The future would provide two important occasions on which the DSN would have good reason to call on these techniques. The first would be to enhance an existing mission, Voyager at Neptune, the other to save a disabled mission, Galileo at Jupiter.

About the time that the PCTA technical studies were being completed in 1982, a broader based study of interagency arraying as a potential benefit to DSN capability for downlink support of planetary missions was initiated. Led by James W. Layland,[38] this study was directed toward determining "which other facilities might be feasibly and beneficially employed for the support of Voyager at Uranus, and examining the Voyager/Neptune Encounter and such other future events and options as might appear." In light of subsequent events connected with the Galileo mission to Jupiter ten years later, this proved to be a very prophetic vision.

At its conclusion in early 1983, the study recommended that the existing plans for support of Voyager at Uranus should be completed. This culminated in the PCTA described above. The study also recommended that the arraying configuration for the Voyager Neptune Encounter should consist of the full array of DSN antennas, plus Parkes, plus the Very Large Array (VLA) at Socorro, New Mexico. There were some qualifications to these recommendations related to budget matters, but they did not affect the course

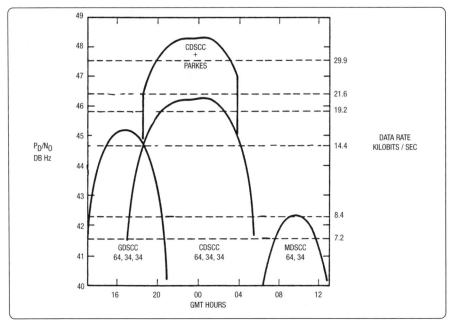

Figure 5-28. Enhanced downlink performance at Uranus.

of history. The comparative downlink performance data on which these recommendations were based are shown in Figures 5-28 and 5-29.

The recommendations from this study set the stage for the continuing DSN interest in interagency arraying and contributed materially to the success of the Voyager Encounter with Neptune in 1989, and eventually to the Galileo Encounter with Jupiter in 1996.

VLA-Goldstone Telemetry Array

Sponsored by the National Science Foundation (NSF) and operated by the National Radio Astronomy Observatory (NRAO), the Very Large Array (VLA) is a premier radio astronomy facility located at Soccoro, New Mexico. It consists of 27 antennas, each 25 meters in diameter, arranged in the form of a "Y" with a 20-km radius. This was the facility that, arrayed with the DSN antennas in Goldstone, California, would be known as the VLA-GDSCC Telemetry Array (VGTA).

The Galileo Era: 1986–1996

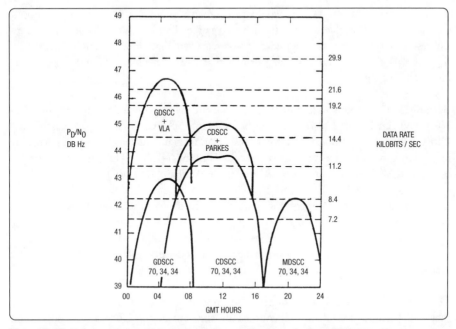

Figure 5-29. Enhanced downlink performance at Neptune. In these figures, each curve represents one of the array options from the study. The left-hand scale represents the performance of the downlink in terms of the ratio of data signal power (P_D) to noise power (N_O) expressed in dB. The right-hand scale shows the Voyager downlink telemetry data rates that correspond to the downlink performance values. These were the downlink data rates which theory predicted would be available 90 percent of the time when weather and other natural effects were taken into account. The reduction in downlink data rate that results from a 6-dB loss in downlink performance, which in turn corresponds to a doubling of Earth-to-spacecraft range, is very evident. The beneficial result of adding a non-DSN antenna to the existing DSN capability is represented by the performance curves. The bottom time scale shows the GMT time at which the Voyager spacecraft would be visible at each longitude during the respective encounters. These figures convey, in a very dramatic way, the essential reason for the strong DSN interest in the technique of antenna arraying.

The VGTA and its Australian counterpart, the PCTA, would be the DSN key to the Voyager Encounter with Neptune in August 1989.

By that time, the range to the spacecraft would have doubled again from 20 AU at Uranus to 40 AU at Neptune. Before that point was reached, however, the DSN 64-m antennas would have been upgraded to 70-m diameter and this, together with the two

34-m antennas and the VGTA, would more than double the downlink capability at Goldstone and offset the loss due to increased range. The PCTA would produce a similar result at the Canberra longitude. It was expected that these enhancements would support data rates at Saturn similar to those that had been used at Uranus, despite the increased range penalty.

By early 1985, a Memorandum of Agreement and Management Plan for a joint JPL-NRAO VGTA Project had been signed by NSF and NASA. The JPL TDA Engineering Office would have responsibility for overall planning and management, supported by implementation and preparation managers at JPL and NRAO. The JPL effort was led by Donald W. Brown, and the NRAO effort by William D. Brundage.[39]

Under the terms of these agreements, NASA agreed to bear the costs of installing low-noise, X-band amplifiers on all 27 VLA antennas, as well as the direct costs of preparing and operating the VLA for Voyager/Neptune support. This would include at least forty monthly tests with Goldstone between late 1984 and early 1989, in addition to forty spacecraft tracks during the Neptune Encounter period in late 1989.

The onsite engineering, installation, test, and operation of the PCTA were managed, as they had been for the Uranus Encounter, by the staff at the CDSCC in cooperation with the staff at Parkes.

A high-level block diagram of the VGTA is shown in Figure 5-30.

In the process of reaching the final operational configuration described above, a number of technical problems that were unique to the VGTA design demanded considerable attention before they were finally resolved.

The problems began with the low-noise front-end amplifiers for the VLA antennas. The baseline plan for the X-band low-noise receivers on the VLA antennas included cooled Field Effect Transistor (FET) low-noise amplifiers. By the time the first three receivers had been installed in 1985, new technology offered the promise of significant improvement in system noise temperature by the use of the High Electron Mobility Transistor (HEMT) low-noise amplifiers. This translated into an increase of 1.5 dB in the expected overall sensitivity of the VLA. As the HEMT amplifiers became available beginning in 1986, tests confirmed the expected improvements, and ultimately all of the VLA antennas were equipped with HEMT amplifiers. It was calculated that when equipped with HEMT receivers, the VLA would equal nearly three 64-m DSN antennas in terms of downlink performance, a valuable resource indeed.

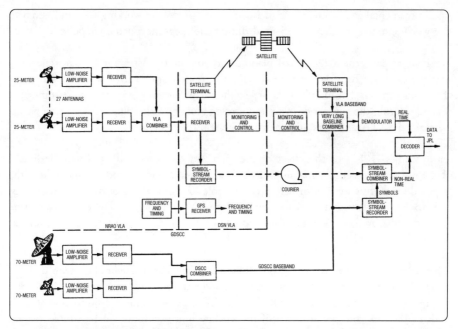

Figure 5-30. **VLA-GDSCC telemetry array.** In concept and operation, the VGTA closely resembles the PCTA described above. The essential features of the VGTA as eventually implemented were 1) X-band reception at the VLA and at GDSCC; 2) full-spectrum combining of 27 separate signals at VLA; 3) carrier demodulation to baseband at both sites; 4) transmission of the baseband signal from VLA to GDSCC via an Earth-satellite link; 5) standard baseband combining of one 70-meter and two 34-meter antennas at Goldstone; 6) baseband combining of the VLA and Goldstone signals in the Very Long Baseline Combiner at GDSCC; 7) convolutional decoding, signal processing, and data transmission to GDSCC to JPL; and 8) symbol stream recording at both sites to back up the real-time system and to allow for symbol stream playback and non-real-time combining at Goldstone.

For the combining (summing) process to work correctly, the signals from the twenty-seven separate VLA antennas must all be in phase with one another. By 1987, with the aid of new, more powerful computers at the VLA, phase measurements on all antenna baseline combinations could be made simultaneously. This allowed adjustments for phase variations along the signal paths to each antenna, caused by fluctuations in the troposphere, to be made in near-real time. The resulting ability to track out the effects of the troposphere, even in the severe summer thunderstorms characteristic of the VLA site, was instrumental in keeping the signal combining procedure working efficiently even in the worst summer weather experienced during the Neptune Encounter period.

An item of concern in the initial planning for use of the VLA for Voyager telemetry was the existence of a data gap in the signal output from each of the VLA antennas. The data gap was 1.6 milliseconds in length and occurred approximately 20 times per second. During the gap period, vital monitoring and "housekeeping" information passed between the VLA antennas and their remote control center.

Early studies (1982) of the effect of the "gap" on Voyager's coding scheme indicated that the error correcting capability of the outer Reed-Solomon code would bridge the gap and yield error-free performance, comparable to standard performance with perhaps 0.5 dB loss in threshold level for the combined VGTA output. These estimates were verified with simulated Voyager data and VGTA hardware and software during tests at JPL in 1985 and 1986.

The procurement and installation of a fully redundant Earth satellite link with dedicated transponders and Earth stations at each site were carried out by a contractor. Following the resolution of initial start-up problems and frequency response adjustments at both ends of the link, the performance of the satellite link was found to be similar to that of the microwave link associated with the PCTA.

During 1985, it became clear that the commercial power supplied to the VLA was too unstable to meet the DSN standards for primary power. Frequent voltage transients and outages occurred during inclement weather conditions. The operational deficiencies of the existing commercial power system were overcome by installing two diesel generators, each of 1,400-kW capacity, to provide primary power for the facility. This was followed by replacement of the deteriorating underground wiring that supplied power to the antennas out to the limit of the 20-km "Y" arrangement.

While an availability requirement of 80 percent was imposed on the PCTA Uranus Encounter configuration, the favorable experience led to an implicit design goal on the order of 90 percent for the entire VGTA for the Neptune Encounter. This required setting much higher reliability standards for the individual elements of the VLA system, including the twenty-seven individual antennas. In the beginning, it was not at all clear that thirty antennas (three DSN plus twenty-seven VLA) could be sustained and operated repeatedly with confidence in their reliability, given the diverse geographical and organizational aspects of the array. Because of these concerns, system reliability was given special attention throughout the implementation phase, and critical elements, such as an online computer and rubidium frequency standard, were provided by the DSN for use at the VLA.

The majority of the equipment supplied by JPL was installed at the VLA and Goldstone in late 1988. As each major assembly was completed, it was tested at the system level before being shipped to VLA or Goldstone for integration with the onsite systems. Monthly tests with the VLA and operational training at both sites continued through the year. These included demonstration tracks with all antennas on the Voyager spacecraft, where the array data was processed through the end-to-end system and delivered to the Voyager project for evaluation of its quality.

By May 1989, the VGTA was ready to provide full operational support to the Voyager mission as the spacecraft began its observatory phase of the mission. It was anticipated that, from this point on, the resolution in the Voyager images would exceed that of the best Earth-based observations. These expectations were completely fulfilled by the performances of the VGTA and the PCTA as the mission progressed. Some examples of the high-quality images that were downlinked at 21.6 kbps from Neptune were shown in Figures 5-7, 5-8, 5-9 of the Voyager section.

All of the careful, pre-Encounter preparations, enhanced by the proficiency of the NRAO and DSN operations personnel at the VLA and Goldstone, paid off handsomely during the Neptune Encounter. Statistics for the period 26 April through 28 September 1989 showed that during the forty Voyager Encounter passes supported in that period, the VLA signal to the DSN at Goldstone was available 99.959 percent of the time. As for the full VGTA and PCTA configurations which included the DSN antennas, data for the same period showed that more than 99 percent of the data transmitted by the spacecraft, from a distance of 40 AU, was captured by the Earth-based antennas of the interagency arrays.

The successful implementation and operation of interagency arrays in Australia and the United States provided a downlink with the capability of supporting telemetry from Neptune at 21.6 kbps for the full view period when the Voyager spacecraft was over Goldstone and Canberra (at the standard DSN 90-percent weather confidence level). Together with the expansion of the 64-meter antennas to 70 meters, interagency arraying had enabled a DSN downlink that effectively doubled the science data return from the Voyager Neptune Encounter.

The X-band Uplink

The arguments that drove the DSN to move from S-band to X-band for operation of their downlinks were not so compelling for the uplink. While the immediate improvement in downlink performance that followed directly from the frequency ratio of S-to X-band was of critical importance to the extension of DSN downlink capability to the outer planets, there were other means by which the S-band uplink capability could be extended. Prior to the early 1980s, improvement in uplink performance was achieved by increasing the power of the DSN S-band transmitters, first from 10 kW to 20 kW, then to 100 kW, and finally to 400 kW. At the same time, the effective radiated power in the uplinks was enhanced by the increased gain that resulted from the addition of new, larger antennas to the Network. Nevertheless, within the DSN Advanced Systems Development Section, a small (but very effective) X-band transmitter development program had been in progress for many years. This program was directed specifically toward supporting the DSN Planetary Radar program.[40] X-band transmitters had also been used to transmit timing signals from Goldstone to Canberra and Madrid by bouncing the time synchronization signal off the Moon.[41] These transmitters were designed for pulsed operation where, although the peak power level was hundreds of kilowatts, the average power was on the order of tens of watts. As distinct from planetary radar, planetary communications required continuous wave (CW) operation at the level of tens of kilowatts. The generation of tens of kilowatts of stable, CW, X-band power, and the power dissipation or cooling problems that were associated with it, was a challenging problem. The DSN needed a powerful rationale to embark upon such a task. It first appeared in 1978 with the gathering interest in the search for gravitational waves.

A proposal was made to detect gravitational waves by means of the Doppler signature imprinted on a CW, two-way radio link between Earth and a spacecraft crossing the interplanetary medium.[42] The key to success for such an experiment lay in the use of 1) a high frequency for the two-way radio path and 2) extremely high stability for the frequency of transmission. Together, this amounted to a requirement for a 20-kW, CW, X-band transmitter with a frequency stability of one part in ten raised to the power of fifteen (1×10^{-15}).

Apart from the potential for collaborating in a major scientific discovery, the DSN was highly motivated to gain the other benefits that were associated with a highly stable X-band uplink. These benefits related principally to improving the quality of the DSN radio metric data used for spacecraft navigation by effectively removing the transmission media effects. Improved accuracy of the DSN data would reduce the level of uncertainty in the orbit determination calculations, on which the spacecraft navigators depended for knowing the precise location of their spacecraft in interplanetary space.

The DSN undertook a phased development program in 1979 to add an X-band uplink capability to the Network. In the first phase of this program, an experimental model of an X-band transmitter and its associated equipment would be built and tested at DSS 13, the DSN's research and development station at Goldstone. In the second phase, a coherent X-band transponder would be added to an existing spacecraft prior to launch, to be used as an inflight technology demonstration of the end-to-end X-band system. The add-on would not interfere with operation of the spacecraft's normal S-band uplink.

By early 1981, a prototype 20-kW X-band transmitter and a highly stable exciter to drive it had been developed and were under test in the DSN labs at JPL. Also, arrangements were in place to add the X-band uplink to the NASA spacecraft, one of two spacecraft planned for the international Solar Polar mission to be launched in 1985.[43] The other spacecraft was to be provided by ESA and would eventually be renamed *Ulysses*.

Despite the cancellation of the NASA spacecraft in 1981, development of the experimental X-band uplink continued. At DSS 13, a full-scale ground system, including the Doppler extractors, X-band to S-band down-converter, and a Block III receiver, was set up and intensively tested. It was then planned to carry out the inflight X-band technology demonstration with the *Galileo* and *Venus Radar Mapper* (later called *Magellan*) spacecraft in 1984, using DSS 13 as the single ground station for the experiment.[44]

In a 3 June 1982 memo to the Assistant Laboratory Director for Telecommunications and Data Acquisition, the DSN Chief Engineer, Robertson Stevens, drew attention to the progress that had been made in the development of a stable X-band uplink at DSS 13. He also identified the advantages that would accrue from a move to X-band for operational use in the DSN. Relative to S-band uplinks, Stevens cited, as examples, improvements in range and Doppler measurements, immunity to signal distortion by charged particles along the transmission path, improved telemetry threshold in the two-way mode of Earth-to-spacecraft communications, and a tenfold improvement in command capability. Benefits not directly related to the uplink performance were to be found in lower cost spacecraft implementation and testing, as well as relief from future radio frequency interference problems.

He concluded by recommending that the operational Network should support the Galileo technology demonstration as a preamble to the initiation of a fully operational X-band uplink capability in the DSN.[45]

As a result, the DSN system engineering offices began writing technical requirements for the antenna, microwave, receiver/exciter, and transmitter subsystems that would be necessary to support the addition of an operational X-band uplink capability to the new 34-m high-efficiency (HEF) antennas, then being planned for implementation in 1984.

Shortly thereafter, Galileo was postponed to 1986 and Magellan to 1988, and the privilege of carrying out the first X-band uplink inflight demonstration fell to Galileo.

Frank P. Easterbrook and Joseph P. Brenkle made persuasive arguments for a joint Gravitational Wave Experiment (GWE) between Galileo (X-band) and Ulysses (S-band only), which provided the incentive needed to establish the GWE as a formal scientific endeavor,[46] as well as a technology demonstration as envisaged by the DSN.

In a carefully worded agreement that reflected the uncertainty of the times with regard to both budget and X-band operational performance, the DSN committed to implementing a 20-kW X-band uplink transmitter and three 34-m HEF antennas to support a technology demonstration which would include Doppler, ranging, two-way radio loss measurements, and commanding techniques.[47] It was believed that particular attention to phase stability would achieve a fractional phase stability approaching 5×10^{-15} to permit the detection of gravitational waves. The capability would be available in Canberra (DSS 45) in January 1987, at Goldstone (DSS 15) in July 1987, and in Madrid (DSS 65) in January 1988.

This plan was in process of being implemented when, as a result of the *Challenger* disaster in January 1986, both Galileo and Magellan were postponed to 1989, and Ulysses to 1990. In the aftermath of this affair, the HEF antennas were completed as planned, but the installation of the X-band uplink was deferred for about three years to meet a new set of requirements for the 1989 Galileo and Magellan missions. The new dates for the availability of an X-band uplink in the DSN were to be DSS 15 in March 1991, DSS 65 in September 1992, and DSS 15 in January 1993. In the meantime, a method was found for controlling the frequency and phase stability of the X-band transmitters to meet the stringent requirements imposed by the GWE.[48] Implementation progressed rapidly.

In fact, the X-band uplink at DSS 15 was completed and declared operational on 22 January 1990. By the end of the year, fully operational X-band uplinks were complete at DSS 45 and DSS 65. Tests were run to prepare them for the start of Galileo X-band operations when the high-gain antenna was unfurled in April 1991 for immediate support of Magellan. Ironically, *Galileo* never did get to use the X-band uplink because of the HGA failure as discussed earlier in this chapter. However, the spectacular suc-

The Galileo Era: 1986–1996

cess of the Magellan mission to Venus firmly established the X-band uplink as a powerful new DSN capability for Doppler, ranging, commanding, navigation, and radio science applications. The stringent requirements for X-band frequency stability were used to great advantage for very precise orbit determination and detection of gravity anomalies on Venus.

The early planning in 1994 for the new 34-meter beam waveguide antennas had made provision for 20-kW X-band transmitters. Budget considerations required these transmitters to include major components that were surplus equipment from tracking stations operated by the Goddard Space Flight Center. Subsequently, it was found that the effective radiated power needed to meet the technical requirements of future flight missions could be achieved with 4-kW, vis-à-vis 20-kW, transmitters on the new BWG antennas. Furthermore, the overall cost of installing new 4-kW transmitters was estimated to be about the same as retrofitting the surplus 20-kW transmitters. On these grounds, it was decided that the new BWG antennas would be implemented with 4-kW X-band transmitters and the work was re-directed accordingly.

By mid-1997, the X-band transmitters were complete at all three 34-m BWG antennas and, after a short period of operational testing, were put into service a few months later. The DSN could now operate two subnets of 34-m antennas with X-band uplink capability.

As the Galileo Era drew to a close in 1996, the X-band transmitters of the DSN were providing uplinks for Mars Global Surveyor, Mars Pathfinder, and Near-Earth Asteroid Rendezvous missions. In 1997, the giant *Cassini* spacecraft depended on X-band uplinks for its mission to Saturn. In 1998, Deep Space 1, the first of the New Millennium missions, and the Mars Surveyor Orbiter and Mars Surveyor Lander missions would also use X-band uplinks. Further into the future, long-term plans within the DSN called for 20-kW X-band transmitters on the 70-m antennas in the first years of the new millennium.

In 1997, 20 years had passed since the DSN, motivated by an interest in the search for gravitational waves, had moved toward X-band for its uplinks. X-band uplinks had become a reality; the improvements cited by Stevens in 1982 had all been fully realized, but the search for gravitational waves using this powerful medium for their detection remained a challenge for the future.

Block V Receiver

The Block V Receiver (BVR) was an all-digital receiver that had been under development as the Advanced Receiver and was characterized by high phase stability and extremely narrow tracking loop bandwidths, both highly desirable features for the Galileo and other weak downlink applications.[49]

The BVR first appeared for initial testing with the Galileo downlink at Goldstone in February 1994. As a result of these tests, some design modifications were made. In a further series of tests at the end of the year, the BVR successfully demonstrated its ability to track the weak Galileo downlink at all modulation settings, including the fully suppressed carrier mode. A telemetry data stream was flowed to JPL, and Doppler data were extracted and delivered to the Metric Data assembly. Repeated tests with the Galileo downlink in early 1995 confirmed the BVR performance under full operational conditions and showed excellent agreement with the performance predicted by theory.

Compared to the existing DSN Block IV receiver, the BVR showed greatly reduced high-frequency noise characteristics; it was virtually free of cycle slips even when the antennas were pointed close to the Sun, and noise in the Doppler data was reduced by at least a factor of three.

The role of the BVR in the development of the DSN Galileo Telemetry Subsystem (DGT) is described later. However, the new capabilities brought to the DSN downlinks by the BVR were not limited to its Galileo application. Its ability to track downlinks in the suppressed carrier mode; the extremely narrow bandwidth (about 0.1 Hz) of its carrier tracking loop; rapid, automatic signal acquisition; and great stability and versatility made it an obvious replacement for the older Block III and Block IV receivers throughout the Network. The time was propitious for the DSN to make such changes. The number of antennas in the Network had more than doubled in recent years, as had the number of simultaneous downlinks now handled by the DSN at each Complex. A separate receiver was necessary for each link, and the existing analog-based Block IV receivers were near the end of their lifespan. New receivers were needed, and the high-performance, multiple-function capability of the all-digital BVRs made them a logical choice for the new era. The 70-m antennas were equipped with BVRs in May 1995, followed by the 34-m HEF and 34-m BWG antennas a year later. A few of the old Block III and Block IV receivers were retained for use with the remaining SDA/SSA units on the 34-m STDN antennas and the 26-m Low-Earth Orbiter antennas.

The BVR not only performed residual carrier tracking, which was the only function performed by the Block III and Block IV receivers, but it also performed suppressed carrier tracking and carried out subcarrier and symbol demodulation functions. It therefore replaced not only the former Block III and IV receivers, but the former SDA/SSA and DSA units as well. In addition, it carried out the functions of signal estimation, Doppler extraction and IF signal distribution, thereby replacing the signal precision power monitors, Doppler extractors and IF distribution equipment that had been an integral part of the Mark IVA DSN.

The BVR accepted an S-band or X-band signal from a low-noise amplifier on the antenna to which it was assigned by the station controllers and, after performing the functions described above, distributed the following outputs for subsequent data processing:

1. two symbol streams to the assigned telemetry channel assemblies (TCAs) for decoding, frame synchronization, and telemetry data delivery;

2. two baseband outputs to the baseband assemblies (BBA) for combining with similar signals from other antennas;

3. two open-loop IF signals to other subsystems, such as radio science and VLBI, for special signal spectrum processing; and

4. two range-modulated signals to the Sequential Ranging Assembly (SRA) for ranging measurements.

Local Area Networks provided for communication and control between the BVR and other subsystems, and between the BVR and station controllers at the Link (LMC) or Complex (CMC) Monitor and Control level.

The DSN Galileo Telemetry (DGT) Subsystem

As it gradually became apparent that the Galileo HGA problem might not be corrected in time to save the mission as originally designed, attention began to turn towards viable options for continuing the mission in a modified form, using the spacecraft low-gain antenna (LGA).

In October 1991, the Telecommunications and Data Acquisition (TDA) Office at JPL chartered a thirty-day study to identify a set of options for improving the telemetry performance of the Galileo downlink at Jupiter, using the LGA, in the event that the HGA could not be recovered. The study team was led by Leslie J. Deutsch, Manager of Technology Development in the TDA Office.

The jobs of tuba player for a Dixieland jazz band and organist for a choir, church, and synagogue might seem unlikely roles for the mathematician most responsible for leading the DSN effort to recover the Galileo mission to Jupiter. However, all three roles were at one time played by Dr. Leslie J. Deutsch. Born and raised in Los Angeles, California, "Les" Deutsch spent much of his early professional life developing electronic music technology for his father's Deutsch Research Laboratories. After earning a doctorate in mathematics from the California Institute of Technology in 1980, Deutsch joined JPL in the Communications Systems Research Section, the section originally led by Walter K. Victor twenty years earlier. Deutsch became manager in 1986.

In 1989, he took charge of the Technology Development program in the Tracking and Data Acquisition Office, and it was while there that he brought the considerable resources of that group to focus on the Galileo problem. The full scope of the research carried out under that program is discussed separately in chapter 6.

A pleasant man of medium build, Deutsch always appeared to be busy. He was as much at ease discussing an abstruse mathematical representation of some physical problem in the DSN as he was executing a complex musical phrase on the pipe organ or tearing off a tuba riff in a jazz band.

In an interesting sidelight to his involvement with Galileo, Deutsch was invited to accompany a small party from NASA and JPL to Padua, Italy, in 1997 to participate in a "Three Galileos" conference. Sponsored jointly by the University of Padua, the German and Italian space agencies, and NASA, the conference honored "Galileo, the Man," "Galileo, the Telescope," and "Galileo, the Spacecraft." Deutsch contributed a technical paper[50] on the Galileo telemetry recovery effort. He was also honored with an invitation to present an organ recital in the Basilica of St. Anthony at Padua that featured music of Galileo's time. Dr. Deutsch continued to manage the program of technology development for the DSN until 1998, when he transferred to another area of JPL to study new technologies for planetary spacecraft.

Deutsch began by establishing a set of assumptions on which the team would base its considerations. Because of the HGA problem, all telemetry was being transmitted from the spacecraft on an S-band carrier through the LGA. The design of the spacecraft precluded the transmission of X-band telemetry through the LGA. If nothing was done to improve the S-band downlink performance by the time the spacecraft arrived at Jupiter in December 1995, the telemetry data rate would be only 10 bits per second, compared to the rate of 134,000 bits per second that had been designed into the original X-band mission.

The question facing the team was, "What could be done by making changes to the DSN or the spacecraft, or both, to improve this situation in the time available, and how much would it cost?"

Because the period of the study coincided with the Galileo Encounter with asteroid Gaspra, technical participation from the spacecraft and mission design areas was limited. Deutsch presented the team's report on 5 November 1991.

Out of the eight options that were identified, the following four were recommended for further evaluation:

Arraying Antennas

> Increase the effective aperture area of the DSN antennas by building new ones or arraying existing DSN antennas with large, non-DSN antennas such as were then in operation in Australia, Japan, Germany, and Russia. In addition, the S-band performance of the existing 70-m antennas could be improved by installing special low-noise feeds called "ultra-cones."

Compression

> Use data compression techniques on the spacecraft to increase the effective telemetry rate by reducing the number of data bits that needed to be transmitted for each image.

Coding

> Improve downlink capability at low data rates by adding more advanced coding to the existing coded telemetry data stream. This would require inflight reprogramming of the spacecraft computers.

Modulation

> Increase the effective power received by the ground antennas by changing the modulation index at the spacecraft transmitter to fully suppress the S-band carrier. This

would have the effect of putting all the transmitted power into the subcarrier sidebands that carried the information content of the downlink.

Estimates of the performance gain, cost, and uncertainty associated with each option were made and evaluated. The study concluded by emphasizing the urgent need for the TDA and the Flight Project Office (FPO) to jointly develop a workable plan to implement the necessary changes in the spacecraft and in the DSN in the limited time available.

The positive results of the largely TDA-driven, thirty-day Galileo Options study, along with its emphasis on the need for timely action, led to the chartering of a further study which would involve both TDA and the Flight Projects Office (FPO). It was to be called the Galileo S-band Mission Study and would run from 9 December 1991 through 2 March 1992. The co-leaders of the study group would be Leslie J. Deutsch, representing TDA, and John C. Marr, Manager of JPL's Flight Command and Data Management Systems Section, representing FPO.

The objectives of that study were to

- assimilate and verify the information from the previous TDA Galileo Options Study;

- solicit additional ideas and assess their performance benefit and feasibility;

- create a conceptual design for the end-to-end telemetry system for the Galileo S-band mission;

- generate a rough cost estimate for this design for submission to the program offices at NASA Headquarters which were responsible for funding TDA (Code O) and FPO (Code S) in time for the Fiscal Year 1994 budget cycle; and

- make specific recommendations for the implementation of an end-to-end data system and hand these over to the organizations that would perform the work.

Whereas the TDA study had addressed only the high-rate science telemetry downlink, this study included real-time, low-rate science and spacecraft engineering data, uplink command and radio metric data, and navigation data. It also stipulated that no improvements to the downlink would be made until the important Probe science data had been successfully returned to Earth using the existing downlink capability.

The basic conclusion of the study was that a viable Galileo S-band mission was indeed feasible, based on a design that could meet a somewhat reduced, but still very acceptable, set of science objectives. Such a design could be implemented in time for the start of the Galileo orbital mission, and the total cost to NASA would be about 75 million dollars spread over seven years.

The results of the study were presented at NASA Headquarters on 20 March 1992 by Deutsch and Marr, supported by the Galileo Project Manager, William J. O'Neil, and the Galileo Project Scientist, Torrence V. Johnson.[51]

O'Neil came straight to the point. There was no way of knowing whether the HGA would eventually be deployed, he said, and NASA should therefore prepare to complete the Galileo mission with the spacecraft LGA. He said that a key design feature of the new LGA mission would limit the data return from the tape recorder to one full tape recorder load for each of the ten targeted satellite encounters. That feature, coupled with the proposed enhancement of the downlink by additions to the spacecraft and DSN, gave reason for great optimism that a worthwhile scientific endeavor could still be successfully accomplished.

In addressing the mission science objectives, Johnson demonstrated that the proposed LGA mission would, in effect, yield 70 percent of the science return expected from the original HGA mission. That new science, he said, would represent a major advance beyond the current level of Jupiter knowledge based on the results of the Voyager mission.

Marr described the reprogramming changes that would have to be made in the spacecraft computers to add the new data compression and encoding algorithms; he also described the costs, risks, and implication of these changes in data format on Galileo mission operations.

The four options recommended for enhancement of the downlink by the DSN were essentially a refined version of those that had originated in Deutsch's original study. Suppressed carrier modulation would yield 3.3 dB improvement; advanced coding, 1.7 dB; ultracones for 70-m antennas, 1.6 dB; and antenna arraying, 0-4 dB depending on which antennas were used. Together with the spacecraft modifications, these enhancements would provide a capability in the DSN to return one full load of tape-recorded data after each satellite encounter. They also satisfied the project requirements for receiving continuous engineering data and low-rate science data. The options for arraying antennas were not limited in this presentation to just the DSN. They included the cost of adding the large antennas at Parkes (Australia), Bonn (Germany), and Usuda (Japan) to the DSN arrays.

Following the presentations, action items to study three alternative approaches were assigned, and the JPL team returned to Pasadena to prepare for a final decision meeting set for 10 April 1991.

The three alternative approaches related to when and how the various arraying options would be implemented in the DSN. This decision largely affected funding profile issues that were of paramount importance in the NASA Headquarters view of the LGA mission.

The final meeting was attended by the newly appointed JPL Director, Edward C. Stone, in addition to the NASA Associate Administrators for Space Communications and Space Science.

While the previous meeting emphasized the technical aspects of the LGA mission and addressed the costs in a rather general way, this meeting concentrated specifically on costs and the time rate of expenditure that would be incurred by the three alternative approaches previously identified. Since support of the LGA mission would require so much 70-m antenna time at the expense of other ongoing and future missions, the issue of DSN loading was also included in the DSN presentation.

The major cost drivers for the DSN were known to be the decoder and decompressor, the ultracones for the 70-m antennas, the signal combiners, communication links between antenna sites, and the cost of renting time on whichever of the non-DSN antennas would be chosen for arraying. Except for the latter, each of the costs was well understood, easily identifiable, and bounded. The arraying costs were not.

At a breakfast meeting on the morning of the presentations, the new JPL director expressed concern over the uncertainty associated with arraying costs and the rather overwhelming number of options available to choose from. His concern prompted Deutsch to propose a last-minute option that he had budgeted out during an impromptu lunchtime session in a vacant office at NASA Headquarters. The "Deutsch" proposal limited all arraying operations to the Canberra site. The Australian antenna at Parkes would be the only non-DSN antenna used for arraying, and DSS 43 would be the only antenna fitted with an ultracone. This approach offered a more manageable proposal on which to base a decision and was included as part of the total DSN presentation. The total DSN-related cost was estimated to be 38.3 million dollars spread over the six fiscal years 1992 through 1997, with the major cost impact in 1994. The spacecraft and mission opera-

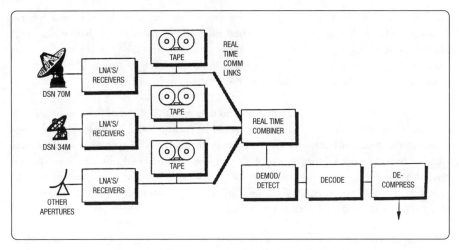

Figure 5-31. The concept of DSN telemetry for the Galileo LGA mission.

tions costs were 31.8 million dollars over the same period. This strategy was successful and was accepted as the way to recover the science data from Jupiter. Taken together, the spacecraft and ground system proposals would provide 70 percent of the original science data value at 1 percent of the original data rate for an overall cost increase of less than 5 percent.

The Galileo LGA mission as eventually approved by NASA satisfied most of the project requirements for real-time data and playback data return. It also provided for the new spacecraft capability but delayed the DSN enhancements until late 1996. It included DSN arraying centered on Canberra, as proposed by Deutsch. Software "hooks" were to be inserted in the design for the possible later inclusion of arraying with Parkes or Usuda, if funds could be identified.

The conceptual arrangement of telemetry proposed for DSN support of the Galileo LGA Mission is depicted in Figure 5-31.

Three types of receive antennas were to be used. These included the DSN 70-m antennas and the DSN 34-m antennas, with the non-DSN antenna at Parkes, Australia, as an option. Each antenna was to be equipped with a low-noise amplifier (LNA). The DSN 70-m antennas and the Parkes antenna would be equipped with ultracones and low-noise maser amplifiers, the others with cooled High Electron Mobility Transistor (HEMPT) amplifiers.

The LNAs were to be followed by receivers that would translate the incoming S-band signals to a lower frequency, filter out the uninteresting parts of the bandwidth, and digitize the resulting signal. The digital signals would be recorded on tape and sent to a central site for combining with the signals from the other antennas. There was also an option to send the digitized signal to the central site in real time.

At the central site, the signals from each antenna would be aligned with one another and combined (summed). The alignment would be accomplished by looking for the signature of the packet synchronization markers in the signals. The summed signal would then be demodulated. During much of the mission, there would not be enough signal strength to perform any demodulation before this summation. In some cases, data could be lost in the aligning process. When this happened, post-processing of the tape-recorded signals would be used to recover the data.

Following demodulation, the composite signal would be decoded in two stages. First, the effective (14, 1/4) convolutional code would be decoded. Then the Reed-Solomon code would be decoded. Finally, the effects of the two spacecraft compression algorithms would be undone in two decompressors. Subsequent processing of the science data would be performed by the Galileo Mission operations organization and the Science Teams.

From a technical point of view, the Galileo LGA Mission represented a remarkable episode of conceptual design in the areas of spacecraft data systems, planetary mission design, and DSN downlink engineering. However, before these concepts could become a reality, they had to be funded and implemented. While the NASA Headquarters people went off to wrestle with the funding problem on the short time scale required, the JPL people returned to Pasadena to initiate the implementation tasks.

Although we are only concerned here with the DSN effort, it needs to be said for the record that this evolution was an intensively interactive process. A satisfactory end result could only have been achieved by keeping all three elements—spacecraft, DSN, and mission operations—focused on the end-to-end performance of the downlink, while the implementation tasks proceeded as separate entities.

The task of implementing the DSN portion of the Galileo LGA mission was assigned to Joseph I. Statman, whose previous experience with developing the Big Viterbi Decoder for the original Galileo mission in 1989 was reported earlier. As Task Implementation Manager, Statman became responsible for transforming the conceptual ideas into hardware and software and installing and testing this equipment at the three DSN sites around the globe.[52] This task was to be accomplished within the cost estimates that had been presented to (and

The Galileo Era: 1986–1996

accepted by) NASA, and completed in time to support multiple Galileo Encounters with the satellites of Jupiter in less than forty-eight months. Although most of the technology was available, none of it had found practical application in this way, or on this scale, before. Some arraying technology had been used in the earlier Voyager encounters with Uranus and Neptune, and better techniques had been the subject of continuous study since then.[53] It had, however, been centered at Goldstone and had not involved intercontinental arraying between Goldstone and Canberra. The application of coding to enhance the performance of telemetry downlinks was well established in the DSN, and an advanced version of concatenated, convolutional coding had been added to the *Galileo* spacecraft downlink prior to launch.[54] This was intended for use at very high data rates (134 kbps) during the Io Encounter on the original X-band mission. Because it could not operate at the low data rates (10–100 bps) which would be typical of the LGA mission, it was of no further use. Nevertheless, the studies that had gone into developing that coding system were to be extended and applied in a new way to the LGA mission.[55]

Under the guidance of Statman, a comprehensive set of system and subsystem requirements and a cost breakdown for the DSN Galileo Telemetry (DGT) Subsystem, as it came to be called, was prepared and presented to a DSN Review Board for approval on 3 December 1992. Five days later, *Galileo* made its second and final close flyby of Earth and was on a direct path to Jupiter, just three years away. There was no time to lose.

In the course of transition from conceptual form to physical form, the DGT had been somewhat modified. Decompression of the received data function had become a Galileo mission operations function, and the arraying function had been limited to the one 70-m and three 34-m DSN antennas at Canberra. The combiner at CDSCC, however, was to make provision for combining the signals from up to seven antennas. This would permit a digitized signal from the DSN 70-m antenna at Goldstone to be added to the Canberra array to create real-time intercontinental arraying during Goldstone-Canberra overlapping view periods. The use of non-DSN antennas was not considered in this review.

The form in which the DGT was implemented at CDSCC is shown in Figure 5-32.

Although it was not considered part of the DGT, the Block V Receiver (BVR), shown in Figure 5-32, was an integral part of the DSN configuration for *Galileo*. Its primary purpose was to track the fully suppressed carrier downlink from a single 70-m antenna to extract telemetry and the two-way Doppler data essential for radiometric-based navigation of the *Galileo* spacecraft. When multiple-antenna arraying was used, the BVR provided Doppler data only and the DGT provided telemetry as described above. Its implementation proceeded in parallel with, but a year ahead of, the DGT.

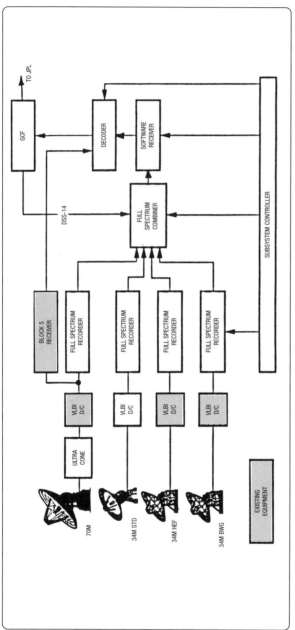

Figure 5-32. Functional diagram of the DSN Galileo Telemetry Subsystem for the Canberra Deep Space Communications Complex. The S-band signals from each of the antennas were first stepped down by 300 MHz to feed the Full Spectrum Recorders (FSR). After recording, the separate digitized signals were summed in a Full Spectrum Combiner (FSC) and delivered to a software receiver, where the detection and demodulation functions were carried out in a Buffered Telemetry Demodulator (BTD).[56] The complex decoding operation on the inner and outer codes was performed by a special Feedback Concatenated Decoder (FCD)[57] that delivered a decoded data stream to the ground communications equipment for formatting and transmission to JPL via the GCF. A system controller performed all the configuration and data management tasks via several local area networks. The arrangement at Goldstone was similar, except that there was no ultracone and no Full Spectrum Recorder. At Madrid, the arrangement was simpler still, consisting only of the decoder and system controller.

The Galileo Era: 1986–1996

Repeated tests with the Galileo downlink in early 1995 confirmed the BVR performance under full operational conditions and it was accepted for future Galileo S-band mission support.

By May 1995, BVRs had been installed at all Complexes and soon demonstrated their ability to track the Galileo downlink in the suppressed carrier mode, deliver telemetry, and extract Doppler data.

In September 1995, based on these impressive results, the Galileo downlink was finally switched to the suppressed carrier mode that, supported with the BVR, would become the standard mode of operation for the rest of the mission. Less than three months later, *Galileo* arrived at Jupiter, relying on the BVR to track its low-powered S-band downlink and recover the most important data of the entire mission, the Probe entry science data. Over the next several months, the BVR lived up to everyone's expectations, and in recovering all of the Probe playback data, it allowed the Galileo project to accomplish its primary mission objective.

The tests conducted at Goldstone in early 1994 to demonstrate the BVR had also been used to demonstrate the proposed full-spectrum recording and full spectrum combining techniques[58] using prototypes of each of these elements of the DGT. The Galileo downlink signals from DSS 14 and DSS 15 were recorded simultaneously by the prototype FSR and were later combined with the prototype FSC.[59] The results were well within the predicted performance limits. Later experiments would verify similar performance in combining the downlinks from the intercontinental array of DSS 43 with DSS 14.

Over the course of the following two years, the operational versions of these units were built, installed, and tested at the Complexes. Since arraying was to be used only at Canberra, that was the only site where a full-spectrum combiner was installed. Also during this period, an ultra-low-noise, receive-only feed system (ultra-cone) was added to the Canberra 70-m antenna to reduce the S-band system noise temperature from its normal value of 15.6 kelvin (K) to 12.5 K. By reducing still further the noise on the downlink, this addition enhanced the performance of the downlink and increased the rate at which it could return data from the spacecraft. The DGT was rapidly taking real form and substance.

Although the DSN had received approval (and funding) from NASA Headquarters for the use of multiple-antenna arraying techniques to support the Galileo LGA mission, this support was limited to DSN antennas only. As design studies matured, however, it became evident that, even with the already-approved downlink enhancements in the DSN, the Galileo science data return would be especially low during the six-month period when the spacecraft was farthest from Earth, November 1996 through April 1997. This was the period when four of the ten satellite encounters would occur. Galileo and DSN concern with this situation in March 1994 led the TDA Executive Committee to recommend that the DSN array configuration planned for Canberra be augmented with the 64-m antenna of the Australian radio telescope at Parkes. The cost of making the necessary modifications to the DSN, making additions to the Parkes antenna, and renting time on this antenna was estimated at 4.3 million dollars. This would be met by funds transferred from the Galileo project. The estimated enhancement in data return was about 10 percent.

NASA approval was quickly followed by an Australian agreement to participate in the Galileo LGA mission, and the DSN began work on the changes needed to add Parkes to the Canberra array. Working arrangements with Parkes were quickly reestablished based on the agreements that had been made for Parkes support of the Voyager mission a few years earlier.

The DGT at Canberra supported the 70-m antenna, plus three 34-m antennas at CDSCC, plus the 64-m antenna at Parkes, in addition to the 70-m antenna at Goldstone whenever overlapping view periods permitted.

While DSN engineers had been focused on implementing the BVR and the DGT in the Network, the Galileo project focused on the changes that were required in the spacecraft to make the end-to-end downlink work. At the same time, the project had to carry the mission forward to the point at which the new flight software could be transmitted to the spacecraft. This point was reached on 13 May 1996, when the DSN began radiating a continuous series of commands to load the spacecraft computers with the new Phase II flight software. The continuous sequence was completed without incident nine days later on 22 May, and the spacecraft was enabled to begin operating in the new mode the following morning.

By that time the DGT, including the FCD decoder, was in place at all three Complexes and ready to handle the new concatenated convolutional (14, 1/4) coded downlink at all bit rates in the suppressed carrier mode.

The Ganymede Encounter on 28 June 1996 was the first critical test of the new flight software and the DGT ability to process it. Since it was not yet ready for full arraying, the DSN used single 70-m antennas to return the Encounter data. As the tape recorder playback data began to arrive over the new downlink for the first time, the Galileo project manager declared the images returned from the Ganymede Encounter to be "absolutely stunning" and a tribute to everyone involved in meeting the challenges of the LGA mission.[60] The end-to-end design of the new downlink had been proven in critical flight operations conditions, and the initial optimism for a successful conclusion to the Galileo Mission was fully vindicated.

The Callisto Encounter in November 1996 was the first occasion on which multiple antenna intercontinental arraying at CDSCC was used in critical real-time operations. The full DSN array capability, augmented with the Parkes antenna, allowed the *Galileo* spacecraft data rate to be raised from 40 bps to 120 bps during overlap periods. This Encounter, too, was extremely successful and demonstrated the full operational capacity of the DGT. Based on the performance of the DGT in supporting the Callisto Encounter, the ensuing recorded data playback, and confidence in the DSN ability to deal with the increased operational complexity of full array operations, the DGT and associated array mode was designated as the standard configuration for all subsequent Galileo operations in the DSN.

The operational complexities and remarkable downlink performance resulting from this arrangement throughout the rest of the Galileo mission were discussed earlier in this chapter.

BEAM WAVEGUIDE ANTENNAS

The weekly edition of "Significant Events in the DSN" for 29 August 1997 reported, "After successfully completing an extended series of mission verification tests on the X-band uplink and downlink capability, the 34-m beam waveguide antenna in Canberra (DSS 34), began operational tracking support this week. This antenna will now be available to provide support [to] such missions as NEAR, Mars Global Surveyor, Mars Pathfinder, and Cassini, in addition to continuing support for the Galileo array." It further reported, "System performance testing began on schedule this week on the new 34-m beam waveguide antenna in Madrid (DSS 54). This will be followed by mission verification testing prior to the antenna being scheduled for operational tracking support on 1 October 1997." And so they were.

Together with DSS 24 at Goldstone, which had been completed in February of 1995, the completion of DSS 34 and DSS 54 gave the DSN a full subnet of three 34-m beam waveguide (BWG) antennas, which represented a most significant increase in uplink and downlink capability.

Based on the successful design of the existing 34-m high-efficiency (HEF) antennas, the new BWG antennas now extended the existing DSN uplink and downlink capabilities on S-band and X-band to include Ka-band (32 GHz).

The design of the 34-m BWG antennas embodied the most advanced principles of antenna and microwave design. It also represented the culmination of nearly ten years of DSN engineering, research, and development, which began with a JPL report on Ka-band downlink capability for Deep Space Communications by Joel G. Smith in December 1986.[61]

In advocating the extension of DSN downlink, and ultimately uplink, capability to Ka-band, Smith argued that, just as the move from L-band (0.96 GHz) to S-band (2.3 GHz) had offered a possible increase in performance by a factor of 5.74 (7.6 dB), and the move from S-band to X band (8.4 GHz) a further increase of 13.5 (11.3 dB), the move from X-band to Ka-band (32 GHz) offered a potential increase in performance of 11.6 dB. These improvements were based on the theory that the capacity of a radio link between two well-aimed antennas is roughly proportional to the square of the operating frequency. In practice, the actual realizable gain is reduced by effects in the propagating media and losses in the various physical components of the link.

However, these latter effects could be controlled more easily at higher frequencies where microwave components, particularly the surface area of the antenna itself, could be smaller, stiffer, and mechanically more precise. Furthermore, measurements had shown that degradation of downlink reception at Ka-band, due to weather effects, was not as severe as had previously been thought.

Based on these considerations, Smith proposed a course of action which would culminate in an operational Ka-band downlink capability in the DSN by 1995. It began with the development of a new research and development antenna at Goldstone to "verify the various approaches to be used in upgrading the existing 34-m and 64-m/70-m antennas to good efficiency at Ka-band."

As Smith pointed out, there were other reasons, independent of the Ka-band decision, for building a new research and development antenna at Goldstone, but there was a strong synergism between the two.

In a companion paper,[62] Smith laid out all of the arguments for a new antenna, including those related to Ka-band, and proposed a specific design based on the existing HEF antennas but modified to incorporate new beam waveguide technology for the transport of microwave energy at low level (receiving) or high level (transmitting) between the antenna and the receiving or transmitting devices.

At that time, 1986, the beam waveguide concept had been around for several years, but it had not been employed in the design of any DSN antennas. It had, however, been used recently in the design of several major non-DSN antennas, most notably the 64-m antenna at Usuda, Japan, and the 45-m radio astronomy antenna at Nobeyama, Japan. R. C. Clauss had made an evaluation of these installations and became a strong advocate for the new technology.[63]

The supporting arguments included significant simplification in the design of high-power water-cooled transmitters and low-noise cryogenic amplifiers, improved accessibility for maintenance and adjustment, and avoidance of performance degradation associated with the accumulation of rain and moisture on the feedhorn cover. Beam waveguide technology allowed these systems to be located in a fixed, nonrotating area beneath the antenna azimuth bearing. As a result, the long bundles of power and signal cables and cryogenic gas lines that were required on the existing antennas could be replaced with short, fixed, nonflexing connections. Better frequency stability, higher reliability, and improved performance could be expected.

A beam waveguide system is shown in conceptual form in Figure 5-33.

The flat reflectors were used to redirect the beam; the curved reflectors were used to refocus the beam. As the antenna rotated in elevation, or azimuth, the transmitting or receiving beam of microwave energy was constrained to the beam waveguide path by the action of the reflecting and refocusing surfaces. The various reflectors were enclosed by a large protective tube or shroud along the beam waveguide path. Although this tube was a highly visible element of a BWG antenna, it played no significant part in the beam waveguide transmission process.

Over the next two years, 1989 and 1990, a design for a new BWG research and development antenna at DSS 13 was completed; funding was obtained; and construction of the new antenna commenced.[64] It would be located about 300 feet south of the existing 26-m antenna at the Venus site at Goldstone, and it was scheduled for completion in mid-1990. It was intended to serve as the prototype for a whole new generation of DSN antennas using the new BWG-type of antenna feed system.

The mechanical and microwave optical components in the BWG path are visible in the diagram of the completed DSS 13 BWG antenna shown in Figure 5-34.

From July 1990 through January 1991, the new BWG antenna was tested as part of its post-construction performance evaluation. Antenna efficiency (gain) and pointing performance measurements at X-band and Ka-band were carried out using unique portable test packages installed at either of the two focal points, F1 and F3. Celestial radio sources

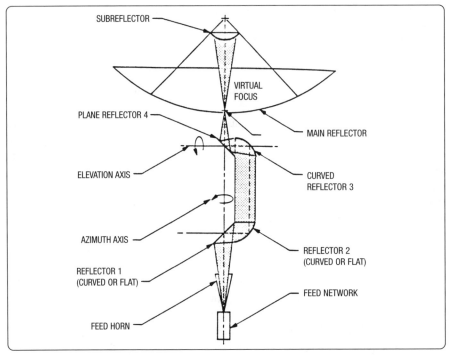

Figure 5-33. Beam waveguide antenna in conceptual form. In this diagram, Reflectors 4 and 3, mounted along the elevation axis of rotation, brought the beam from the subreflector on the main antenna down to the alidade structure. There, Reflectors 2 and 1, along the vertical axis, brought the beam to a stationary equipment room below the alidade that housed the transmitting and receiving equipment.

of known flux density were used in the calibration process. The values of peak efficiency of 72.38 and 44.89 percent, at X-band and Ka-band respectively, measured at the beam waveguide focus, met the functional requirements for antenna performance and agreed well with the predicted design values.[65] Further testing followed, as frequency stability, noise temperature, and the G/T ratio were evaluated and optimized.

Because system noise temperature is a critical parameter in deep space communication systems, the system noise temperature of the new antenna was of special concern. In the early tests, the system noise temperature was found to be higher than expected due to the spill-over losses of the BWG mirrors having a greater effect than previously thought.[66]

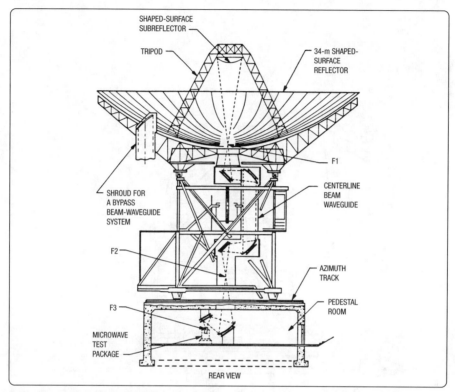

Figure 5-34. DSS 13 beam waveguide antenna.

New feedhorns were designed and tested and new methods for determining noise temperature in beam waveguide systems were developed.[67]

While the construction and testing of the prototype antenna at DSS 13 was in progress, submissions for funding for a fully operational version of a BWG at Goldstone (DSS 24) were being made. At this time, the Galileo mission to Jupiter was progressing normally and the high-gain antenna had not been deployed. The planned close flyby of Io in December 1995 over Goldstone demanded the full X-band support of the 70-m and HEF antennas at Goldstone, leaving little or no resources for other flight projects. Using this as justification, further funding for two more BWG antennas was requested from NASA Headquarters and, at the expense of some institutional compromises, was awarded. Now there was funding for a cluster of three BWG antennas at Goldstone, DSS 24, DSS 25, and DSS 26. These were placed on contract with an American construction

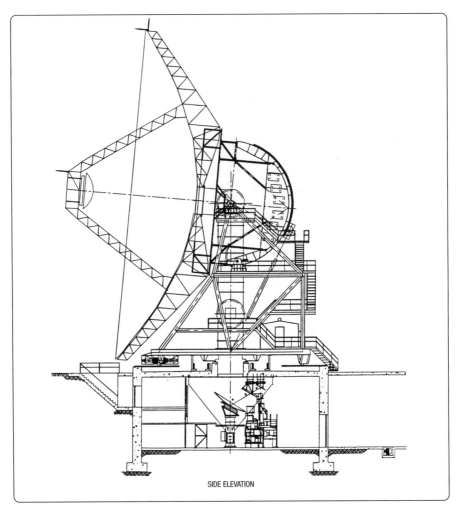

Figure 5-35. Design for operational 34-meter beam waveguide antenna. In the operational version of the 34-m beam waveguide antenna, the alternative bypass microwave system was eliminated and the alidade structure was strengthened to add stiffness and to accommodate the additional drive and ancillary services.

company, TIW Systems, Inc. All three antennas would be located in the Apollo Valley at Goldstone, and work began immediately.

By the time construction began, the original design for the DSS 24 antenna had been modified to incorporate many of the lessons learned from the DSS 13 experience and the intensive program of performance measurements that had been conducted on that antenna. In particular, the antenna structure had been modified to improve its "gravity performance" by controlling the antenna distortion for best fit to the ideal parabolic shape, as the antenna moved through the full range of elevation angles. Other improvements related to the azimuth cable wrap-up and to the substitution of nonmechanical "RF Choke" bearings for the mechanical tube bearings used at the rotating junctions of the waveguide shroud. Also, the four large (30-inch diameter) wheels which supported the alidade structure on the azimuth track had been "surface hardened" to improve wearability, based on operational experience with the existing HEF antennas. After a short period of use, these wheels cracked and had to be replaced with "through hardened" wheels to stand up to the loads and wear induced by continuous use under operational conditions. The design features of the operational version of the DSN 34-m BWG antenna are illustrated in Figure 5-35.

Construction of DSS 24 began in February 1992 with the blasting of a large hole, 30 feet deep and 60 feet in diameter, into solid bedrock to accommodate the pedestal room with its 24-ft- high ceiling. Because this was the first antenna to be built, it revealed the presence of several unsuspected problems in design and planning and, acting somewhat as a pathfinder for the others, was not completed until May 1994.

Microwave performance testing of the new antenna was carried out during the summer of 1994 using the same techniques that had been employed at DSS 13. Antenna efficiency, pointing calibrations, and system noise temperature were all included at S-band, X-band, and Ka-band.[68] In addition, microwave holography was used to adjust the reflector panels on the main antenna for optimum efficiency.

The values for gain and efficiency obtained from these tests are tabulated below:

DSS 24 Peak Gains and Efficiencies

Band, Frequency (GHz)	Gain (dBi)	Efficiency (percent)
S, 2.295	56.79	71.50
X, 8.45	68.09	71.10
Ka, 32.00	78.70	57.02

The results of these measurements confirmed the soundness of the basic design and the improvements that had accrued from the modifications mentioned above.

Experience gained from the DSS 24 task enabled the remaining two antennas to be built in a parallel fashion to save time. They were completed in July and August 1996, respectively, about 18 months after construction started in 1994.

Unlike DSS 13, which was built for deep space communications research and development purposes, the other BWG antennas were intended for operational use in the Network. Consequently, after the construction phase was completed, they were subjected to an intensive program of testing under operational conditions before they were declared operational and accepted for tracking flight spacecraft as an element in an array or in a stand-alone mode. The "operational" dates for all the BWG antennas are given in the table below, and they generally followed the completion of the antennas by several months.

DSN plans for BWG antenna construction in Spain and Australia were significantly affected by the dramatic events, described earlier, that occurred on the *Galileo* spacecraft in April 1991. To save the mission, DSN and Galileo engineers had proposed an alternative mission using the spacecraft's S-band, Low-Gain Antennas and an array of S-band antennas in Australia. This situation provided the DSN with the high-profile justification needed for adding a fourth BWG antenna at Canberra (DSS 34). With design and specification details already available from the Goldstone task, funding for construction of this antenna was quickly approved.

International bidding was opened for the erection of a fourth BWG antenna in Australia with an option for two more when funding became available. (It was recognized that significant savings could be realized by building the antennas in groups of three). The contract was won by the Spanish firm Schwartz-Hautmont, and construction began at the Canberra site in July 1994.

Initially, problems with water percolation in the large excavation for the pedestal room, along with difficulties with the contract and Australian unions, caused significant delay to the construction schedule. Eventually these problems were dealt with, work resumed, and the antenna was completed in November 1996, very close to the original schedule. After a short period of operational testing, the antenna was accepted for operational support and immediately began supporting Galileo on S-band as part of the Canberra array. Later, the X-band uplink was added, and after a period of mission verification testing, it began operational tracking support with X-band uplink and downlink in August 1997, as mentioned above. A photograph of the completed antenna is shown in Figure 5-36.

The Galileo Era: 1986–1996

Figure 5-36. DSS 34 BWG antenna at the Canberra Deep Space Communications Complex, Australia, 1996.

Two years after the Galileo high-gain antenna problem changed the course of BWG history in the DSN, another spacecraft event occurred which once again changed the course of the program.

In August 1993, eleven months after launch and three days before insertion into Mars orbit, communication with the *Mars Observer* spacecraft was lost and never recovered. In the aftermath of this loss, NASA hastily proposed and funded a new mission to Mars to be called *Mars Global Surveyor* (MGS). Its arrival at Mars toward the end of 1997 would be just in time to create a conflict for X-band uplink support over Madrid with the *Cassini* spacecraft scheduled for launch in October of that year. Once again there was a strong programmatic need for an additional BWG antenna in the Network, this time at the Madrid site. But this time, there was no funding immediately available. Fortunately, the DSN was able to reprogram sufficient money saved from the contracts for the first and second groups of three antennas to fund the option for the fifth BWG antenna at Madrid. The contractor, Schwartz-Hautmont, began construction in June 1995. Despite a water problem similar to the one that had occurred during construction in Australia, the antenna was completed by August 1997. System and mission verification tests followed, and the antenna was accepted for operations support in October 1997.

To a large degree, the disposition of S-band and X-band capability among the different BWG antennas was driven by the funding available for procurement of the necessary electronics packages. Although all of the antennas were capable of operation at Ka-band, the use of Ka-band for spacecraft telecommunications links, except for *Cassini*, had not matured beyond the need for a demonstration capability at one BWG antenna by the end of 1997. The need for a full subnet capability was expected to develop in the years ahead. A Ka-band downlink capability existed only at DSS 25.

In late 1994, two 34-m BWG antennas that had formerly belonged to the U.S. Army and were situated near the Venus site at Goldstone were transferred as surplus property to JPL. On transfer to the DSN, they became identified as DSS 27 and DSS 28 and were designated as High Speed Beam Waveguide (HSB) type. These antennas were azimuth/elevation mounted and, before they could be put into service, had to be stiffened and strengthened to meet DSN standards for reliability and stability under conditions of continuous use. DSS 27 was equipped to provide backup for the one remaining 26-m at Goldstone, DSS 16, and to supplement DSN capability for high-Earth orbiter missions like Infrared Space Observatory (ISO) and Solar Heliospheric Observatory (SOHO). DSS 27 thus became a 34-m BWG that had the performance of a 26-m S-band antenna with one multifunction receiver and a low-power, 200-W transmitter. Like DSS 16, it was operated remotely from the SPC located some ten miles away from the antenna site. Activation of DSS 28 was postponed for later, when funding for suitable electronics would become available. The photograph in Figure 5-37 shows a cluster of three 34-m BWG antennas at Goldstone in 1995.

The times at which the various BWG antennas came into operational service in the Network and their uplink and downlink operating bands as of December 1997 are summarized in the table below.

The Galileo Era: 1986–1996

Figure 5-37. Cluster of three 34-m BWG antennas at Goldstone, 1995.

DSN 34-Meter Beam Waveguide Antennas

DSS	In Service	Uplinks	Downlinks
13*	July 1990	S, X, Ka	S, X, Ka
24	Feb. 1995	S	S, X
25	Aug. 1996	X	X, Ka
26	Aug. 1996	X	X
27	July 1995	S	S
28	Oct. 2000	N/A	N/A
34	Nov. 1996	S, X	S, X
54	Oct. 1997	S, X	S, X

* DSS 13 for research and development use only.

N/A: Electronics not available as of January 1998.

With the completion of DSS 54 in 1997, the DSN finally had a complete subnetwork of three 34-m BWG antennas, all of which were capable of operation at S, X, and Ka-band. By then, the trend for future missions was towards shorter passes and higher data rates, a natural application for Ka-band uplinks and downlinks, as Joel G. Smith pointed out in the 1988 papers described earlier. In this context, the DSN was well positioned to cope with mission requirements in the immediate future. The possibility of a justification for more antennas seemed unlikely, and in terms of antennas, the DSN had reached a plateau that appeared to extend well into the foreseeable future.

SIGNAL PROCESSING CENTER UPGRADE TASK

Background

The progressive enhancement of uplink and downlink capabilities through the Mark II, Mark III, and Mark IVA models of the DSN has been traced in previous chapters. By 1993, the major goals of the Mark IVA configuration had been achieved. New antennas had been added to the DSN, Networks had been consolidated, and the Signal Processing Centers (SPCs) had been integrated. Centralized control was emerging as the established process for conducting mission operations at the SPCs. Spacecraft data received at the Deep Space Communication Complexes (CDSCCs) could now be returned to JPL at higher data rates and with fewer errors, thanks to complementary improvements in the NOCC and GCF. Fewer errors meant less need for lengthy data recalls and faster delivery of the final data records to the flight projects. Despite a few remaining functional and operational problems associated with the introduction of these changes, the Network was running smoothly and at maximum capacity.

In 1993, and for the remainder of the Galileo Era, implementation of new uplink and downlink capability in the DSN proceeded more on the basis of individual tasks highly focused on specific objectives, rather than as part of a general overall enhancement of the entire Network, as had been the case in the past. Three such tasks, the Galileo telemetry, Beam Waveguide Antennas, and X-band uplinks, are examples of this change in approach and have already been discussed.

While these tasks were in progress, however, a fourth major task involving changes and additions at the subsystem level in most of the major systems of the DSN was being carried out at the SPCs. It was called the SPC Data Systems Upgrade and covered the period 1990 through 1993.

Signal Processing Centers Upgrade

The drivers for the SPC upgrade were the future Mars Observer and International Solar Terrestrial Physics (ISTP) missions. Between them, these two future missions placed heavy demands for 26-m support, for X-band uplinks, and for additional telemetry downlinks, which exceeded the capability of the existing DSN.

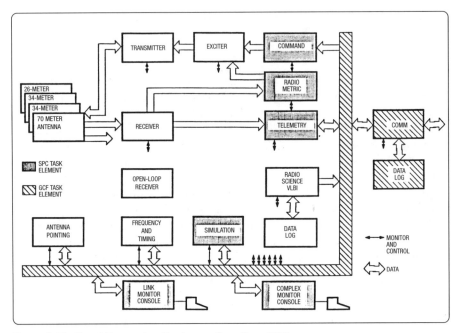

Figure 5-38. Signal Processing Center configuration, 1993.

There were four main objectives for this task:

1. Upgrade the telemetry system to support data rates up to 2.2 Mbps and provide Reed-Solomon decoding capability at the SPCs.

2. Add the 26-m Earth orbiter antennas to the existing SPC capability for centralized monitor and control of all Complex subsystems.

3. Increase the number of telemetry and command links to five per complex and expand the capability to control them by adding two more Link Monitor and Control (LMC) computers and consoles.

4. Replace the obsolete Modcomp II computers with new Modcomp 7845 machines, which had significant growth potential.

The scope of the SPC upgrade task is identified in the SPC configuration diagram shown in Figure 5-38.

In the planning and execution of this task, great care was taken to avoid impacting the SPC ability to continue full-scale mission operations throughout the entire period. This constraint led to a phased implementation plan that began in 1990 and ended in 1993. The SPC implementation plan was carefully integrated with similar and interlocking plans for the GCF and the NOCC. However, other events soon overtook the original plans, so by 1993, the original objectives had been extended. It was not until about 1995 that the task of upgrading the SPCs was completed.

SPC Telemetry

As part of the SPC data systems upgrade, the DSN Telemetry System, which began the Galileo Era with four groups (strings) of data processing equipment, was expanded to five strings, each capable of handling two data streams from a single spacecraft at selected data rates.[69] Each group was configured, monitored, and controlled by a Group Controller that set the spacecraft data channels, monitored their performance, logged the data, and formatted it for transmission to JPL via the GCF.

Groups 1 and 2 were composed of the original Mark IVA assemblies, subcarrier demodulator (SDA), Symbol Synchronizer (SSA), Maximum Likelihood Convolutional Decoder (MCD), and Telemetry Processor (TPA). Their convolutional decoding capability was limited to constraint lengths of 24 or 32 at rate 1/2 for sequential decoding at data rates of 3 bps to 5 kbps, and a constraint length of 7 at rate 1/2 or 1/3 for data rates 10 bps to 134 kbps, for maximum likelihood decoding. Uncoded data in the range 2 bps to 15 kbps was processed by the TPA.

These capabilities were expanded in Groups 3 and 4 by the addition of new functions and more advanced assemblies. For combining signals from up to eight receivers simultaneously, a baseband combiner assembly (BBA) was added. In addition to the combining function, the BBA also performed subcarrier demodulation and symbol synchronization and various combinations of all three functions. A more advanced version of the MCD, MCD II extended the data rate for maximum likelihood decoding to 2.2 Mbps, while frame synchronization at the frame transfer level was performed by a new frame synchronizer assembly (FSA). For telemetry data streams employing concatenated coding, a Reed-Solomon decoder for data rates of 2 bps to 2.2 Mbps was provided. Control and monitoring functions for each Group were performed by a new TPA, which also extended the upper limit for uncoded data from 15 kbps to 2.2 Mbps and carried out sequential decoding functions on either channel in the range 3 bps to 5 kbps.

With the exception of baseband combining, the capabilities of Groups 3 and 4 were replicated in Group 5.

Although the older Mark IVA assemblies in Groups 1 and 2 were eventually replaced with equipment which was equivalent to that in Group 5, the former data rates and formats were retained in these two Groups.

Over the next several years, improvements were made as the need arose. Responding to requests from the DSCCs, a self-test capability was added to the Telemetry Group Controller (TGC) and Telemetry Channel Assembly (TCA) software in 1995 to reduce the time required by the station operators to carry out pre-track readiness checks.

Following a long period of complex development and an exhaustive series of tests at Goldstone, new Block III Maximum Likelihood Decoders (B3MCD) were added to the telemetry subsystem at each Complex in 1997. The B3MCD could be selected as an alternative to the standard MCD in one channel of each of telemetry Groups 3, 4, and 5. The B3MCD responded to the requirements of *Mars Pathfinder* and *Cassini* for maximum likelihood decoding of convolutional codes of constraint length 15 and rate 1/6, in addition to the standard (K = 7, R = 1/2) decoding provided by the DSN. The addition of the B3MCD provided 1 dB to 2 dB improvement in bit signal-to-noise ratio compared to the existing (7, 1/2) MCD.

The B3MCD represented not only a major increase in downlink capability, but also a major accomplishment in new VLSI technology. It grew out of a 1988 concept by J. I. Statman for a powerful new decoding device for the long constraint length (15), high-rate (1/4) convolutional code to be employed on the Galileo 1989 X-band mission for returning Io Encounter science at a high data rate (134 kbps) from Jupiter.[70] The new Big Viterbi Decoder (BVD) grew out of two contemporary developments in the DSN's Advanced Systems program—the successful search for a long constraint length code with a "2 dB additional coding gain,"[71] and the burgeoning expertise in Very Large Scale Integrated (VLSI) technology.[72]

Early progress was rapid, and by mid-1991 a prototype had been completed and run successfully. Using experience gained from the prototype and improved VLSI chips, a new "single board" decoder was being constructed for actual implementation in the DSN. The new BVD would be programmed to decode convolutional codes having constraint length of 2 to 15 at rates of 1/2 to 1/6, and it would handle the highest Galileo telemetry data rate of 134 kbps.[73]

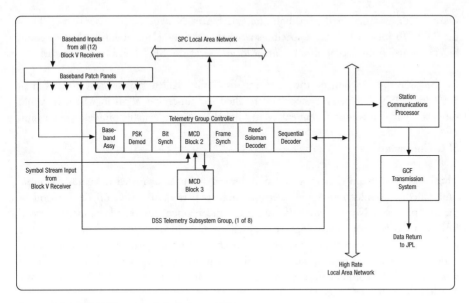

Figure 5-39. DSN Telemetry Subsystem, 1997.

But the great expectations that were held for the BVD would not be fulfilled. In the wake of the Galileo HGA problem in 1991, further work on the BVD was suspended, and Statman was redirected to manage the DGT task in an effort to recover downlink data from the Galileo S-band mission at data rates of 10 bps to 100 bps.

Later, the need for an improved decoder for the Telemetry System prompted the revival of the BVD concept. However, it was an entirely new development task using even more advanced VLSI technology that resulted in the design for Block 3 Decoders that were eventually produced for implementation at all three complexes in 1997.

This brought the DSS Telemetry Subsystem to the generic configuration shown in functional form in Figure 5-39.

The number of telemetry groups at each complex varied according to the number of antennas to be supported, eight being the maximum, at Goldstone. A Baseband Patch panel provided flexibility for switching or combining multiple inputs from up to twelve Block V receivers. Each group contained the assemblies necessary for performing the telemetry-related functions of baseband or symbol stream combining, demodulation, bit and frame synchronization, and convolutional, Reed-Solomon, or sequential decoding.

Output data was delivered to a high-rate Local Area Network for transport to JPL via the GCF. To form a link for a specific spacecraft pass, the elements of any group could be configured from a centralized control location via the SPC Local Area Network.

By the time the DSN made the transition from the Galileo Era to the Cassini Era in 1997, the DSN Telemetry System had reached the functional form in which it would support the new missions of that era, Mars Global Surveyor, Mars Pathfinder, and Cassini.

SPC Command

The DSN Command System also was improved to meet increased demands for more simultaneous uplinks with higher performance, as part of the SPC upgrade period of the mid-nineties.[74] Not only were the number of independent command channels increased from four to five, but switching arrangements were added to allow station controllers greater flexibility in switching any command channel to any exciter-transmitter combination. Command data rates and subcarrier frequencies were increased, and provision was made for the command system to be operated directly from the NOCC or a remote Mission Operations Center (MOC).

At each complex, each of the five groups in the command subsystem comprised a Command Processor Assembly (CPA), Command Modulation Assembly (CMA), and Command Switching Assembly (CSA).

The CPA contained software that provided electrical interfaces with all other related subsystems in the Complex and with the NOCC and MOC. It accepted the serial, pulse-code-modulated (PCM) command data stream from the MOC and delivered it to the CMA for modulation on to a subcarrier.

The CMA generated the desired subcarrier in the frequency range of 100 Hz to 16 kHz and modulated it with the command data received from the CPA. Phase Shift Keyed (PSK) or Frequency Shift Keyed (FSK) modulation was available at a command data rate ranging from 1 bps to 2,000 bps.

The CSA connected the command-modulated output of the CMA to a designated exciter for generating the S-band or X-band carrier, which was finally radiated by the transmitter and assigned antenna. These functions are shown in Figure 5-40, which illustrates the DSCC Command System in 1996.

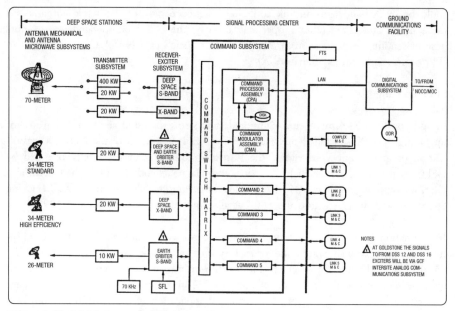

Figure 5-40. DSCC Command System, 1996.

In 1997, an additional 34-m antenna, the BWG, with a 4-kW X-band transmitter, was added to each complex and integrated with the command system by means of the Command Switch Matrix.

In the upgraded SPC, assignment of command groups to specific exciters, transmitters, and antennas was carried out by a single "complex" operator. The resulting "uplink" was then configured by a "link" operator to support a specific spacecraft in the next scheduled tracking and commanding period. After the correct configuration and operation of the link had been validated, it was turned over to the MOC for use in transmitting the desired commands to the spacecraft. Because the DSN Command System had the potential to cause unintended, undesirable, or irreversible actions in a spacecraft, particular care was taken to ensure that the command data radiated by the transmitter was identical to the command data input to the CPA. This assurance was provided by verification and confirmation functions as an integral part of the command system.

After the final installation of CPA/CMA software in September 1995, each complex had the capability to support five active command uplinks simultaneously, two of which could be on X-band. Any uplink could be assigned to any of the active antennas on the complex and could (after proper validation by the DSN) be operated for the purpose of sending commands from a remote MOC (such as a NASA Center other than JPL, or an agency other than NASA).

SPC Tracking

In addition to the telemetry and command subsystems discussed above, DSN plans for upgrading the SPCs included the tracking subsystem. As a result of the Network consolidation program some years earlier, the 9-m and 26-m antennas, formerly operated by GSFC as part of the Spaceflight Tracking and Data Network (STDN), had become a DSN responsibility. DSN plans in 1990 called for them to be added to the DSCC Tracking Subsystem and integrated into the SPCs at all complexes by May or June of 1995.[75]

In a functional sense, the Tracking Subsystem consisted of four basic groups of equipment (hardware and software), one for each type of antenna—namely, 70-m, 34-m STD, 34-m HEF, and 9-m/26-m—where the 9-m/26-m antennas were represented by a single type. The principal application of the 70-m/34-m group was in tracking deep space or high-Earth orbiting spacecraft. The principal application of the 9-m/26-m group was in initial acquisition of all spacecraft for which the DSN was to provide tracking and data acquisition services, and in tracking low-Earth orbiting spacecraft.

All 70-m/34-m groups were identical and contained a Metric Data Assembly (MDA) and a Sequential Ranging Assembly (SRA). Their primary functions were to generate radio metric data consisting of Doppler, range, differenced range versus integrated Doppler (DRVID) and uplink tuning data. Performance monitoring and data validation were secondary functions. The primary functions were applied to S-band and X-band uplinks and downlinks as determined by the capability of the antennas to which they were assigned. Each group was permanently associated with a specific transmitter-exciter-receiver combination and, like all other subsystems in the SPC, could be configured and controlled from the centrally located DSCC Monitor and Control (DMC) station via a Local Area Network.

With the exception of the MDA, which was replaced by a Metric and Pointing Assembly (MPA), the composition of the 9-m/26-m groups was similar to the 70-m/34-m groups. Their primary functions, however, were very different. The main purpose of the MPA was to point the 9-m/26-m antenna, a function which was carried out by a special Antenna Pointing Subsystem on the other antennas.

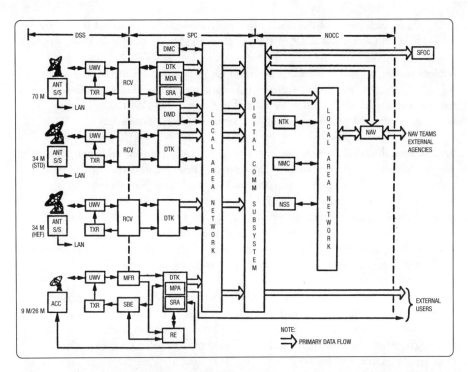

Figure 5-41. DSCC Tracking Subsystem, 1995.

Using trajectory data unique to its specific site and spacecraft of interest supplied by an external entity (generally GSFC), the MPA generated spacecraft signal acquisition parameters in the form of topocentric angles, Doppler frequency, and range delay. These data were used to point the antenna in various modes of tracking and to validate the Doppler and ranging performance of special ranging equipment used on these antennas for support of older Earth-orbiting missions. The MPA also used angle and acquisition data to generate predictions for any colocated antenna. This capability was frequently used to point the 34-m antennas during the initial downlink signal acquisition sequence following a spacecraft launch, when the spacecraft trajectory was not known with sufficient precision to point the narrow-beam 34-m antennas. The MPA controlled and monitored the SRA in a manner identical to the MDA.

The functions just described for all four types of antennas are shown in block diagram form in Figure 5-41.

In this figure, each group of the DSCC Tracking Subsystem is represented by the designator, DTK. A fifth group was added in 1997 to accommodate the new 34-m BWG antennas.

Implementation of the new MDA at the SPCs in Madrid, Canberra, and Goldstone was completed in early 1993 without significant problems. This was followed a few months later by the MPA, so that by the end of the year, the DSCC Tracking Subsystem as originally planned was in place and supporting mission operations.

By that time, however, several major events had occurred that would materially affect the just-completed DSCC Tracking Subsystems. These events were the failure of the Galileo HGA and subsequent emergence of the DGT and BVR to support a Galileo S-band mission, encouraging results from development tests of an X-band exciter and approval of plans for another set of 34-m antennas (BWG). The BVR would replace the existing Block IV receivers, the new exciter would bring X-band uplinks to the DSN, and the BWG antennas would increase the number of tracking system groups to five. In all three instances, new MDA software and hardware was required.

Although an early version of the BVR appeared at Goldstone at the beginning of 1993 for testing with the Galileo downlink, it was not until May of 1995 that it was implemented in full operational form at all complexes and ready to begin acceptance testing with the new MDA software. These tests revealed problems with missing Doppler data between the BVR and the MDA and required more changes to the MDA software. The DSN was to switch to the BVR for Galileo downlink mission operations on 18 September 1995. Also at that time, the Galileo downlink would be switched permanently to the suppressed carrier mode. *Galileo* was approaching Jupiter, and good Doppler data was vital to the success of the Galileo Jupiter Orbit Insertion sequence. Under pressure of these impending changes, the new MDA software was developed; checked in the DSN test facility at JPL, DTF 21; tested at Goldstone (DSS 14); shipped to the other complexes; and installed in eight weeks. It was soon followed by a new software package for the SRA, which, together with a new hardware board, allowed it to generate ranging data with the BVR. It was this software that was used to support the highly successful Galileo mission operations at Jupiter in December 1995.

Subsequently, more changes were made to the MDA software to add operability enhancements requested by the users at the Complexes, to improve the Monitor interfaces, and to provide an interface to the Block V X-band exciter. Although this latter capability was not driven by *Galileo*, it would be required by other spacecraft: *Mars Global Surveyor*, *Mars Pathfinder*, and *Cassini*, all to be launched within the next year or two.

By 1997, when the new beam waveguide antennas came into operation, the DSCC tracking subsystem had been greatly expanded at each Complex to accommodate the additional X-band uplink and downlink.

In addition to the DSCC Tracking Subsystem, there was another important component of the overall DSN tracking system located at each complex. This was called the DSCC Media Calibration Subsystem (DMD). The two principal functions of the DMD were to make continuous measurements of ground weather parameters and to make continuous measurements of ionospheric conditions at the site. These data were formatted and transmitted to the NOCC for use by the Navigation Team in correcting radio metric data for media effects before it could be used for orbit determination purposes.

The local ground weather observations were made by a meteorological monitoring assembly and had gradually been improved over the years. However, the ionospheric data had for many years been supplied to the DSN by other agencies and were based on ad hoc observations of an aging Earth satellite (ETS-6) from several remote-observing sites. This unsatisfactory method of determining total electron content (TEC) was replaced in 1992 by a system that measured TEC continuously at the local site. Based on Global Positioning Satellite (GPS) technology, the new GPS Receiver/Processor Assembly (GRA) could automatically acquire and track at least four GPS space vehicles when they were above the local horizon without requiring operator intervention. The GRA computed the differential group delay and differential phase delay between the L1 and L2 carriers from the space vehicles and estimated the slant TEC. The DMD made provisions for monitoring, validating, and recording the TEC data prior to its transmission to JPL.

NETWORK OPERATIONS CONTROL CENTER (NOCC)

Located at JPL, the NOCC furnishes coordination, control, and monitoring services to the Network and serves as the single data and communications interface between the Complexes and the external users of the DSN. Although these functions were not changed, they were significantly affected by the SPC upgrade, and necessitated an upgrade to existing NOCC capability.

Concurrent and interlocked with the upgrade of the SPC, the task of upgrading the NOCC was directed toward complementing the new capabilities of the SPC and improving the reliability and operating efficiency (operability) of the Center. The primary functions of the NOCC—providing a control interface between the Complexes and external users such as the Flight Projects, and providing a central facility for the control and monitoring of the DSN—were not changed.

The upgraded NOCC gave its operators better information with which to monitor the performance and status of the tracking stations, resulting in a reduced need for verbal interaction with the station operators. The upgraded SPC gave the station operators more information with which to identify and correct problems independently of the NOCC operators. The new ability of the SPCs to perform frame synchronization and decommutation of telemetry data and to calculate the deviation of actual performance from predicted values was a key factor in this area.

Other areas of activity in the NOCC saw similar benefits. A manual procedure for reporting frequency and timing offsets was replaced by an automatic procedure, and the need for a member of the radio science team to be present at a station during radio science activity was eliminated. Faster identification of operational errors and the reduction of command, tracking, and monitor communications traffic between the NOCC and the complexes led to significant improvement in the quality of the VLBI navigation data generated by the Network.

The NOCC upgrade was carried out as planned and, when completed in 1993, represented a significant advance in the efficiency of centralized control of the Network. Except for a change in its function as an interface between external users and the DSN, the NOCC continued to function as an integral part of the DSN without further modification through the end of the Galileo Era.

Ground Communications Facility

In order to make the changes that were planned for the SPC, it was necessary to upgrade the existing communication facilities at the three Signal Processing Centers (SPCs), as well as those at the Central Communication Terminal (CCT) at JPL. This became the Ground Communication Facility (GCF) Upgrade Task, and it ran concurrently with, and with the same constraints as, the other upgrade tasks.

Prior to the GCF upgrade, the interchange of data between the subsystems at each SPC was carried on a single Ungerman-Bass Local Area Network (LAN). To handle the additional telemetry, command, and tracking equipment that came with the SPC upgrade, the GCF task provided two new LANs, each connected to a new Station Communications Processor (SCP). One LAN was linked to the old UBI LAN via a "gateway" and was designated as the SPC LAN. The other was designated as the Hi-rate LAN. Interactive traffic between the Complex/Link Controllers on the UBI LAN and the telemetry, command, and tracking subsystems on the SPC LAN flowed through the "gateway." Traffic between these subsystems and the NOCC moved through the SCP. The Hi-rate LAN handled only telemetry, transporting the output from all five telemetry channels to the SCP. The SCP provided digital recording and formatting functions for the data on both LANs before transmitting it to the Central Communications Terminal (CCT) at JPL via wideband satellite and ground communication circuits provided by NASCOM.

The CCT at JPL was the terminal for all voice and data circuits between the NOCC and each of the complexes. Located at the CCT, the Central Communications Processor (CCP) could be viewed as a mirror image of the Station Communications Processor with the GCF transmission subsystem, and it included the far-reaching communication channels of NASCOM, forming the link between them. A LAN connected the CCP with a digital recording subsystem, a monitor and control subsystem, and three "gateways." It was through these "gateways" that the GCF digital communications subsystem connected the Complexes and the NOCC to the outside world. The NOCC gateway provided for all NOCC traffic. An "External User" gateway conducted traffic with remote Mission Operations Control centers such as GSFC or space agencies in foreign countries, while the third "SFOC" gateway handled the data flow between the SFOC* and the DSN.

A simplified diagram of the subsystems of the GCF as it was configured in 1992 is given in Figure 5-42.

The Big Pipe and Little Pipe elements of the GCF transmission system were descriptors for the NASCOM communication channels. They delivered data to the CCT at 672 kbps and 64 kbps, respectively, from Madrid and Canberra, and 1.2 Mbps and 224 kbps from Goldstone.

Implementation of these changes was to be phased in with incremental changes in SPC capability and was in progress when further changes were imposed by the requirements of the Galileo mission redesign in 1993. To meet the arraying requirements of

---

* At various times, the facility at JPL from which JPL inflight missions were controlled was known by different names that reflected the changing scope of its activities. Some of these names include the Space Flight Operations Facility (SFOF), the Mission Control and Computing Center (MCCC), the Space Flight Operations Center (SFOC), and the Multimission Ground Data System (MGDS). At the time of the GCF upgrade, it was called the SFOC.

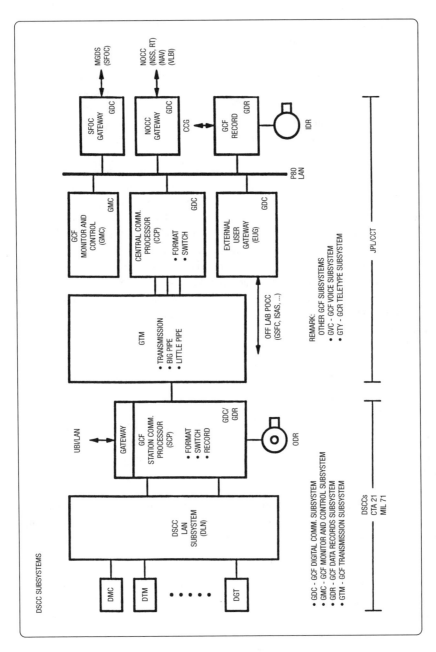

Figure 5-42. Ground Communications Facility, 1992.

the DGT, a special channel having a capacity of 400 kbps was established between Goldstone and Canberra. This channel, which consisted of two standard intersite circuits, was routed through the CCT for monitor and control purposes and was specified to require an error rate better than 2×10^{-7} with a 1-way, 2-hop delay that would not exceed 0.75 second.

As part of the original SPC upgrade, new capability had been added to all DSN subsystems to allow them to transfer data files among each other. This was extended in 1993 to include the SFOC by providing suitable hardware and software in the GCF digital communications subsystem for routing and monitoring non-real-time data file transfers. This service eventually came into regular use for automatic post-pass transfer (PPT) of data files between the complexes and the flight project mission control centers, and it proved to be of great value in streamlining the conduct of routine DSN operations.

A SUCCESSFUL CONCLUSION

The successful conclusion of the Galileo Prime Mission in December 1997 brought the Galileo Era to a close. The effort devoted to conducting mission operations, for both the spacecraft and the DSN, was scaled back but continued, at a much reduced level, as the Europa Continuation Mission. By that time, the Galileo project had been on the DSN "books" in various forms for nearly twenty years. Also, NASA had recently announced a new initiative for "faster, better, cheaper" spacecraft. *Cassini*, the last of the planetary "mega-spacecraft," had already been launched and was on its way to Saturn.

The Cassini Era was at hand.

Endnotes

1. Bruce Murray, *Journey Into Space* (New York: W. W. Norton, 1989), pp. 227, 228, 321.

2. C. Stelzried, L. Effron, and J. Ellis, "Halley Comet Missions," TDA Progress Report PR 42-87, July-September 1986 (15 November 1986), pp. 240–42.

3. C. T. Stelzried and T. Howe, "Giotto Mission Support," TDA Progress Report PR 42-87, July-September 1986 (15 November 1986), pp. 243–48.

4. N. A. Mottinger and R. I. Premkumar, "Giotto Navigation Support," TDA Progress Report PR 42-87, July-September 1986 (15 November 1986), p. 249–62.

5. R. Linfield and J. Ulvestad, "Source and Event Selection for Radio-Planetary Frame-Tie Measurements Using the Phobos Landers," TDA Progress Report PR 42-92, October/December 1987 (15 February 1988), pp. 1–12.

6. K. M. Cheung and F. Pollara, "Phobos Lander Coding System: Software and Analysis," TDA Progress Report PR 42-94, April/June 1988 (15 August 1988), pp. 274–86.

7. P. H. Stanton and H. F. Reilly, Jr., "The L-/C-band Feed Design for the DSS 14 70-Meter Antenna (Phobos Mission)," TDA Progress Report PR 42-107, July/September 1991 (15 August 1991), pp. 364–83.

8. J. Withington, "DSN 64-Meter Antenna L-band (1688-MHz) Microwave System Performance Overview," TDA Progress Report PR 42-94 April/June 1988 (15 August 1988), pp. 294–300.

9. C. E. Hildebrand, B. A. Iijima, and W. M. Folkner, "Radio-Planetary Frame Tie from Phobos-2 VLBI Data," TDA Progress Report PR 42-119, July-September 1994 (15 November 1994).

10. A. L. Berman, "Deep Space Network Preparation Plan for Magellan Project," DSN Document 870-73, JPL D-4852 (Pasadena, California: JPL, 15 March 1989).

11. A. L. Berman, D. J. Mudgway, and J. C. McKinney, "The 1986 Launch of the Galileo Spacecraft via the Space Transportation System," TDA Progress Report PR 42-72, October-December 1982 (15 February 1983), pp. 186–99.

12. Jet Propulsion Laboratory, "Magellan Spacecraft Final Report," Internal Report MGN-MA-011 (Pasadena, California: Martin Marietta Technologies, Inc., January 1995), pp. 50–51.

13. Rick Gore, "Neptune; Voyager's Last Picture Show," *National Geographic Magazine* (August 1990): 35–64.

14. Bradford A. Smith, "Voyage of the Century," *National Geographic Magazine* (August 1990): 48–65.

15. W. J. O'Neill et al., "Project Galileo; Mission and Spacecraft Design," JPL D-0518, Jet Propulsion Laboratory, January 1983 (compilation of papers presented at AIAA 21st Aerospace Sciences Meeting, Reno, Nevada, 10–13 January 1983).

16. A. L. Berman, D. J. Mudgway, and J. C. McKinney, "The 1986 Launch of the Galileo Spacecraft via the Space Transportation System," TDA Progress Report PR 42-72, October-December 1982 (15 February 1983), pp. 186–99.

17. S. Dolinar, "A New Code for Galileo," TDA Progress Report PR 42-93, January/March 1988 (15 May 1988), pp. 83–96.

18. D. J. Mudgway, Jet Propulsion Laboratory, Telecommunications and Data Acquisition Office, interoffice memorandum to R. J. Amorose, from "Galileo Convolutional Coding Proposal," 6 July 1987.

19. J. W. Layland et al., "Galileo Array Study Team Report," TDA Progress Report PR 42-103, July/September 1990 (15 November 1990), pp. 161–69.

20. T. K. Peng, J. W. Armstrong, J. C. Breidenthal, F. F. Donivan, and N. C. Ham, "Deep Space Network Enhancement for the Galileo Mission To Jupiter," IAF-86-304 (Innsbruck, Austria: International Astronautical Federation, October 1986).

21. W. J. O'Neil, "The Galileo Messenger," Issue 31 (Pasadena, California: Galileo Project Office, JPL, February 1993), p. 9.

22. L. Deutsch and J. Marr, "Low Gain Antenna S-band Contingency Mission," Galileo Project Document 1625-501 (Pasadena, California: JPL, 10 April 1992).

23. J. I. Statman, "Optimizing the Galileo Space Communication Link," TDA Progress Report PR 42-118 (15 February 1994), pp. 114–20.

24. P. E. Beyer, D. J. Mudgway, and M. M. Andrews, "The Galileo Mission to Jupiter: Interplanetary Cruise; Post-Earth Encounter through Jupiter Orbit Insertion," TDA Progress Report PR 42-125, March/April 1996 (15 May 1996), pp. 1–16.

25. P. E. Beyer, B. G. Yetter, R. G. Torres, and D. J. Mudgway, "Deep Space Network Support for the Galileo Mission to Jupiter: Jupiter Orbital Operations From Post-Jupiter Orbit Insertion Through the End of the Prime Mission," TDA Progress Report PR 42-133, March/April 1998 (15 May 1998), pp. 1–23.

26. Willis G. Meeks and Ed. B. Massey, "Ulysses: An Investigation of the Polar Regions of the Heliosphere," personal communication, July 1997.

27. European Space Agency, "The Ulysses Data Book," European Space Agency Document, ESA-BR 65 (Noordwijk, Netherlands: ESTEC, June 1990).

28. D. M. Enari, "DSN Preparation Plan: Ulysses Project," JPL Document D-1508 (Pasadena, California: JPL, 1 August 1985).

29. M. R. Traxler, "DSN Support of Mars Observer," TDA Progress Report PR 42-113, January-March 1993 (15 May 1993), pp. 118–22.

30. T. A. Rebold, A. Kwok, G. E. Wood, and S. Butman, "The Mars Observer Ka-band Link Experiment," TDA Progress Report PR 42-117, January-March 1994 (15 May 1994), p. 250.

31. Earl J. Montoya and Richard O. Fimmel, "Space Pioneers and Where They Are Now," NASA EP-264 (Washington, District of Columbia: NASA Educational Affairs Division, 1987).

32. Richard, P. Rudd, "Voyager Approach to Maintaining Science Data Acquisition for a 30-Year Extended Mission," Voyager Project Web site, *http://jpl.nasa.gov/voyager/30plan.html* (Pasadena, California: JPL, November 1997).

33. N. A. Fanelli, "JPL Emergency Support of TDRSS and Compatible Satellites," TDA Progress Report PR 42-82, April-June 1985 (15 August 1985), pp. 120–24.

34. N. A. Fanelli to D. J. Mudgway, "Emergency Cross Support Agreements," private communication, 11 November 1997.

35. B. Wood, "Report on the Mechanical Maintenance of the 70-Meter Antennas," DSN 890-257, JPL D-10442 (Pasadena, California: JPL, 31 January 1993), pp. 5, 207.

36. Donald H. McClure, "Rehabilitation of the DSN 70-Meter Antennas: The DSN Antenna Rehabilitation Team Final Report," DSN Document 890-226 (Pasadena, California: JPL, 1 November 1989).

37. D. W. Brown, H. W. Cooper, J. W. Armstrong, and S. S. Kent, "Parkes-CDSCC Telemetry Array: Equipment Design," TDA Progress Report PR 42-85, January-March 1986 (15 May 1986), pp. 85–110.

38. J. W. Layland et al., "Interagency Array Study Report," TDA Progress Report PR 42-74, April-June 1983 (15 August 1983), pp. 117-48.

39. D. W. Brown, W. D. Brundage, J. S. Ulvestad, S. S. Kent, and K. P. Bartos, "Interagency Arraying for Voyager-Neptune Encounter," TDA Progress Report PR 42-102, April-June 1990 (15 August 1990), pp. 91–118.

40. Andrew J. Butrica, *To See the Unseen; A History of Planetary Radar Astronomy*, NASA SP-4218 (Washington, District of Columbia: NASA, NASA History Office, 1996).

41. J. R. Paluka, "100-kW, X-band Transmitter for FTS," Technical Report TR 32-1526, Vol. VIII, January/February 1972 (15 April 1972), pp. 94–98.

42. A. L. Berman, "The Gravitational Wave Experiment: Description and Anticipated Requirements," TDA Progress Report PR 42-46, May/June 1978 (15 August 1978), pp. 100–08.

43. T. A. Komarek, J. G. Meeker, and R. B. Miller, "ISPM X-band Technology Demonstration, Part 1. Overview," TDA Progress Report PR 42-62, January/February 1981 (15 April 1981), pp. 50–62.

44. J. G. Meeker and C. T. Timpe, "X-band Uplink Technology Demonstration at DSS 13," TDA Progress Report PR 42-77, January-March 1984 (15 May 1984), pp. 24–32.

45. R. Stevens, Jet Propulsion Laboratory, Telecommunications and Data Acquisition Office, Interoffice Memorandum to P. T. Lyman, "X-band Uplink Experiment for Galileo" (3 June 1982).

46. Frank B. Easterbrook, "Gravitational Wave Searches with Ground Tracking Networks," IAF-85-386 (paper presented at the 36th Congress of the International Astronautical Federation, Stockholm, Sweden, October 1985).

47. D. J. Mudgway, "1986 Galileo Project NASA Support Plan," JPL Document D-585 (Pasadena, California: JPL, 15 November 1983).

48. R. M. Perez, "Improvements in X-band Transmitter Phase Stability Through Klystron Body Temperature Regulation," TDA Progress Report PR 42-109, January-March 1992 (15 May 1992), pp. 114–20.

49. S. Hinedi, "A Functional Description of the Advanced Receiver," TDA Progress Report PR 42-100, October-December 1989, (15 February 1990), pp. 131–49.

50. University of Padua, "The Galileo Spacecraft: A Legacy for Future Space Flight" (proceedings of the Three Galileos Conference, Padua, Italy, 1997).

51. Jet Propulsion Laboratory, Telecommunications and Mission Operations Directorate, "Galileo Low Gain Antenna S-band Contingency Mission," presentation to NASA Headquarters, JPL Document 1625-497 (20 March 1992).

52. J. I. Statman, "Optimizing the Galileo Space Communication Link," TDA Progress Report PR 42-116, October-December 1993 (15 February 1994), pp. 114–20.

53. J. W. Layland et al., "Galileo Array Study Team Report," TDA Progress Report PR 42-103, July/September 1990 (15 November 1990), pp. 161–69.

54. A. Mileant and S. Hinedi, "Overview of Arraying Techniques in the Deep Space Network," TDA Progress Report PR 42-104, October-December 1990 (15 February 1991), pp. 109–15.

55. S. Dolinar, "A New Code for Galileo," TDA Progress Report PR 42-93, January-March 1988 (15 May 1988), pp. 83–96.

56. H. Tsou, B. Shah, R. Lee, and S. Hinedi, "A Functional Description of the Buffered Telemetry Demodulator (BTD)," TDA Progress Report PR 42-112, October-December 1992 (15 February 1993), pp. 50–73.

57. S. Dolinar and M. Belongie, "Enhanced Decoding for the Galileo S-band Mission," TDA Progress Report PR 42-114, April-June 1993 (15 August 1993), pp. 96–111.

58. S. Million, B. Shah, and S. Hinedi, "A Comparison of Full Spectrum and Complex-Symbol Combining Techniques for the Galileo S-band Mission," TDA Progress Report PR 42-116, October-December 1993 (15 February 1994), pp. 128–49.

59. T. T. Pham, S. Shambayati, D. E. Hardi, and S. G. Finley, "Tracking the Galileo Spacecraft with the DSCC Galileo Telemetry Prototype," TDA Progress Report PR 42-119, September/October 1994 (15 November 1994), pp. 221–35.

60. Jet Propulsion Laboratory, "The Galileo Messenger," Issue 39, Galileo Project Office, JPL (July 1996).

61. J. G. Smith, "Ka-band (32 GHz) Downlink Capability for Deep Space Communications," TDA Progress Report PR 42-88, October/December 1986 (15 February 1987), pp. 96–103.

62. J. G. Smith, "Proposed Upgrade of the Deep Space Network Research and Development Station," TDA Progress Report PR 42-88, October/December 1986 (15 February 1987), pp. 158–63.

63. R. C. Clauss and J. G. Smith, "Beam Waveguides in the Deep Space Network," TDA Progress Report PR 42-88, October-December 1986 (15 February 1987), pp. 174–82.

64. T. Veruttipong, W. Imbriale, and D. Bathker, "Design and Performance Analysis of the DSS-13 Beam Waveguide Antenna," TDA Progress Report PR 42-101, March/April 1990 (15 May 1990), pp. 99–113.

65. S. D. Slobin, T. Y. Otoshi, M. J. Britcliffe, L. S. Alvarez, S. R. Stewart, and M. M. Franco, "Efficiency Calibration of the DSS 13 34-m Diameter Beam Waveguide Antenna at 8.45 and 32 GHz," TDA Progress Report PR 42-106, April-June 1991 (15 August 1991), pp. 283–89.

66. M. S. Esquival, "Optimizing the G/T ratio of the DSS 13 34 m Beam Waveguide Antenna," TDA Progress Report PR 42-109 January-March 1992 (15 May 1992), pp. 152–61.

67. W. Imbriale, W. Veruttipong, T. Otoshi, and M. M. Franco, "Determining Noise Temperature in Beam Waveguide Systems," TDA Progress Report PR 42-116, October-December 1993 (15 February 1994), pp. 42–52.

68. L. S. Alvarez, M. J. Britcliffe, M. M. Franco, S. R. Stewart, and H. J. Jackson, "The Efficiency Calibration of the DSS 24 34-Meter Diameter Beam Waveguide Antenna," TDA Progress Report PR 42-120 (15 February 1995), p. 174.

69. D. L. Ross, "DSCC Subsystem Functional Requirements; Telemetry Subsystem 1992–1995," JPL Document D-6111, 824-35 (15 May 1989).

70. J. Statman, G. Zimmerman, F. Pollara, and O. Collins, "A Long Constraint Length VLSI Viterbi Decoder for the DSN," TDA Progress Report PR 42-95, July-September 1988 (15 November 1988), pp. 134–42.

71. C. R. Lahmeyer and K. M. Cheung, "Long Decoding Runs for Galileo Convolutional Codes," TDA Progress Report PR 42-95, July-September 1988 (15 November 1988), pp. 143–52.

72. J. Statman, J. Rabkin, and B. Siev, "Big Viterbi Decoder Results for (7,1/2) Convolutional Code," TDA Progress Report PR 42-99, July-September 1989 (15 November 1989), pp. 122–23.

73. I. M. Onyszchuk, "Testing Interconnected VLSI Circuits in the Big Viterbi Decoder," TDA Progress Report PR 42-106, April-June 1991 (15 August 1991), pp. 175–82.

74. B. Falin, "DSCC Subsystem Functional Requirements: Command Subsystem 1992–1996," JPL Document D-6395, 824-36 (1 August 1989).

75. Jet Propulsion Laboratory, "DSCC Subsystem Functional Requirements: DSCC Tracking Subsystem 1992–1996," JPL Document D-7335, 824-38 (1 July 1990).

CHAPTER 6

THE CASSINI ERA: 1996–1997

THE CASSINI ERA

Winds of Change

Throughout its life, the form of the DSN at any single point in time was shaped by many forces acting on it simultaneously and often in different directions. These forces included budget constraints from the NASA Office of Tracking and Data Acquisition (Code O), the requirements for support of new missions from the NASA Office of Space Science (Code S), the drive of new technology, the pressure of real-time operations, and the concerns of foreign agencies and competing requirements for limited antenna time from radio astronomy and other space science constituencies. Dealing with these often conflicting currents had become a way of life at all levels within the DSN structure, and the remarkable progress recorded in the forthcoming pages attests to the high degree of success that was achieved.

By 1995, however, the winds of change were blowing more strongly than ever before throughout NASA, and by 1997 their effect was being keenly felt in the DSN. The changes would affect the established purpose, scope, and functions of the DSN, and

Cassini Era Mission Set	
Deep Space Missions (Launch)	Earth-Orbiter Missions
Galileo (1989)	ISTP-Soho
Voyager (1977)	ISTP-Polar
Ulysses (1990)	ISTP-Wind
Pioneer 10 (1972)	ISTP-Geotail
N-E Asteroid Rndv. (2/17/96)	IR Space Observatory
Mars Global Surveyor (11/7/97)	ASTRO-D
Mars Pathfinder (12/04/96)	TOMS-EP
Cassini (10/15/97)	YOHKOH Solar-D
	Space Tech. Res. Vehicle A
	Radar Satellite
	Roentgen Satellite
	SURF Satellite-1
	SSTI-Lewis
	Hotbird-2,-3
	HALCA

Figure 6-1. Cassini Era mission set, 1997.

The Cassini Era: 1996–1997

they first appeared in the Telecommunications and Mission Operations Directorate (TMOD) Implementation Plan for Fiscal Year 1998.[1] This plan is discussed fully in a later chapter.

This chapter will review the DSN and the missions it was supporting as the Galileo Era came to an end and the Cassini Era opened to an environment of change.

The Cassini Era Mission Set

A snapshot of the DSN operational tracking schedule in 1997 would have revealed the portfolio of deep space and Earth-orbit missions shown in Figure 6-1. This represented the DSN mission set for the first year or two of the Cassini Era.

Except for Ulysses, which was an ESA mission, all of the deep space missions were of NASA origin. Half of the missions, Galileo, Voyager, Ulysses, and *Pioneer 10*, were older missions that had been in the DSN for many years and have been discussed in previous chapters. The Near-Earth Asteroid Rendezvous, Mars Global Surveyor, Mars Pathfinder, and Cassini missions were new in 1996 and 1997, and brought with them new challenges and changes for the DSN.

DEEP SPACE MISSIONS

GENERAL

Through the end of the Galileo Era and the beginning of the Cassini Era, the ongoing deep space missions, Galileo, Ulysses, *Pioneer 10*, and *Voyagers 1* and *2*, presented no new challenges for the DSN. They continued to return science and engineering data, each according to its own remaining capability, and were provided with minimal tracking station support, sufficient only to maintain their viability as scientific missions. The changes and upgrades that had been incorporated into the DSN since these missions began had always maintained a capability to handle their diminishing requirements for telecommunications and data acquisition support.

GALILEO

Amongst the aging missions, Galileo was the most active. As described in the previous chapter, *Galileo* had completed its primary mission in December 1997 and continued at a much reduced pace to carry out further observations of Europa under the name Galileo Europa Mission. It remained entirely dependent upon arraying of the DSN 70-m/34-m antenna for its downlinks, since the Parkes antenna was no longer available to support the arrayed configurations.

ULYSSES

As the *Ulysses* spacecraft embarked on its second 6.2-year orbit over the poles of the Sun, its science instruments continued to perform normally. The north and south polar passes were planned for 2000 and 2001, respectively. On this orbit, the properties of solar winds in high solar latitudes during the maximum of the solar activity cycle were to be investigated. Through 1997, Earth-pointing maneuvers and instrument calibrations were made regularly, in preparation for observations during the fifth solar opposition in March 1998. Daily passes on the 34-m antennas at S-band or X-band, and telemetry running at 512 bits per second to 2,048 bits per second, were typical of this period.

VOYAGERS 1 AND 2

The continuation of the *Voyager 1* and *2* missions beyond the outer planets was called the Voyager Interstellar Mission (VIM) and would continue through the year 2019. Through 1996 and 1997, *Voyager 1* and *Voyager 2* maintained a presence on the DSN tracking schedules with several passes per week for each spacecraft on the 34-m antennas, with occasional support on the 70-m antennas.

At approximately 2:10 P.M. Pacific time on 17 February 1998, according to VIM project manager Ed B. Massey, *Voyager 1* reached an Earth-to-spacecraft range of 10.4 billion km (6.5 billion miles), exceeding the record held by Pioneer 10 for 25 years, and became the most distant human-made object in space. Almost 70 times farther from the Sun than Earth (70 AU), with the radio signal taking 9 hours and 36 minutes to travel from the spacecraft to Earth, *Voyager 1* was then at the very edge of the solar system. The downlink data rate was 160 bits per second or 600 bits per second on the 34-m antennas and could be increased to 1.4 kilobits per second on the 70-m antennas when required for tape recorder playback. At a smaller distance of 52 AU, *Voyager 2* was being supported by the southern hemisphere stations in Canberra and could receive downlink data at 7.2 kilobits per second.

JPL marked the event with a special press release,[2] "*Voyager 1* Now Most Distant Human-Made Object In Space." The article explained the scientific significance of this remarkable achievement:

> Having completed their primary explorations, *Voyager 1* and its twin *Voyager 2* were studying the environment of space in the outer solar system. Although they were beyond the orbit of all the planets, the spacecraft remained well within the boundary of the Sun's magnetic field, called the heliosphere. Science instruments on both spacecraft sensed signals that scientists believed were coming from the outermost edge of the heliosphere, known as the heliopause. The heliosphere results from the Sun's emission of a steady flow of electrically charged particles called the solar wind. As the solar wind expands supersonically into space in all directions, it creates a magnetized bubble—the heliosphere—around the Sun. Eventually, the solar wind encounters the electrically charged particles and magnetic field in the interstellar gas. In this zone, the solar wind abruptly slows down from supersonic to subsonic speed, creating a termination shock. Before the spacecraft travel beyond the heliopause into interstellar space, they will pass through this termination shock. Dr. Edward C. Stone, Voyager project scientist and director of JPL, said, "The data coming back from *Voyager* now suggest that we may pass through the termination shock in the next three to five years. If that's the case, then one would expect that within ten years or so, we would actually be very close to penetrating the heliopause itself and entering into interstellar space for the first time."

Reaching the termination shock and heliopause would be major milestones for the mission because no spacecraft had been there before, and the *Voyagers* would gather the first direct evidence of their structure. Encountering the termination shock and heliopause had been a long-sought goal for many space scientists. Exactly where these two boundaries were located and what they were like remained a mystery.

Although these were significant long-range telecommunications records for the DSN, uplink and downlink communications with both spacecraft remained routine matters. With plenty of performance margin still remaining, the DSN expected to be able to support the VIM for the next twenty years.

Pioneer 10

After an active life of 25 years, the *Pioneer 10* mission came to an end on 31 March 1997, with the last track at DSS 63. Launched on 2 March 1972 and tracked by the DSN ever since, *Pioneer 10* support had almost become a permanent feature of routine DSN activity. Although the downlink was still running at 8 bps, the signal was very weak, and it had been determined that the science value of the data being returned was no longer sufficient to justify further continuation of the mission. Pioneer operations had been conducted from the mission control center at the Ames Research Center in Sunnyvale, California, for many years, and the mission had many notable scientific achievements to its credit. By 1983, it had become the first spacecraft to travel beyond the orbits of Neptune and Pluto. Following a solar system escape trajectory away from the Sun, it continued to move further outward in search of the heliopause, the region of the solar system where the influence of the Sun itself finally ends and true interstellar space begins. Now, following a trajectory in the opposite direction to that of *Pioneer 10*, the pursuit of that goal would be taken over by Voyager. In the immediate future, the *Pioneer 10* spacecraft would be used by the DSN operations group at JPL to train new station controllers for future missions.

NEAR-EARTH ASTEROID RENDEZVOUS

The Near-Earth Asteroid Rendezvous (NEAR) mission was the first in NASA's Discovery program of lower cost, highly focused planetary science missions. It was developed and managed by the Johns Hopkins University Applied Physics Laboratory, Laurel, Maryland.[3]

The primary objective of the NEAR mission was to inject a spacecraft into a 50-km orbit around the near-Earth asteroid Eros for the purpose of making scientific observa-

The Cassini Era: 1996–1997

tions to determine its size, shape, mass, density, and spin rate. It could also determine the morphology and mineralogical composition of the surface.

Launched in February 1996, the NEAR mission used an Earth gravity-assist flyby in early 1998 to deliver the spacecraft to the asteroid Eros in January 1999. Orbital operations were to begin immediately following arrival and were planned to be about one year in duration. In the early part of its mission, June 1997, NEAR flew by the asteroid Mathilde. It found Mathilde to be composed of extremely dark material with numerous large impact craters, including one nearly six miles deep.

DSN support for the NEAR mission in 1996 and 1997 was conducted primarily on the 34-m HEF or BWG antennas, and the X-band uplinks and downlinks carrying telemetry, command, and ranging data were well within existing capabilities. Initial acquisition and a trajectory correction maneuver in March 1997 were accomplished by the DSN without incident.

During the critical phases of the mission and mission operations at Eros, NEAR required continuous tracking support (three 8-hour passes per day) from the 34-m stations, with occasional use of the 70-m antennas. At other periods during the cruise phase, the tracking support was reduced to one or two passes per day.

In preparation for an Earth swing-by maneuver in early 1998, tracking support for NEAR was increased to the level of three passes per day in December 1997.

Mars Global Surveyor

Mars Global Surveyor (MGS) was the first mission in a new program of Mars exploration called the Mars Surveyor Program. It was also the first of NASA's new "faster, better, cheaper" family of missions. MGS was designed to deliver a single spacecraft to Mars for an extended orbital study of the surface, atmosphere, and gravitational and magnetic fields of the planet. A search for gravitational waves and a demonstration of Ka-band downlink communications technology was to be conducted during the flight to Mars. The spacecraft was to be launched during the November 1996 Mars opportunity, using a Delta launch vehicle with a PAM D upper stage. The transit time to Mars would be about ten months. MGS was to be the first U.S. spacecraft to orbit Mars since the Viking Orbiters twenty years before and the first to use aerobraking rather than propulsive maneuvers to adjust its orbit upon arrival. The use of aerobraking techniques had been demonstrated in mid-1993 during the final stages of the Magellan mission to Venus and was considered to be a promising technique for reducing the

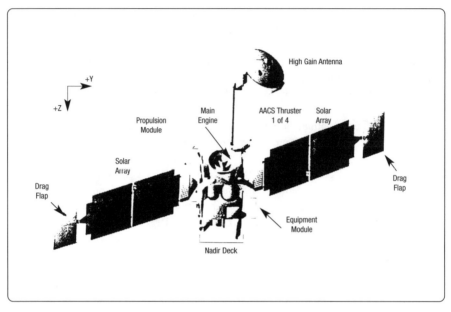

Figure 6-2. *Mars Global Surveyor* spacecraft in mapping configuration. The spacecraft was built and operated by Lockheed Martin Astronautics, JPL's industrial partner on the mission, in Denver, Colorado.

spacecraft fuel expended in circularizing the spacecraft orbit around a planet. Repetitive observations of the Mars surface and atmosphere were to be carried out from a nearly circular, low-altitude (378 km) orbit over a period of one Martian year (687 Earth days).

An overview of the major components of the spacecraft is shown in Figure 6-2.

Like the earlier *Mars Observer* spacecraft, MGS uplinks and downlinks were both on X-band, and it also carried a Ka-band downlink beacon as a new downlink technology demonstration. The downlink telemetry would employ Reed-Solomon and convolutional ($K = 7$, $R = 1/2$) encoding and would range from 10 bits per second to 32 kilobits per second for engineering data, and 4 kilosymbols per second to 85 kilosymbols per second for Reed-Solomon encoded science data. In coding terminology, a symbol is essentially a Reed-Solomon encoded data bit. The MGS spacecraft transmitted 250 encoded bits to represent every 218 bits of its raw science data, a ratio of 1.147 to 1.0.

The Cassini Era: 1996–1997

The command uplink would operate at a nominal rate of 125 bits per second, although several other rates were available. Radio metric data consisting of two-way Doppler and sequential ranging data would be generated from the coherent X-band uplink and downlink carriers, and a noncoherent signal from the spacecraft's ultra-stable oscillator (USO) would be used to generate project data during the frequent occultation experiments.

The MGS mission required the tracking stations to be able to acquire the X-band downlink and achieve telemetry "in-lock" within five minutes of spacecraft view at the station, for data rates of 18.6 ksps or greater. Reacquisition of the telemetry signal following each Earth occultation during the mapping phase was to be accomplished within one minute of spacecraft view. These requirements, which posed some problems for the DSN, could not be solved before launch but were satisfied by the time the spacecraft reached Mars.

The DSN was required to deliver at least 95 percent of the science data transmitted from the spacecraft during the mapping phase to the Advanced Multimission Operations System (AMMOS) at JPL. Because the Mars Orbiter camera instrument made extensive use of data compression, the imaging science data could not tolerate significant gaps in the "packets" of telemetry data delivered by the DSN. This produced a further requirement on the DSN to deliver science data "packets" containing no more than one packet gap or error within 10,000 packets. This was equivalent to one error per 100 million bits (1×10^{-8}) and would require some additional backup recording equipment in the GCF to meet the stringent requirement.

Following initial acquisition of the downlink signal after launch by the 26-m antenna at Canberra, the mission was to be supported by the 34-m HEF antennas, DSS 15, 45, and 65. The 34-m BWG antennas would also provide support as they became available in 1997. The 70-m antennas were to support the critical Mars Orbit Insertion sequence and to be available to support any other situations which were declared to be "critical" by the MGS flight operations manager.

DSN readiness to support the launch and cruise phase of the MGS mission was presented to a Telecommunications and Mission Operations Directorate (TMOD) Review Board in October 1996. For the first time, it was part of a broad-based review which included DSN and AMMOS as an integrated service of the TMOD. It was this integrated service (which included its multimission operators) that would be used by the MGS flight operations manager to conduct the mission after launch.

The DSN presentation satisfied the Review Board that it had verified telecommunications compatibility with the spacecraft, had adequately trained its operations personnel, had developed an Initial Acquisition Plan covering all available launch windows, and that all of the facilities necessary for launch and early cruise were in place. The uplink and downlink requirements for MGS were well within the capabilities of the DSN systems that had been implemented earlier under the SPC upgrade program. Because the issues of downlink acquisition time and telemetry data accountability were related to operations at Mars, they were not included in this review. The DSN, together with AMMOS, was ready to support the MGS launch.

At noon on 7 November 1996, *Mars Global Surveyor* lifted off launch pad 17A at Cape Canaveral Air Station, Florida, on a three-stage Delta II launch vehicle bound for Mars. All launch vehicle sequences were completed as planned and ended with separation from the spacecraft about 50 minutes after launch. An anomaly that occurred during deployment of the spacecraft solar panels was not a cause for concern at the time, since it was believed it could be corrected the following day. About 70 minutes after launch, just after the spacecraft came in view of Canberra, the X-band downlink was very quickly acquired, first by the 26-m antenna (DSS 46), and then by the 34-m HEF (DSS 45), using the offsets from DSS 46. The DSN settled into routine operations as MGS cruise support began.

Over the next several weeks, spacecraft engineers continued to evaluate various solutions to the problem created by the solar array anomaly, although it posed no immediate threat to the mission.

In January 1997, the spacecraft was turned to point its high-gain antenna toward Earth, and with a stronger downlink available, the spacecraft immediately began transmitting telemetry at the higher data rates. During a period of low activity in May, a program of observations directed to the search for gravitational waves was carried out. This was the long-awaited opportunity that had eluded the project investigators when the Galileo X-band mission failed. The high-stability X-band exciter and X-band uplink and downlink were key factors in reaching the sensitivity needed to detect the presence of gravity waves by their characteristic three-perturbation signature in the two-way Doppler data. There was no immediate answer to whether the observations revealed the presence of gravitational waves, since analyses of these data would take many months to complete.

Ka-band technology demonstration passes with DSS 15 and DSS 13 were conducted over a two-week period beginning on 21 July. The Ka-band downlink was received at DSS 13 in a coherent mode with the X-band uplink from DSS 15. After a period of good ranging data had been recorded, DSS 15 transferred its uplink to DSS 34 to generate comparative data over the Goldstone/Australia path.

The Cassini Era: 1996–1997

In July, with the spacecraft rapidly approaching Mars, mission controllers and DSN operators engaged in several simulation exercises to prepare for orbital operations. During the simulations, the spacecraft radio transmitter was turned on and off over the course of a six-hour period to simulate three actual orbits. Because capturing the project data and acquiring the telemetry data under these rapidly changing conditions would be an operationally demanding task for the tracking station operators and project team, these simulations were used to familiarize them with the necessary operational procedures.

About three weeks before the actual MOI, the DSN team reviewed its state of readiness to support the rapidly approaching orbit insertion and orbital operations phase of the MGS mission. As in the Launch Readiness Review, the review included the status of the AMMOS as well as that of the DSN.

The DSN team presentation focused on the steps that had been taken to meet the telemetry delivery (data gaps) and acquisition (lock-up time) requirements that had been of concern to the DSN prior to launch. It was shown that the implementation of reliable network servers in the GCF and the correction of a known antenna-pointing anomaly at the BWG stations had accounted for most of the difference between actual measured performance of the DSN and the performance required by the MGS specifications. The remaining errors would be attributable to data gaps induced by downlink data rate changes on the spacecraft. Although lab measurements at JPL and past performance at the stations confirmed that telemetry lock-up time would meet the requirement, some uncertainty remained until the actual downlink became available and performance was verified under fully operational conditions. An operations support plan for the MOI sequence; availability of the necessary antennas, communications, and facilities; and a demonstrated level of crew training completed the DSN presentation. There were no outstanding issues or concerns. The DSN's Telecommunications and Mission Services (TMS) manager, John C. McKinney, assessed the status as Network-ready to support the Mars Orbit Insertion and Aerobraking events.

The DSN configuration used to support *Mars Global Surveyor* orbit insertion and orbital operations is shown in Figure 6-3.

The spacecraft executed the orbit insertion sequence perfectly, beginning with a 22-minute retro-engine burn on 11 September 1997, during the mutual view period of Canberra and Madrid. Twelve minutes after the start of the engine burn, the first Mars occultation occurred, providing the tracking station operators with their first opportunity to demonstrate the benefits of the simulated acquisition training. When the spacecraft emerged from behind the planet four minutes after the retro-engine burn had been com-

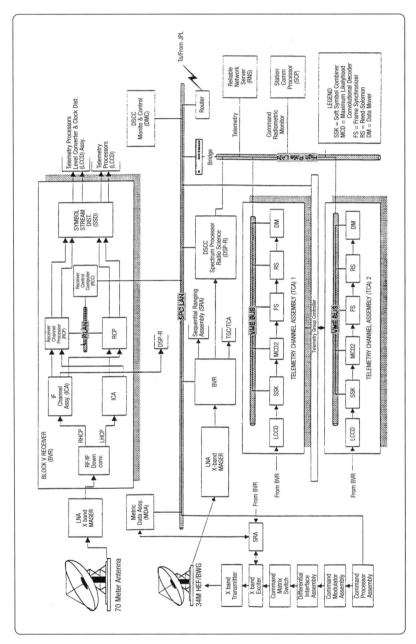

Figure 6-3. DSN configuration for *Mars Global Surveyor*. This was the standard DSN configuration for the Cassini Era and has been discussed earlier as part of the SPC upgrade task. It was used in various forms for all the missions in 1996 and 1997.

pleted, the downlink was acquired within seconds. Doppler data indicated that the spacecraft had entered a highly elliptical orbit within one minute of the intended 45-hour orbit. The spacecraft was about 250 km above the Martian surface (periapsis) and about 56,000 km above the surface at the farthest point (apoapsis).

A few days after the spacecraft began orbiting the planet, the spacecraft magnetometer detected the existence of a planet-wide magnetic field. This discovery was hailed by Vice President Gore as "another example of how NASA's commitment to faster, better, cheaper Mars exploration is going to answer many fundamental questions about the history and environment of our neighboring planet."

The first aerobraking maneuvers began with retro-engine burns at apoapsis on 17, 20, 22, and 24 September. As each periapsis passage began to take the spacecraft into the upper reaches of the Mars atmosphere, atmospheric drag provided additional aerobraking effect. While the orbit slowly decreased, data taking progressed to the point where the MGS Science Team was able to hold a press conference at JPL on 2 October to report their findings from all six science experiments. "The spacecraft and science instruments are operating magnificently," reported Dr. Arden Albee, the MGS project scientist at the California Institute of Technology, Pasadena, California. "The initial science data we've obtained from the walk-in phase of aerobraking are remarkable in their clarity, and the combined measurements from all of the instruments over the next two years are going to provide us with a fascinating new global view of the planet."

Suddenly, during the fifteenth periapsis passage on 6 October, significant movement was observed in the damaged solar array panel. On 12 October the orbit was temporarily raised to reduce the stress on the solar panel at each periapsis passage while the operations teams at JPL and Lockheed Martin Astronautics investigated this potentially alarming situation. Two weeks later, with the spacecraft in a 35-hour orbit and closest approach to the surface of 172 km, aerobraking was resumed at a much slower rate than before to avoid putting undue stress on the unlatched solar panel. This decision would extend the aerobraking phase by eight to twelve months and would change the final mapping orbit. However, it would not significantly affect the ability of MGS to accomplish its primary mission's science objectives.

Apart from the faulty solar panel, the spacecraft was operating perfectly, and with DSN support available from the BWG antennas, science data collection continued unabated. An image of the giant volcano Olympus Mons taken during this period is shown in Figure 6-4.

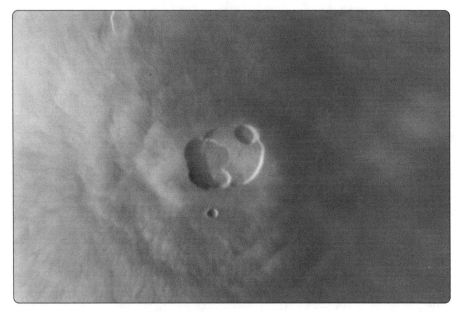

Figure 6-4. ***Mars Global Surveyor* image of Olympus Mons.** *Mars Global Surveyor* obtained this spectacular wide-angle view of Olympus Mons from an altitude of 900 km (500 mi) above the surface on its 263rd orbit around the planet on 25 April 1998. More than three times the height of Mt. Everest, this giant volcano is almost flat, with its flanks having a gentle slope of three to five degrees.

MGS completed its 100th orbit around Mars in January 1998. Assisted by the calm state of the Martian atmosphere, aerobraking operations were proceeding satisfactorily. The spacecraft had reached a 19-hour orbit around the planet, with a high point of about 28,000 km and a low point of about 120 km.

The target date for the start of mapping operations was March 1999, and it was hoped that the duration of the mapping phase would still be one Martian year. The DSN was providing support with the 34-m HEF antennas on a routine basis with typical downlink data rates of 2,000 bits per second for engineering and 21,333 symbols per second for playback of science data.[4]

The Cassini Era: 1996–1997

Mars Pathfinder

About one month after *Mars Global Surveyor* lifted off the launch pad at Canaveral Air Force Station bound for Mars, the second spacecraft in the Mars Discovery Program was launched from the same pad with the same destination, but with a different objective. *Mars Pathfinder* (MPF) was primarily an engineering demonstration of key technologies and concepts for eventual use in future missions to Mars employing scientific Landers. MPF also delivered a scientific instrument package to the surface of Mars to investigate the elemental composition of Martian rocks and soil and the structure of the Mars atmosphere, surface meteorology, and geology. In addition, MPF carried a free-ranging surface Rover, which was deployed from the Lander to conduct technology experiments and to serve as an instrument deployment mechanism. At launch, the mass of MPF was 890 kilograms, compared to 1,060 kilograms for MGS. MPF would arrive at Mars on 4 July 1997, seven months after launch, compared to 11 September 1997, ten months after launch, for MGS.

The MPF spacecraft comprised three major elements: the cruise vehicle, the deceleration systems, and the Lander vehicle that contained the Rover.[5]

The cruise vehicle provided the major spacecraft functions of power generation, propulsion, and attitude control prior to entry into the Martian atmosphere. The deceleration subsystem comprised an aeroshell, parachute, tether, retrorockets and inflatable airbags. The tetrahedral Lander structure enclosed the science instruments, the Rover, and the engineering subsystems necessary for cruise and surface operations. The radio transponder and transmitter, along with attitude control and data handling electronics and software, were also located on the Lander and were connected to the cruise stage via a detachable cable harness. The Lander structure was self-righting, with three side petals that opened to establish an upright configuration on the surface from which the Rover could be deployed. Power was provided by solar arrays mounted on the surface panels.

The MPF Lander with the Rover vehicle in a deployed configuration on the Mars surface is depicted in Figure 6-5.

The duration of the primary mission was defined to be 30 days beginning at the time of landing. Pathfinder requirements for tracking and data acquisition lay within current (1997) DSN capabilities, and the Telecommunicatons and Mission Services (TMS) manager Dennis M. Enari concluded that no special implementation was needed. Both uplink and downlink for MPF were on X-band and would be used for command, telemetry, and generation of radio metric Doppler and range data. There were no requirements for

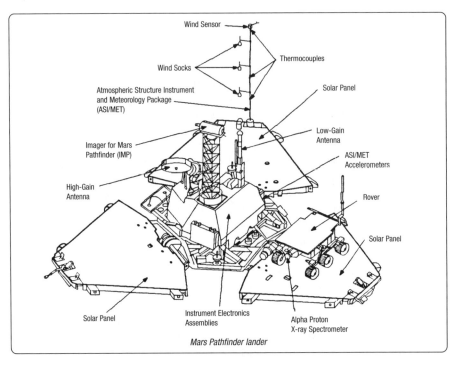

Figure 6-5. Pathfinder Lander deployed on the surface of Mars.

project observations, although the DSCC project equipment would be used to record open-loop spectral data from the X-band downlink during the rapid and critical entry, descent, and landing sequence. Descent Doppler profiles and key engineering telemetry data would be reconstructed from this data after the event. Telemetry data rates would range from 40 bits per second in the cruise phase to 22,120 bits per second during surface operations. All data would be convolutionally coded at either (R = 7, K = 1/2) or (R = 15, K = 1/6). There would be no uncoded data. Command data rates ranging from 7 to 500 bits per second would be used on the X-band uplink.

During the entry, descent, and landing phase, both the 70-m and the 34-m HEF antennas at Madrid would be tracking the MPF spacecraft simultaneously. At other times, the tracking support would be provided by the 34-m HEF and 34-m BWG antennas, with occasional use of the 70-m antennas for the downlink only. The 26-m antennas at DSS 46 and DSS 16 would provide support for the initial acquisition events only.

The Cassini Era: 1996–1997

Mars Pathfinder blasted into space on 4 December 1996, the third day of its launch window. The launch, ascent, orbital injection, and spacecraft separation events were all normal. About 70 minutes after lift-off, the spacecraft made its first appearance over the Goldstone radio horizon. Stations DSS 16 and DSS 15 quickly locked up their receivers on the downlink signal and, less than two minutes later, began flowing telemetry data to the mission controllers at JPL. On the evidence of these data, the spacecraft team reported that all critical spacecraft systems, such as power, temperature, and attitude control, were performing well. "Everything looks really good and we're very happy," said Tony Spear, Pathfinder project manager at NASA's Jet Propulsion Laboratory.

An event of some significance to the DSN occurred on 18 January 1997 when the new Block 3 Maximum Likelihood Decoder (MCD3) was used for the first time to process live inflight data from DSS 15 Goldstone.

The MCD3 grew out of the technology that had been used by Statman in developing the Big Viterbi Decoder for the original Galileo X-band mission in 1989. Although that effort was dropped when the spacecraft high-gain antenna problem forced a redesign of the Galileo mission in 1991, the enhancement in downlink performance available from newer, more powerful convolutional codes remained viable. Although the encoding process in the spacecraft was a relatively simple process, the new codes required much bigger and extremely complex decoders in the DSN to take advantage of the improvement in performance. In the intervening years, DSN engineers had pursued the development of such a machine and, by the time of the MPF mission, were ready to demonstrate the first machine of its kind in a real, inflight situation.

In this demonstration, the MPF telemetry downlink consisted of a convolutionally coded ($R = 15$, $K = 1/6$) data stream, running at 4,070 bits per second. In light of the importance of this machine to later MGS, MPF, and Cassini mission support, its first successful demonstration under operational conditions was a matter of considerable satisfaction to the DSN.

The cruise phase passed quietly without the occurrence of any significant incidents. Spacecraft and science instrument status checks were made routinely, and several trajectory correction maneuvers were successfully executed. The DSN operations team carried out several rehearsals to improve proficiency in the procedures that would be used in July for the entry, descent, and landing sequence in July.

On 30 June, with all spacecraft systems in excellent operating condition and commanded by a sequence from its onboard computer, the MPF spacecraft began the entry, descent, and landing phase of its mission. On the morning of 4 July, an update of the MPF orbit based on the latest DSN Doppler and range data indicated that MPF was heading straight for the center of its predetermined 60-mile by 120-mile landing ellipse and would enter the upper atmosphere at an entry angle just 0.75 degrees less than its original design value of 14.2 degrees. A few hours later, at 10:07:25 A.M. Pacific time, 4 July 1997, the MPF spacecraft landed successfully on the surface of Mars, marking NASA's return to the surface of the Red Planet after more than twenty years. A low-power transmission from an independent antenna on one of the petals confirmed the landing and indicated that the craft had landed on its base petal in an upright position.

The first low-gain antenna transmission was received on schedule about four hours later. It contained preliminary information about the health of the Lander and Rover; the orientation of the spacecraft on the surface; the entry, descent, and landing; and the temperature and density of the Martian atmosphere. This was soon followed by a high-gain antenna transmission which contained the first images from the Lander. The next day, 5 July, after clearing a problem with the petal that the Rover needed as a ramp to reach the surface and rectifying a Rover-to-Lander communication link problem, the mission control team moved the Rover vehicle (named Sojourner) down the ramp under its own power and onto the Martian soil. Images of the Rover from the Lander and vice versa, plus engineering data from both, confirmed that both vehicles were in their proper positions and in good operating condition. With the downlink telemetry data rate set to 6,300 bps, science activities from the surface began in earnest.

During the first two weeks of surface operations, uplink and downlink communications were interrupted several times due to Lander computer resets that unexpectedly switched the low-gain and high-gain antennas, and to operational errors related to the timing of the limited duration downlink sessions. Once these initial problems were cleared up, the data rate was raised to 8,300 bps and the return of high-quality data from the MPF Lander became a matter of routine operation for the DSN. Downlink sessions with the Lander were of short duration, generally about ninety minutes, during which about 60 megabits of science and engineering data were returned. By the time MPF had completed its primary 30-day mission on 3 August, it had returned 1.2 gigabits of data, including 9,669 images of the Martian landscape.

The designers of the spacecraft and the mission had every reason to be proud of their efforts. "This mission demonstrated a reliable and low-cost system for placing science payloads on the surface of Mars," said Brian Muirhead, Mars Pathfinder project manager at JPL. "We've validated NASA's commitment to low-cost planetary exploration,

The Cassini Era: 1996–1997

Figure 6-6. The MPF Lander camera views the Sojourner Rover in operation near Yogi Rock on the surface of Mars, July 1997.

shown the usefulness of sending microrovers to explore Mars, and obtained significant science data to help understand the structure and meteorology of the Martian atmosphere, and to understand the composition of the Martian rocks and soil."

The downlink telemetry data from Mars, delivered by the DSN stations to JPL in real time were subsequently processed into images by the MPF scientists and made available to the media and on the World Wide Web. The images returned from both Lander and Rover were remarkable indeed, but it was the robust, semi-autonomous Rover, Sojourner, that captured the imagination of the public. To accommodate the swell of public interest in following the mission via the World Wide Web, JPL engineers, in cooperation with several educational and commercial institutions, constructed twenty Pathfinder mirror sites. Together, these MPF sites recorded 565,902,373 hits worldwide during the period 1 July–4 August. The highest number occurred on 8 July, when a record 47 million hits were logged, more than twice the number received by the official Web site for the 1996 Olympic Games in Atlanta, Georgia.

A typical high-quality MPF image transmitted from Mars to Earth over the DSN telemetry downlink is shown in Figure 6-6. The inherent capability of the DSN downlink contributed to the remarkable detail observed in this and other MPF images.

Downlink communications with the Lander continued with no sign of trouble until 27 September, when DSS 15 at Goldstone was unable to detect the presence of a downlink at the scheduled transmission time. Declaration of a spacecraft emergency by the Pathfinder mission director authorized Enari to negotiate the release of the 70-m antennas from Galileo and MGS support to assist with DSN attempts to recover the MPF downlink. At the time, it was surmised that the downlink problems were most likely related to depletion of the spacecraft battery and uncertainties in the status of the onboard clock. While the spacecraft team investigated various scenarios to explain what might have happened to the spacecraft, the DSN went into emergency mode on a daily basis for all MPF passes.

While these downlink recovery efforts were going on, however, there was a great deal of excitement in the science community as the scientific results of the mission became available. A press release on 9 October 1997 reported that "Mars was appearing more like a planet that was very Earth-like in its infancy, with weathering processes and flowing water that created a variety of rock types, and a warmer atmosphere that generated clouds, winds, and seasonal cycles."

At an 8 October 1997 press briefing at JPL, Mars Pathfinder project scientist Dr. Matthew Golombek observed, "What the data are telling us is that the planet appears to have water-worn rock conglomerates, sand, and surface features that were created by liquid water. If," he added, "with more study, these rocks turn out to be made of composite materials, that would have required liquid water flowing on the surface to round the edges in pebbles we see on the surface or explain how they were embedded in larger rocks. That would be a very important finding."

Golombek also stressed the amount of differentiation—or heating, cooling, and recycling of crustal materials—that appeared to have taken place on Mars. "We're seeing a much greater degree of differentiation—the process by which heavier elements sink to the center of the planet while lighter elements rise to the surface—than we previously thought, and very clear evidence that liquid water was stable at one time in Mars'[s] past. Water, of course, is the very ingredient that is necessary to support life," he added, "and that leads to the $64,000 question: Are we alone in the universe? Did life ever develop on Mars? If so, what happened to it and, if not, why not?"

The Cassini Era: 1996–1997

Despite the most intense efforts by the DSN and spacecraft controllers to detect the presence of a downlink, however weak or off-frequency it may have been, no downlink could be found. There was conjecture that, without the heat generated by the battery-powered transmissions, the spacecraft temperature would fall below its operating limits and the spacecraft computer that controlled the spacecraft transmission times would no longer operate correctly. Whatever the cause, the downlink was never recovered, and further efforts were discontinued in mid-October.

At the time of the last downlink, the Lander had operated for nearly 3 times its design lifetime of 30 days, and the Sojourner Rover had operated for 12 times its design lifetime of 7 days. After the 4 July landing, it returned 2.6 billion bits of data, which included more than 16,000 images from the Lander and 550 images from the Rover, as well as more than 15 chemical analyses of rocks and extensive data on winds and other weather factors. All MPF requirements of the DSN for telecommunications and data acquisition support had been fulfilled.

The cost of designing and building the Pathfinder had been $171 million; that of the Sojourner Rover, $25 million. Together, their combined costs would have been a mere round-off error in the $3 billion (1997 dollars) cost of NASA's previous mission to Mars, the Viking Landers in 1976. In that sense, the MPF mission had accomplished one of its primary objectives. As Brian Muirhead, Mars Pathfinder project manager at JPL, declared at the end of the mission, "This mission has demonstrated a reliable and low-cost system for placing science payloads on the surface of Mars. We've validated NASA's commitment to low-cost planetary exploration."

NASA seemed well satisfied. A few weeks later, NASA Administrator Daniel S. Goldin recognized the efforts of all involved with the mission in a formal press release: "I want to thank the many talented men and women at NASA for making the mission such a phenomenal success. It embodies the spirit of NASA and serves as a model for future missions that are faster, better, and cheaper. Today, NASA's Pathfinder team should take a bow, because America is giving them a standing ovation for a stellar performance."

At the same time that it was dealing with the highly visible Mars Pathfinder events described above, the DSN was also handling the arrival and aerobraking maneuvers of the *Mars Global Surveyor* and arrayed operations for the Galileo encounter with the Jupiter satellite Callisto. While these dramatic events were unfolding at JPL and around the Network, another deep space mission was being readied for launch at Cape Canaveral. The Cassini Saturn Orbiter with the Huygens Titan Probe had been successfully mated with the Titan IV/Centaur at the Cape Canaveral Air Force Station, and launch was planned for 13 October.

Cassini

The Cassini mission to Saturn and Titan was a cooperative endeavor by NASA, the European Space Agency (ESA), and the Italian Space Agency (ASI). Its genesis can be traced to the early 1980s, when the international science community agreed upon a combined Saturn Orbiter/Titan Probe mission as the next logical step in the exploration of the solar system, following the Galileo mission to Jupiter. It was seen as a follow-on mission to the brief reconnaissance of Saturn, which had been carried out by the *Pioneer 11* spacecraft in 1979 and the *Voyager 1* and *2* encounters of 1980 and 1981. It would be an enterprise that would span nearly thirty years from its initial vision to its completion in the year 2008.

The Cassini mission was designed to make a detailed scientific study of Saturn; its icy moons, magnetosphere, and rings; and the satellite Titan. Observations relative to the first four science objectives were to be made by twelve science instruments carried on an orbiting spacecraft (*Cassini*). The Titan observations would be made by six science instruments onboard a probe (Huygens), which would be carried to the vicinity of Saturn by *Cassini* before being released to descend to the surface of the satellite. In addition to these experiments, the DSN Project System would be used to conduct at least three gravitational wave searches en route to Saturn and to carry out numerous occultation experiments with the planet, moons, and rings during the orbital phase of the mission.

Scheduled for launch in October 1997 using a Titan IV/Centaur launch vehicle, the spacecraft would require a unique set of four gravity-assist maneuvers to arrive at Saturn almost seven years later, in July 2004. The gravity-assists would be provided by close flybys of Venus in April 1998 and June 1999. It would then fly by Earth in August 1999 and Jupiter in December 2000. The *Cassini* interplanetary trajectory with the four gravity-assist maneuvers is shown in Figure 6-7.

To reduce costs, the development of many mission operations capabilities was deferred until after launch. Except for the three gravitational wave experiments, there were no plans to acquire science data during the cruise phase or gravity-assist maneuvers. Spacecraft operations during the interplanetary cruise would be centralized at JPL. During the orbital phase of the mission, a system of distributed science operations would allow scientists to operate their instruments from their home institutions with the minimum interaction necessary to collect their data.

The Cassini Era: 1996–1997

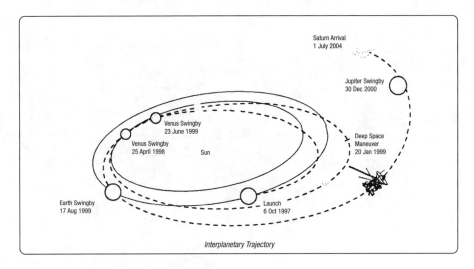

Figure 6-7. *Cassini* interplanetary trajectory.

Cassini was a three-axis-stabilized spacecraft, the main body of which was formed by a stack consisting of a lower equipment module, a propulsion module, an upper equipment module, and the HGA. The various science instruments were mounted on pallets and, together with the Huygens Probe and RTGs, were attached to the stack at appropriate points. The spacecraft contained twelve engineering subsystems in addition to its complement of twelve science instruments.

At launch, the combined weight of *Cassini*, the Huygens Probe, fuel, and the launch vehicle adaptor was 5,712 kilograms, making it the heaviest interplanetary spacecraft ever launched by NASA. More than half of the launch weight was contributed by the propellant needed for the 94-minute main engine burn that would inject *Cassini* into Saturn orbit. The spacecraft stood 6.8 meters high, with a maximum diameter of 4 meters due to the HGA. The major components of the spacecraft are identified in the diagram shown in Figure 6-8.

The design of the *Cassini* spacecraft was the end result of extensive tradeoff studies that considered cost, mass, reliability, the availability of hardware, and past experience. Moving parts were eliminated from the spacecraft wherever possible. Science instruments and the high-gain antenna, which replaced the deployable type of antenna used on *Galileo*, were permanently attached to the spacecraft, their pointing functions performed by rotation of the entire spacecraft. Tape recorders were replaced with solid-state recorders

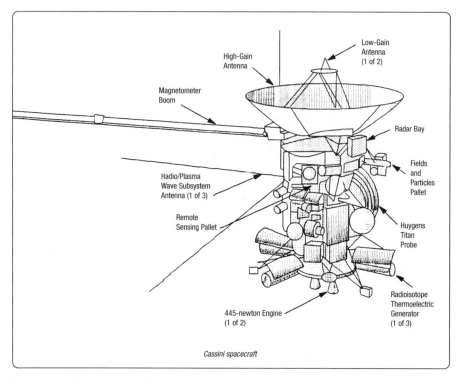

Figure 6-8. *Cassini* spacecraft with Huygens Probe.

and mechanical gyroscopes were replaced with hemispherical resonator gyros. The high-gain antenna would be used for the short-duration (2.5 hours), S-band radio-relay link between *Cassini* and the Huygens Probe, as well as for the permanent radio link between *Cassini* and Earth.

Uplink and downlink communications on X-band were provided by the Radio Frequency Subsystem. A Radio Frequency Instrument Subsystem provided unmodulated RF carriers on the Ka-band and S-band for project experiments only during the cruise and orbital phases. The X-band uplink carried command data in the range 7.8 bps to 500 bps and sequential ranging modulation. The Ka-band uplink was unmodulated. Both Radio Frequency Subsystems included redundant Deep Space Transponders (DSTs). Originally developed as a joint DSN/Flight project task at JPL in 1990 for use on X-band only, the DST was first flown on the NEAR and *Mars Pathfinder* spacecraft in 1996.[6] It was later upgraded to include Ka-band and was flown for the first time in that configuration on *Cassini*.[7]

The X-band downlink carried telemetry data with a wide range of data rates, from 5 bps to 249 kilobits per second. The telemetry data was encoded with Reed-Solomon (255, 223) coding concatenated with Viterbi (7, 1/2) or (15, 1/6) coding prior to transmission. It could also be transmitted in uncoded form. In addition to telemetry, the downlink carried ranging modulation.

The versatile design of the DST provided X-band, Ka-band and S-band downlinks that could be either coherent or noncoherent with the X-band uplink. The Ka-band downlink could be coherent with, or noncoherent with, the Ka-band uplink. This allowed the downlinks to be optimized for the various project investigations under a wide range of conditions.

The Antenna Subsystem included the 4-meter-diameter, narrow-beam-width HGA that was designed for transmitting and receiving on all four communication bands, and two wide-beam-width, low-gain antennas that could receive and transmit on X-band only.

Working with Cassini mission designers, the Telecommunications and Mission Services (TMS) manager, Ronald L. Gillette, negotiated tracking and data acquisition services for the mission that generally fell within the existing (1997) or planned (2002) capability of the Network. The 70-m, 34-m BWG, and 34-m HEF subnets would be required singly or arrayed at various times during the mission. The initial acquisition and early cruise requirements, when the spacecraft would be using its low-gain antennas, would be met without the need for any new implementation. By 2001, however, the 20-kW X-band uplinks on the 34-m HEF antennas would not provide adequate uplink margin for commanding and X-band transmitters, and exciters would be needed on the three 70-m antennas. The DSN agreed to provide this new capability. The DSN also agreed to provide a Ka-band uplink and downlink at DSS 25 for the project experiments, particularly the first gravitational wave experiment of 40 days' duration in 2001. Where the research and development installation at DSS 13 had been equipped with an 80-watt transmitter, this first fully operational Ka-band capability in the Network would have a 4-kW transmitter and provide simultaneous X-band and Ka-band downlink capability.

A new and innovative approach to orbital operations planning was expected to effect economies in the scale of DSN coverage required to support this critical phase of the mission. For the purposes of operational simplicity, DSN coverage was classified into two categories: "high activity," when science opportunities would be most intensive, such as targeted satellite flybys and Saturn periapses; and "low activity," when fewer science opportunities would permit lower data return volume. The science community had agreed that a data volume return of 4.0 gigabits per day for high activity periods and 1.0 gigabit per day for low activity periods would be adequate to accomplish their science objectives. Because Saturn would be above the plane of the ecliptic during the period of orbital operations (2004–08), the daily view periods for the DSN antennas in the southern hemisphere would be of shorter duration, and the received downlink signal-to-noise ratio about 1.5 dB less, than they would be for the DSN antennas in the Northern Hemisphere. For this reason, the DSN agreed to provide coverage, most of the time, for antennas in the Northern Hemisphere, Goldstone, and Madrid. High-activity data return would be accommodated by the 70-m antennas and 70-m/34-m arrayed antennas in the Northern Hemisphere. Low-activity periods would be covered by the 34-m Northern Hemisphere antennas or the 70-m and 34-m Southern Hemisphere antennas, if necessary.

The sustainable data rates on the X-band downlink during Saturn orbital operations also depended on the position of Earth in its orbit around the Sun and on the change in elevation of the spacecraft as it made its daily pass over the tracking station. When all these variables were considered, it was estimated that the downlink data rates would vary over the range 14 to 166 kilobits per second.

In a joint effort to reduce the complexity and cost of daily orbital operations, the factors just described were used by the DSN and Cassini mission planners to develop a data return strategy based on an average pass length of about 9 hours. This strategy optimized the downlink capability with the defined science activity period while limiting the data rate changes on the telemetry downlink to two or three per station pass. Once developed, each strategy would remain in effect for a fixed period, such as 90 days, during which time data rates and pass lengths would not be changed. This arrangement promised significant economies in mission planning and operations effort for both DSN and the Cassini projects.

After the White House gave approval to proceed, *Cassini* was launched without incident from Cape Canaveral, Florida, on 15 October 1997. White House approval to launch was required by presidential directive due to the use of RTG units to power the *Cassini* spacecraft. This matter had been the subject of a number of environmental protest demonstrations at Cape Canaveral in the weeks immediately preceding launch.

The Titan IV/B and Centaur Launch Vehicle functioned perfectly. "Right on the money," observed Cassini program manager Richard J. Spehalski. The sequence of post-launch events, including initial acquisition of the downlink by Canberra stations DSS 46 and DSS 45, was

The Cassini Era: 1996–1997

executed precisely as planned. First reports from the Cassini mission controllers indicated that the angular deviation of the trajectory from its design value was insignificant at better than 0.004 degrees. Mission plans had called for an expected adjustment in the post-launch trajectory of 26 meters per second, but flight data derived from the DSN radio metric data in the first few hours of two-way tracking showed that a correction of only one or two meters per second would be necessary to correct for the small injection error. Early DSN telemetry from CDSCC also showed that all subsystems on the spacecraft, including the solid-state recorder, were operating normally. "I can't recall a launch as perfect as this one," said Cassini mission director Chris Jones; "everything we see is within predictions, with no failures."

With the spacecraft high-gain antenna pointed toward the Sun and the X-band uplink and downlink established through the low-gain antennas, the DSN began the first 30 days of continuous tracking on the 34-m antenna subnet at the rate of 21 passes per week. During this time, mission controllers took advantage of the high data rate (14.2 kbps) available on the downlink, due to the short spacecraft-to-Earth range and the relatively quiescent (no science) state of the spacecraft, to carry out a series of engineering and science instrument maintenance activities. These activities included a checkout of the Huygens Probe about one week after launch. Because of the near-perfect post-launch injection conditions, the first trajectory correction maneuver on 9 November required only a 2.7-meters-per-second adjustment in spacecraft velocity to fine-tune its flight path. Real-time telemetry at 948 bps from DSS 54 allowed the spacecraft controllers to observe the 34.6-second main-engine burn in progress and provided reassuring evidence that this critical subsystem was functioning correctly.

By early January 1998, the *Cassini* flight team had completed all spacecraft activities planned for the early phase of the mission. Onboard software, stored in the command and data subsystem, had been directing spacecraft activities as planned, and the spacecraft thrusters were maintaining the spacecraft in its correct attitude in space. For the next 14 months, *Cassini* would fly with the HGA facing the Sun to shield the spacecraft from the intense solar radiation characteristic of the inner solar system. Throughout this period, the uplinks and downlinks with the DSN would be maintained through either of the two low-gain antennas, depending on the relative geometry of Earth, the Sun, and the spacecraft. Downlink data rates and weekly DSN tracking schedules were adjusted to meet the predefined, returned-data volume requirements. With all its subsystems working perfectly, and periodic instrument maintenance and spacecraft housekeeping activities dominating its routine schedule, *Cassini* moved steadily around its orbit toward an appointment with Venus on 26 April 1998, the first milestone on its seven-year voyage to Saturn.[8]

EARTH-ORBITING MISSIONS

GENERAL

By 1997, almost two-thirds of the missions that composed the total DSN mission set were missions of the Earth-orbital type. These are listed in the mission set for the Cassini Era, shown in Figure 6-1. Together with the Deep Space missions, these accounted for a Network operations load of 30 to 35 mission events being handled by the NOCC in a typical 24-hour period in 1997. In addition to the real-time mission events, a typical operations day for the NOCC would include two to five data playback sessions with the complexes.

Some of these spacecraft, like the Solar Heliospheric Observatory (Soho), Polar, Wind, and Geotail, were part of the continuing International Solar and Terrestrial Physics (ISTP) program and had been supported by the DSN for several years. Others, like Hotbird, were supported by the DSN only for the launch and early orbit period, a few days at most. Some were technology satellites; others were communications satellites; and some were solely for scientific purposes. Except for the cooperative Space VLBI mission HALCA and the reimbursable ESA mission Hotbird, all were NASA missions. However, the cooperative and reimbursable options for DSN support of future, non-NASA missions remained open. There were a number of other NASA missions, not listed in this set, for which the DSN assumed responsibility for providing tracking support in the event of a spacecraft emergency. Appropriate authorities and procedures to invoke support of this kind were permanently in place in the NOCC.

Spacecraft in low-Earth orbit (apogee less than 12,000 km) were supported on the 26-m or 34-m antennas, depending on their requirements for downlink data reception and handling. Those in high-Earth orbit usually required 34-m HEF or 34-m BWG antennas.

The HALCA spacecraft was a special case that warrants further discussion because of the new technology associated with its entry into the DSN. It was developed for the Japanese VLBI Space Observatory Program (VSOP) and was known by that name prior to its actual launch. Following a successful launch in February 1997, it became HALCA and participated in an international cooperative Space VLBI program for which the DSN provided some of the Earth-based antenna support.

The Cassini Era: 1996–1997

Space VLBI Observatory Program

The Space Very Long Baseline Interferometry (SVLBI) program was a two-mission, cooperative venture among NASA, Japan's Institute of Space and Astronautical Sciences (ISAS), and the Russian Astro Space Center (ASC). NASA participated in both missions with funding for the DSN; the National Radio Astronomy Observatory (NRAO) Very Long Baseline Array at Socorro, New Mexico; and the NRAO tracking station at Green Bank, West Virginia.

As early as 1983, Gerry S. Levy and his JPL associates had shown that the resolution of an array of Earth-based VLBI antennas could be greatly increased by means of a special spacecraft which would act as an extension of the Earth-based "radio baseline."[9] The SVLBI program applied and extended these basic ideas to improve the resolution of radio-astronomy images of celestial radio sources by extending the use of VLBI technology into space. The SVLBI missions therefore had a space component and a ground component, both of which heavily involved the DSN.

DSN involvement in the space component of the SVLBI project included the implementation of a new subnetwork of 11-meter antennas with Ku-band uplink and downlink capability. This effort is described later in this chapter. DSN participation in the ground component of SVLBI as a co-observer with the DSN 70-meter antennas is discussed in the Radio Astronomy section of this book.

The space component of the SVLBI program was to consist of two spacecraft: the VSOP (VLBI *Space Observatory Program*), provided by ISAS and launched in 1997, and the *RadioAstron*, provided by ASC and due to be launched several years later. Each spacecraft would carry a radio-telescope antenna and suitable receiving equipment for making radio-astronomy observations from a high-Earth orbit. The DSN's 11-meter antennas would provide continuous tracking support to supplement the limited tracking resources of the other participants in the program.

The VSOP antenna was capable of receiving signals in the three standard radio astronomy bands, L-band (1.6 GHz), C-band (5.0 GHz), and Ku-band (22 GHz). The digitized radio astronomy data, together with time synchronization signals, was to be transmitted to Earth over a Ku-band downlink at 128 Mbps. A Ku-band uplink carried time and frequency reference signals derived from a Hydrogen-maser at the Earth-based receiving stations to the spacecraft for use as a frequency reference. Spacecraft command and control functions and generation of radio metric tracking data used an S-band uplink and downlink that was completely independent of the Ku-band link. These functions were exercised from the Kagoshima Control Center and were not part of the DSN responsibility.

The VSOP spacecraft was successfully launched on 12 February 1997, on the then-new ISAS Mark-V launch vehicle, from the Kagoshima Space Center in Japan.[10] The satellite was renamed HALCA after launch. Following initial acquisition of the S-band downlink by the 26-m station at Goldstone, the Ku-band uplinks and downlinks were activated and communications were established with the 11-meter subnet consisting of DSS 23 (Goldstone), DSS 33 (Canberra), and DSS 53 (Madrid). Three perigee-raise maneuvers, done from 15 to 21 February, were used to place HALCA in a high elliptical orbit with an apogee of 21,400 km, perigee of 560 km, and orbit period of 6.3 hours. This high elliptical orbit had in inclination of 31 degrees (the latitude of the Kagoshima Space Center) and would enable VLBI observations on baselines up to three times longer than those achievable with Earth-based antennas. First interference fringes between HALCA and ground telescopes were found at the Mitaka, Japan, correlator in May, and at the Penticton, Canada, and Socorro, New Mexico, correlators in June 1997.

Tracking station passes varied in length from 10 minutes to as much as 5 hours, depending on the position of the spacecraft in its orbit at the time of the station pass. The short-period orbit provided two or three passes per day at each complex. The DSN was not required to perform any processing of the downlink data. It was simply recorded on special wide-band VLBI-compatible tapes and shipped to designated locations in the United States, Japan, and Canada. Special wide-band correlators at these centers processed the spacecraft observations together with corresponding data from the ground observatories, including the DSN 70-meter antennas, to produce the desired images of celestial radio sources.

The Cassini Era: 1996–1997

THE NETWORK

Complexes and Antennas

By 1997, when the Cassini Era became a reality in the DSN, the total number of antennas operated by the DSN around the globe had increased to twenty-four, twelve at Goldstone and six at each of the Madrid and Canberra Complexes. The DSN also engaged in spacecraft mission operations with many other institutions around the world, each of which had one or more antennas of its own.

The sites of the DSN antennas and Signal Processing Centers were identified according to their geographical location, as shown in Figure 6-9. The sites of related antennas such as Weilheim, Germany; Usuda, Japan; and Parkes, Australia, were identified in the same manner.

It was convenient to describe the Network in terms of several subnets according to the metric diameter of the antennas composing it. There were a 70-meter subnet, two 34-m subnets, a 26-m subnet, an 11-m subnet, and several unique antennas such as DSS 13 and DSS 17 at Goldstone. An inventory of antennas in the DSN in 1997 is given in Figures 6-10, 6-11, and 6-12 for the Goldstone, Canberra, and Madrid complexes, respectively.

These tables show the various uplinks and downlinks that were available in 1997 to support space operations on each of the antennas. Comprehensive though this listing was, substantial additions were planned for it in the immediate future.

Location Identifier for DSN and Related Sites

Number Range	Geographic Description and Position*
00–29	JPL/Goldstone and the VLA, Socorro, NM
30–49	Pacific Basin: Between L = 180 and L = 270
50–69	European Basin: Between L = 345 and L = 90
70–89	Atlantic Basin: Between L = 270 and L = 345
90–99	Miscellaneous
100–999	GPS and Future Use

*Note: L = Longitude in degrees, measured east from zero longitude.

Figure 6-9. Location identifiers for DSN and related sites.

Goldstone Deep Space Communications Complex, 1997

DSS	Diam., Type	Uplinks	Downlinks	Operational
12	34 m, STD	none	none	Decom'd. 2/96
13	34 m, BWG, R&D	S, X, Ka	S, X, Ka, Ku	6/90
14	70 m	S	S, X	64 m/70 m, 5/88
15	34 m, HEF	X	S, X	8/84
16	26 m	S	S	1/67
23	11 m, OVLBI	X	X, Ku	2/96
24	34 m, BWG	X, S	S, X	2/95
25	34 m, BWG	S, X	S, X, Ka	8/96
26	34 m, BWG	X	X	8/96
27	34 m, BWG, HS	S	S	7/95
28	34 m, BWG, HS	S	S	10/00

Figure 6-10. Goldstone Deep Space Communications Complex, 1997.

Canberra Deep Space Communications Complex: 1997

DSS	Diam., Type	Uplinks	Downlinks	Operational
33	11 m, OVLBI	X, Ku	X, Ku	6/96
34	34 m, BWG	X	S, X	11/96
42	34 m, STD	S	S, X	Decom'd. 12/99
43	70 m	S	S, X	9/87
45	34 m, HEF	X	S, X	12/84
46	26 m	S	S	12/83

Figure 6-11. Canberra Deep Space Communications Complex, 1997.

Madrid Deep Space Communications Complex, 1997

DSS	Diam., Type	Uplinks	Downlinks	Operational
53	11 m, OVLBI	X, Ku	X, Ku	6/96
54	34 m, BWG	X	S, X	10/97
61	34 m, STD	S	S, X	Decom'd. 12/99
63	70 m	S	S/X	64 m/70 m, 7/87
65	34 m, HEF	X	S/X	4/87
66	26 m	S	S	12/84

Figure 6-12. Madrid Deep Space Communications Complex, 1997.

The 70-m subnet would be provided with X-band uplinks for first use with *Cassini*, beginning with DSS 14 in 2000. Already, plans were being made to install an X-/Ka-band dichroic plate on the DSS 25 BWG antenna. This would be the first step in providing an operational Ka-band downlink for use on the Deep Space-1 mission and for project use with *Cassini* in 2000. The DSS 26 BWG antenna was being modified to demonstrate the more stringent antenna pointing capability required for Ka-band operation.

Driven by the need to reduce maintenance and operations costs, the closure of the 26-m subnet by the end of Fiscal Year 2001 had been under consideration since 1996. Several alternatives, including the automation of routine telemetry, command, and monitoring functions, the transfer of low-Earth-orbiting spacecraft to NASA's Wallops Island facility, and eventually to a commercial ground network, were being actively pursued in 1997.

Ways of making more effective use of DSN antenna time were demonstrated in several areas in 1996 and 1997. The "Multiple Spacecraft Per Antenna" concept was demonstrated effectively at Goldstone in September 1997 with the successful tracking of *Mars Global Surveyor* and *Mars Pathfinder* from a single antenna. Both spacecraft were at Mars and both lay within the beamwidth of a single 34-m BWG antenna (0.063 degrees at X-band). Several hundred commands were transmitted to each spacecraft, and good telemetry and two-way Doppler was obtained from each spacecraft. The operational advantages of this mode of operation were clearly demonstrated, particularly those related to switching the uplink from one frequency to another, and subsequent acquisition of individual spacecraft receivers. The DSN planned to develop this technique for future use as the Mars Exploration initiative would provide more opportunities for effecting economy in usage of DSN antennas.

For many years, it had been the practice to make regular, biweekly Earth motion and clock synchronization measurements between complexes by using two 70-m antennas simultaneously to observe a celestial radio source. Comparison of time off-sets in the observables provided a measure of the differences in time and frequency between the complexes. In 1997, with the addition of updated VLBI equipment and precision data from GPS receivers at the complexes, it became possible to accomplish the same biweekly measurements with two 34-m HEF antennas rather than two 70-m antennas.

In November 1997, the Future Programs Council at JPL heard a briefing on the future of deep space communications. The scope of the briefing included optical as well as conventional radio communications, but in the context of existing DSN antennas, it recommended that "to meet the increased needs of a growing mission set, NASA should rapidly deploy Ka-band capability throughout the existing DSN as a low[-]cost path which would allow future missions to double their data return while cutting their (antenna) contact time in half."

In time, by 1997, that statement seemed to indicate the direction of future development for antennas in the Network.

Ka-band Downlink

Along with design concepts for beam waveguide antennas, the advantages of Ka-band communication links and the application of these ideas to future uplinks and downlinks for the DSN were discussed by Joel G. Smith, Robert C. Clauss, and James W. Layland in three companion reports in 1987.[11] These studies showed that a spacecraft-to-Earth downlink on Ka-band (32 GHz) could carry telemetry data at 3 to 10 times the data rate possible with an X-band downlink, given the same spacecraft transmitter weight, antenna size, and power requirements. This enhancement in downlink performance derived from increased antenna gain at higher frequencies (shorter wavelengths), but was reduced somewhat by other factors such as atmospheric and antenna losses and susceptibility to wet or foggy weather. A more indepth study of the tradeoffs associated with Ka-band operation in the DSN was carried out in 1988. This study described a scenario for establishing Ka-band as the primary Deep Space communications frequency of the future and proposed a baseline plan for reaching that goal.[12]

The Cassini Era: 1996–1997

As the first two steps toward realization of Ka-band communication links in the future DSN, it was proposed that a Ka-band receiving capability should be added to the new research and development beam-waveguide antenna being built at DSS 13, and that a Ka-band beacon experiment (KABLE) should be planned for the *Mars Observer* spacecraft due to be launched to Mars in September 1992.[13]

This methodology had been used to introduce the X-band into the Network in 1973 as an S-/X-band experiment on the *Mariner 10* missions to Venus and Mercury. The Ka-band transponder for the *Mars Observer* spacecraft was the first step in a parallel development plan for spacecraft device and component technology proposed by Arthur L. Riley.[14]

The BWG antenna at DSS 13 was completed by mid-1990, as described in the preceding chapter, and its Ka-band capability was evaluated over the period July 1990 through January 1991.[15] The microwave feed package used on this antenna for KABLE consisted of an X/Ka-band dichroic plate and two low-noise amplifiers (LNAs), one for X-band and one for Ka-band. While the X-band LNA used a simple high-electron mobility transistor (HEMT) to achieve a system temperature of 37 kelvin, the ultra-LNA for Ka-band required the development of JPL's first Ka-band cavity maser to achieve a system temperature of 5 kelvin.[16] The Ka-band and X-band downlink signals, amplified by the LNAs, were converted to intermediate frequency and delivered to separate digital Advanced Receivers (ARX) for tracking, demodulation, and delivery to the telemetry decoders. A data-handling terminal carried out the telemetry recording and delivery tasks. Of the two advanced receivers, one was the unit that was being developed for the Galileo application in the DGT; the other was specially built for use at DSS 13 and would remain at the station. A special Doppler tuner was required for this unit to handle the Doppler jerk (rate of acceleration) expected on the Ka-band downlink when MGS would be in orbit around Mars.

The *Mars Observer* spacecraft was equipped with a special 33-milliwatt Ka-band transmitter and used the concave surface of the subreflector on the spacecraft's high-gain antenna as an antenna for the Ka-band downlink. To keep the Ka-band downlink coherent with the X-band uplink and downlink, its frequency, 33.66 GHz, was obtained by simply multiplying the X-band downlink frequency (8.4150 GHz) by four. Telemetry and ranging modulation were provided on both downlinks. At this time, fiber-optic cables were being introduced for intra-site communications and transmission of highly stable reference and timing signals between antennas on each complex. Full advantage was taken of this new facility to connect DSS 13 to the SPC 10 Control Room some 20 miles away, where the existing ranging and Doppler equipment was located.

The major objectives for KABLE included a telemetry demonstration, a ranging demonstration, and a tracking experiment. The telemetry demonstration would prove the feasibility of a telemetry link at Ka-band by collecting a minimum of one million bits of data, at 250 bits per second, with minimal errors. The ranging demonstration would follow the telemetry demonstration and illustrate the correct demodulation of DSN ranging codes at Ka-band. The tracking experiment would develop a database of Ka-band downlink statistics including weather, system noise, temperature, and antenna performance for use in the future development of Ka-band capability in the Network.

By an unfortunate coincidence, the telemetry demonstration took place during some of the worst weather conditions ever experienced in the Goldstone area. Heavy rains and dense fog were daily occurrences. This type of weather has a major degradation effect on Ka-band link performance and proved to be a limiting factor in reaching the data collection goal for telemetry. Although the ranging demonstration was limited to two passes only, due to various equipment problems and to the weather, it yielded a sufficient number of good ranging points for favorable comparison with theoretical performance. Equipment problems and bad weather also limited the amount of tracking data that was gathered. Nevertheless, sufficient Ka-band tracking data was retrieved to show a potential improvement of 4 to 7 dB over X-band performance under equivalent conditions.

Ironically, on 21 August 1993, just when the weather was improving and the equipment problems were being resolved, uplink and downlink communications with the *Mars Observer* spacecraft were lost. Later investigations led to the conclusion that the spacecraft had been destroyed during the engine pressurization sequence prior to the maneuver for Mars orbit insertion.

Despite its abrupt ending, sufficient data was obtained to declare the KABLE effort partially successful. The data that had been collected was thoroughly analyzed, and it ultimately provided a strong justification for a repeat experiment as soon as another opportunity presented itself.[17]

That opportunity would not be long in coming. In November 1996, when the *Mars Global Surveyor* spacecraft was launched on a mission similar to that of the ill-fated *Mars Observer*, it carried with it a much-improved version of the original Ka-band technology experiment called KABLE-II. Although KABLE-II was identical in concept to the earlier experiment, it benefited from several changes to the spacecraft and ground elements of the link. The Ka-band transmitter power had been increased to approximately 1 watt, and a change had been made to the way in which the frequency of the downlink Ka-band signal was derived in the spacecraft transponder. This change effectively placed the downlink signal frequency in the center of the receiving pass-band, rather than to one side of it, as was the case in Kable-I. Since the microwave components in the DSS 13 BWG antenna, particularly the dichroic plate, were optimized to receive at this band-center frequency, substantial loss in the downlink signal margin was avoided. At DSS 13, a new seven-element Ka-band array feed was used to receive the downlink and point the 34-m antenna with millidegree precision; Block V receivers replaced the earlier ARX developmental versions; and numerous other improvements related to operational reliability were made.

When the MGS spacecraft was pointed to Earth in mid-January 1997, strong signals were immediately received at DSS 13 and simultaneous X-/Ka-band tracking was carried out on six occasions thereafter. The array feed tracked the spacecraft within 1 millidegree in angle, and both carriers were tracked simultaneously to generate Doppler data by an experimental tone-tracker. Measurements were in good agreement with theory. The telemetry experiment began in February with an improved low-noise tracking feed which not only incorporated a multimode coupler to generate pointing error signals for driving the antenna pointing system, but also was cooled to reduce the system noise temperature.

On 21 March, the KABLE-II experimenters reported that a major objective had been successfully accomplished. Telemetry data received over the 1 watt, Ka-band link at 2 kbps had been compared bit-by-bit with data received over the 25-watt X-band downlink. All 12 million contiguous bits were in agreement. There were no errors. This event marked the first error-free telemetry reception at Ka-band in the DSN.

At this time, the Ka-band array feed, which had been replaced by the multimode-coupler feed at DSS 13, was installed and tested on the DSS 14 antenna in preparation for the start of Ka-band efficiency tests of the 70-m antenna in May. This capability would be needed for Cassini in the years ahead.

In mid-April, real-time telemetry data received from MGS over the Ka-band and the X-band links was presented to the MGS project operations team at JPL for evaluation. Dual-frequency tracking and ranging measurements were also made at this time. The experimental Ka-band downlink was received at DSS 13, while an operational X-band uplink and downlink was maintained at DSS 15. Both downlinks had been converted to 300-MHz IF frequency and transferred over fiber-optic circuits to SPC 10 for telemetry and radio-metric data processing and transmission to JPL.

The KABLE-II telemetry demonstrations with MGS at DSS 13 continued as scheduled through 8 May 1997. The successful demonstrations of Ka-band technology occasioned by the two KABLE experiments lent impetus to its transfer to the operational Network beginning with DSS 14 and DSS 25 at Goldstone.

The first two missions that would be carrying Ka-band downlinks to demonstrate an operational rather than an experimental capability in the DSN were the New Millennium mission DS-1 and the Cassini mission to Saturn.

DS-1 was a technology demonstration mission with secondary scientific objectives. Primary communications would be conducted on X-band, and DSS 25 was to provide a Ka-band receive capability as an operational technology demonstration. DSS 13 would also provide some analytical support. DS-1 would not be launched before the middle of 1998.

Cassini also planned to use X-band for its primary telemetry and radio metric support, but, additionally, it planned an extensive project program that would be accomplished using a combination of S-, X-, and Ka-band downlinks and X-band and Ka-band uplinks. The Ka-band uplink and downlink were required to support the Cassini Gravitational Wave Experiment in 2001.

While the DSN telecommunications engineers had been pursuing a path leading to an operational Ka-band capability in the Network, spacecraft telecommunications engineers had been pursuing a somewhat parallel path in developing an X-band/Ka-band transponder for future spacecraft. The NASA Deep Space Transponder (DST) had been in development since 1991 and was intended for first use with *Cassini*, at that time planned for launch in 1996. It would replace the existing S-band transponders, which, with an external S-band-to-X-band down-converter to provide the X-band capability, had been used on all previous planetary spacecraft. Using conventional spacecraft technology, the DST contained an automatic phase-tracking receiver for X-band uplink only and an X-band exciter to drive redundant downlink X-band transmitters. It provided appropriate reference frequency signals to devices external to the DST for the generation of independent Ka-band and S-band downlink signals.

The *Cassini* spacecraft was launched in October 1997, carrying the DST into space for the first time. *Cassini* would depend on the DST for its primary spacecraft communications links with the DSN; uplink and downlink on X-band for telemetry, command, and radio metric data; and downlink only on Ka-band and S-band for project and gravity-wave experiments.

The Cassini Era: 1996–1997

The Ka-band transponder, also known as the Small Deep Space Transponder, had been under development since 1995 and was designed specifically for Ka-band uplink and downlink operation.[18] It included a selectable X-band capability that could be switched in or out as required, as well as several state-of-the-art components, including sampling mixers, a Ka-band dielectric resonator oscillator, and microwave monolithic integrated circuits (MMICs) to perform the functions of up and down frequency conversion, modulation, and amplification. It was proposed for first use as a telecommunications evaluation experiment on DS-1, to be launched in 1998.

Ka-band was definitely on the move in the DSN in 1997, but its arrival as an operational capability lay several years in the future.

Orbiting VLBI Subnetwork

As discussed earlier in this chapter, Space Very Long Baseline Interferometry (SVLBI) was a cooperative project among NASA, ISAS, and ASC, representing space agencies from the United States, Japan, and Russia. ISAS provided the VSOP (HALCA) spacecraft, ASC was to provide the RadioAstron spacecraft, and the Russian tracking stations and NASA provided the SVLBI subnet of four tracking stations. One of these stations was the existing 14-meter NRAO Green Bank station in West Virginia. The other three antennas were built by NASA specially for this cooperative project and, at their completion in 1996, became part of the Deep Space Network. One antenna, with its unique electronics, was installed at each complex. Within the DSN, they were referred to collectively as the Orbiting VLBI (OVLBI) 11-meter subnet and designated DSS 23, DSS 33, and DSS 53 for Goldstone, Canberra, and Madrid, respectively. However, from the overall project viewpoint, they remained part of the SVLBI subnet.

The main functions of 11-meter stations were to automatically acquire and track the spacecraft X-band and Ku-band downlinks, generate and provide two-way integrated Doppler data to the DSN Navigation Subsystem; provide uplink to the spacecraft with a phase-stable reference frequency tied to the DSN Hydrogen Maser Frequency standard at each site; and receive, demodulate, and record the 144 Mbps downlink telemetry data in special VLBA and/or S-2 data format. The spacecraft data would be delivered to the project for later correlation with data from the co-observing ground telescopes. These functions were to be performed autonomously, without the need for operator intervention except to change recording tapes, at a frequency of 4 passes per day, 7 days per week. Interfaces with the existing DSN systems would provide the new antennas with frequency and time references; monitor and control functions; and the necessary schedule, frequency prediction, and antenna pointing and orbital state vector information to enable autonomous operation.[19]

Initial studies in 1990 had shown that considerable cost savings to NASA would result from the use of commercially available resources, rather than JPL "in-house" resources, to provide the antennas. A comprehensive set of technical design requirements was prepared in 1991, and after due process, a contract for the design and implementation of three high-precision 11-meter antennas and associated electronics and software as a "turn-key" package was awarded to Scientific-Atlanta (S-A) Corporation in September 1991.

JPL was responsible for selecting and preparing the sites and providing access roads, antenna foundation, buried cable conduits, air-conditioned buildings, power, water, and related facilities.

Although the design was based on an S-A standard, the 11-meter production antenna, a great deal of new development was necessary to meet the extremely tight design requirements for frequency and phase stability required for the intended OVLBI application.[20]

The functional block diagram in Figure 6-13 shows the interfaces between the 11-m antenna in the OVLBI subnet and various systems of the Network.

Each station in the new subnetwork contained two new subsystems: the OVLBI Tracking Subsystem (OTS) and the OVLBI Data Subsystem (DTS).

The OVLBI Tracking Subsystem included the 11-meter-diameter antenna, on an azimuth/elevation mount, with a third "tilt" axis to allow for a direct overhead pass without loss of tracking data due to the keyhole effect inherent in AZ/EL mounts. Concentric, five-horn monopulse feeds in a cassegrain subreflector configuration handled uplink and downlink signals at X-band, while the center horn also acted as a common receiving aperture for Ku-band. The Ku-band monopulse error signals were derived from a TE21 mode coupler at the base of the horn. Low-noise amplifiers, microwave components, tracking receivers, antenna-control servos and data receivers and demodulators provided for antenna pointing and downlink data reception and demodulation. Frequency converters and local oscillators driven from computers in the Data Tracking Subsystem provided the important Doppler compensation and signal coherence functions. The uplink for RadioAstron was provided by a small 5-watt solid-state X-band transmitter, while the uplink for VSOP was provided by a 0.5-watt Ku-band transmitter.

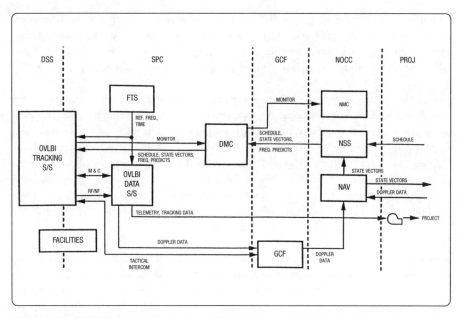

Figure 6-13. OVLBI Subnetwork functional block diagram. The following abbreviations are used in this diagram:

FTS	Frequency and Timing System
DMC	DSN Monitor and Control
NOCC	Network Operations Control Center
NMC	NOCC Monitor and Control
NSS	NOCC Support Subsystem
NAV	NOCC Navigation
DSS	Deep Space Station
SPC	Signal Processing Center
GCF	Ground Communications Facility
PROJ	Flight Project Mission Operations Teams
S/S	Subsystem

On these antennas, special measures were taken to minimize the effect of temperature variations on the phase stability of the X-band and Ku-band signals. Transmission of these signals between the antenna and the Control Room was carried on fiber-optic cables, and all frequency conversion equipment was located in the temperature-regulated Control Room. The temperature of microwave and optical equipment mounted on the antenna itself was maintained to within ±1.0 degrees of normal by an independent, proportional air-conditioning system. To further enhance phase stability, the fiber-optic cables between the antenna and the Control Room were buried six feet below ground level. Except for routine maintenance and changing of recording tapes, the station was fully operated by a ground station control computer as part of the OTS.

The OVLBI data subsystem contained various components that were peculiar to the VSOP and RadioAstron missions. A Doppler compensation computer drove the local oscillators in the Tracking Subsystem, which programmed out the expected Doppler frequency shift on both the uplink and downlink. A Doppler extractor integrated the Doppler frequency shift and provided this data to the DSN Navigation Subsystem for calculating the precise spacecraft position. After decoding, the telemetry data containing the VLBI observations from the spacecraft were recorded digitally in a special Very Long Baseline Array (VLBA) format. An additional interface to provide the VLBI data to the Canadian S-2 recording subsystem was added later.

In August 1995, four years after the award of the contract, factory acceptance testing of the first two systems for Goldstone and Canberra was successfully completed at Scientific-Atlanta. The hardware performed well, and performance against critical requirements relating to antenna gain-to-system noise temperature (G/T), phase stability, Doppler compensation, tracking accuracy, and telemetry bit error rate was well within specification. There were, however, a significant number of software anomalies which would be corrected over the following weeks. The Goldstone system was dismantled and prepared for shipping, while the Canberra system was retained on the antenna test range to provide a test-bed for clearing the software anomalies. The Canberra system was shipped a few weeks later, but the system for Madrid was delayed by shipping and other problems and did not reach the Madrid Complex until May 1996.

By then, the Goldstone system (DSS 23) was acquiring and tracking the SURFSAT spacecraft automatically without operator intervention at X-band, as well as generating two-way Doppler on a regular basis. The SURFSAT tracks were continued throughout the year to gain experience, while the Doppler data was evaluated for quality and accuracy and the last of the software problems were worked on.

A photograph of DSS 23 at the Goldstone Apollo Site with the DSS 24 BWG antenna in the background is shown in Figure 6-14.

In May and June, while the Canberra system (DSS 33) was completing its onsite system acceptance testing, assembly of the antenna for DSS 53 began at the Madrid site. By the end of 1996, all three 11-m antennas had been completed and most major prob-

Figure 6-14. DSS 23, 11-meter OVLBI antenna at Goldstone, 1996.

lems with the DPRG and the Doppler and phase generation software had been solved. Full end-to-end capability between a ground station (DSS 33) and the VLBI correlator at Socorro, New Mexico, was demonstrated for the first time in December, using X-band data from the SURFSAT spacecraft. At that time, none of the 11-m stations had yet been exposed to a Ku-band downlink, and the VSOP launch, set for February 1997, was then just over two months away.

In mid-January 1997, the SURFSAT spacecraft was commanded to switch its downlink from X-band to Ku-band to give the new 11-m stations some experience in routine tracking at the frequency they would be seeing for the VSOP downlink. For the next four weeks, all 11-m stations engaged in routine tracking of the SURFSAT Ku-band downlink, while final software deliveries were installed and preparations were made for the start of VSOP inflight operations.

The VSOP launch took place on 12 February 1997, and the first Ku-band track of HALCA, as it became known after launch, was made at DSS 23 on 20 March. With minor exceptions, 11-m subnet support for HALCA operations in 1997 was satisfactory and is described earlier in this chapter. In the longer term flight operations environment, however, the 11-m subnet required significant operations and engineering effort from JPL and the Complexes to meet most of its design specifications.

EMERGENCY CONTROL CENTER

Each of the facilities that compose the DSN had a unique set of policies and procedures to cover emergencies and disasters that might occur in its own particular locality. The response of each facility to an emergency situation is determined by its own set of standard emergency procedures. Should the capability of the overall Network be affected by the local emergency or the steps taken to deal with it, the Network situation would be handled by the Operations Chief at the NOCC using whatever alternate resources were at his disposal at the time. However, the loss of the NOCC at JPL, for whatever reason, would critically impair the ability of the entire Network to continue support for inflight spacecraft.

Aware of this situation, the Galileo project had, in 1995, already developed a capability to conduct the critical "insertion into Jupiter orbit" phase of its mission from Goldstone in the event that a natural disaster precluded operations at JPL. A key feature of this fairly basic emergency center was the ability to send commands to the *Galileo* spacecraft from any DSN 70-m antenna, independent of support from the regular Galileo mission control center at JPL.

As a consequence of the growing perception of Network vulnerability to the occurrence of a disaster at JPL, NASA Headquarters soon requested that DSN develop and publish an official Network Disaster Plan. The Network Disaster Plan recognized the potential for disablement of the NOCC due to earthquake, bomb threat, fire, or industrial dispute, and established appropriate procedures to be followed by Network personnel. The procedures involved the relocation of basic NOCC and mission control activities to an alternate location at the Goldstone Deep Space Communications Complex (GDSCC), located about 120 air-miles northwest of JPL. There, in 1996, a permanent new facility called the Emergency Control Center (ECC) was implemented, with voice and basic data communication capabilities, specifically for the purpose of continuing Network and mission control activities on a limited scale in the event of a disaster at JPL.[21]

In its original concept, the ECC was intended to accommodate only the DSN personnel involved with Network control. It quickly became evident, however, that the spacecraft and mission control facilities that were accommodated in the same Space Flight Operations Facility (SFOF) building at JPL as the NOCC were equally vulner-

The Cassini Era: 1996–1997

able to the effects of a disaster. Under the direction of the Telecommunications and Mission Operations Directorate (TMOD), the design was broadened to include these latter capabilities in addition to those required for Network control, and the facility became the TMOD ECC.

In its basic form, the TMOD ECC consisted of a central router which could be connected, on demand, to the Goldstone Signal Processing Center (SPC) via an ethernet link and to the SPCs at Canberra and Madrid via permanently available and dedicated voice/data communication circuits. By design, the circuits to both of the overseas sites bypassed the high-earthquake-risk Los Angeles area. In the event of an emergency, the Complexes would be instructed to patch their communications circuits to alternate communications routers which would redirect their communications traffic to the ECC at Goldstone rather than to the NOCC at JPL.

Within the ECC, an ethernet LAN connected the central router to several Hewlett-Packard and Sun Microsystems high-capacity workstations arranged in a 10-base T-hub configuration. Most of these were for the use of spacecraft and mission controllers in carrying out navigation, command, telemetry data handling, and sequence generation functions. Network control functions and the generation and dissemination of prediction data products for the antenna pointing, radio metric, telemetry, and frequency control subsystem were carried out at a large VAX workstation. A DSN data base was kept current by frequent downloading of current data from the active Network support subsystem in the NOCC. Likewise, flight project software for the spacecraft and mission workstations was updated on a regular basis to reflect all changes and revisions.

Funding for implementation of the ECC task was in place by mid-1996, and responsibilities for development, planning, and implementation were assigned. The ECC was to be located at the Echo site at Goldstone in the vacant control room of the DSS 12 antenna, which had been decommissioned from DSN service earlier in 1996. Under pressure to be ready prior to the Cassini launch in October 1997, progress was rapid. By early 1997, the workstations were installed in the ECC and testing had begun. Late deliveries of the remaining equipment and a "freeze" on all new implementation activity in the Network during the *Mars Pathfinder* landing and Mars surface operations in July delayed the completion of the work until mid-August, when Cassini and Ulysses mission controllers were able to begin evaluation of the ECC for flight operations support. The first tests with Canberra and Madrid used 28-kilobits-per-second communication circuits routed through JPL, since the special 512-kilobits-per-second direct circuits between the ECC and the overseas sites had not yet been made available to JPL.

Meanwhile, the Cassini project had developed a new launch "constraint" requirement for an independent back-up for the existing Cassini Launch Operations Center at JPL. Fortunately, the 28-kilobits-per-second circuits could be used to satisfy the newly imposed the Cassini launch criteria in the event of further delay in the availability of the wideband 512-kilobits-per-second circuits. A few weeks later, these circuits did become available, just in time to fully test the system with the overseas sites and to satisfy the new Cassini requirements.

The TMOD Emergency Control Center was reported to be operational for all JPL flight projects on 6 October 1997. *Cassini* was successfully launched nine days later, with the ECC providing back-up for its prime Launch Operations Center at JPL.

The ECC was first used for full in-flight mission support by *Galileo* at the end of October 1997. A complete track, including planning tasks and spacecraft command functions, was run successfully from the ECC by Galileo flight operations personnel. Similar demonstrations were planned to familiarize Ulysses, Voyager, and Mars Global Surveyor operations personnel with the capabilities (and limitations) of the new facility. In due course, these demonstrations came into regular scheduled use as a means of keeping the ECC in a state of readiness and exercising its potential users in emergency awareness.

The Cassini Era: 1996–1997

OTHER ASPECTS

The years 1996–1997 brought to a close the traditional form of mission operations that the DSN had employed to support all of NASA's planetary missions, and many of its lunar and Earth-orbiting missions, for almost forty years. New, more efficient, and less costly ways of conducting mission operations, including a move to Ka-band frequencies, were being introduced in the Network to match NASA's call for "faster, better, cheaper" missions. In the opening years of the Cassini Era, Mars Global Surveyor and Mars Pathfinder, the first planetary missions designed to new guidelines, were launched on a fast track to Mars. *Cassini*, the last planetary spacecraft based on the old principles, began its long mission to Saturn.

While the proper business of the DSN was always the support of NASA's inflight spacecraft missions, "mission operations" represented only one aspect of DSN history over the period covered by this book. Technology, science, and organization, although not directly involved with "mission operations", were also essential to the growth, scientific stature, and fiscal well-being of the Network during that time.

To this point, the narrative has emphasized "mission operations" in terms of five major eras of planetary missions from 1957 though 1997. With that completed through the Cassini Era, we turn now to consider other aspects of DSN history, beginning with the growth of technology in the DSN.

Endnotes

1. Jet Propulsion Laboratory, Telecommunications and Mission Operations Directorate, "Bridging the Space Frontier" (30 January 1998). *http://deepspace.jpl.nasa.gov/920/public/922_strat_plan/*.

2. Voyager Web site: *http://www.jpl.nasa.gov/releases/98/vgr217.html*.

3. NEAR Web site: *http://near.jhuapl.edu/*.

4. Mars Global Surveyor Web site: *http://mars.jpl.nasa.gov/mgs/*.

5. Mars Pathfinder Web sites: *http://www.jpl.nasa.gov/marsnews*, *http://mpfwww.jpl.nasa.gov/mpf/engineering.html*.

6. N. R. Mysoor, J. D. Perret, and A. W. Kermode, "Design Concepts and Performance of a NASA X-band (7162 MHZ/8415 MHz) Transponder for Deep-Space Spacecraft Applications," TDA Progress Report PR 42-104, October-December 1990 (15 February 1991), pp. 247–56.

7. N. R. Mysoor, J. P. Lane, S. Kayalar and A. W. Kermode, "Performance of a Ka-band Transponder Breadboard for Deep-Space Applications," TDA Progress Report PR 42-122, April-June 1995, (15 August 1995), pp. 175–83.

8. Cassini Web site: *http://www.jpl.nasa.gov/cassini*.

9. VSOP Web site: *http://www.vsop.isas.ac.jp*.

10. J. S. Ulvestad, C. D. Edwards, and R. P. Linfield, "Very Long Baseline Interferometry Using a Radio Telescope in Earth Orbit," TDA Progress Report PR 42-88, October-December 1986 (15 February 1987), pp. 1–10.

11. J. G. Smith, "Ka-band (32-GHz) Downlink Capability for Deep Space Communications," TDA Progress Report PR 42-88, October-December 1986 (15 February 1987), pp. 96–103; J. W. Layland and J. G. Smith, "A Growth Path for Deep Space Communications," TDA Progress Report PR 42-88, October-December 1986 (15 February 1987), pp. 120–25; R. C. Clauss and J. G. Smith, "Beam Waveguides in the Deep Space Network," TDA Progress Report PR 42-88, October-December 1986 (15 February 1987), pp. 174–82.

12. J. W. Layland, "Ka-band Study-1988, Final Report," 890-212, JPL Document D-6015, (Pasadena, California: Jet Propulsion Laboratory, 15 February 1989).

13. J. G. Smith, "Proposed Upgrade of the Deep Space Network Research and Development Station," TDA Progress Report PR 42-88, October-December 1986 (15 February 1987), pp. 158–63; A. L. Riley, D. M. Hansen, A. Mileant, and R. W. Hartop, "A Ka-band (32 GHz) Beacon Link Experiment (KABLE) With Mars Observer," TDA Progress Report PR 42-88, October-December 1986 (15 February 1987), pp. 141–47.

14. A. L. Riley, "Ka-band (32 GHz) Spacecraft Development Plan," TDA Progress Report PR 42-88, October-December 1986 (15 February 1987), pp. 164–73.

15. S. D. Slobin, T. Y. Otoshi, M. J. Britcliffe, L. S. Alvarez, S. R. Stewart, and M. M. Franco, "Efficiency Calibration of the DSS 13 34-m Diameter Beam Waveguide Antenna at 8.45 and 32 GHZ," TDA Progress Report PR 42-106, April-June 1991 (15 August 1991), pp. 283–89.

16. J. Shell and R. B. Quinn, "A Dual-Cavity Ruby Maser for the Ka-band Link Experiment," TDA Progress Report PR 42-116, October-December 1993 (15 February 1994), pp. 53–70.

17. T. A. Rebold, A. Kwok, G. E. Wood, and S. Butman, "The Mars Observer Ka-band Link Experiment," TDA Progress Report PR 41-117, January-March 1994 (15 May 1994), pp. 250–82.

18. N. R. Mysoor, J. P. Lane, S. Kayalar, and A. W. Kermode, "Performance of a Ka-band Transponder Breadboard for Deep-Space Applications," TDA Progress Report PR 42-122, April-June 1995 (15 August 1995), pp. 175–83.

19. J. Ovnick, "DSN Orbiting VLBI Subnet, Task Implementation Plan," JPL Document D-7787, 803-120, Vol. 1 (August 1990).

20. J. Ovnick, "DSN Orbiting VLBI Subnet, Design Requirements," JPL Document D-9619, DM515606 (November 1993).

21. Jet Propulsion Laboratory, "Deep Space Network Standard Operations Plan," DSN Document 841, Vol. 1, DSN Operations (31 July 1996), p. 14-1.

Chapter 7

The Advance of Technology in the Deep Space Network

THE DSN TECHNOLOGY PROGRAM

No history of the DSN would be complete without a full appreciation of its contribution of advanced technology to the successful development of the Network. The wellspring of new and innovative ideas for increasing the existing capability of the Network; for improving reliability, operability, or cost effectiveness; or for enabling recovery from potential mission-threatening situations has resided, from the very beginning of the Network's history, in a strong program of advanced technology, research, and development. Known by various names through the years, the numerous elements of the advanced technology program were formally identified as a complete entity and placed under the direction of the newly established Technology Office of the Telecommunications and Mission Operations Directorate (TMOD) in 1997. It was known as the TMOD Technology Program. The program scope was further expanded to include technologies relevant to the full end-to-end Deep Space Mission System (DSMS). This expansion included flight components of the physical communications links (data services) and added mission services activities like protocol developments, mission planning and execution tools, science data visualization, and the merging of tools and autonomy.

These disciplines covered almost every aspect of DSN-related technology. Theoretical work in any of them could always be complemented with experimental work in well-equipped laboratories and machine shops at JPL and verified with field measurements at DSS 13, which was maintained primarily for research and development (R&D) purposes.

Prior to 1997, the program was known most generally as the DSN Advanced Systems Program and covered antennas, low-noise amplifiers, Network signal processing, frequency and timing, radio metric tracking, navigation, Network automation, atmospheric propagation. It also involved the evolution of the program research and development tracking station, DSS 13, at Goldstone. These were all elements of ground-based systems. In 1985, a second Systems Development Program was added to cover telecommunications systems analysis, spacecraft radio communication systems, and inflight demonstrations of new uplink and downlink communications systems.

Due in no small part to the unremitting efforts of Nicholas A. Renzetti to ensure that the results of these efforts were properly documented, a continuous record of the early work carried out under the DSN Technology Program was published by JPL in a series of progress reports. Starting in 1969 as Volume II of the JPL Space Programs Summary 37-XX series, it became the DSN Progress Reports in 1971, TDA-PRs in 1980, and the TMO-PRs in 1998. In August 1994 the series began publication on the World Wide

The Advance of Technology in the Deep Space Network

Web. An index of all articles published after 1970 also became available online at that time. (See the appendix at the back of the book.)

The technology underlying many of the major engineering changes, discussed earlier in the context of DSN operations support for flight missions, was described by James W. Layland and Lawrence L. Rauch in 1997 (see section 1 of the appendix). As those authors made clear, most of the progress in antenna design; uplink and downlink performance improvement; and the application of radio metric techniques to spacecraft navigation, radio science, radio astronomy, and radar astronomy originated in the DSN program of advanced research. These new technologies were later made available to the DSN engineering groups for implementation into the operational Network.

This phase of DSN endeavor is illustrated by tracing the advancement of technology in those key areas. The material that follows is derived from the work of Layland and Rauch and is intended for the general reader. The technical reader is referred to the appendix at the back of the book, which cites significant publications associated with each topic.

The Great Antennas of the DSN

A photograph of the NASA Deep Space Communications Complex near Canberra, Australia, is shown in Figure 7-1. The largest antennas in the photo are quasi-parabolic reflector antennas, one with a diameter of 70 meters; the others, 34 meters. These were used for deep space mission support, while the other, smaller antennas, 26-meter or 9-meter antennas, provided tracking support for Earth-orbiting missions.

Each of the antennas has what is termed a Cassegrain configuration with a secondary reflector mounted on the center axis just below the focal point of the primary reflector "dish." The secondary reflector serves to relocate the focal point closer to the surface of the main dish and thus establish a more convenient location for the low-noise amplifiers, receivers, and powerful transmitters.

The efficiency with which parabolic antennas collect radio signals from distant spacecraft is degraded to some extent by radio noise radiated by the Earth terrain surrounding the antenna. This form of radio noise is known scientifically as "black body radiation" and is a physical characteristic of all material with a temperature above absolute zero (-273 degrees Celsius or 0 kelvin). The magnitude of noise power radiated by a material body depends on its temperature. While it is incredibly small at the temperature of typical Earth surfaces, it is enormous at the temperature of the Sun, for instance. All parabolic

Figure 7-1. The Deep Space Communications Complex in Canberra, Australia

radio antennas have sidelobes in the beam pattern, and the magnitudes of those sidelobes can increase as the antennas deviate from the ideal shape. The shape of the beam and its sidelobes is essentially the same whether the antenna is used for transmitting or receiving signals. The sidelobes are analogous to the circles of light surrounding the main beam of a flashlight when it is held close to a reflecting surface.

When an antenna is used for transmitting a strong signal to a spacecraft, the sidelobes are of no great consequence. However, when the antenna is receiving a weak signal from a distant spacecraft, particularly at or near the horizon, the "Earth noise" picked up by the sidelobes is sufficient to obscure the spacecraft signal in the extremely sensitive receivers used on the DSN antennas. This interference produces errors in the data stream being delivered to the spacecraft engineers and scientists, and it may cause the DSN receivers and antennas to lose the spacecraft signal altogether, in which case the data stream is completely lost.

The earliest antennas in the DSN (Figure 7-2) were of commercial design and were parabolic in shape. Then, as now, the actual efficiency of the antenna represented a compromise between maximum signal-gathering capability and minimum susceptibility to radio noise picked up from the surrounding Earth.

The Advance of Technology in the Deep Space Network

Figure 7-2. An early parabolic design of a DSN antenna.

Improved technology that could reduce the sidelobes and increase the signal collection capability (gain) of future DSN antennas appeared in the Advanced Systems Program in the early 1970s. The new technology was based on a "dual-shape" design wherein the surface shapes of both the primary and secondary reflectors were modified to illuminate the slightly reshaped "quasi-parabolic" surface of the main reflector more uniformly. However, it was not until the 1980s, when the first 34-meter high-efficiency antennas were built, that the new "dual-shape" design saw operational service in the Network (Figure 7-3).

These antennas were needed by the DSN for support of the Voyager spacecraft in their tour of the outer planets. At the time, the DSN was in transition from the lower, less capable S-band (2.3 GHz) operating frequency, for which the early spacecraft and antennas were designed, to X-band (8.4 GHz), a higher, more capable operating frequency. When X-band technology became available in the DSN, largely as a result of work in the Advanced Systems Program, all the later spacecraft and DSN antennas were designed to operate at X-band frequencies. These antennas were therefore the first to be optimized for performance at X-band.

Figure 7-3. The quasi-parabolic 34-m high-efficiency (HEF) antenna, DSS 15.

As the *Voyager 2* spacecraft headed outward toward Neptune, it was recognized that an increased signal-collecting area was needed on Earth to effectively support this unique science opportunity. The DSN's largest antennas at the time were 64-m parabolas of the original design. Calculations showed that the best investment of scarce construction funds would be to modify these antennas, using the dual-shape design, to expand their diameter to 70 m. It was also apparent that the upgraded large antennas would benefit the planned Galileo and Magellan missions.

Completed in time for support of *Voyager 2* at Neptune, the 70-m enhancement project (Figure 7-4) resulted in an increase of more than 60 percent in the effective collecting area of these large antennas. Fully half of the increase was attributed to the dual-shape design, a product of the DSN Advanced Systems Program.

Figure 7-4. A 70-meter antenna with dual-shape reflector design.

By the end of the century, several new 34-m antennas employing the dual-shaped reflector design in conjunction with beam waveguide (BWG) techniques had been constructed for operational use in the Network. The dual-shaped reflector design enhanced the radio performance of the antenna, while the beam waveguide configuration greatly facilitated maintenance and operation of the microwave receivers and transmitters. Using a series of additional secondary reflectors to relocate the focal point into a stationary room below the main dish, the BWG design feature enabled these critical components to be mounted in a fixed environment rather than in the more conventional type of moving and tipping enclosure mounted on the antenna itself.

Beam waveguide antennas had been used for many years in Earth communications satellite terminals where ease of maintenance and operation outweighed the consideration of losses introduced into the microwave signal path by the additional microwave-reflecting mirrors. For deep space applications, however, where received signal power levels were orders of magnitude smaller, any losses in the signal path were a matter of great concern, and the losses associated with BWG designs kept such antennas out of consideration for DSN purposes for many years. Researchers in the DSN Advanced Systems Program nevertheless pursued the idea of BWG antennas for the DSN and, by 1985, were ready to conduct a collaborative experiment with the Japanese Institute for Space and Aeronautical Sciences (ISAS) using its new 64-m beam waveguide antenna at Usuda, Japan. Using one of the DSN's low-noise microwave receivers installed on the Usuda antenna to receive a signal from the *International Cometary Explorer* (ICE) spacecraft, the researchers made very precise measurements of the microwave losses, or degradation, of the downlink signal.

The results of the experiment were very surprising. The measured losses, attributable to the BWG design, were much smaller than expected, exhibiting similar performance at zenith and better performance at low-elevation angles; they confirmed the efficacy of the BWG configuration.

Encouraged by this field demonstration, researchers sponsored by the Advanced Systems Program moved forward with the construction of a prototype BWG antenna for potential application in the Network.

This new prototype BWG antenna was built at the Venus site at Goldstone and replaced the aging 26-m antenna that had served for many years as a field test site for technology research and development (R&D) programs. The designers used microwave optics analysis software, an evolving product of the Advanced Systems Program, to optimize the antenna for operation over a wide range of current and future DSN operating fre-

The Advance of Technology in the Deep Space Network

Figure 7-5. A 34-meter beam waveguide antenna at Goldstone Technology Development Site, DSS 13.

quency bands. When completed, the antenna successfully demonstrated its ability to operate effectively at S-band, X-band, and Ka-band (approximately 2, 8, and 32 GHz, respectively). Figure 7-5 shows the completed BWG antenna, and Figure 7-6 shows the interior of the equipment room below the antenna structure.

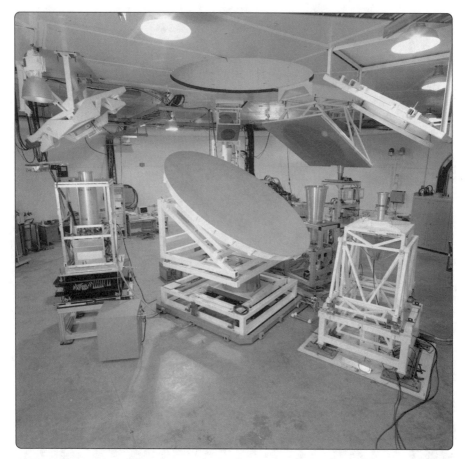

Figure 7-6. Stationary equipment room below the BWG antenna.

The various frequencies and modes of operation for the BWG antenna were selected by rotating the single microwave mirror at the center of this room. Lessons learned by Advanced Systems Program personnel in the construction and evaluation of this antenna were incorporated into the design of the operational BWG antennas for the Network, with the result that the performance of these somewhat exceeded that of the prototype, especially at the lower frequencies. A selection of technical references related to great antennas of the DSN can be found in section 2 of the appendix at the end of the book.

The Advance of Technology in the Deep Space Network

Forward Command/Data Link (Uplink)

The large antennas of the DSN are used for transmission of radio signals carrying instructions and data to the spacecraft, as well as for reception of signals. Getting data to distant spacecraft safely and successfully requires that substantial power be transmitted from the ground and directed in a narrow beam at the spacecraft. For most "normal" situations, the compatible design of spacecraft and the DSN is such that power of about 2 kW to 20 kW is adequate. However, situations in space are not always normal. Unexpected events can redirect a spacecraft's main antenna away from Earth, leaving only a low-gain or omnidirectional antenna capable of receiving anything from Earth. Transmitter power of up to 400 kW at S-band can be sent from the 70-m antenna during attempts to regain contact with a spacecraft in such an emergency situation.

The initial design and evaluation of R&D models of the high-power transmitters and their associated instrumentation was carried out under the Advanced Systems Program. Much of the essential field testing was carried out as part of the planetary radar experiments. This cooperative and productive arrangement provided a realistic environment for testing without exposing an inflight spacecraft to an operational unqualified uplink transmission. Later, DSN engineers implemented fully qualified operational versions of these transmitters in the Network at all sites.

The pointing of the narrow forward link signal to the spacecraft is critical, especially when making initial contact without having received a signal for reference, as is typical in emergency situations. The beamwidth of the signal from the 70-m antenna at S-band is about 0.030 degrees, while that of the 34-m antenna at X-band is about 0.017 degrees. Achieving blind pointing to that precision requires a thorough understanding of the mechanics of the antenna, including the effects of gravity and wind on the dish and specifics of the antenna bearing and positioning mechanisms, as well as knowledge of the spacecraft and antenna positions, atmospheric refraction, and other interferences.

Forward link data delivered to a spacecraft, if incorrectly interpreted, have the potential for causing that spacecraft to take undesirable actions, including some that could result in an emergency situation for the spacecraft. To guard against that possibility, the forward link signal is coded with additional redundant data that allow the spacecraft data system to detect or correct any corruption in that signal. Operating on the presumption that it is always better to take no action than to take an erroneous one, the forward link decoding accepts only data sets for which the probability of error is extremely small, and it discards those that cannot be trusted. A selection of technical references related to the forward command/data link (uplink) can be found in the section 3 of the appendix at the end of the book.

Return Telemetry/Data Link (Downlink)

Throughout the Network, the stations use the same antennas for both the forward link and the return data-link signals. Because the strength of a signal decreases as the square of the distance it must travel, these two signals may differ in strength by a factor of 10^{24} in a single DSN antenna. Isolating the return signal path from interference by the

Figure 7-7. Dichroic (frequency-selective) reflector developed under Advanced Systems Program.

much stronger forward signal poses a significant technical challenge. Normally, these two signals differ somewhat in frequency, so at least a part of this isolation can be accomplished via dichroic or frequency-selective reflectors. These reflectors (Figure 7-7) consist of periodic arrays of metallic/dielectric elements tuned for the specific frequencies that either reflect or pass the incident radiation. These devices must not only be frequency-selective, but they must also be designed to minimize the addition of extraneous radio noise picked up from the antenna and its surroundings, which would corrupt the incredibly weak signals collected by the antenna from the desired radio source in deep space.

The DSN Advanced Systems Program developed the prototypes for almost all the reflectors of this type in current use in the Network. As an adjunct to this work, powerful microwave analytical tools that can be used to affix design details for almost any conceivable dichroic reflector applicable to the frequency bands of the DSN were also developed under the Program.

Low-Noise Amplifiers

The typical return data link signal is incredibly small and must be amplified before it can be processed and the data itself reconstructed. The low-noise amplifiers that reside in the antennas of the DSN are the most sophisticated in the world and provide this amplification while adding the least amount of noise of any such devices.

Known as traveling-wave masers (TWMs), the quietest (in terms of adding radio noise) of these operational devices amplify signals that are propagated along the length of a tuned ruby crystal. Noise in a TWM depends upon the physical temperature of the crystal, and those in operation in the DSN operate in a liquid helium bath at 4.2 kelvin. (Zero kelvin is equivalent to a temperature of minus 273.18 Celsius. Therefore, the temperature of the helium bath is equivalent to approximately minus 269 Celsius.) The practical amplifiers for the DSN were invented by researchers at the University of Michigan, and early development of these amplifiers was carried out under the DSN Advanced Systems Program, as were many improvements throughout the Network's history. The quietest amplifiers in the world today (Figure 7-8), which operate at a physical temperature of 1.2 kelvin, were developed by the DSN Advanced Systems Program and demonstrated at the Technology Development Field Test Site, DSS 13.

Some of the low-noise amplifiers in the DSN today are not TWMs, but a special kind of transistor amplifier (Figure 7-9) using high-electron mobility transistors (HEMTs) in amplifiers cooled to a physical temperature of about 15 kelvin.

Developed initially at the University of California at Berkeley, such amplifiers were quickly adopted by the scientific community for radio astronomy applications. This, in turn, spawned the JPL development work that was carried out via collaboration involving JPL and the DSN Advanced Systems Program, radio astronomers at the National Radio Astronomy Observatory (NRAO), and device developers at General Electric. This work built upon progress in the commercial sector with uncooled transistor amplifiers. In the 2-GHz DSN band, the cooled HEMT amplifiers are almost as noise-free as the corresponding TWMs, and the refrigeration equipment needed to cool the HEMTs to 15 kelvin is much less troublesome than that for the TWMs. Primarily for this reason, current development efforts in the DSN Advanced Technology area are focused on improving the noise performance of the HEMT amplifiers for the higher DSN frequency bands.

Figure 7-8. Ultra-low-noise amplifier (ULNA) at DSS 13.

The first DSN application of the cooled HEMT amplifiers came with the outfitting of the NRAO Very Large Array (VLA) in Socorro, New Mexico, for collaborative support of the Voyager-Neptune Encounter. The VLA was designed for mapping radio emissions from distant stars and galaxies and consists of 27 antennas, each 25 meters in diame-

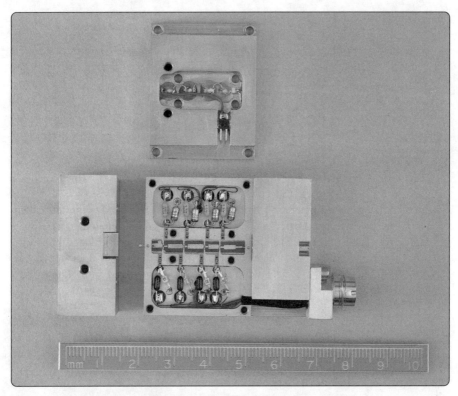

Figure 7-9. High-electron mobility transistor (HEMT) low-noise microwave amplifier.

ter, arranged in a tri-axial configuration. Within the funding constraints, only a small part of the VLA could be outfitted with TWMs, whereas HEMTs for the entire array were affordable and were expected to give an equivalent sensitivity for the combined full array. In actuality, technical progress with the HEMTs under the Advanced Systems Program during the several years taken to build and deploy the needed X-band (8-GHz) amplifiers resulted in better performance for the fully equipped VLA than would have been possible with the VLA partially equipped with the more expensive TWMs. Since that time, many of the DSN operational antennas have had cooled HEMT amplifiers installed for the 2-GHz and 8-GHz bands. A selection of technical references related to return telemetry (downlink) can be found in section 4 of the appendix at the end of the book.

Phase-Lock Tracking

Once the first stages of processing in the low-noise amplifiers are completed, there are still many transformations needed to convert the radio signal sent from a spacecraft into a replica of the data stream originating on that spacecraft. Some of these transformations are by nature analog and linear; others are digital with discrete quantification. All must be performed with virtually no loss in fidelity in order for the resultant data stream to be of practical use.

Typically, the downlink signal consists of a narrow-band "residual carrier" sine wave, together with a symmetric pair of modulation sidebands, each of which carries a replica of the spacecraft data. (Specifics of the signal values vary greatly, but are not essential for this general discussion.) If this signal is cross-correlated with a pure identical copy of the residual carrier, the two sidebands will fold together, creating a low-frequency signal that contains a cleaner replica of the spacecraft data than either sideband alone. Of course, such a pure copy of the carrier signal does not already exist; it must be created, typically via an adaptive narrow-band filter known as a phase-locked loop. The recreated carrier reference is thus used to extract the sidebands. The strength of the resultant data signal is diminished to the extent that this local carrier reference fails to be an identical copy of the received residual carrier. Noise in the spectral neighborhood of the received residual carrier and dynamic variations in the phase of the carrier itself limit the ability to phase-lock the local reference to it.

These dynamic variations are due predominantly to the Doppler effect in play between a distant spacecraft and the DSN antenna on the surface of a spinning Earth. The variations interfere with the return data link process, but they themselves provide for a radio location function. Over the years, the DSN Advanced Systems Program has contributed significantly to the design for the phase-locked loops and to the knowledge of phase-coherent communications, and thus to the performance of the operational DSN. A selection of technical references related to phase-lock tracking can be found in section 5 of the appendix at the end of the book.

Synchronization and Detection

Further steps in converting a spacecraft signal into a replica of the spacecraft data stream are accomplished by averaging the signal over brief intervals of time that correspond to each symbol (or bit) transmitted from the spacecraft and by sampling these averages to create a sequence of numbers, often referred to as a "symbol stream." These averages

The Advance of Technology in the Deep Space Network

must be precisely synchronized with the transitions in the signal as sent from the spacecraft so that each contains as much as possible of the desired symbol and as little as possible of the adjacent ones. Usually, a subcarrier, or secondary carrier, is employed to shape the spectrum of the spacecraft signal, and it must be phase-tracked and removed prior to the final processing of the data itself. The Network contains several different generations of equipment that perform this stage of processing. Designs for all of these have their roots in the products of the DSN Advanced Systems Program. The oldest current equipment is of a design developed in the late 1960s by a partnership between the DSN Advanced Systems Program and the DSN implementation programs. This equipment is mostly analog in nature and, while still effective, is subject to component value shifts with time and temperature, and thus requires periodic tending and adjustments to maintain desired performance.

As digital devices became faster and more complex, it became possible to develop digital equipment that could perform this stage of signal processing. Digital demodulation techniques were demonstrated by the Advanced Systems Program in the early 1970s in an all-digital ranging system. Similar techniques were subsequently employed for data detection in the second generation of the Demodulator-Synchronizer Assembly. A selection of technical references related to data synchronization and detection can be found in section 6 of the appendix at the end of the book.

A Digital Receiver

Rapid evolution of digital technology in the 1980s led researchers to explore the application of digital techniques to various complex processes found in receiving systems such as those used in the Network. The processes of filtering, detection, and phase-lock carrier tracking, formerly based on analog techniques, were prime candidates for the new digital technology.

In this context, the Advanced Systems Program supported the development of an all-digital receiver for Network use. Known as the Advanced Receiver (ARX), the developmental model embodied most of these new ideas and demonstrated capabilities far exceeding those of the conventional analog receivers then installed throughout the Network.

Encouraged by the performance of the laboratory model, an engineering prototype was built and installed for evaluation in an operational environment at the Canberra, Australia, Complex. Tests with the very weak signal from the *Pioneer 10* spacecraft, then approaching the limits of the current DSN receiving capability, confirmed the designer's performance and, incidentally, significantly extended the working life of that spacecraft.

As a result of these tests, the DSN decided to proceed with the implementation of a new operational receiver for the Network that would be based on the design techniques demonstrated by the ARX. The new operational equipment, designated the Block V receiver (BVR), would include all of the functions of the existing receiver in addition to several other data processing functions, such as demodulation and synchronization, formerly carried out in separate units.

As the older generation receivers were replaced with the all-digital BVR equipment, the Network observed a general improvement in weak signal tracking performance and operational reliability. The receiver replacement program was completed throughout the Network by 1998. A selection of technical references related to digital receivers can be found in section 7 of the appendix at the end of the book.

Encoding and Decoding

The data generated by science instruments must be reliably communicated from the spacecraft to the ground, despite the fact that the signal received is extremely weak and that the ground receiver corrupts the signal with additive noise. Even with optimum integration and threshold detection, individual bits usually do not have adequate signal energy to ensure error-free decisions. To overcome this problem, structured redundancy (channel encoding) is added to the data bit-stream at the spacecraft. Despite the fact that the individual "symbols" resulting from this encoding have even less energy at the receiver, the overall contextual information, used

Figure 7-10. The Advanced Receiver (ARX) as used to track the *Pioneer 10* spacecraft.

properly in the decoding process on the ground, results in more reliable detection of the original data stream.

High-performance codes to be used for reliable data transfer from spacecraft to DSN were identified by research performed under the Advanced Systems Program and adopted for standard use in the Network while the search for even more powerful and efficient codes continued. New, more efficient block codes, which made better use of the limited spacecraft transmitter power by avoiding the need to transmit separate synchronizing signals, were developed under the DSN Advanced Systems Program and first demonstrated on the *Mariner 6* and *7* spacecraft in 1969. By putting the extra available spacecraft transmitter power into the data-carrying signal, the new block code enabled the return of the Mars imaging data at the astonishing (for the time) rate of 16,200 bits per second, an enormous improvement over the 270-bps data rate for which the basic mission had been designed. Of course, conversion of the encoded data stream back to its original error-free form required a special decoder. The experimental block decoder, developed under the same program for this purpose, formed the basis for the operational block-decoders implemented in the Network as part of the Multimission Telemetry System shortly thereafter.

While the JPL designers of the Mariner spacecraft were pursuing the advantages of block-coded data, the designers of the *Pioneer 9* spacecraft at the Ames Research Center (ARC) were looking to very complex convolutional codes to satisfy their scientists. The scientists agreed to accept intermittent gaps in the data caused by decoding failure in exchange for the knowledge that successfully decoded data would be virtually error-free. In theory, a convolutional code of length k = 25 would meet the requirement, but it had a most significant drawback. The decoding process was (at the time) extraordinarily difficult. Known technically as "sequential decoding," this was a continuous decoding operation rather than the "one block at a time" process used by the DSN for decoding the *Voyager* data.

The original plan was to perform the decoding operation for *Pioneer 9* in non-real-time at ARC, using tape-recorded data provided by the DSN. However, Pioneer engineers working in conjunction with the DSN Advanced Systems Program explored and demonstrated the potential for decoding this code in real time via a very-high-speed engineering model sequential decoder. With the rapid evolution in capability of small computers, it became apparent that decoding Pioneer's data in such computers was both feasible and economical. Subsequent implementation of sequential decoding in the Network was done via microprogramming of a small computer, guided by the knowledge gained via the efforts of the Advanced Systems Program. The subsequent *Pioneer 10* and *11* spacecraft flew with a related code of length k = 32 and were supported by the DSN in a computer-based decoder.

The DSN standard code, flown on *Voyager* and *Galileo*, consisted of a short convolutional code that was combined with a large block-size Reed-Solomon code. The standard algorithm for the decoding of convolutional codes was devised in consultation with JPL researchers and demonstrated by simulations performed under the Advanced Systems Program. Prototypes of the decoding equipment were fabricated and demonstrated at JPL, also with the support of the Advanced Systems Program.

The application of coding and decoding technology in the DSN was paced by the evolution of digital processing capability. At the time of the Voyager design, a convolutional code of length k = 7 was chosen as a compromise between performance and decoding complexity, which would grow exponentially with code length. Equipment was implemented around the DSN to handle this code from *Voyager* and subsequently from *Magellan*, *Galileo*, and others. Modern digital technology has permitted the construction of much more complex decoders, and a code of length k = 15 was devised with the support of the Advanced Systems Program. This code was installed as an experiment on the *Galileo* spacecraft shortly before its launch. The corresponding prototype decoder was completed soon afterward. Though not used for *Galileo* because of its antenna problem, the more complex decoder was implemented around the Network for support of the Cassini and subsequent missions.

Efforts of the Advanced Systems Program provided the understanding of telemetry performance to be expected with the use of these codes. Figure 7-11 displays the reliability of the communication (actually, the probability of erroneous data bits) as it depends upon the spacecraft signal energy allocated to each data bit for uncoded communication and three different codes.

Research on new and even more powerful coding schemes, such as turbo codes, continued to occupy an important place in the Advanced Systems Program. Turbo codes are composite codes made up of short-constraint-length convolutional codes and a data stream interleaver. The decoding likewise consists of decoders for the simple component codes, but with an iterative sharing of information between them. These codes, which push hard on the fundamental theoretical limits to signal detection, can result in almost a full decibel of performance gain over the best previous concatenated coding systems. A selection of technical references related to data encoding and decoding can be found in section 8 of the appendix at the end of the book.

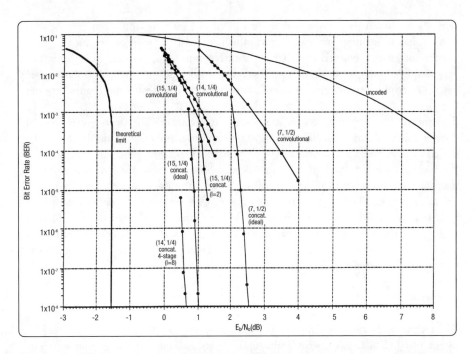

Figure 7-11. Telemetry communication channel performance for various coding schemes. The first set of curves shows the *Voyager* k = code, both alone and in combination with the Reed-Solomon code. The second set of codes illustrates the k = 15 code, which was to be demonstrated with Galileo's original high-rate channel, shown alone and in combination with the Reed-Solomon code, either as constrained by the *Galileo* spacecraft data system (l = 2) or in ideal combination. The third set shows the k = 14 code, devised by the Advanced Systems Program researchers for the actual Galileo low-rate mission, both alone and in combination with the selected variable-redundancy Reed-Solomon code and a complex four-stage decoder. The added complexity of the codes, which has its greatest effect in the size of the ground decoder, clearly provides increased reliability for correct communication.

Data Compression

Source encoding and data compression are not typically considered a part of the DSN's downlink functions, but the mathematics that underlie coding and decoding are a counterpart of those that guide the development of data compression. Simply stated, channel encoding is the insertion of structured redundancy into a data stream, while data compression is the finding and removal of intrinsic redundancy. Imaging data are often highly redundant and can be compressed by factors of at least two, and often four or more, without loss in quality. For *Voyager*, two influences led to a factor-of-two increase in the number of images returned from Uranus and Neptune. The first was the effecting of a very simplified image-compression process constrained to fit into available onboard memory. The second improvement involved corresponding changes to the channel coding.

The success of data compression technology in enhancing the data return from the *Voyager* missions firmly established the technique as an important consideration in the design of all future planetary downlinks. The original telecommunication link design for the *Galileo* spacecraft used data compression to almost double the amount of imaging data that the spacecraft could transmit from its orbital mission around Jupiter. The failure of the spacecraft's high-gain antenna prior to *Galileo*'s arrival at Jupiter prompted an intense effort to find even more complex data compression schemes that would recover some of the Jupiter imaging data that otherwise could not have been returned. A selection of technical references related to data compression can be found in section 9 of the appendix at the end of the book.

Arraying of Antennas

The technique of antenna arraying, as practiced in the Deep Space Network, made use of the physical fact that a weak radio signal from a distant spacecraft that is received simultaneously by several antennas at different locations is degraded by a component of radio noise that is independent of each receiving station. By contrast, the spacecraft signal itself is dependent, or coherent, at each receiving site. In theory, therefore, the power of the signal, relative to the power of the noise, or signal to noise ratio, (SNR) could be improved by combining the individual antennas in such a way that the coherent spacecraft signals were reinforced, while the independent or non-coherent noise components were canceled out.

In practice, this involved a complex digital process for compensating for the time, or phase, delays caused by the different distances between each station and the spacecraft. It also called for compensating for differing distances between the various antenna locations and the common station where the combining function was carried out. This technique became known as antenna arraying, and the digital processing function that realized the theoretical "gain" of the entire process was called "signal combining."

By 1970, conceptual studies had described and analyzed the performance of several levels of signal combining and two of these schemes, carrier and baseband combining, were of potential interest to the Network. Both techniques involved compensation for the phase delays caused by the various locations of the arrayed antennas. The difference lay in the frequency at which the combining function was performed. "Carrier" combining was done at the carrier frequency of the received signal, while "baseband" combining was carried out at the frequencies of the subcarrier and data signal that modulated it. Each had its advantages and disadvantages, but baseband combining proved easier to implement and was, obviously, tried first.

The "arraying and signal combining" concept was first developed and demonstrated in 1969 and 1970 by J. Urech, a Spanish engineer working at the Madrid tracking station. Using signals from the *Pioneer 8* spacecraft and a microwave link to connect two 26-m stations located 20 km apart (DSS 61 and DSS 62), he succeeded in demonstrating the practical application of the principle of baseband combining in the Network for the first time. Because of the low baseband frequency of the *Pioneer 8* data stream (8 bits per second), as well as the close proximity of the antennas, no time-delay compensation was necessary.

Within the bounds of experimental error, this demonstration confirmed the R&D theoretical estimates of performance gain and encouraged the Advanced Systems researchers to press forward with a more complex form of baseband combining at a much higher data rate (117 kilobits per second) in real time.

The demonstration took place at Goldstone in September 1974, using the downlink signals from the *Mariner-Venus-Mercury* (MVM) spacecraft during its second encounter with the planet Mercury. Spacecraft signals from the two 26-m antennas, DSS 12 and DSS 13, were combined in an R&D combiner with signals from the DSS 14 64-m antenna in real time at 117 kbps. The less-than-predicted arraying gain obtained in this demonstration (9 percent versus 17 percent) was attributed to small differences in performance between key elements of the several data-processing systems involved in the test. Although this experience demonstrated both the practical difficulty of achieving full theoretical gain of an antenna arrayed system and the critical effect of very small variations in the performance of its components, it also established the technical feasibility of baseband arraying of very weak high-rate signals.

In 1977, with the lessons learned from these demonstrations as background, the DSN started to develop an operational arraying capability for the Network. The *Voyager 1* and *2* Encounters with Saturn in 1980 and 1981 would be the first to use the arraying in the Network. A prototype baseband real-time combiner (RTC), based on the analysis and design techniques developed by the earlier R&D activity, was completed in the fall of 1978. Designed to combine the signals from DSS 12 and DSS 14 at Goldstone, it was used with varying degrees of success to enhance the signals from the *Voyagers* at Jupiter in March and July of 1979 and the *Pioneer 11* Encounter of Saturn in August and September of that year.

Like the previous demonstration, this experience emphasized the critical importance of having all elements of the array-receivers, antennas, and instrumentation operating precisely according to their specified performance capabilities. With this very much in mind, the DSN proceeded to the design for operational versions of the RTC for use at all three complexes to support the *Voyager 1* and *2* Encounters of Saturn. The operational versions of the RTC embodied many improvements derived from the experience with the R&D prototype version. By mid-August 1980, they were installed and being used to array the 64-m and 34-m antennas at all three complexes as *Voyager 1* began its far-Encounter operations. During this period, the average arraying gain was 0.62 dB, about 15 percent greater than that of the 64-m antenna alone. While this was good, improvement came slowly as more rigorous control and calibration measures for the array elements were instituted throughout the Network. By the time *Voyager 2* reached Saturn in August 1981, these measures, supplemented with additional training and calibration procedures, had paid off. The average arraying gain around the Network increased to 0.8 dB (approximately 20 percent), relative to the 64-m antenna alone. This was clearly a most satisfactory result and the best up to that time. Antenna arraying had become a permanent addition to the capability of the Network.

While researchers working within the Advanced Systems Program continued to explore new processes for arraying antennas, engineers within the DSN took advantage of the long flight time between the Voyager Saturn and Uranus Encounters to refine the existing RTC configuration. Over the next five years, the formerly separate data-processing functions of combining, demodulation, and synchronization were integrated into a single assembly. This integration facilitated improvements in performance, stability, and operational convenience. By the time *Voyager 2* approached Uranus in 1985, the new Baseband Assemblies (BBAs), as they were called, had been installed at all three Complexes. In addition, a special version of the basic four-antenna BBA was installed at the Canberra Complex. This provided for combining the Canberra array of one 64-m and two 34-m antennas with signals from the 64-m Parkes Radio Telescope, 200 km distant (Figure 7-12).

In January 1986, this arrangement was a key factor in the successful return of Voyager imaging data from the unprecedented range of Uranus. But even greater achievements in antenna arraying lay ahead.

The Advance of Technology in the Deep Space Network

Figure 7-12. The 64-m antenna of the Radio Astronomy Observatory, Parkes, Australia.

In 1989, the DSN used a similar arrangement with great success to capture the Voyager imaging data at a still greater range—from Neptune. This time the Goldstone 70-m and 34-m antennas were arrayed with the 27 antennas of the Very Large Array (VLA) (Figure 7-13) of the National Radio Astronomy Observatory at Socorro, New Mexico.

Figure 7-13. The Very Large Array (VLA) of the National Radio Astronomy Observatory (NRAO) at Socorro, New Mexico.

DSN support for the Voyager Encounter of Uranus was further augmented by the Canberra-Parkes array in Australia, which the DSN had reinstated with the addition of new BBAs, new 34-m antennas, and the upgraded 70-m antenna.

The success of these applications of the multiple-antenna arraying technique provided the DSN with a solid background of operational experience. The DSN drew heavily on this experience a few years later, when it was called upon to recover the science data from *Galileo* after the failure of the spacecraft's high-gain antenna in 1991. Together with the data compression and coding techniques discussed earlier, the Network's Canberra/Parkes/Goldstone antenna arrays succeeded in recovering a volume of data that, according to the Galileo project, was equivalent to about seventy percent of the original mission.

With time, arraying of multiple antennas within Complexes, between Complexes, or between international space agencies came into general use as a means to enhance the downlink capability of the Network. In the latter years of the century, most of the enhancements to the arraying in the DSN were driven by implementation and operational considerations rather than by new technology, although the Advanced Systems Program continued to explore the boundaries of performance for various alternative arraying architectures and combining techniques. A selection of technical references related to antenna arraying can be found in section 10 of the appendix at the end of the book.

Radio Metric Techniques

In addition to being able to exchange forward and return link data with an exploring spacecraft, it is equally important to know the precise location of the spacecraft and its velocity (speed and direction). Information about the position and velocity of the spacecraft can be extracted from the one-way or two-way radio signals passing between the spacecraft and the DSN. When these data are extracted by appropriate processing and further refined to remove aberrations introduced by the propagation medium along the radio path between spacecraft and Earth, it can be used for spacecraft navigation.

Radio metric techniques similar to those used for spacecraft navigation can also be used for more explicit scientific purposes, notably radio science, radio astronomy, and radio interferometry on very long baselines (VLBI).

Since its inception, the DSN Advanced Systems Program has worked to develop effective radio metric tools, techniques, observing strategies, and analysis techniques that furthered the DSN pre-eminence in these unique fields of science. In more recent times, the Advanced Systems program demonstrated the application of Global Positioning System technology to further refinement of radio metric data generated by the DSN. A selection of technical references on radio metric tools can be found in section 11 of the appendix at the end of the book.

Doppler and Range Data

If the Earth and the spacecraft were standing still, the time taken for a radio signal to travel from the Earth to the spacecraft and back would be a measurement of the distance between them. This is referred to as the round-trip light time (RTLT). However, since the Earth and the spacecraft are both in motion, the RTLT contains both position and velocity information, which must be disentangled through multiple measurements and suitable analysis. The precision at which such measurements can be obtained is limited by the precision of the time-tag marker to the radio signals, and by the strength of the signal in proportion to the noise mixed with it, or by the signal-to-noise ratio (SNR).

Precise measurements of changes to this light time are far easier to obtain via observing the Doppler effect resulting from the relative motions. Such measurements are mechanized via the phase-locked loops in both spacecraft and ground receivers using the spacecraft's replica of the forward link residual carrier signal to generate the return link signal, and counting the local replica of the return link residual carrier against the original carrier for the forward link signal. The raw precision of these measurements is comparable to the wavelength of the residual carrier signal, e.g., a few centimeters for an X-band signal (8 GHz). Numerous interesting error sources tend to corrupt the accuracy of the measurement and the inferred position and velocity of the spacecraft, and they have provided significant technical challenge for work under the Advanced Systems Program.

The observed Doppler contains numerous distinct components, including the very significant rotation of Earth. As Earth turns, the position of any specific site on the surface describes a circle, centered at the spin axis of the Earth, falling in a plane defined by the latitude of that site. The resultant Doppler component varies in a diurnal fashion with a sinusoidal variation, which is at its maximum positive value when the spacecraft is first observable over the eastern horizon. Its corresponding negative value occurs at approach to the western horizon. A full-pass Doppler observation from horizon to horizon can be analyzed to extract the apparent spacecraft position in the sky, although the determination is somewhat weak near the equatorial plane. Direct measurements of the RTLT are useful for resolving this difficulty.

Three distinct generations of instruments designed to measure the RTLT were developed by the Advanced Systems Program and used in an ad hoc fashion for spacecraft support before a hybrid version was designed and implemented around the DSN. The third instrument designed, the Mu-II Ranging Machine, was used with the Viking Landers in a celestial mechanics experiment, which provided the most precise test, up to that time, of the general theory of relativity.

These devices function by imposing an additional "ranging" modulation signal on the forward link, which is copied on the spacecraft (within the limits imposed by noise) and then imposed on the return link. The ranging signal is actually a very long period-coded sequence that provides the effect of a discrete time tag. The bandwidth of the signal is on the order of 1 MHz, giving the measurement a raw precision of a few hundred meters, resolvable with care to a few meters. Among other features, the Mu-II Ranging Machine included the first demonstrated application of the digital detection techniques that would figure strongly in future developments for the DSN.

Timing Standards

The basic units of measurement for all radio metric observations, Doppler or range, derive from the wavelength of the transmitted signal. Uncertainties or errors in knowledge of that wavelength are equivalent to errors in the derived spacecraft position. The need for accurate radio metrics has motivated the DSN Advanced Systems Program to develop some of the most precise, most stable frequency standards in the world. While the current suite of hydrogen maser frequency standards in the DSN field sites was built outside of JPL, the design is the end product of a long collaboration in technology development, with research units being built at JPL under the DSN Advanced Systems Program and elsewhere.

Continued research under the Advanced Systems Program for improved frequency standards resulted in the development of a new linear ion trap standard (Figure 7-14) that

Figure 7-14. The new linear ion trap (LIT) standard.

offered improved long-term stability of a few parts in 10^{16}, as well as simpler and easier maintenance than that required by the hydrogen masers.

Work was under way to implement the LIT standard in the DSN, while research efforts continued for improvements that could be transferred to field operation in the future. A selection of technical references related to timing standards can be found in section 12 of the appendix at the end of the book.

Earth Rotation and Propagation Media

Radio metric Doppler and range data enable the determination of the apparent location of a spacecraft relative to the position and attitude of the rotating Earth. Earth, however, is not a perfectly rigid body with constant rotation, but contains fluid components as well, which slosh about and induce variations in rotation of perhaps a few milliseconds per day. Calibration of Earth's attitude is necessary so that the spacecraft's position in inertial space can be determined—a necessary factor in navigating the spacecraft toward a target planet. Such calibration is available via the world's optical observatories and, with greater precision, via radio techniques, which will be discussed further in the sections entitled "VLBI and Radio Astronomy" and "Global Positioning System."

The interplanetary media along the signal path between Earth and the spacecraft affect the accuracy of the Doppler and range observations. The charged ions in the tenuous plasma spreading out from the Sun, known as the solar wind, bend and delay the radio signal. Likewise, the charged ions in Earth's own ionosphere and the water vapor and other gases of the denser lower atmosphere bend and delay the radio signal. All of these factors are highly variable because of other factors, such as intensity of solar activity, season, time of day, and weather. All factors must be calibrated, modeled, or measured to achieve the needed accuracy; over the years, the DSN Advanced Systems Program has devised an increasingly accurate series of tools and techniques for these calibrations. A selection of technical references related to Earth rotation and propagation can be found in section 13 of the appendix at the end of the book.

Radio Science

Radio science is the term used to describe the scientific information obtained from the intervening pathway between Earth and a spacecraft by the use of radio links. The effects of the solar wind on the radio signal path interfere with our efforts to determine the location of the spacecraft, but if the relative motions of Earth and the spacecraft are modeled and removed from the radio metric data, much of what remains is information about the solar wind and, thus, about the Sun itself. Other interfering factors are also of scientific interest.

In some situations, the signal path passes close by a planet or other object, and the signal itself is bent, delayed, obscured, or reflected by that object and its surrounding atmosphere. These situations provide a unique opportunity for scientists to extract information from the signal about object size, atmospheric density profiles, and other factors not otherwise observable. Algorithms and other tools devised to help calibrate and remove interfering signatures from radio metric data for use in locating a spacecraft often become part of the process for extracting scientific information from the same radio metric data stream. The precision frequency standards, low-noise amplifiers, and other elements of the DSN derived from the Technology Program are key factors in the ability to extract this information with a scientifically interesting accuracy. Occasionally, engineering models developed by the Program are placed in the Network in parallel with operational instrumentation for ad hoc support of metric data-gathering for some unique event.

The effects of gravity can also be observed by means of the radio link. Several situations are of interest. If the spacecraft is passing by or in orbit about an object that has a lumpy, uneven density, that unevenness will cause a variation in the spacecraft's pathway that will be observable via the radio metric data. If the radio signal passes near a massive object such as the Sun, the radio signal's path will be bent by the intense gravity field, according to the theories of general relativity. And in concept, gravitational waves (a yet-to-be-observed aspect of gravity field theory) should be observable in the Doppler data from a distant spacecraft. All of these possibilities depend upon the stability of the DSN's precision frequency standards for the data to be scientifically interesting. A selection of references related to radio science can be found in section 14 of the appendix at the end of the book.

VLBI and Radio Astronomy

The technical excellence of the current DSN is, at least in part, a result of a long and fruitful collaboration with an active radio astronomy community at the California Institute of Technology (Caltech) and elsewhere. Many distant stars, galaxies, and quasars are detectable by the DSN at radio frequencies. The furthest of these are virtually motionless and can be viewed as a fixed-coordinate system to which spacecraft and other observations can be referenced. Observations relative to this coordinate set help to reduce the distorting effects of intervening material in the radio signal path and uncertainties in the exact rotational attitude of Earth during spacecraft observations.

Little precise information can be extracted by observing these objects one at a time and from a single site, but concurrent observation at a pair of sites will determine the relative position of the two sites referenced to the distant object. The observing technique is known as very long baseline interferometry (VLBI) and was developed by the research of many contributors, including substantial work by the DSN's Advanced Systems Program. If three sites are used in VLBI pairs and multiple objects are observed, the positional attitude of Earth and the relative positions of the observed objects can be determined. If one of the observed is a spacecraft transmitting a suitable signal, its position and velocity in the sky can be very accurately defined. A demonstration of this technique via the Advanced Systems Program led to operational use for spacecraft such as *Voyager* and *Magellan*.

VLBI can also be used in conjunction with conventional radio metric data types to provide the calibration for the positional attitude of Earth. Such observations can be made without interfering with spacecraft communication except for the time utilization of the DSN antennas. In addition to determining Earth's attitude, the observations measure the relative behavior of the frequency standards at the widely separated DSN sites, and thus help to maintain their precision performance. Again, demonstration of this capability via the Advanced Systems Program led to routine operational use in the DSN.

Design and development of the DSN equipment and software needed for VLBI signal acquisition and signal processing (correlation) was carried out in a collaboration involving the Advanced Systems Program, the operational DSN, and the Caltech radio astronomy community. Tools needed to produce VLBI metric observations for the DSN were essentially the same as those for interferometric radio astronomy. Caltech received funding from the National Science Foundation for this activity, and both Caltech and the DSN shared in the efforts of the design while obtaining products that were substantially better than any that they could have been obtained independently.

Another area of common interest between the DSN and the radio astronomy community is that of precision wideband spectral analysis. Development efforts of the Advanced Systems Program produced spectral analysis tools that have been employed by the DSN in spacecraft emergency situations and in examining the DSN's radio interference environment, and they have served as pre-prototype models for equipment for the DSN. Demonstration of the technical feasibility of the very-wide-band spectral analysis and preliminary observations by a megachannel spectrum analyzer fielded by the Advanced Systems Program helped establish the sky survey planned as part of the former SETI (Search for Extraterrestrial Intelligence) Program.

The Advance of Technology in the Deep Space Network

Another technique (one similar to the use of VLBI for a radio metric reference) is used if two spacecraft are flown to the same target; the second can be observed relative to the first, providing better target-relative guidance once the first has arrived at the target. Techniques for acquiring and analyzing such observations have been devised by the Advanced Systems Program. A selection of technical references related to VLBI and radio astronomy can be found in section 15 of the appendix at the end of the book.

The Global Positioning System

The Global Positioning System (GPS) is a constellation of Earth-orbiting satellites designed (initially) to provide for military navigation on Earth's surface. Research under the Advanced Systems Program showed that these satellites could provide an excellent tool to calibrate and assist in the radio metric observation of distant spacecraft. GPS satellites fly above the Earth's atmosphere and ionosphere in well-defined orbits so that their signals can be used to measure the delay through these media in a number of directions. With suitable modeling and analysis, these measurements can be used to develop the atmospheric and ionospheric calibrations for the radio path to a distant spacecraft.

Additionally, since the GPS satellites are in free orbit about Earth, their positions are defined relative to the center of mass of Earth, and not its surface. They provide another method to observe the uneven rotation of Earth.

GPS techniques can also be used to determine the position of an Earth-orbiting spacecraft relative to the GPS satellites, as long as the spacecraft carries a receiver for the GPS signals. The potential of this technique was initially demonstrated by the Advanced Systems Program. GPS was subsequently used by the TOPEX/POSEIDON Project for precise orbit determination and a consequent enhancement of its scientific return. A selection of technical references related to the GPS can be found in section 16 of the appendix at the end of the book.

Goldstone Solar System Radar

The Goldstone Solar System Radar (GSSR) is a unique scientific instrument for making observations of nearby asteroids, the surfaces of Venus or Mars, the satellites of Jupiter, and other objects in the solar system. Although the GSSR makes use of the DSN 70-m antenna for its scheduled observing sessions, its receiving, transmitting, and data-processing equipment is unique to the radar program. The GSSR is a product of many years of development by the DSN Advanced Systems Program. In the early days of the DSN, the Advanced Systems Program took ownership of the radar capability at the DSN's Goldstone, California, site and evolved and nurtured it as a vehicle for developing and demonstrating many of the capabilities that eventually would be needed by the Network.

Scientific results abounded as well, but they were not its primary product. Timely development of DSN capabilities was the major result. Preparations for a radar observation at the DSN Technology Development Field Site bore many resemblances to those for a spacecraft planetary encounter, since the radar observations could only be successful during the few days when Earth and the radar target were closest together.

In the conventional formulation of the radar sensitivity equations, that sensitivity depends upon the aperture, temperature, power, and gain of the system elements. Here, aperture refers to the effective size, or collecting area and efficiency, of the receiving antenna; temperature is a way of referring to the noise in the receiving system, where a lower temperature means a lesser noise; power refers to the raw power level from the transmitter; and gain is the effective gain of the transmitting antenna, which depends in turn upon its size, its surface efficiency, and the frequency of the transmitted signal. Where the same antenna is used both to transmit and to receive, the antenna size and efficiency appear twice in the radar equations.

Significant improvements to the DSN's capability for telemetry reception were to come from the move upward in frequency from S-band (2 GHz) to X-band (8 GHz) on the large 64-m antennas. Performance of these antennas at the higher frequencies and the ability to successfully point them were uncertain, however, and these uncertainties would best be removed by radar observations before spacecraft with X-band capabilities were launched. The radar had obvious benefit from the large antenna and the higher frequency. The first flight experiment for X-band communication was carried out on the 1973 Mariner Venus Mars mission. Successful radar observations from the Goldstone 64-m antenna demonstrated that the challenge of operating the large antennas at the higher X-band frequency could be surmounted.

High-power transmitters were needed by the DSN for its emergency forward link functions but were plagued by problems such as arcing in the waveguide path when power densities became too high. High-power transmitters were essential for the radar to "see" at increased distances and with increased resolution. Intense development efforts at the DSN Technology Development Field Site could take place without interference or risk to spacecraft support in the Network. Successful resolution of the high-power problems for the radar under the Advanced Systems Program became the successful implementation of the high-power capability needed by the Network for uplink communications.

The Advance of Technology in the Deep Space Network

Low-noise amplifiers were needed by the DSN to increase data return from distant spacecraft. Low-noise amplifiers were essential for the radar to enable it to detect echoes from increasingly distant targets or to provide for increased resolution of already detectable targets. The synergistic needs of both the radar system and the Network led to the development of the extremely low-noise maser amplifiers that became part of the standard operational inventory of the DSN.

Digital systems technology was rapidly evolving during this period and would play an increasing role in the developing DSN. Equipment developed by the Advanced Systems Program for its radar application included 1) digital encoders to provide for spatial resolution of parts of the radar echo, 2) computer-driven programmable oscillators to accommodate Doppler effects on the signal path from Earth to target to Earth, and 3) complex, high-speed digital signal processing and spectrum analysis equipment. Much of the digital technology learned this way would transfer quickly to other parts of the signal processing work under the Advanced Systems Program and eventually into the operational DSN. Some of the elements would find direct application, such as the programmable oscillators, which became essential for maintaining contact with the *Voyager 2* spacecraft following a partial failure in its receiver soon after launch. And the signal analysis tools would be called on many times over the years to help respond to spacecraft emergencies.

Some of the products of the early radar observations (see Figure 7-15) were both scientific in nature and essential for providing information for the planning and execution of NASA's missions.

One notable "first" was the direct measurement of the astronomical unit. (One astronomical unit (AU) is equal to 1.5×10^8 km, the mean distance between Earth and the Sun.) It sets the scale size for describing distances in the solar system. The measurement was made in support of preparations for sending *Mariner 2* to Venus and provided a correction of 66,000 km from conventional belief at that time. It also enabled corrections that brought the mission into the desired trajectory for its close flyby of the planet. The GSSR was also used in qualifying potential Mars landing sites for the Viking Landers, and it continues to provide information about the position and motion of the planets, which is used to update the predicted orbits for the planets of the solar system. A selection of technical references related to the Goldstone solar system radar can be found in section 17 of the appendix at the end of the book.

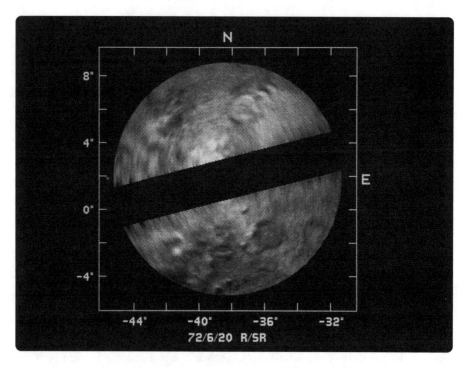

Figure 7-15. First high-resolution radar image of Venus.

Telecommunications Performance of the Network

The progress of deep space communications capability over the period of forty years since the inception of the Network is illustrated in Figure 7-16.

In interpreting the data presented in Figure 7-16, it will be observed that the logarithmic scale that displays the data rate gives an impression that the early improvements are more significant than the later improvements. This is because the steps represent fractional or percentage increases, rather than incremental increases. The latter would show the actual data rate increases, which are much larger in the later improvements. If the value was proportional to the amount of data, then the display of the incremental increases would be more meaningful than the logarithmic display.

The Advance of Technology in the Deep Space Network

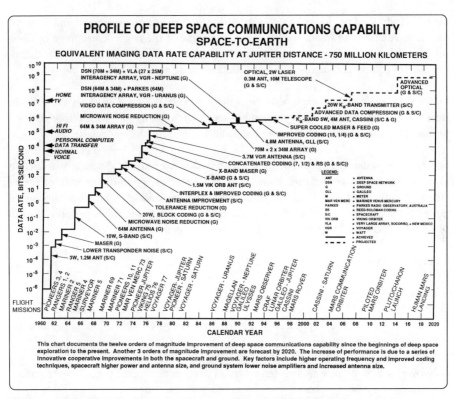

Figure 7-16. **Profile of deep space communications capability.** The timeline on the horizontal axis of the figure covers the first forty years of actual Network operational experience through the close of the century and extends for a further twenty years to forecast the potential for future improvements through the year 2020. The vertical axis displays the growth in space-to-Earth downlink capability of the Network. The downlink capability is given in units of telemetry data rate (bits per second) on a logarithmic scale and represents the equivalent imaging data rate capability for a typical spacecraft at Jupiter distance (750 million kilometers). Significant events in the history of deep space telecommunications and deep space exploration are appropriately annotated on both axes.

Presented this way, however, the figure clearly shows that, from inception though 1997, the downlink capability of the Network grew from 10^{-6} to 10^6 bits per second, equivalent to twelve orders of magnitude.

This remarkable progress is not, of course, solely due to improvements in the Network. Many of the steps result from "cooperative" changes on the part of both the DSN and

493

the spacecraft. Coding, for example, is applied to the data on the spacecraft and removed on Earth. A change in frequency has resulted in some of the larger steps shown by causing the radio beam from the spacecraft to be more narrowly focused. Such change necessitates equipment changes on both the spacecraft and Earth.

Other steps represent advances that are strictly spacecraft-related, such as increases in return-link transmitter power or increases in spacecraft antenna size, which improves performance by more narrowly focusing the radio beam from the spacecraft.

Still other steps depict improvements strictly resulting from the DSN, such as reduction in receiving system temperature, increase in the size of the ground antennas, or use of arrays of antennas to increase the effective surface area available for collecting signal power. A selection of technical references related to telecommunications performance can be found in section 18 of the appendix at the end of the book.

COST-REDUCTION INITIATIVES

With one exception, the program continued to pursue the same broad themes in this period, December 1994 to December 1997, that had characterized its earlier work. The new theme that began to appear in the program in 1994 was directed at the reduction of Network operations costs.

A Network automation work area was set up to develop automated procedures to replace the extremely operator-intensive work of running a spacecraft "pass" over a DSN tracking station. This effort soon produced demonstrable results. A fully automated satellite tracking terminal that could reduce operations costs for near-Earth satellites was demonstrated in 1994. A software prototype that reduced the number of manual inputs for a typical 8-hour track from 900 to 3 was installed at DSS 13 and used to support Ka-band operations at that site. Eventually this technology would find its way into the operational Network. A contract for a new, small deep space transponder offering lower size and power needs, and most importantly lower production costs, was initiated with Motorola in July 1995. This became an element in JPL's future low-cost micro-spacecraft. Development of a space-borne micro-GPS receiver, which offered the promise of low-cost orbit determination for most low-Earth orbiting spacecraft, was also initiated in 1995.

Prototypes of a new class of low-cost, fully automated, autonomous ground stations that would simplify implementation and operation and reduce the life-cycle cost of tracking stations in the DSN were introduced in 1995, 1996, and 1997. The first of these terminals was designed for tracking spacecraft in low-Earth orbit and was named LEO-T. It was enclosed in a radome and mounted on the roof of a building at JPL, where it accumulated over two years of unattended satellite tracking operations without problems. Prompted by the success of LEO-T, the program undertook a fast-track effort to develop a similar automated terminal for deep space applications. It would be called DS-T. The prototype DS-T was to be implemented at the 26-m BWG antenna at Goldstone. Automation technology was carried one step further into the area of Network operations in 1997 with the introduction of Automated Real-time Spacecraft Navigation (ARTSN). In addition to the antenna system, the DS-T included an X-band microwave system, a 4-kW transmitter, and an electronics rack containing commercial-off-the-shelf equipment to carry out all baseband telemetry downlink, command uplink, and Doppler and ranging functions. It was planned to demonstrate DS-T with *Mars Global Surveyor* early in 1998 and to use this technology (autonomous uplink and downlink) in the Network with the New Millennium DS-1 spacecraft later that year. A selection of technology references related to cost-reduction initiatives can be found in section 19 of the appendix at the end of the book.

KA-BAND DEVELOPMENT

In the past, the major improvements in the Network's deep space communications capabilities were made by moving the operating frequency to the higher frequency bands. Recognizing this, research and development at Ka-band (32 GHz) was started in 1980. Initial efforts were directed toward low-noise amplifier development and system benefit studies. However, it was also clear that the performance of existing antennas (which were designed for much lower frequencies) would severely limit the improvement in performance that could be realized from the higher operating frequency. Accordingly, in 1991, a new antenna specifically designed for research and development at Ka-band was installed at the DSS 13 Venus site at Goldstone. It would be used as the pathfinder for development of large-aperture beam waveguide antennas that would, in due course, be implemented throughout the Network.

Small imperfections in the surface of an antenna cause larger degradations at Ka-band than at lower frequencies, and, because of the narrower beam width, small pointing errors have a much larger effect. In 1994, improvements in antenna efficiency and in antenna pointing were made on the research and development Ka-band antenna at DSS 13 and on the new operational antenna at DSS 24. These improvements were effected by the use of microwave holography for precise determination of antenna efficiency, along with a special gravity-compensation system to counteract the effect of gravitational sag as a function of antenna elevation angle. In a search for further downlink improvement, a new feed system consisting of a maximally compact array of seven circular Ka-band horns was designed and tested at DSS 13. Each horn was connected to a cryogenically cooled low-noise amplifier, a frequency down-converter, an analog-to-digital converter, and a digital signal processor. The signals from each horn were optimally combined in a signal processor and presented as a single output with a quality equivalent to that of a signal from an undistorted antenna. The measured gain in downlink performance was 0.7 dB.

The technology program continued to develop operational concepts that would eventually lead to the adoption of Ka-band for deep space missions. Tradeoff studies between X-band and Ka-band showed the overall advantage of Ka-band, taking due account of the negative effects inherent in its use, to be about a factor of four, or 6 dB, in data return capability. Obviously, end-to-end system demonstrations were needed to instill confidence in the new technology. In 1993, the *Mars Observer* spacecraft carried a nonlinear element in its transmission feed to produce a fourth harmonic of the X-band signal. The demonstration (called KABLE for Ka-band Link Experiment) provided a weak Ka-band signal for the tracking antenna at DSS 13. A second demonstration

(KABLE II) was conducted using the Mars Global Surveyor mission in 1996. This experiment would be used to characterize Ka-band link performance under real flight conditions and validate the theoretical models derived from the studies mentioned above.

KABLE-II required additions to the MGS spacecraft radio-transponder to generate a modulated Ka-band downlink from which the improvements that had been made to the DSS 13 antenna to reduce pointing errors and improve performance while tracking a spacecraft at Ka-band could be evaluated. While the main objective of KABLE-II was to evaluate Ka-band for future operational use, it also served as a testbed for new Ka-band technology applications in both the flight systems and the ground systems.

In the course of transition to simultaneous X-/Ka-band operation, the DSN needed the capability to support various combinations of X- and Ka-band uplinks and downlinks. These included Ka-band receive only; X-/Ka-band simultaneous receive, with or without X-band transmit; and full X-band transmit/receive simultaneously with Ka-band receive/transmit. The technology program developed new microwave techniques using frequency-selective surfaces and feed junction diplexers to provide the frequency and power isolation necessary to realize the performance required by a practical device. In late 1996, a demonstration at DSS 13 succeeded in showing that these four different modes of operation could coexist on a single beam waveguide antenna within acceptable performance limits. This work provided a viable solution to the problem of simultaneous X-/Ka-band operation on a single antenna; this solution would be needed by the operational network to support the Cassini radio science experiment (search for gravitational waves) in 2000.

Recognizing the need for a cheaper, smaller, less power-consuming radio-transponder to replace the existing device on future deep space missions, the advanced development program embarked on a joint program with other JPL organizations to develop the Small Deep Space Transponder (SDST). The concept employed Microwave Monolithic Integrated Circuits in the RF circuits and Application Specific Integrated Circuit techniques to perform digital signal-processing functions, with a RISC microprocessor to orchestrate overall transponder operation. The transponder would transmit coherent X-band and Ka-band downlinks and receive an X-band uplink. Besides minimizing production costs, the principal design drivers were reduction in mass, power consumption, and volume. The Small Deep Space Transponder was flown in space for the first time aboard the DS-1 New Millennium spacecraft in July 1998.

With a view to providing a better understanding of the performance of Ka-band links relative to X-band links from the vantage point of a spaceborne radio source, the DSN technology program engaged in the development of a small low-Earth-orbiting spacecraft called SURFSAT-1. Launched in 1995, the experiment provided an end-to-end test of Ka-band signals under all weather conditions and DSS 13 antenna elevation angles as the spacecraft passed over Goldstone. The SURFSAT data was also used for comparison with the KABLE data received from MGS. Later, the SURFSAT X-band and Ka-band downlinks were used to great advantage to test and calibrate the DSN's new 11-meter antennas prior to their support of the VSOP (HALCA) Orbiting VLBI mission in 1996. A selection of technical references related to Ka-band development can be found in section 20 of the appendix at the end of the book.

OTHER TECHNOLOGIES

While reduction of Network operating costs and the introduction of Ka-band into the Network were important features of the Technology Program in the late 1990s, they were not the only areas of activity. Interagency agreements allowed Wide-Area Differential GPS techniques, originally developed under the technology program for NASA spacecraft orbit determination and DSN Earth platform calibrations, to be made available for applications within the Department of Transportation and Federal Aviation Administration.

In addition to these other technologies, further advances were made in the areas of photonics and optical communications. A selection of technical references related to other technologies can be found in section 21 of the appendix at the end of the book.

OPTICAL COMMUNICATIONS DEVELOPMENT

Beginning in 1980, the Technology Program supported theoretical analyses that predicted, under certain system and background light conditions typical of deep space applications, the ability to communicate at more than 2.5 bits of information per detected photon at the receiver. Laboratory tests later confirmed these theoretical predictions. However, detection power efficiency was only one of the many factors that needed to be studied to bring optical communications to reality. Others included laser transmitter efficiency, spatial beam acquisition, tracking and pointing, link performance tools, flight terminal systems design, definition of cost-effective ground stations, and mitigation of Earth's atmospheric effects on ground stations.

Several system-level demonstrations were carried out as this work progressed. The first, carried out in December 1992, involved the detection of a ground-based pulsed laser transmission by the *Galileo* spacecraft during its second Earth fly-by. The second demonstration was carried out over the period November 1995 to May 1996 with the Japanese Earth-orbiting satellite ETS VI.

Both experiments yielded important observational data in support of theoretical studies and encouraged the further development of optical communications technology with supportive flight demonstrations. A selection of technical references related to optical communications development can be found in section 22 of the appendix at the end of the book.

DSN Science

Science and technology have always been closely coupled in the DSN. Since the very beginnings of the DSN, its radio telescopes had provided world-class instruments for radio astronomy, planetary radar, and radio science. Many technology program achievements were of direct benefit to these scientific endeavors, and DSN science activities frequently resulted in new techniques that eventually found their way into the operational Network.

In the period reported here, the program supported radio astronomy investigations related to the formation of the stars and to the study of microwave radio emissions from Jupiter, as well as radio science measurements of the electron density in the solar plasma outside the plane of the ecliptic (Ulysses). It also supported a program of tropospheric delay measurements which would be of direct benefit to the Cassini gravitational waves experiment. The Goldstone Solar System Radar (GSSR) continued its highly successful series of Earth Crossing Asteroid (ECA) observations, which began with images of Toutatis (asteroid 4179) in 1992 and continued with Geographos (asteroid 1620) in 1994 and Golevka (asteroid 6489) in 1995. This work was expected to increase in the years ahead as new and improved optical search programs enabled the discovery of more ECAs. A selection of technical references related to DSN science can be found in section 23 of the appendix at the end of the book.

CHAPTER 8

THE DEEP SPACE NETWORK AS A SCIENTIFIC INSTRUMENT

THE TDA SCIENCE OFFICE

In addition to their principal function of tracking and acquiring data from distant spacecraft, the unique capabilities of the Deep Space Network were also used for scientific research in the fields of radio science, radio astronomy, radar astronomy, crustal dynamics, and the Search for Extraterrestrial Intelligence (SETI). With the completion of the first 64-m antenna at Goldstone in 1966, radio astronomy, which explores the origin and location of natural sources of radiation in the galaxies, and radar astronomy, which investigates the surface and shape of planetary bodies by studying the characteristics of a reflected radar return, became of much greater significance in the overall function of the Network.[1] Originally, both radio astronomy and radio science were included in the term "radio science." As they commanded more attention and more resources from the DSN, it became necessary to identify and manage them independently of each other. They are discussed separately in later sections of this chapter.

From 1967 through 1995, the vigorous efforts of Nicholas A. Renzetti brought the potential of the DSN as an instrument for scientific research to the attention of the scientific community throughout the United States, Germany, France, Italy, Australia, and Japan. It was his advocacy that led to the negotiation of international cooperative experiments in radio science, as well as radio and radar astronomy, and to funding for U.S. interagency support agreements for technology development in the fields of crustal dynamics and SETI.

While the work in radio science, radio astronomy, and radar astronomy directly involved the DSN in development, engineering, and the use of the DSN antennas and facilities, the work in crustal dynamics, except in its very earliest stages, was independent of the mainstream of DSN activity. Other than its need for observing time on the 70-m antennas, the SETI program was also independent of the DSN. It was the strong cross-coupling between the technologies used by both these latter programs and those of the DSN that justified their residence at JPL and their association with the DSN. As the following paragraphs will show, DSN association with the crustal dynamics and SETI programs was relatively short-lived, while radio science, radio astronomy, and radar astronomy became permanent features of the DSN mission set.

In 1983, Peter T. Lyman, then Director of Telecommunications and Data Acquisition at JPL, created an office to coordinate and properly manage the several streams of science activity with which the DSN was becoming involved. The TDA Science Office would guide and coordinate all of JPL's efforts on behalf of NASA in the fields of radio science,

The Deep Space Network as a Scientific Instrument

SETI, crustal dynamics, radio astronomy, and planetary radar astronomy (Goldstone Solar System Radar). Nicholas A. Renzetti was appointed as its first manager.

Nicholas Renzetti had already established a successful career in research and development when he came to JPL in 1959. Reared in the Bronx, New York, as the son of an Italian immigrant family with a will to succeed, Renzetti had, by 1940, developed those humble beginnings into a Ph.D. from Columbia University, New York. During World War II, Renzetti was engaged in research on countermeasures for acoustic and magnetic mines for the Navy's Counter-Mine Warfare Service. In the decade that followed, Renzetti was involved with research on the ballistics of rocket projectiles and development of test facilities and optical and electronic instrumentation for missiles at the Naval Ordnance Test Station, China Lake, California. The years 1954 to 1959 saw Renzetti working as a physicist, conducting research on the physical and chemical processes of the Los Angeles atmosphere under the auspices of the Air Pollution Foundation, chaired by President Lee A. Dubridge of Caltech.

When Eberhardt Rechtin hired Renzetti in 1959 to manage the actual operation of the new Network, he brought a wealth of in-the-field experience to the position. As manager of the new DSIF Communications and Operations Section in Rechtin's Telecommunications Division, Renzetti's experience served him well in the years to follow, when the Network expanded and the spacecraft it was called upon to support increased in number, complexity, and institutional affiliation.

Stocky in build, Renzetti always dressed in a sport shirt and bolo tie (thin leather cord fastened with a silver or turquoise bolo). He made an exception on formal occasions, when he favored a bright bow tie. A cigar or curved Sherman pipe completed the picture. He was a gruff man, loud of voice, and inclined to shout for emphasis when he was excited or agitated. Renzetti's phone conversations or exchanges at staff meetings were frequently audible to people in nearby offices or corridors. Behind that rather intimidating exterior lay an extraordinary gentleness that became evident when, on occasion, he invited his engineers to join his family Christmas celebrations.

From the outset, Renzetti determined to leave a written record of the work and accomplishments of the DSN. He insisted that his people produce publishable accounts of all of the spaceflight operations in which they were involved, and he saw to it that all research and development work associated with the DSN, even though it was performed in areas beyond his responsibility, was likewise properly reported. The formal documents that he established for publishing these reports were known as the DSN (or later, TDA)

Progress Reports. They remain a testament to his success and a permanent record of the accomplishments of the Network and those associated with it.

Always promoting the potential of the DSN for purposes other than tracking planetary spacecraft, Renzetti left the area of DSN "operations" in 1983 to head the DSN Science Office. There, he led the DSN effort in radio astronomy, radar astronomy, geodynamics, and the Search for Extraterrestrial Intelligence. The value of his continuous support for and promotion of the Goldstone Solar System Radar for Planetary Radar Astronomy was recognized by the International Astronomical Union in the naming of Asteroid Renzetti in 1997.

He retired from JPL in 1996 and died at his home in San Marino, California, in 1998.

The combination of sound scientific leadership and NASA's liberal policy for support of ground-based radio science in the Deep Space Network proved very productive. A quantitative measure of its productivity is depicted in Figure 8-1, which shows the number of papers published in the scientific literature by investigators using the facilities of the DSN for radio astronomy, radio science, and radar astronomy for over thirty-five years, from 1962 to 1997.

Note: In the absence of a better estimate, the data shown in Figure 8-1 for Goldstone radar astronomy represents fifty percent of the total number of publications in the literature for the years 1962 to 1987.[2]

At the end of 1995, Renzetti moved to a technical staff position in the TDA director's office, and management of the TDA Science Office passed to M. J. Klein. By that time, the scope of the effort in the office had diminished somewhat. Two of the major science programs described later in this chapter, crustal dynamics and the Search for Extraterrestrial Intelligence Project (SETI), were no longer under the purview of the TDA Science Office. The work on crustal dynamics had moved out of JPL to the National Geodetic Survey, and SETI had been canceled. A third program, radio science, became more the province of the flight projects as *Galileo* arrived at Jupiter and commenced an intense two-year campaign of Radio Science experiments with multiple Jovian satellite encounters.

Nevertheless, radio astronomy continued to flourish under the strong scientific leadership of Thomas B. Kuiper, while radar astronomy, which had been brought to prominence in the scientific world by Richard M. Goldstein, expanded the scope of its scientific investigations under the direction of Steven J. Ostro.

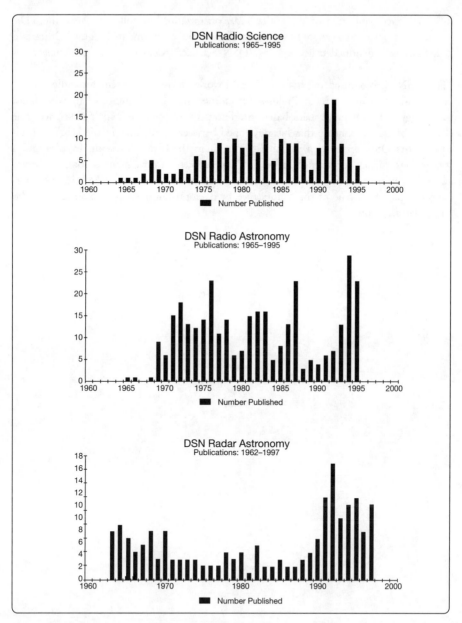

Figure 8-1. DSN science-related publications, 1962–97.

In a major reorganization of the entire TDA organization the following year, the TDA Science Office was incorporated into a new Plans and Commitments Program Office and its activity was identified collectively as "DSN Science." Klein remained as manager.

"The DSN is recognized internationally as a world[-]class instrument for radio astronomy," said Klein in 1997, "and allows astronomers to take advantage of the very latest technology in Earth[-] and space-based interferometry to produce first[-]rate science. The close symbiotic relationship that has developed between technology in the DSN and science in the DSN serves as a valuable stimulant to the benefit of both. In addition, a strong constituency for science in the DSN promotes improvements in both technical and cost performance for the DSN, and it enhances the image of NASA in the international scene by virtue of the significance of its contributions to the sciences of radio and radar astronomy."[3]

RADIO SCIENCE

By the time the DSIF was established in 1958, radio astronomy had been in existence as a scientific discipline for many years. Radio astronomers, supported largely by the NSF, generally directed their attention to extragalactic radio sources, leaving the exploration of the bodies and media of the solar system to optical astronomers.

Thus, the availability of radio-equipped spacecraft traversing the immediate vicinity of the inner planets encouraged scientists to explore ways in which the downlinks from these spacecraft might be used to yield new data about the planets, the Sun, and the interplanetary medium. Furthermore, this type of science required no additions to the spacecraft science payload and, for the most part, made use of the signals already generated by the DSN for its own use in spacecraft navigation. The new science became known as radio science and was always associated with a particular spacecraft and the total science inventory of the flight project to which it belonged.[4] Radio science investigations for the flight projects formed the major science activity in the DSN for the first half of the 1960s. Using only the Doppler and range data generated by the tracking system, a great deal of significant new planetary science was generated in this period.

In 1981, Renzetti and Allen L. Berman published the first well-documented description of radio science in the DSN.[5] Later, Sami W. Asmar, then manager of radio science in the Network, expanded this work and brought it up to date through 1993.[6]

NOTE: The DSN published later versions of the Radio Science System online at http://radioscience.jpl.nasa.gov.

It was Sami Asmar who explained the basic technique of radio science. "By examining small changes in the phase and/or amplitude of the radio signal received from a planetary spacecraft, radio science investigators are able to study the atmospheric and ionospheric structure of planets and satellites, planetary gravitational fields, shapes and masses, planetary rings, ephemerides of planets, solar plasma, magnetic fields, cometary dust, and several aspects of the theory of general relativity."

The early radio science investigators relied on the Doppler and range data generated by the DSN Tracking system, primarily for spacecraft navigation purposes, to provide the minute changes in phase and amplitude of the radio signal that were clues to the science they were seeking from a distant planet or satellite.

Doppler data represents the relative motion (velocity) between the spacecraft and the Earth station. It is derived from the measured frequency difference between the transmitted uplink signal and the received downlink signal. Range data represents the distance (range) between the spacecraft and the Earth station. It is derived from the measured time delay between a precision range code transmitted on the uplink and its delayed image received from the spacecraft on the downlink. Special radio transponders carried on the spacecraft receive the uplink from Earth, add telemetry data, and retransmit the signal back to Earth without introducing any unwanted changes in the phase characteristics of the signal. Such a downlink frequency is said to be "coherent" with the uplink frequency; therefore, any phase changes observed by the receiving station are attributed to the "relative velocity" of the spacecraft.

The spacecraft also carries a very stable radio-frequency oscillator of known frequency that can be used to supply the downlink frequency when no uplink is present. In this case, "one-way" Doppler is derived from the measured difference in frequency between the known frequency transmitted by the spacecraft and the actual frequency received by the Earth station.

Before it can be used by radio scientists or spacecraft navigators, Doppler and range data must be refined (to remove perturbations introduced by the interplanetary media though which it has traveled) and calibrated (to remove errors introduced by the radio circuitry in the spacecraft, ground receivers, and transmitters).

The tracking system is the element of the overall DSN architecture that generates Doppler and range data from the coherent uplink and downlink between spacecraft and tracking station. It provides these data, along with appropriate time tags and ancillary data, in a suitable format for use in spacecraft navigation, orbit determination, and radio science.

To generate radio metric data of great precision and stability, the tracking system requires an ultra-stable clock as a reference, or yardstick, for its uplink and downlink frequency measurements. In the early years, this was provided by an atomic clock at each Complex. These were rubidium and cesium frequency standards with stability characteristics of one part in 10^{12} (rubidium) or one part in 10^{13} (cesium).

NOTE: To appreciate the significance of these numbers in the time domain, a simple calculation will show that one part in ten raised to the thirteenth power (1×10^{13}) is approximately equivalent to three seconds per one million years.

The rubidium and cesium atomic clocks were eventually superseded by even more stable clocks driven by hydrogen masers (1×10^{-14}). Equally stringent requirements for phase and frequency stability applied to the spacecraft radio system where the range and Doppler "turnaround" functions were performed.

Radio metric data generated by the tracking system was used for many applications in radio science. Doppler data from spacecraft on planetary flyby or orbiting missions provided information regarding the orbit, mass, gravitational, and physical properties of the planetary bodies for celestial mechanics investigations. It also provided planetary atmospheric information as the downlink signal was refracted through planetary neutral atmospheres. When dual-frequency downlinks (S-band and X-band) became available (first on *Mariner 10* (Venus/Mercury) in 1973), it enabled measurements of columnar electron density changes in the solar wind and planetary ionospheres.

Range data was used to measure planetary distances, as well as the signal time delay due to the solar wind (electron density) and gravitational (relativistic) bending near the Sun.

In 1976, the various components (such as open-loop receivers, stable oscillators, and recorders) that had been used for the previous decade or so to make radio science observations were incorporated into a new DSN system designed to handle the unique needs of radio science for all flight projects. Known as the Radio Science System, it generated radio science data as a data product for delivery to the Flight Project Science Team in the same way as telemetry or radio metric data was considered to be a "deliverable." Together with the tracking system and ancillary calibration data, the Radio Science System generated all of the DSN radio science data for the planetary missions after this time.

As described by Berman, the essential feature of the new DSN Radio Science System was its ability to reduce the amount of radio science data to be recorded (digitally) for each occultation by a factor of 10 to 100, compared to the fixed tuned analog recording technique previously employed by the Network. This was accomplished by automatically tuning the open-loop receivers in conformance with a predicted Doppler frequency profile based on precise knowledge of the planet and spacecraft orbits, along with a model of the planetary atmosphere. Depending on the accuracy of the "predicts" and the nature of the data itself, the output of the open-loop receiver would then become a new "baseband" frequency lying within a few kHz of the predicted frequency and could easily be digitized and written on computer-compatible tape at the observing station. In addition to the "baseband" frequency, the digital recording contained a copy of the frequency profile used to generate the "baseband" data with appropriate timing information. This information was then shipped to the experimenter, transmitted to JPL via

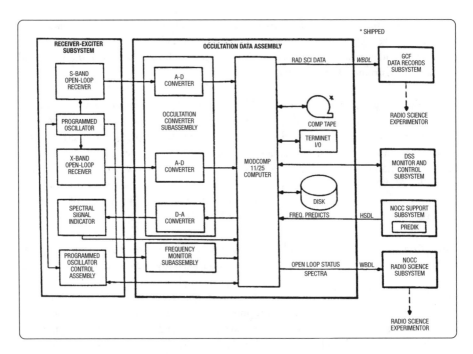

Figure 8-2. Real-time bandwidth reduction in the radio science Occultation Data Assembly.

the GCF, or both. By combining the "predict" frequency with the "baseband" frequency, the experimenter could recover the actual received frequency. The new system also provided real-time spectral data for display on Signal Spectrum Indicators (SSIs) at the tracking station and in the Radio Science Team area at JPL. The SSIs were used for monitoring performance, calibration, and analysis purposes, as well as for quick-look science evaluation.

The new capability, referred to as "real-time bandwidth reduction," was delivered to the Network for first operational use with the Pioneer Venus mission in December 1978. The Occultation Data Assembly, a part of the Radio Science Subsystem in which the bandwidth reduction function described above was performed, is illustrated in functional form in Figure 8-2.

As an integral part of the DSN, the Radio Science System was upgraded from time to time to reflect the introduction of new technology, processes, and capability in the overall DSN. In 1996, the Radio Science System acquired a capability that allowed the Radio

Science Team at JPL to share control of the Radio Science System at the Complexes with the station operators. This placed the burden of operating the complicated Radio Science System back on the user group and reduced the responsibility and workload of the DSN tracking stations. At last, the Radio Science Teams would have the same flexibility and control of their instrument as the other science experiment teams enjoyed with instruments carried by the spacecraft. Remote control for radio science was first used for Galileo orbital operations in late 1996 and, after some initial integration problems, proved to be of great value to both the investigators and the DSN in dealing with the heavy load of radio science through the end of the Galileo Mission.

CELESTIAL MECHANICS

Renzetti and Berman suggested that the DSN (then DSIF) radio metric capabilities should be first used for radio science purposes by Anderson and Warner for celestial mechanics experiments with the 1962 *Mariner 2* mission to Venus. Using only the Doppler navigational data from this flight, the experimenters were able to make order-of-magnitude improvements in the determination of the masses of the Moon and Venus. All through the Mariner Era, the *Mariner 4* (Mars), *Mariner 5* (Venus), and *Mariner 9* (Mars) missions were used for celestial mechanics experiments to determine the masses of Venus and Mars.

Doppler measurements from the Lunar Orbiter and Apollo missions led to the discovery of mass concentrations (mascons) on the Moon, and analysis of the *Mariner 10* Doppler data in 1973 and 1974 enabled the determination of the mass and shape of Mercury.

In the Voyager Era, celestial mechanics investigators used the DSN radio science data from the Pioneer and Voyager missions to Jupiter and Saturn to produce a wealth of new knowledge on the physical properties of those planets and their satellites.

Venus had received intense study by celestial mechanics investigators since the time of the Mariners. Data from long-term orbiting spacecraft, most notably *Pioneer Venus Orbiter* and *Magellan Radar Mapper*, produced a data volume equivalent to thousands of single flybys, making the gravity-field of Venus almost as well understood as that of Earth.

Celestial mechanics studies of Mars also had the benefit of orbiting spacecraft, starting with *Mariner 9* in 1971 and the two *Vikings* in 1976. These studies would have been extended with *Mars Observer* in 1993 had the spacecraft not been lost just prior to entering Mars orbit. They were, however, resumed with the Mars Global Surveyor mission in 1997.

Prior to the Galileo mission to Jupiter in 1995, knowledge of the gravity fields of Jupiter and its satellites was based on the combined results of celestial mechanics experiments conducted during the flybys of *Pioneers 10* and *11* and *Voyagers 1* and *2*. This had provided a good estimate of the Jupiter mass but contributed little to existing knowledge of the satellites. Although the *Galileo* orbits of Jupiter were too high to contribute much to gravity field measurements of the planet itself, the close Encounters of Io, Ganymede, Callisto, and Europa provided an immense amount of data for analyses of the internal structure of these Jovian satellites.

During the period July 1996 to December 1997, investigators carried out twenty-three celestial mechanics experiments while the *Galileo* spacecraft was in orbit around Jupiter. The purpose of this campaign was to refine the knowledge of the position of Jupiter in the plane of the sky which, in turn, led to a significant improvement in knowledge of the ephemeris of Jupiter. To obtain the increased accuracy required for these observations, the DSN generated ranging data from several stations simultaneously during the spacecraft overlap view period between Complexes, with the spacecraft in the "one-way" mode. Reduction of these data to extract the simultaneous spacecraft-to-station, path-length differences from the known station locations yielded data from which a very precise determination of the spacecraft orbit could be made. Investigators were then able to refine their knowledge of the position of Jupiter from its effect on the very precisely known orbit of the spacecraft. The range differencing technique used in these experiments was known as Delta Differenced One-way Ranging (Delta DOR).

The *Voyager 1*, *Voyager 2*, and *Pioneer 11* Encounters of Saturn contributed radio science data for celestial mechanics estimates of the masses and gravity fields of Saturn and its satellites Rhea, Titan, Tethys, and Iapetus. A previously unknown moon, S13, was first discovered by its gravitational perturbations of the Doppler data from the *Voyager 2* Saturn Encounter. Its presence was later verified by optical imaging from the spacecraft.

Further celestial mechanics investigation of Saturn must await the arrival of *Cassini* with the Huygens Titan Probe in 2004. Then radio science data at X-band and Ka-band from multiple *Cassini* orbits through the Saturnian system will provide investigators with additional information on masses, densities, and interactions between the planet and its satellite Titan.

Celestial mechanics information on Uranus was derived from the radio science data taken during the *Voyager 2* Encounter of the planet in 1986. Combined with imaging data of the satellites, the Doppler perturbations enabled great improvement in determination of the masses of five of the major satellites and of the planet itself.

The two-way Doppler data obtained by the DSN Radio Science System during the *Voyager 2* Close Encounter of Neptune in 1989 provided celestial mechanics investigators with basic data from which the masses of Neptune and its satellite Triton could be estimated to an accuracy of 0.0003 percent and 0.3 percent respectively.

Solar Corona and Solar Wind

In 1965, as the *Mariner 4* (Mars) spacecraft passed within 0.6 degrees of the solar disk, Richard M. Goldstein used the not-yet-fully-operational 64-m antenna (DSS 14) at Goldstone with an "open-loop" receiver to observe the spectral spreading of the spacecraft signal as it passed through the solar corona. This marked the first use of the DSN for Solar Corona research. Berman, and later Woo, were able to interpret the observed results in terms of the relationship between spectral broadening and integrated coronal electron density. Subsequently, similar experiments with the *Helios* and *Pioneer* spacecraft would confirm and expand the original measurements and the underlying theories.

In addition to spectral broadening, the radio signal from a planetary spacecraft experiences a second effect as it passes through the Solar Corona. The RF signal experiences a small, but measurable, rotation of its plane of polarization. Known as "Faraday Rotation," this effect was linked to both electron density and the solar magnetic field in an experiment carried out at DSS 14 by Charles T. Stelzried in 1968.[7] To support this experiment, DSS 14 had been equipped with a rotatable, linear microwave feed system driven by a closed-loop polarimeter that would automatically track the orientation of the polarization of the received signal. Using the linearly polarized signal from the *Pioneer 6* spacecraft as it was occulted by the Sun in November 1968, these observations yielded valuable information on both transient and steady-state effects in the solar corona.

Similar experiments were carried out at the other 64-m stations with Helios and Pioneer spacecraft in 1975 and 1977. An ambitious Faraday Rotation experiment, planned for the Galileo mission to Jupiter in 1995, was eliminated in 1991 when the failure of the *Galileo* high-gain antenna limited the downlink to S-band right circular polarization only. However, successful Faraday Rotation experiments using the *Magellan* spacecraft at Venus were carried out by Asmar and Bird in 1991 at both S-band and X-band.

Range data generated by the DSN tracking system also provided a powerful tool for determination of electron density in the solar corona, provided that all other sources of error could be properly accounted for. Duane O. Muhleman, John D. Anderson, and others first used this method during the solar conjunctions of *Mariners 6* and *7* in 1970, using ranging data acquired at DSS 14.

During the 1975 solar conjunctions of *Pioneer 10, Pioneer 11*, and *Helios 1*, Allen L. Berman and Joseph A. Wackley of JPL, working in collaboration with Jose M. Urech at the Madrid Complex, were able to demonstrate that the fluctuations observed in the Doppler signature of a single S-band downlink provided an excellent measure of turbulence in the solar wind. Since Doppler fluctuations (or "noise") were automatically computed for monitoring the performance of the DSN tracking system, it represented a "free" source of scientific data and was compared favorably with the more accurate dual-frequency methods. Electron density measurements based on dual-frequency S- and X-band Doppler were not degraded with many of the sources of error present in single-frequency measurements. Measurements with dual-frequency Doppler downlinks on the *Mariner 10* spacecraft in 1974 and dual-frequency ranging on the *Viking* spacecraft in 1977 and 1979 yielded excellent results.

Observations of Doppler scintillation from *Pioneer Venus* were compared with independent observations of solar plasma transients from *Helios 1* by Woo and Schwen in 1991. They showed that several Doppler transients, observed by *Pioneer Venus* at 15 to 18 solar radii, were observed about 48 hours later as solar plasma shocks by *Helios 1* at 170 to 174 solar radii.[8]

Solar wind scintillation experiments played a large part in the Galileo radio science program. The experiments were designed to gather data for the study of the solar wind within 0.3 AU of the Sun. The scintillation experiments were originally intended to complement solar corona experiments but were eliminated by the loss of the HGA and the dual-frequency downlinks. Five experiments were conducted successfully, on S-band only, during superior conjunctions between December 1991 and February 1997.

Radio Propagation and Occultation

Almost every planetary mission, past and future, is concerned with the investigation of planetary atmospheres, ionospheres, rings, and magnetic fields. Such investigations are based on the perturbation of the phase and amplitude of the spacecraft downlink by the extended atmosphere of the planet as the radio signal passes through it during periods of occultation. These perturbations are contained in the data captured by the open-loop receivers of the Radio Science System and can be converted into a refractivity profile for the planetary atmosphere after the data are corrected for the contribution due to the geometrical flight path of the spacecraft. Information regarding electron distribution in the ionosphere, the temperature-pressure profile in the neutral atmosphere, or particle size distribution in the ring material can be derived from the refractivity profile.

According to Asmar and Renzetti, the method was first proposed by V. R. Eshelman in 1962. Independently, D. L. Cain, leading a JPL team engaged in analyzing the effect of refraction in Earth's atmosphere on the accuracy of Doppler data, recognized the possibility of applying the sensitivity of Doppler phase measurements to the study of atmospheres and ionospheres of other planets. The theoretical analysis was developed by G. Fjelbo in 1964.

The occultation experiments required special receivers to extract the Doppler data in addition to the normal DSN receivers. The normal receivers in the DSN tracking system were of the "closed-loop" type; that is, they employed a "phase-locked tracking loop" to keep them tuned to the frequency of the downlink carrier signal. As the frequency of the downlink changed as a result of the Doppler effect of the spacecraft's motion relative to Earth, the "phase-lock loop" caused the receiver to follow the changes and thereby "maintain lock." Doppler data was generated by comparing the received downlink frequency with the uplink transmitter frequency (two-way), or a precise estimate of the frequency of the Ultra Stable Oscillator (USO) carried by the spacecraft (one-way).

When a spacecraft entered occultation by a planet, the downlink signal was cut off and the receivers dropped lock. When the spacecraft exited occultation and the signal reappeared, valuable time was lost while the receivers reacquired the downlink. Thus, the important radio science data was only available for a short period immediately before entry and immediately after exit as the signal transited the thin planetary atmosphere. Therefore, in 1964, special wide-bandwidth "open-loop" receivers were added to the DSN tracking system to ensure that the downlink would appear in the "open-loop" receiver passband the instant it emerged from occultation. The "open-loop" receivers were tuned by a programmable local oscillator, which was driven by a computer program based on the spacecraft orbit and other pertinent parameters. The open-loop output was recorded on analog magnetic tape recorders at the station and subsequently shipped to JPL. There it was digitized and delivered to the experimenters for analysis.

Occultation experiments using open-loop receivers in the DSN were made for the first time in 1965, when *Mariner 4* flew by Mars. These experiments were highly successful; they showed that the Martian atmosphere was predominantly carbon dioxide and that the surface pressure was less than one percent of that of Earth, an order of magnitude less than had previously been thought.[9] Measurements of the electron density in the Martian ionosphere were also obtained. During the late 1960s and early 1970s, occultation experiments were carried out by the Mariner and Viking missions to the inner planets (Mercury, Venus, and Mars).

The Voyager Era was prolific in producing occultation radio science for the outer planets. This began in 1973 with the *Pioneer 10* flyby of Jupiter, repeated a year later by *Pioneer 11*. *Pioneer 11* went on to make the first occultation measurements of the atmosphere and ionosphere of Saturn in 1979. Occultation experiments were conducted with *Voyagers 1* and *2* at Jupiter in 1979 and at Saturn in 1980 and 1981, with *Voyager 2* at Uranus in 1986 and at Neptune in 1989. These experiments determined atmospheric pressure levels, temperatures, and constituents, as well as the characterized ionospheres of Jupiter, Saturn, and Triton.[10] The theory of radio occultation techniques was extended in 1980 to include ring occultation experiments during the *Voyager 1* Encounter of Saturn.

Ulysses experimenters used occultation radio science during their spacecraft gravity assist flyby of Jupiter in February 1992 to derive the columnar electron density of the Io plasma torus. This experiment was enhanced by the presence of a dual-frequency (S- and X-band) downlink from the spacecraft.

In 1991 and 1992, Magellan experimenters obtained highly accurate profiles of atmospheric refractivity and absorptivity of Venus using occultation radio science. These profiles were used to deduce ionospheric density, atmospheric temperature, pressure, and sulfuric-acid abundance on Venus. The dual frequency, combined with the higher power of the Magellan downlinks, enhanced the end result by eliminating errors and making it possible to penetrate the atmosphere to greater depth than would otherwise have been possible.

The orbital part of the Galileo mission to Jupiter provided many opportunities for occultation radio science. In 1996 and 1997, twenty-two occultations were observed. These included a single occultation by Callisto and multiple occultations by Jupiter, Io, Ganymede, and Europa. The information derived from these data substantially increased the existing body of knowledge about the ionospheric characteristics of all of those bodies.

The Cassini mission to Saturn in 2004 will include several occultation flybys of Saturn and Titan, which will be observed with S-band, X-band, and Ka-band downlinks. Data obtained from observations in each of these bands has advantages in the study of dispersive media like the atmospheres of Saturn and Titan and also makes it possible to determine the particle size distribution in Saturn ring structure.

Relativistic Time Delay

Depending on the Earth-Sun-spacecraft geometry, range measurements from Earth to a planetary spacecraft passing through solar conjunction can enable a test of a portion of the General Theory of Relativity. Einstein's theory predicts that electromagnetic waves passing close to massive bodies, such as the Sun, are influenced by the Sun's gravitational field in such a way that the signal path between spacecraft and Earth is "bent" by the intense gravitational force of the Sun. The effect of the "bending" is to introduce a time delay along the signal path between spacecraft and Earth. This time delay can be measured by the DSN tracking system. Using the measured signal time delay and knowledge of the Earth-Sun-spacecraft geometry at the time the measurement was taken, radio scientists are able determine a value for a key parameter in the mathematical equations by which the theory is described. This parameter, known as the "Schwartzchild metric," is denoted by the Greek symbol gamma. In the theory, the value of gamma is 1.0. For a relativistic experiment, therefore, the closer the results approach a value of 1.0 the better its agreement with theory.

To measure signal time delay, the DSN used the ranging system which forms a part of the DSN tracking system described earlier.

The first major attempt to measure the relativistic bending of a spacecraft signal was made by John D. Anderson and others during the solar conjunction of *Mariners 6* and *7* in the mid-1970s.[11] The results obtained from this experiment verified the value of gamma to about 6 percent, i.e., gamma = 1.00 ± 0.06. The major sources of error were attributed to uncertainty in the accuracy of the spacecraft orbit and in the measurement of time delay due to scattering effects of free electrons in the solar corona. In time, these errors were reduced, and a later experiment with *Mariner 9* brought the uncertainty in the value of gamma down to 2 percent.

A very significant improvement in accuracy resulted from the relativistic experiments carried out in 1976 and 1977 by Shapiro, Cain, Reasonberg, and others with the Viking Mars Orbiter and Lander spacecraft. Benefiting from significant improvements in the calibrations of the tracking system, the use of dual-frequency downlinks, and precise knowledge of the position of the landed spacecraft, independent experimenters at MIT and JPL determined the value for gamma of 1.00 ± 0.002 the most accurate test of the theory to that time. Repeating the experiment again during the *Voyager 2* solar conjunction in December 1985, Anderson and others obtained a value of 1.00 ± 0.03 for gamma.

Improved accuracy is expected from two relativistic experiments to be conducted during the solar conjunction periods of the Cassini mission to Saturn in 2003 and 2002. The greater accuracy will derive from the use of Ka-band, which, with its wavelength of less than one centimeter, is less affected by interplanetary media than the longer wavelength signals used previously.

GRAVITATIONAL WAVES

The General Theory of Relativity predicted that the violent collapse of stellar bodies into supermassive black holes would generate gravitational waves in the form of spatial strains propagating through the solar system at the speed of light. These waves were expected to slightly alter the distance between two separated free masses, which would be detectable as a proportionate fractional-frequency shift in ultra-precise Doppler data from a space-based detector using Earth and a planetary spacecraft as free test masses. Because gravitational waves are very weak and propagate at very low frequencies, space-based systems offered the best chance for their detection.

The use of ultra-precise two-way Doppler from a distant spacecraft (greater than 1 AU) to detect very-low-frequency gravitational waves was proposed by Estabrook and Wahlquist in 1975.[12] Driven by an ultra-stable frequency standard, the DSN tracking system generates precision two-way Doppler from a comparison of the uplink and coherent downlink frequencies from the spacecraft and, in doing so, continuously measures the fractional frequency shift due to relative Earth-spacecraft motion. A gravitational wave passing through the system causes a unique "three-pulse" signature to appear in the Doppler data due to a combination of buffeting effects on the Earth and the spacecraft and speed-up of the tracking station clocks (frequency and timing systems). The characteristics of the "three-pulse" signature depend on the relative positions of Earth and the spacecraft and the direction of the gravitational wave. These effects are illustrated very simply in Figure 8-3.

Experiments conducted through 1990 using *Viking*, *Voyager*, and *Pioneer* spacecraft failed to detect the existence of gravitational waves. However, the sensitivity of these tests was limited by noise due to plasma propagation.

The advent of hydrogen masers as frequency standards for the DSN and the addition of X-band uplinks to the existing S- and X-band downlinks improved the sensitivity of

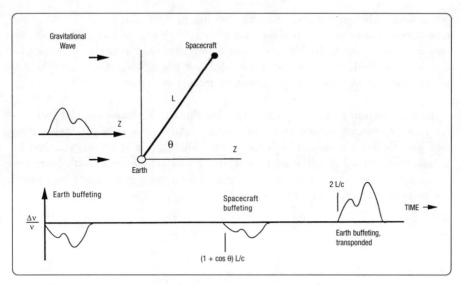

Figure 8-3. Characteristic three-pulse signature of a gravitational wave in two-way Doppler data.

the tracking system for detection of the gravity waves by a factor of about 10. With the launch of *Galileo* in 1989, carrying X-band uplink in addition to the standard S-band uplink and S- and X-band downlinks, expectations were high for the detection of gravity waves in the Galileo Era. A particularly interesting coincidence experiment involving *Mars Observer*, *Galileo*, and *Ulysses* was planned for early 1993. The sensitivity of the tracking system had been set at 5×10^{-15} as offering the best chance for successful detection of the elusive gravitational waves. This experiment afforded several unique opportunities for immediate confirmation of the detection of a gravity wave and determination of the direction of the incident wave from the Doppler signatures of the three spacecraft. The joint experiment ran for three weeks in the spring of 1993. Two further searches for gravitational waves with the Galileo spacecraft were planned for May 1994 and June 1995 during the solar opposition periods.

Unfortunately, the loss of the X-band uplink and downlink, as a consequence of the failure of the Galileo High Gain antenna in April 1991, reduced the sensitivity of the radio science system for gravity wave detection to that of the residual S-band mission, deemed marginal at best. None of the experiments succeeded in detecting the presence of gravitational waves; nevertheless, interest in pursuing the search for gravity waves persisted. The launch of *Cassini* in 1997, with Ka-band and X-band uplinks and downlinks, as well as more stable ground and spacecraft elements in the tracking system, offered the hope of observations at a much higher sensitivity level. It was estimated that the sensitivity could be as high as 10^{-16} for about one month of tracking near each of three solar oppositions in 2001, 2002, and 2004, provided that tropospheric effects could be sufficiently removed with water vapor radiometer techniques.

Looking to the future for gravitational wave searches, S. W. Asmar believed that sensitivity improvements beyond this level would require moving the tracking station itself into Earth orbit so that both test masses would be in space, unperturbed by the many unmodeled geophysical effects that are associated with an Earth-based station. In fact, this concept is to be employed in future spaceborne, laser-based, gravitational wave experiments.

RADIO ASTRONOMY

The origins of the DSN owe much to the science of radio astronomy. By the time of its establishment in 1958, radio astronomy had been in existence as a scientific discipline for many years. As discussed in the early chapters of this book, the original 26-m-diameter antennas of the DSIF were adaptations of existing, commercially available radio astronomy dishes and mounts. The original 64-m-diameter antenna built at Goldstone in 1966 was based on the design of a radio telescope of similar size operated by the Australian Scientific and Industrial Research Organisation at Parkes, Australia.

It came as no surprise when the radio astronomy community showed early interest in the substantial new global facility then being implemented by NASA for exploration of the solar system by means of planetary spacecraft. However, the National Science Foundation, which was the principal sponsor of radio astronomy research in the United States, tended to promote research in regions of the cosmos beyond the solar system. It was more interested in sponsoring studies associated with astrophysics, the galaxies, and stars than those related to furthering the existing knowledge base of the solar system. Thus, the early capabilities of the DSN, designed for in situ observations of the inner solar system with planetary spacecraft, offered little of interest to radio astronomers dependent on the NSF for their sponsorship.

In 1966, the first of the DSN 64-m antennas was put into operation at Goldstone. With that event, radio astronomy, as a scientific endeavor distinct from that associated with planetary spacecraft, became a new capability for the DSN. The new antenna, over six times more sensitive than the existing 26-m antennas, more than doubled the range of the Network and, in doing so, provided the best instrument available for radio astronomy at that time. With such an instrument, radio astronomers could now direct their research to regions beyond the solar system with a capability that was comparable to, and in many ways better than, that of the other large radio telescopes throughout the world. Prompted by a strong advocacy from Nicholas A. Renzetti, who was managing tracking operations for the DSN at that time, the NASA Office of Tracking and Data Acquisition (OTDA) approved the allocation of a limited amount of time on the new antenna for radio science observations on a noninterference basis with the antenna's prime task of tracking planetary spacecraft. Official notice of the availability of the new facility brought an immediate response from the radio astronomy community.

In 1967, the DSN was approached by David S. Robertson of the Australian Space Research Group with the first proposal for a long baseline interferometry measurement of extragalactic radio sources. This was to be done using the baseline between Goldstone and

DSN antennas at Woomera and Tidbinbilla in Australia, which would make a resolving power of 2×10^{-3} arc-second possible. With Allen T. Moffet as a coexperimenter, the first observations were made in 1967 and continued at intervals for several years thereafter.[13]

In 1968 and 1969, DSS 14 was used extensively by Moffet of Caltech and Ekers and Downs of JPL to study pulsar emissions. By 1969, the volume of requests for observing time on the antenna had increased to the point where it became necessary to evaluate and select the most appropriate and worthy non-NASA proposals to compete for the limited amount of available antenna time. A Radio Astronomy Experiment Selection Panel (RAES), composed of distinguished radio astronomers nationwide, was established for this purpose, and a DSN radio astronomy support group was created to facilitate the technical and logistical interfaces between the astronomers and the operational DSN. These organizations were announced to the radio science community in the technical literature at the end of 1969.

The next two years saw a substantial increase in the demand for radio astronomy time at DSS 14 and DSS 13 as pulsars, very long baseline interferometry, Jupiter radio emissions, brightness, temperature, and rotation period continued to attract the attention of radio astronomers.[14]

Management policy for support of "Ground-Based Radio Science" in the Deep Space Network was formalized by NASA in 1971 with an appropriate NASA Management Instruction setting forth the requirements for submission of requests for experiment support and the levels of approval required. A level of support was determined in terms of percentage of available antenna time and assigned to the RAES panel, and its subsequent equivalent for distribution amongst approved experimenters. To distinguish between the radio experiments involving spacecraft and those involving celestial bodies, NASA adopted the following definitions.

"Radio science" would refer to the acquisition and extraction of information from spacecraft-originated signals that had been affected by celestial bodies or the propagation media. "Radio astronomy" would refer to the acquisition and extraction of information from signals that had been emitted by or reflected from natural (non-spacecraft) sources.

In the past, all experiments had been loosely referred to as "radio science," but from this point forward, the DSN would observe these new conventions in its management and support for radio science and radio astronomy. Although at that time the important planetary radar observations being made at Goldstone came within the definition of radio astronomy, it would later evolve into a separate activity as the Goldstone Solar System Radar (GSSR).

For the next decade and a half, radio astronomy continued to play an important role in the total sum of DSN activity. As the DSN capability increased with the construction of 64-m antennas in Australia and Spain, and later with their enlargement to 70-m diameter, so did the capability for radio astronomy, and the scientists were quick to use them to advantage.

The level of radio astronomy activity in the DSN over the period 1971 through 1987 is summarized below:

Radio Astronomy in the Deep Space Network: 1967–87*

1967–71
    Various Sponsors, NSF, CIT, JPL, Australia
        Twenty experiments, various subjects
1972–87
  RAES-Sponsored Research
        Sixty discrete experiments in various subjects
  NASA-Sponsored Research
        Five continuous experiments in the following:
                Interstellar microwave spectroscopy
                Planetary radio astronomy
                Hipparcos VLBI
                Quasar and galactic nuclei VLBI
                Southern Sky Survey with Tidbinbilla Interferometer

**The details of this work were collected and published by Renzetti and others in 1988.*

INTERNATIONAL COOPERATION

While individual radio astronomers in the United States continued to carry out their work through the scheduling and interface organization in the DSN Operations Office at JPL, associations of a more international nature were also formed during this period.

In 1983, the Goldstone 70-m antenna cooperated with the National Radio Astronomy Observatory 43-m antenna and the 22-m antenna of the Center for Astrophysics in the Crimea, USSR, in an interferometric network to study extragalactic sources at a wavelength of 3 cm with high resolution. Thus was forged the DSN's first close cooperative association with radio astronomers from NRAO, Cornell, Caltech, and the Soviet Union. This association would continue to be productive in the field of radio astronomy research in the years that followed.

Tidbinbilla Interferometer

In 1972, Samuel Gulkis of JPL, in association with David L. Jauncey and Michael J. Yerbury, then of Cornell University, put forward a proposal that would use the two existing DSN antennas at the Canberra Deep Space Communication Complex in Australia to create a high-sensitivity, phase-stable interferometer that could be operated in near-real time. The two antennas (26 m and 64 m in 1972, later enlarged to 34 m and 70 m) were separated by 195 meters on a north-south baseline and, together with available low-noise maser amplifiers and a common operating frequency at 2.3 GHz, provided a unique opportunity to implement a high-sensitivity interferometer with a positional capability of approximately 2.0 arc-seconds at minimal cost. It would be the only instrument in the Southern Hemisphere capable of carrying out rapid measurements of weak radio source positions with arc-second accuracy and flux densities to the millijansky level. These data would be used for unambiguous identification of radio sources with cataloged infrared and optical celestial objects. Earlier, whole (southern) sky surveys made with the Parkes Radio Telescope at 2.7 GHz had listed over 12,000 radio sources, but these data did not have sufficient positional accuracy to permit confident identification with visible objects and left unanswered many questions of interest to radio astronomers. The proposed interferometer would be used to supplement these data and provide a complete southern sky survey.

The Tidbinbilla two-element interferometer was constructed and tested in the years 1977 to 1979 and began radio astronomy observations at 2.3 GHz in 1980.[15] It was upgraded to operate simultaneously at 2.3 GHz and 8.4 GHz in 1986.

By 1988, over 1,500 of the radio sources listed in the Parkes survey had been measured, and many had been identified optically. A great deal of attention was paid to the study of quasars. The identification in 1982 by Jauncey and others of the most distant quasar discovered to that time was attributed solely to the accuracy of the Tidbinbilla interferometer.[16] A sensitive search of the brighter stars in the Southern Hemisphere for possible radio emissions was initiated in 1986, depending on the narrowness of the interferometer beam to reduce confusion with the background galactic sources of radiation. Data like these would be used by the Space Telescope and other astronomical spacecraft (Hipparchus) as well as for navigation of conventional spacecraft and station location purposes.

The Tidbinbilla interferometer continued to operate as a significant capability for radio astronomy in Australia under the Host Country program described below.

Orbiting VLBI

The feasibility of using a spacecraft in Earth orbit as one element of a very long baseline interferometer for radio astronomy observations was demonstrated by Gerry Levy and his team in 1983.[17] Building on these ideas, James Ulvestad, Roger Linfield, and Chad Edwards applied these principles to make successful observations of Extra Galactic Radio Sources (EGRSs) in August 1986. The enormous increase in baseline length (several Earth diameters) thus obtained would provide a proportionate increase in resolving power, greatly exceeding that of an interferometer constrained by Earth-based antennas. The experiment was conducted at a frequency of 2.3 GHz using the 4.9-m antenna of the geosynchronous NASA Tracking and Data Relay Satellite System (TDRSS) as the space-borne element. The DSN 64-m antenna at Canberra, Australia, and the ISAS 64-m antenna at Usuda, Japan, formed Earth-based elements. In January 1987, 23 EGRSs were detected on a baseline as long as 2.15 Earth diameters.

These successful experiments demonstrated that a ground-based frequency reference could be accurately transferred to an orbiting spacecraft and that observations of radio sources acquired by an orbiting spacecraft could be transmitted and recorded on Earth with the necessary precision to achieve correlation with similar data from Earth-based antennas. The success of these early experiments had positive implications for the future of orbiting VLBI.

Within the next ten years, the DSN would become a participant in the Space VLBI cooperative program involving NASA (U.S.A.), ISAS (Japan), and ASC (Russia). The first of two space VLBI missions, VSOP (HALCA), would be launched in 1997 and a special subnetwork of 11-m tracking antennas, operating at Ka-band, would be added to the Network. The DSN 70-m antennas would also participate as co-observers in the Space VLBI project with many other radio astronomy observatories around the world.

Host Country Programs

The original intergovernmental treaties that were executed between the United States, Spain, and Australia for the implementation and operation of Deep Space Communications Complexes in those countries made explicit provision for the conduct of independent radio astronomy research on these facilities by Host Country agencies on a noninterference basis with the prime NASA business of spacecraft tracking. In the DSN resource planning process, "Host Country" programs were allotted a specified amount of antenna time on an annual basis, and it was left to the local complex directors to assign the time to the individual investigators as appropriate.

In Spain, the prime cooperating agencies were the Instituto Geografico National and the Instituto Nacional Tecnica Aeroespacial, while in Australia, the prime cooperating agency was the Commonwealth Scientific and Industrial Research Organisation (CSIRO). In both countries, the DSN antennas participated in interferometric geodetic surveys of the continents and observation of pulsars under the Host Country program. In 1992, the Tidbinbilla interferometer was brought into the Host Country program and continued to share antenna time with other Australian radio astronomy investigations at the CDSCC.

CROSS-SUPPORT AGREEMENTS

The important role played by the Parkes Radio Astronomy antenna in supporting the Voyager Neptune Encounter as an element in the Parkes Canberra Telemetry Array (PCTA) was discussed in an earlier chapter. The NASA/CSIRO negotiations for this support in September 1986 resulted in a Memorandum of Understanding. This document provided time on the DSN 70-m antenna for the use of CSIRO radio astronomers on a quid pro quo basis for the time provided by Parkes for the Voyager Encounter. This "cross-support" agreement was a productive source of radio astronomy science, although the over-subscription of the DSS 43 antenna for spacecraft support in later years made it difficult for the DSN to meet the original expectations for "payback."

GUEST OBSERVER PROGRAM

Under the Guest Observer Program, DSN facilities were made available to non-JPL scientists for radio astronomy research, subject to peer review and approval of their proposals by the U.S. National Radio Astronomy Observatory (NRAO). The purpose of the review was to establish the scientific merit of the proposal and verify its full dependence on the unique capabilities of the Deep Space Network for success. Similar review processes applied to the use of DSN facilities for cooperative VLBI experiments with other agencies in the United States and Europe.

ANTENNA UTILIZATION FOR RADIO ASTRONOMY

The numerous agreements, directives, and requirements for radio astronomy support, accumulated through the intervening years, were brought up to date by Larry N. Dumas, director of the TDA office at JPL in 1990. With the agreement of the program office at NASA Headquarters, Dumas established new limits for the DSN antenna time that would be applied to radio astronomy throughout the Network.

NASA- and RAES-approved experiments would not exceed 3 percent of the antenna time available at any operational DSN station. NASA/CSIRO cross-support experiments using DSS 43 would receive a minimum of 108 hours per year through August 1999, with a 34-m antenna-hour equivalence of two for one. Host Country experiments would not exceed 3 percent of the time available at any one of the stations available at each complex (approximately 250 hours per year).

The new guidelines formed the basis for scheduling DSN resources for radio astronomy support thereafter.

A summary of antenna utilization for the ensuing ten years, given in the table below, reveals how well the guidelines were followed in the operational environment that prevailed through the Network in those years.

The figures represent hours of observing time on the 70-m and 34-m DSN antennas and represent the fulfillment of the 1988 guidelines discussed above.

Radio Astronomy in the Deep Space Network: 1987–97

(Observing Antenna Hours)

Year	JPL/NASA	Guest	Other*	Total
1997	1,999	253	712	2,964
1996	2,735	129	4,691	7,555
1995	1,884	115	1,201	3,200
1994	1,196	531	1,316	3,043
1993	1,379	234	1,086	2,699
1992	1,947	217	1,496	3,660
1991	1,733	128	887	2,748
1990	1,189	147	573	1,909
1989	1,491	57	199	1,747
1988	793	19	13	825
1987	667	35	203	905

* Included Host Country, Space Geodesy, CSIRO Cross-Support

The unusually large amount of time devoted to other institutions and JPL/NASA programs in 1996 reflects a substantial JPL/CSIRO effort devoted to the development of a new wideband spectrum analyzer for radio astronomy at Canberra and Goldstone, in addition to normal observation time.

The DSN Also Benefits

Without question, the science of radio astronomy has benefited enormously from the resources made available to it by the Deep Space Network since the advent of the first 64-m antenna in 1966.

However, radio astronomy was not the only beneficiary of this association with the DSN. There was a less obvious, but no less substantial, benefit that accrued to the DSN from its association with radio astronomy.

In a 1994 white paper on the role of radio astronomy in the DSN, Thomas B. Kuiper, then DSN Program Manager for radio astronomy in the DSN Science Office, addressed this subject.[18] He cited the following examples of technology, methodology, and operational innovation that, currently used in various forms by the DSN, were attributable to the use of the DSN by radio astronomers:

- digital spectrometers (Goldstein, 1963),
- very long baseline interferometry (Moffet, Ekers, Robertson, 1967),
- antenna calibration with natural radio sources (Klein, 1972),
- connected element interferometry (Gulkis, Jauncey, 1979),
- antenna pointing models (Peters, Ruis, 1982),
- K-band masers (Gulkis, Kuiper, Jauncey, 1982), and
- Internet service to deep space Complexes (Kuiper, 1993).

In addition to its scientific merit, radio astronomy, by its association with other prestigious scientific institutions and the technical challenges it brought with it, stimulated the interest of many talented professionals in all areas of the DSN. DSN capabilities were extended; seldom-used functions were exercised and maintained; and new developments were field-tested in the less demanding risk environment of radio astronomy. It provided an intellectually rewarding, nonoperational environment in which to explore new ideas and new technological development. As Kuiper pointed out, "Radio astronomers would accept lower reliability to achieve performance improvement, thus giving the DSN the valuable opportunity of evaluating and developing experience with new technologies in noncritical operations."

At that time (1994), Kuiper continued, "the key DSN technologies of antennas and receivers and radio astronomy were nearing maturity, and the next dramatic innovations would come from other technologies, such as information management, automation, and networking. As applied to radio astronomy in the DSN, these technologies could lead to unattended observations, remote operation and monitoring, and noninterference with tracking operations. Because of lower reliability requirements, these new techniques could be rapidly integrated into the radio astronomy capability of the DSN for evaluation, before adapting them for operational use." That appeared to be the pathway to the future for radio astronomy in the DSN, and so it proved to be.

By 1997, the DSN had created a world-class facility for radio astronomy research.[19] When operated at their highest potential frequencies, the DSN antennas equaled or exceeded the capabilities of the best radio telescopes of the time. A well-documented site on the World Wide Web encouraged potential investigators worldwide with a plethora of technical descriptive material, along with interactive instructions to propose experiments using the radio astronomy resources of the Deep Space Network.

NOTE: *The DSN resources available to potential investigators in radio astronomy were described online at* http://dsnra.jpl.nasa.gov/.

As an instrument for radio astronomy research, the DSN was highly regarded by the international science community and had become a visible tribute to NASA's continuing commitment to international science. Encouraged by the radio astronomers that used it, the DSN continued to improve in the knowledge that radio astronomy research would quickly adapt new opportunities in deep space communications technology to further its scientific objectives.

SIGNIFICANT ADVANCES

In the role of a scientific instrument for radio astronomy research, the DSN contributed much to the field over the years 1967 to 1997. As a measure of this contribution, the DSN lead radio astronomer, Thomas Kuiper, selected the following published papers as representing significant advances in science over the period 1967 to 1997.[20] Incidentally, these examples reflected not only advances in scientific knowledge, but advances in the DSN's capability to collect and process the data on which such achievements were based.

Significant advances in radio astronomy using DSN antennas include the following:

1967: First intercontinental very long baseline interferometry (VLBI).
Fine details in the structure of radio sources can be resolved by combining the signals from distantly separated radio telescopes (VLBI). The greater the separation, the finer the details. Gubbay and others performed VLBI measurements using baselines of intercontinental dimensions. Prior to this, VLBI had been limited to telescopes in North America.[21]

1969: Pulsar observations at 2,295 MHz (13-cm wavelength), highest frequency to date.
Moffet, Ekers, and others used the Goldstone 64-m antenna at S-band (2,295 MHz) to determine the polarization characteristics of a number of pulsars. At higher frequencies, pulsar signals are less affected by the interstellar medium through which they pass on their way to Earth.[22]

1969: Detection of the first jump in the period of the Vela pulsar.
Reichley, Downs, and others detected a sudden increase in the period of the Vela pulsar while monitoring the period of a set of pulsars with the Goldstone 64-m antenna. They attributed the jump to the settling of the matter of the neutron star, similar to an earthquake.[23]

1971: Discovery of superluminal motions in 3C273 and 3C279.
By comparing the images made at two different times using the VLBI technique (see 1967), Clark and others showed that matter in these objects appeared to be moving with a speed greater than the speed of light. Later, this was understood as an illusion due to the orientation of matter flow and the effect of relativity.[24]

1973: Best upper limit to date on cosmic background small-scale anisotropy.
The currently clumpy distribution of matter in the universe, in stars and galaxies, and clusters of galaxies evolved from a very smooth initial distribution according to radio astronomer Carpenter. Explaining this remained (in 1997) a challenge for the cosmologists.[25]

1982: Most distant quasar discovered with Tidbinbilla interferometer.
Using a radio interferometer consisting of two antennas at NASA's Canberra (Tidbinbilla) Complex, Batty, Jauncey, and others determined precise positions for a large number of radio sources. This allowed these sources to be associated with optical objects which were then identified. One of them turned out to be the most distant quasar known to that time.[26]

1986: First space-based VLBI with TDRSS antenna.
As a proof of concept, Linfield, and others used the DSN antennas with a NASA Tracking and Data Relay Satellite to make VLBI (see 1967) observations of three quasars. In addition to obtaining finer detail than ever before, this demonstrated that VLBI baselines could be larger than the diameter of Earth.[27, 28]

1992: First speckle hologram of the interstellar plasma.
   Desai, Gwinn, and others used an array of Southern Hemisphere antennas, including two at NASA's Canberra Complex and two former DSN antennas (Hartebeesthoek, South Africa, and Hobart, Australia) to observe the Vela pulsar. Correlations in the time variation of the Vela data from the various stations were analyzed to obtain information about the structure of the interstellar medium.[29]

1995: First images of motion in the nearest active galactic nucleus.
   Over a period of eight years, Tingay, Jauncey, and others carried out a series of VLBI observations using telescope arrays including NASA 70-m antennas. Through these observations it was learned that jets of plasma flow out of the core of Centaurus A. The behavior of this core was thought to be typical of other, more distant active galactic nuclei.[30]

1995: First image of infall of cloud material onto a protostar.
   A protostar is a star that is still growing by accreting matter from the cloud out of which it formed. Vellusamy, Kuiper, and Langer, by combining data from the Goldstone 70-m antenna with data from the Very Large Array, made images of the material falling onto the protostar B335. Neither telescope by itself could have produced such images.[31]

1996: First evidence for the formation of a preprotostellar core by coagulation.
   By comparing Goldstone 70-m data with VLA data, and comparing the results with those obtained with other telescopes, Kuiper, Langer, and Vellusamy found the cloud core L1498 to be growing by the process of accreting material. The cloud was not collapsing like B335 (see above), but was thought to be growing towards a stage when collapse would occur. Prior to this, it was not known whether prestellar cores grew by slow accretion of surrounding material.[32]

1996: Nano-arcsecond resolution of the Vela pulsar.
   Using interstellar scattering data derived from further analysis of the speckle data obtained in 1992 (see above), Gwinn and others determined the size (approximately 500 km) and approximate shape of the emitting region of the Vela pulsar.[33]

1997: First VLBI astrometric detection of an unseen companion star.
   Guirado, Reynolds, and others used the 70-m antenna at NASA's Canberra Complex, combined with the (former DSN) 26-m radio astronomy antenna at Hobart, Australia, and the 64-m radio astronomy antenna at Parkes, Australia, to make precise measurements of the position of the star AB Doradus. A wob-

bling in the positions of the star revealed the presence of a dark companion with about a tenth of the mass of the Sun.[34]

1997: First images from the HALCA Space VLBI mission.
The Japanese HALCA space telescope was the first dedicated orbiting VLBI antenna. Its orbit was chosen to provide the longest baselines to that date. The NASA 70-m antennas, because of their size and geographic locations, played an important role as a ground element in the HALCA mission to produce extremely sensitive and detailed images of compact radio sources.

THE SEARCH FOR EXTRATERRESTRIAL INTELLIGENCE (SETI)

In the late 1970s, against a background of serious scientific interest in the detection of extraterrestrial intelligence dating back to the early 1960s and strong advocacy by scientists at Jet Propulsion Laboratory and the Ames Research Center, NASA initiated a formal program to search for evidence of extraterrestrial intelligence. Under the acronym SETI (Search for Extraterrestrial Intelligence), the program was to be managed by the Astrophysics Division of the NASA Headquarters Office of Space Sciences (OSS).

SETI was to be a research and development program aimed at demonstrating the necessary technology to search a well-defined volume of sky and frequency space, within a finite time, for evidence of intelligent signals of extraterrestrial origin. It was to use existing antennas augmented with new spectrum analysis and data-processing systems.

In June 1980, TDA progress reports by Robert E. Edelson and Gerry S. Levy[35] describing the telecommunications technology associated with SETI, and by Berman[36] describing the SETI observational plan, reflected the growing involvement of the TDA Office in the new program. Both of these reports drew attention to the dependence of the proposed SETI program on the use of existing large antennas, ultra-sensitive microwave receiving systems, identification of terrestrial sources of radio frequency interference (RFI), and wideband spectrum analysis and data management technology, most of which was to be found in the existing Deep Space Network. Although not explicitly stated, there was also perceived to be a symbiotic relationship between the new technology that would be needed for SETI and that which the DSN would use for wideband signal acquisition, and RFI and spectrum management. These were the principal factors that established a basis for the DSN association with the SETI program.

Supported by funding from the NASA Office of Space Science and, to a lesser extent by funding from the NASA Office of Tracking and Data Acquisition, the SETI program at JPL was carried forward as a parallel but independent program to that of the Deep Space Network. Within JPL, the SETI program was managed by the TDA Science Office under the direction of Nicholas A. Renzetti.

In the overall plan, two parallel but complementary strategies were to be used to conduct the SETI observations—a "targeted survey" and an "all-sky survey." The targeted survey would target approximately 800 to 1,000 candidate stars within 100 light-years of Earth. It would search for pulsed or continuous signals in the 1- to 3-GHz frequency band and would require high-sensitivity, narrow-bandwidth data acquisition systems that would be used with the 64-m DSN antennas and the NRAO 305-m antenna at Arecibo, Puerto

Rico. The "all-sky survey" would trade lower sensitivity for the ability to survey 99 percent of the sky that was not covered by the targeted search, and it would span a wider frequency range, from 1 to 10 GHz. It would be conducted on the DSN 34-m antennas.

Working within these broad objectives and guided by a SETI Science Working Group that had been set up by NASA at the outset of the SETI program, engineers at JPL began to develop the elements of hardware and software based on Fast Fourier Transform (FFT) techniques that would be required for both strategies.[37] It should be made clear at this point that the engineering groups doing this work for the SETI program were, at the same time, doing similar work for the DSN. Although the funding sources were different, the organizations were one and the same, and the benefits of the DSN technology base obviously accrued to the SETI program.

The objective of this effort was to be a field demonstration at Goldstone, using the DSS 13 antenna and known sources of radiation such as NASA planetary spacecraft.

The first step in this direction was to assess the radio frequency interference (RFI) environment in which the sky survey would be carried out. The resulting data base would reveal the incidence of false signals and the presence of interfering signals and noise—information that would be essential to the design of the final signal-detection, analysis, and data-processing equipment for SETI.

For this purpose, a Radio Spectrum Surveillance System was built, mounted in a transportable van with a 1-meter-diameter antenna, and deployed at Goldstone in 1984.[38] The system comprised seven low-noise Field Effect Transistors (FETs) to cover the range of frequencies 1–10 GHz; a commercially available, programmable, swept-frequency spectrum analyzer; and a floppy disc storage device. These were all controlled by a BASIC software program. In addition to its primary use by the SETI program, the RFI van proved to be a remarkably effective facility that was subsequently used for many years by DSN spectrum management engineers to continuously monitor and identify RFI in the deep space bands at Goldstone.

The initial field demonstrations for SETI, using FFT-based hardware and software, began at DSS 13 in March 1985.[39] The field-test instrumentation consisted of the DSS 13 26-m antenna with low-noise receiving systems for the 2,200- to 2,290-MHz and 8,400- to 8,500-MHz bands. It also included a 65,000-channel FFT spectrum analyzer, designed and built by JPL and housed in the RFI van, which was parked nearby. A 72,000-channel Multichannel Spectrum Analyzer, designed and built for ARC at Stanford University for the targeted search program, was also part of the field demonstration. The detection strategy and configuration used for the sky survey field tests is shown in Figure 8-4.

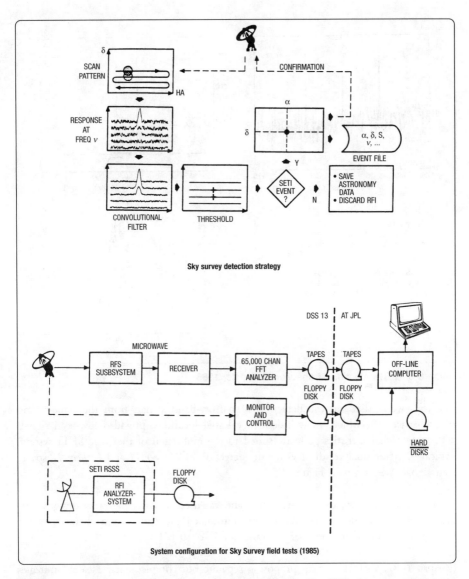

Figure 8-4. Detection strategy and system configuration for the SETI sky survey field tests, 1985.

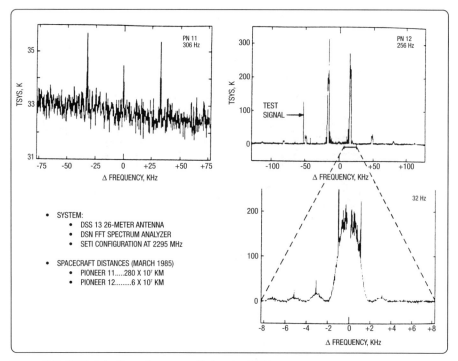

Figure 8-5. SETI detection of spacecraft signals, 1985.

The tests were based on the detection of narrow-band signals from the *Voyager* and *Pioneer* spacecraft, which, in well-defined celestial locations, provided low-level signals with known characteristics perfectly suited to the evaluation of the basic SETI system. Typical spectrum analyzer displays for the detection of *Pioneer 11* and *Pioneer 12* spacecraft signals are shown in Figure 8-5.

For the next two years, the field-test program was used to verify hardware designs, further develop signal-detection strategy and automated procedures, and characterize the RFI environment at the site over the range of 1 to 10 MHz.

In April 1987, a program plan proposing separate but complementary search strategies was presented to NASA by representatives of the SETI program from ARC and JPL. The two Field Centers jointly proposed to develop and operate observing systems using high-speed signal-processing equipment that could be used in conjunction with existing ground-based radio astronomy and deep space communication facilities. The targeted search was to be the responsibility of ARC, while JPL would be responsible for the all-sky survey. ARC would act as the Lead Center for the program, which would be managed by the Life Sciences Division of OSS with important technical and financial support from the Office of Space Communications (OSC, formerly OTDA). The DSN would provide support on its 34-m antennas for the all-sky search and time on its 70-m antennas for those parts of the targeted search that could not be provided by the Arecibo antenna or the NRAO antenna. With this decision made, the JPL team focused its efforts on the development of equipment and techniques that were optimized for the all-sky survey. Henceforth, the newly directed program would be known as the High Resolution Microwave Survey (HRMS).

Building on search strategies and FFT techniques that had been developed while the initial demonstration field tests were in progress at Goldstone,[40] a more refined signal detection strategy for an operational HRMS all-sky survey evolved, and a much larger, wide-band, high-resolution spectrum analyzer (WBSA) had been designed.[41] These and the observing procedures derived from the field tests experience at Goldstone would be assembled into a prototype system and used for a limited proof-of-concept search program to be conducted in the early 1990s. Experience gained from these proof-of-concept tests would be used to improve the design of the full-scale operational system, which was then scheduled for completion in 1996.

The objective of the all-sky survey was to search the entire sky over the frequency range from 1.0 to 10.0 GHz for evidence of narrow band signals of extraterrestrial origin. Frequency resolutions as narrow as 20 Hz were required. The objectives of the survey, as presented by Coulter Klein and others are summarized in the table below.

Characteristics of the All-Sky Survey for a 34-m Antenna

Spatial coverage	4 pi steradians (full sphere)
Frequency coverage	1–10 GHz with higher spot bands
Duration of survey	5–10 years (7-year goal)
Dwell time per source	0.5 to 10.0 seconds
Frequency resolution	10–30 Hz
Sensitivity, (allowing for spatial uniformity and variation with frequency)	Approx. 10^{-23} watts/meter^2
Polarization	Simultaneous dual circular
Signal type	Continuous Wave (CW)

To accommodate the limited observing time that would be available on the DSN 34-m antennas for HRMS, an antenna scan pattern of "racetrack" form was developed. The racetrack scan pattern utilized the available antenna time efficiently, maintained maximum sensitivity with uniformity across each sky frame, and facilitated the signal detection process. Specific features of this pattern could also be used to discriminate against certain types of RFI.

In collaboration with the DSN Advanced Systems Program, an engineering development model of a wideband spectrum analyzer (WBSA) was built to serve as a prototype for the final operational model. This machine was key to the design of the HRMS system.

The WBSA featured a pipelined FFT architecture that transformed 40 MHz of input bandwidth into 20-Hz wide bins in the frequency domain. To accommodate the wide bandwidth in real time, the WBSA performed 4.5 billion operations per second and used two input channels to search simultaneously for left and right polarized signals. The output of the WBSA was passed to a signal-detection module that performed baseline estimation, filtering, and thresholding functions. Although the prototype system bandwidth was limited to 20 MHz, plans were made to increase this to 320 MHz in the final operational design in order to complete the survey in 7 years. It would have taken 100 years to complete the survey with a 20-MHz bandwidth.

Progress in developing the two parallel systems at JPL and at ARC was such that on 12 October 1992, the observational phase of HRMS was inaugurated with simultaneous, coordinated observations of a selected sky frame by the all-sky system at DSS 13, and a star within that frame by the targeted-search system at Arecibo. The occasion was marked with appropriate pomp and circumstance, and keynote addresses were delivered by SETI pioneers Philip Morrison (at Arecibo) and Carl Sagan (at Goldstone). The use of the recently completed 34-m BWG antenna at DSS 13 added to the significance of the event.

A view of the eastern sky as seen from Goldstone at noon on that day is shown in Figure 8-6. The shaded areas define the boundaries of the first two sky frames of the HRMS Sky Survey. The location of the first star observed by the Targeted Search Team at Arecibo is shown in sky frame number 1.

The initial surveys were conducted near 8.5 GHz to characterize the system operating performance and determine the RFI environment and the system's ability to discriminate against it. Later, repeated observations of a set of three sky frames at other, lower frequencies provided data of interest to the radio astronomy community, as well as the

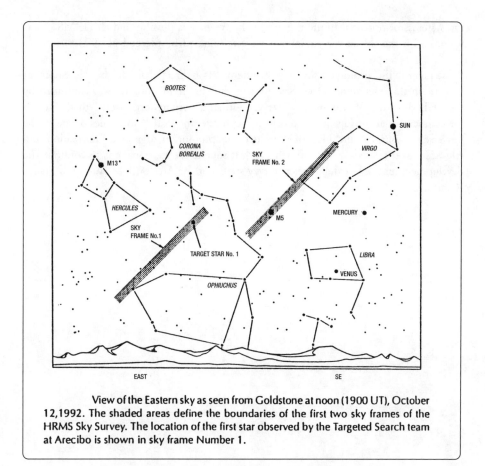

View of the Eastern sky as seen from Goldstone at noon (1900 UT), October 12, 1992. The shaded areas define the boundaries of the first two sky frames of the HRMS Sky Survey. The location of the first star observed by the Targeted Search team at Arecibo is shown in sky frame Number 1.

Figure 8-6. The first HRMS sky survey, Goldstone, 12 October 1992.

HRMS program, and allowed further improvement in the efficiency of the signal-processing algorithms that reduced the effects of RFI. During the first year of observations, no signals from beyond our solar system were detected by either the sky-survey or the targeted-search systems. However, many valuable lessons had been learned and were about to be incorporated into the full operational system. Confidence in and enthusiasm for the HRMS program were high.

Suddenly, in October 1993, the U.S. Congress unexpectedly terminated the program on the grounds of financial expediency. Apart from preserving the advanced signal-processing systems for potential future use, no further work was possible and all HRMS observations with the sky-survey equipment were terminated by the end of 1993.[42]

Subsequently, the prototype sky survey system was redirected for further development and use in the operational Deep Space Network. The SETI initiative was continued by the SETI Institute of Mountain View, California, a corporation devoted to the search for extraterrestrial intelligent life and to research relevant to the disciplines of life in the universe. Its purpose included the conduct and promotion of public information and education as it related to SETI. To this end, it sought private donations to continue the development and use of the targeted survey system at the Arecibo Observatory in Puerto Rico.

CRUSTAL DYNAMICS

By the early 1970s, very long baseline interferometry (VLBI) was well established in the DSN as a technique for spacecraft navigation, radio astronomy observations of the structure and position of Extra Galactic Radio Sources (EGRS), and the determination of baseline vectors between fixed reference points on Earth (DSN tracking stations). Success in these areas gave rise to an increasing interest by the science community in applying VLBI techniques to accurate measurements in the fields of geodynamics and geodesy.

However, the fixed antennas of the DSN were not suited to the conduct of an extensive geodetic monitoring program that would require frequent and accurate measurements along many reference baselines. For this purpose, a mobile facility, which employed the same VLBI techniques but possessed an added capability for rapid deployment to numerous selected sites of geodetic interest, was required.

To this end, in 1971, the NASA Office of Applications funded a research program to demonstrate the capabilities of portable, independent-station radio interferometry. Because of its relationship to well-established DSN technology and its dependence on the 64-m DSN antenna at Goldstone as a reference point, the program was assigned to JPL and was managed by the Telecommunications and Data Acquisition Office, with Nicholas A. Renzetti as the Program Manager. It was named Astronomical Radio Interferometric Earth Surveying (ARIES) and was led by P. F. MacDoran.

By refurbishing a surplus 9-meter transportable antenna and implementing the necessary instrumentation in trailer-mounted vans, a transportable ARIES facility was rapidly assembled and, by the end of 1973, was ready to start field operations at Goldstone.

The initial series of experiments were performed by Ong, MacDoran, and others, on a short 307-meter baseline over the period from December 1973 to June 1974, using ARIES as one observing station and DSS 14 as the other. The arrangement is illustrated in Figure 8-7.

Use of a short baseline for the initial tests enabled a comparison of conventional survey methods with few-centimeter accuracy to minimize interferometry errors due to transmission media effects, source locations, and Earth orientation parameters. The results, representing approximately 28 hours of data, were in excellent agreement with the survey baseline in all dimensions within the formal uncertainty of 3 cm.[43] Over the next five years, ARIES was moved to various locations in the western United States to refine

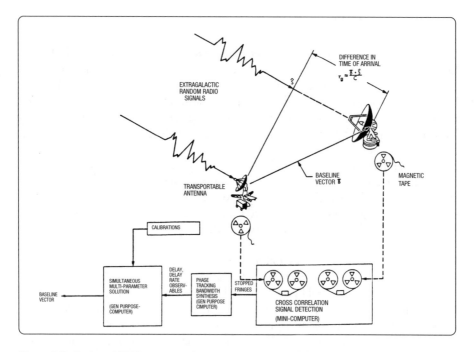

Figure 8-7. Project ARIES transportable antenna interferometer system.

and demonstrate the feasibility of the mobile VLBI concept and to initiate studies of crustal motion on a regional scale.

The success of this program eventually led to a requirement for two additional mobile VLBI units and a dedicated base station for the Crustal Dynamics project. The design and construction of the mobile units and implementation of the Mojave Base Station at Goldstone were carried out by Brun, Wu, and others as part of the ORION project by the DSN in 1982.[44, 45]

The Mojave Base Station utilized a redundant 12.2-meter antenna and Control Room built in 1962 to support a communications satellite program operated by the Goddard Space Flight Center. In June 1983, after major refurbishment by a joint JPL and GSFC engineering team, the Mojave Station, operated by Bendix contractor personnel, began to support the Crustal Dynamics project. The completion of this station relieved DSN stations DSS 14 and DSS 13 from the burden of supporting the Crustal Dynamics field observing program. The two mobile VLBI (MV) units, which used much smaller antennas and were more readily transportable than the original ARIES, started observing in 1980 and 1982, respectively.

Between 1980 and 1985, the three MV units, in conjunction with Mojave and several other fixed base stations in the western United States, were operated as part of a surveying program to determine the relative motions and regional strain fields near tectonic plate boundaries in California and Alaska. The results of this effort, and the difficulties encountered in realizing the necessary accuracy to deliver viable data, are described by J. M. Davidson and D. W. Trask, who led the effort for the TDA Office at JPL.[46] They reported that "baseline measurements utilizing the current Mobile VLBI systems had attained an accuracy of 2 cm, or better, in the horizontal plane" and observed that "since average geological rates of horizontal motion are on the order of 5 cm/yr across the plate boundary regions being studied, it was likely that crustal motion would be detected within the next few years, provided [it was] occurring at the geological rates."

The mobile VLBI project at JPL had reached the point where it could be transferred to its ultimate user, the National Geodetic Survey (NGS)/National Oceanic and Atmospheric Administration (NOAA) and Goddard Space Flight Center (GSFC). The transfer was completed in 1984 and 1985.

Even as the more economical, portable VLBI stations of the ORION task were under construction, technologists were experimenting with a new, potentially low-cost method for measuring Earth crustal deformation. This method utilized the L-band signals transmitted by the Global Positioning System (GPS), a constellation of 24 Earth-orbiting satellites under development in the 1980s by the Department of Defense. Unlike the weak quasar signals recorded by VLBI, the GPS signals broadcasted at 1.2 and 1.5 GHz were relatively strong and did not require 9-m antennas and low-noise receivers for their acquisition. A system based on GPS technology, therefore, offered the promise of high performance, low cost, and much greater mobility than the MV system. It would, in addition, have very significant application to the existing DSN for spacecraft navigation and other purposes related to station locations. With this in mind, the Geodynamics Program funded JPL for an experimental demonstration of the new GPS technique. It started with the Satellite Emission Range Inferred Earth Survey (SERIES) project in 1979, also under the leadership of Peter F. MacDoran.[47]

Under the SERIES project, two GPS receivers were built at JPL for proof-of-concept measurements on short baselines. These receivers measured the differential group delay of signals transmitted at two different frequencies from the GPS satellites. This information could be used not only to determine the location of the receiver on Earth's surface to a level of precision approaching that of the mobile VLBI system, but also to calibrate the ionosphere in terms of the total electron content (TEC) along the line of sight between the receiver and the satellite. The former function and other data derived from it was of particular interest to the Geodynamics Program for short-baseline geodetics measurements, while the latter function was of particular interest to the DSN for adding media calibrations to the radio metric data used for spacecraft navigation.

While the SERIES project successfully demonstrated the potential of the new GPS technology, it also showed the need for a more compact receiver with enhanced capability for deployment at multiple field sites. At that time, receivers that had a capability for parallel processing of signals from four (or more) satellites simultaneously and were sufficiently portable for field applications were not available commercially. It was this situation that led to the development of the SERIES-X receivers at JPL, starting in 1982 under the direction of William G. Melbourne.

The SERIES-X design was also very promising as an instrument for continuous, precise ionosphere calibrations at each DSN Complex. The Faraday Rotation satellites (ATS-6), which had been used for this purpose for many years, were rapidly being decommissioned. The GPS technique offered a cost-effective, more accurate means of calibrating radio metric tracking data. The DSN proposed to implement GPS receivers at each Complex and to develop the data retrieval system and software required to produce the calibrations for the inflight projects. An upgrade to the SERIES-X design was developed for this purpose. The receivers that were actually installed at the three complexes were built by a commercial partner, Alan Osborne Associates (AOA), and were referred to as the "Rogue Receivers."

Under the auspices of the NASA Geodynamics program, the Rogue Receivers were used for geodetics measurements and, as part of the JPL Crustal Dynamics project, to demonstrate the concept of a Fiducial Network.[48, 49] Within the DSN, they were used specifically to determine the total electron content of the ionosphere[50] for radio metric data calibration purposes.

Following the implementation of the Rogue Receivers in the DSN, further development of GPS receivers shifted to the Geodynamics program. The Rogue Receivers were produced commercially by AOA and were used in a number of field campaigns in the mid- to late 1980s. The first campaigns to use the GPS system were conducted in 1985 with baselines in Mexico and the Caribbean regions. Data from the DSN Rogue Network were used to establish a terrestrial reference frame for these experiments. It was these experiments which clearly demonstrated the economics of GPS Geodesy and sparked further development of GPS technology for the Geodynamics program.

As the SERIES-X program got under way, another sponsor for GPS technology emerged in the form of the TOPEX project. Managed by JPL and planned for launch in 1989, TOPEX was an Earth-orbiting satellite dedicated to the study of ocean topography. TOPEX carried, amongst other scientific instruments, a highly accurate radio altimeter for the measurement of ocean surface height (sea altitude). To meet the specified goal of 10-cm altitude accuracy, a very precise knowledge of the spacecraft orbit was required, and the use of the system then available would have resulted in an accuracy of ±14 cm.

In a proposal to the TOPEX Project in July 1984, the TDA Science Office described a system by which the accuracy could be increased to ±2.4 cm using GPS.[51] The proposal called for a flight-qualified GPS receiver and antenna aboard the TOPEX spacecraft, a SERIES-X (Rogue) ground receiver at each DSN Complex, and at least three ground receivers at other remote locations. It also included a data communications network and a central data processing facility to provide the TOPEX precision orbit determination functions. The underlying concept of the demonstration was that an inflight GPS receiver could, in conjunction with precisely located GPS ground stations, be used for highly accurate determination of orbiting satellites in an Earth-based reference frame.

Orbit determination for TOPEX, replacement of the Mobile VLBI stations for Crustal Dynamics, and ionosphere calibration for the DSN became the three principal drivers for further GPS-based development, and funding was shared accordingly by the respective program offices.

GPS-based technology was formally introduced into the operational capability of the DSN in November 1986 with a Functional Design Review for a new Media Calibration Subsystem for the DSCCs.[52] At each Complex, the new DSCC Media Calibration Subsystem (DMC) would perform continuous measurements of local ground weather parameters in addition to generating differential P-code group delay and carrier phase from the GPS satellites. These data were to be recorded and transmitted from the Complexes to the NOCC at JPL for the use of the Navigation Team in calibrating the radio metric data delivered by the DSN Tracking system. A description of the GPS system is beyond the scope of this book, but the essential elements as they applied to the DMC in 1986 are shown in Figure 8-8.

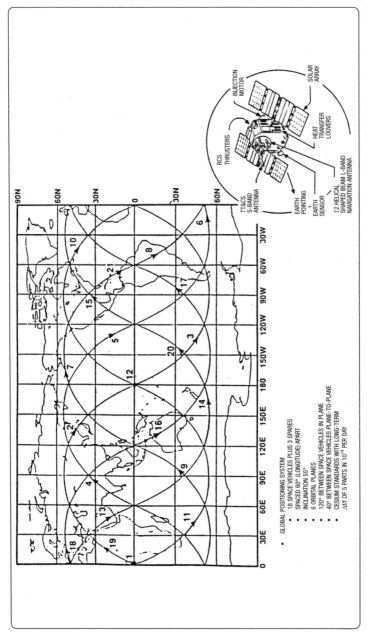

Figure 8-8. **Global Positioning System, 1986.** The GPS functions of the DMC were carried out by SERIES-X (Rogue) receivers, which acquired and simultaneously tracked the two GPS L-band carriers from at least four satellites, extracted the ephemeris data from each, and measured the differential P-code group delay and the carrier phase. Because of the sensitivity of the military to the use of the NAVSTAR system for other than classified purposes, these operations were performed without an explicit knowledge of the classified P-codes themselves.

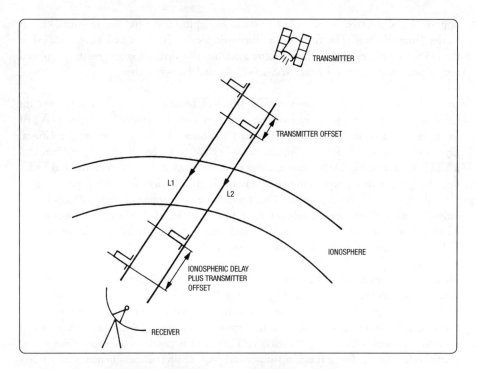

Figure 8-9. Schematic view of GPS-based ionospheric calibration.

The basic principle of GPS-based ionospheric calibration is illustrated schematically in Figure 8-9.

The remarkable performance of the receivers is reflected in the following estimates for their accuracy in determining TEC:

Performance of the DMC for Ionosphere Calibration Using GPS

Differential delay	Precision (el/m^2)	Accuracy (el/m^2)
P-code groups	1×10^{16}	5×10^{16}
L2-L1 carrier phase	0.02×10^{16}	0.02×10^{16}

NOTE: Total Electron Content (TEC) of the ionosphere is measured in units of electrons per square meter (el/m^2).

The DMC was put into operation at each of the three complexes in 1989 and immediately became the DSN standard source for ionosphere calibration data. Eventually, the original Rogue Receivers were replaced by smaller, laptop computer-size receivers, known as the Turbo-Rogues. The first of the Turbo-Rogues to be produced were installed at the DSN Complexes, and as they became available, the Turbo-Rogues gradually replaced the original receivers within the NASA GPS Global Network also.

By 1988, GPS technology sponsored by JPL had become a significant new tool for Geodynamics research on short baselines and as a complementary technique to VLBI for measurements on long (intercontinental) baselines. At the same time, the Crustal Dynamics project at GSFC was operating a Satellite Laser Ranging (SLR) System and a VLBI System (using DSN antennas as well as international VLBI stations) and had a large budget for that purpose. By comparison, the JPL budget for GPS applications to Crustal Dynamics was very small. With time, it became apparent that GPS could not only determine short baselines almost as accurately as VLBI, but was also approaching VLBI performance on the long baselines, and some reassessment of the relative value of GPS to the NASA Geodynamics program was called for.

To this end, a conference of all parties involved in the Geodynamics program was convened by the NASA program manager at Coolfont, West Virginia, in July 1989. For Geodynamics Solid Earth studies, the Coolfont conference recommended the establishment of a backbone consisting of a multicountry, intercontinental network of VLBI stations colocated with fixed GPS stations. The VLBI network would maintain a strong inertial reference for Earth rotation studies and provide precise measurements over intercontinental distances for the study of tectonic motion. The VLBI stations would also be complemented with a multicountry network of roving GPS stations for making measurements on short baselines for crustal motion studies.

Thus, Coolfont represented a major turning point for the Geodynamics program in general and the Crustal Dynamics project at JPL in particular. As a consequence of the agreements reached there, the Crustal Dynamics project at JPL was ended and GPS became a permanent and major part of the NASA Geodynamics program with JPL as the Lead Center for GPS technology. The VLBI technology was pursued by GSFC and involved JPL only to the extent that the VLBI program needed support from the DSN antennas.[53]

Under the cognizance of the NASA International Geodynamics Program, the field observing program thus evolved into two components, one based on VLBI and managed by GSFC, the other based on GPS and managed by JPL. The international VLBI network (which included but was not limited to the DSN 70-m antennas) comprised a set of permanent sites, each of which had a large radio astronomy antenna colocated with a fixed GPS receiver installation. The international GPS network was made up of a Network of portable GPS receivers installed and operated by the observing scientists themselves in areas of their specific interest.

So ended the Crustal Dynamics Project, the techniques for which had been founded and nurtured by the TDA Science Office for fifteen years. Out of this work had grown a burgeoning GPS technology, which, in addition to its proliferation in the Geodynamics program, was finding increasing applications in the operations of the DSN. The synergy that had developed between the Crustal Dynamics project and the DSN had given rise to significant improvements in the orbit determination processes on which the DSN, and users of the DSN, depended for spacecraft navigation purposes. Ionosphere and troposphere calibrations, Earth-orientation measurements, and intercomplex time synchronization were areas in which VLBI techniques, complemented by GPS techniques derived from the Crustal Dynamics work, had been most effective.

In recent years, the extension of GPS technology to DSN- and spacecraft-related problems, and to non-NASA applications, was sponsored by the TDA Technology Office and is reported in the previous chapter.

PLANETARY RADAR

In the context of science in the DSN, it can be stated unequivocally that "in the beginning, there was planetary radar." Planetary radar came before radio astronomy, before radio science, and before SETI or crustal dynamics became initiatives in the Deep Space Network. As a matter of fact, planetary radar preceded planetary spacecraft in the DSN. As we shall see, the first planetary radar experiment (Venus) took place on 10 March 1961, while the first successful planetary spaceflight (*Mariner 2*) took place on 27 August 1962.

Early in 1960, plans were being made at JPL for spaceflights to the inner planets, beginning with the Mariner mission to Venus. The DSIF, as it was known at that time, comprised four antennas with diameters of 26 meters (85 feet): two at Goldstone, California; one at Woomera, Australia; and one under construction at Johannesburg, South Africa. Teletype, telephone, and undersea cable communications links tied each of the antenna sites to a rudimentary Control Center at JPL.

The planetary missions presented many new challenges to existing technology, among which was a requirement for hitherto unprecedented accuracy in spacecraft navigation. To a large degree, spacecraft navigation depends for its accuracy upon the quality of the Doppler data provided by the tracking stations and a knowledge of the ephemeris of the planet and the value of the Astronomical Unit (AU). The DSIF was working on the Doppler data issue, but the accuracy of the Venus ephemeris and the value for AU, based on optical astronomy observations, left much to be desired for planetary navigation purposes. With this situation in mind, Eberhardt Rechtin, Walter K. Victor, and Robertson Stevens proposed to carry out a bold experiment during the 1961 conjunction of Venus, using existing facilities at Goldstone to refine the value of the Astronomical Unit and improve the ephemeris of Venus. Using radar techniques, the round-trip time required for a radio signal transmitted from Earth and reflected back from the surface of the planet to return to Earth would be measured very precisely. From these and ancillary data, the desired ephemeris and value for AU could be determined.

The experiment began on 10 March 1961, in the bistatic mode with one of the Goldstone 26-m antennas transmitting (Echo site) and the other receiving (Pioneer site). On 16 March, the Director of the DSIF, Eberhardt Rechtin, announced the success of the Goldstone Venus Radar Experiment in a message to his colleagues at the MIT Lincoln Laboratory, which had for several years previously been attempting to do the same thing, "HAVE BEEN OBTAINING REAL TIME RADAR REFLECTED SIGNALS FROM VENUS SINCE MARCH 10 USING 10 KW CW AT 2388 MC AT A SYSTEM

TEMPERATURE OF 55 DEGREES." The reply from Lincoln Laboratory offered hearty congratulations but reported no success with their own efforts to that time. Once unambiguous daily contacts with Venus were well established with the ranging system turned on, measurements of the Earth-Venus distance began and continued through mid-May 1961. Based on these results, Victor and Stevens announced their results to the scientific world in *Science*, July 1961.[54] The value for the AU, based on data from the first planetary radar experiment, was determined to be 149,599,000 with an accuracy of ±1,500 km. At that time, this represented a major improvement in accuracy compared to that obtained from optical and other radar-based experiments. After further refinement by D. O. Muhleman and others, the uncertainty was reduced to ±500 km by August and to ±250 km in the following year.

NOTE: *The International Astronomical Union adopted the value of 149,600,000 km in 1964, based on the radar method first demonstrated successfully at Goldstone in 1961. For a detailed account of this event, and of planetary radar in general, the reader is referred to* To See the Unseen *by Andrew J. Butrica.*[55]

The new value of the AU and the refined ephemeris resulting from the Venus radar experiment were immediately put to use for Mariner orbit determination. Later, it was estimated that without the improved value for the AU and ephemeris improvement, *Mariner 2* would have passed Venus at too great a distance to have returned any useful data. (*Mariner 1* failed on launch.)

In the opinion of historian Andrew Butrica, "The JPL experiment succeeded because it did not depend on . . . prior knowledge of the AU. On the other hand, Lincoln Laboratory and Jodrell Bank based their experiments on an assumed, yet commonly accepted, value for AU and consequently for the distance between Earth and Venus during inferior conjunction." The experiment also owed much of its success to the substantial improvements in uplink and downlink performance that were made to the existing equipment built to support the Echo Balloon experiment in August the previous year. Because the power in the return signal from Venus was expected to be 50 dB to 60 dB (100,000 to 1,000,000 times) weaker than had been received from the Echo balloon, an improved receiving system working with maximum transmitter power would be required to detect the Venus signal.

To meet these requirements, an S-band, low-noise maser and parametric amplifier was added to the receiver at the Pioneer antenna.[56] The single-cavity, three-level ruby maser, the first of its kind, had been developed by Walter Higa at JPL specifically for this application.[57] Together with an improved microwave antenna feed, this combination reduced the system noise temperature from the value of 1,570 K, which had been used for Echo, to about 60 K for Venus. When installed at Goldstone and brought into operation for the Venus experiment, the Pioneer receiver was believed to be the most sensitive operational receiving system in the world. At the time, nothing could be done to increase the output of the existing S-band transmitter at Echo. However, it was operated for the Venus experiments at its maximum power output level of 13 kW.

With this extraordinary instrument, and supported by the strong technical motivation of Rechtin, Victor, and Stevens, the two young JPL astronomers Duane O. Muhleman and Richard M. Goldstein had succeeded where two other distinguished scientific institutions, Lincoln Laboratory and Jodrell Bank, which had conducted similar experiments on Venus, had not.

Between 10 March and 10 May 1961, some 238 hours of scientific data were recorded during radar experiments conducted almost daily.

So began the DSN association with planetary radar.

Soon after the Venus radar experiment, Muhleman left JPL for Harvard, and Rechtin, Victor, and Stevens turned their attention to the rapidly evolving Network and its support for the NASA planetary spacecraft missions. Goldstein, however, continued his close association with the Goldstone radar and its further development for many years.

The first Venus radar experiment had been conducted amid growing concern at JPL about the need for two operational antennas at Goldstone to support the rapidly expanding lunar and planetary programs, which would need tracking support from the DSIF. Added to this concern was the urgent need for an antenna which could be dedicated to research and development activities for the engineering side of the DSIF. The Echo antenna, being of the Azimuth/Elevation (Az-El) type, was not entirely suitable for tracking planetary spacecraft where horizon-to-horizon passes at a constant declination were generally required. It was therefore decided to build a second, 26-m polar-mount antenna at the Echo site and to move the existing Az-El antenna to a new site at Goldstone, where it would be used principally for engineering development for the Network and, incidentally, for planetary radar. The second operational antenna was completed in May, and in a monumental feat of transportation, the 26-m Az-El antenna and its pedestal were moved as a complete unit six kilometers across the desert to a new site to be named, obviously, Venus, in June 1962.

As the new polar-mount antenna at Echo prepared to support the *Mariner 2* flight to Venus in August 1962, as part of the operational Network, the Az-El antenna was refurbished at the new Venus site and resumed Venus observations as a research and development station.

The 1961 Venus radar had been a bistatic radar, that is, it used two antennas, one for transmitting (Echo) and another for receiving (Pioneer). The 1962 Venus radar was a monostatic radar using a single antenna (the former Echo antenna) for both transmitting and receiving. This required the addition of a high-power/low-loss duplexer to accommodate both the transmitting and receiving functions on the one antenna. The transmitter could be operated in a continuous-wave (CW) mode, an amplitude modulation (AM) mode, or a keyed (on/off) mode. A CW receiver and an AM receiver provided open-loop or closed-loop reception for any transmitter mode. Signal level and power spectra data were obtained using the open-loop receive configurations, while the Earth-Venus velocity and Earth-Venus range data were obtained using the closed-loop configurations. Several other changes were made to the Doppler and range systems to accommodate the various modes of operation.[58]

A functional block diagram of the 1962 Venus radar is shown in Figure 8-10.

With this powerful new instrument, Goldstein and others continued further successful radar observations of Venus during the 1962 and 1964 conjunctions and made the first detections of Mars and Mercury in 1963.

Despite these scientific successes, however, the radar depended for its continued existence on the research and development program which the DSN maintained to support its operational Network. It benefited from the development work that the DSN carried out under its Advanced Systems Development Program but was never formally established and funded as a NASA program in its own right. Because much of the technology required for a planetary radar was the same as that required for deep space communications, the Goldstone radar received, in its early years, a great deal of support from the DSN research and development teams. For them it offered a less critical environment in which new technology could be evaluated before being adapted to meet the (mission) critical needs of the operational Network. There were many areas in which this commonality of technology was used to the mutual advantage of the Goldstone radar and the DSN research and development programs. These areas include:

- very high-power transmitters,
- very low-noise microwave amplifiers,

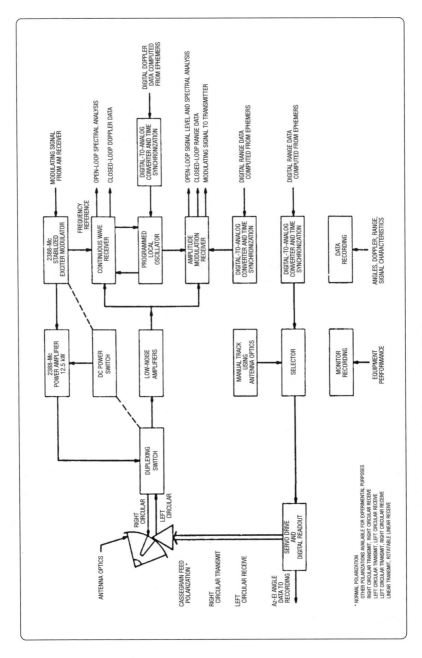

Figure 8-10. Functional block diagram of the Venus radar, 1962.

- high-efficiency microwave antenna feed systems,
- advanced ranging/Doppler systems,
- data acquisition and signal processing, and
- planetary ephemeris prediction.

In addition to providing a realistic demonstration or test bed for new technology in the DSN, the Goldstone radar, under the guidance of R. M. Goldstein, continued to produce good scientific data and to develop a presence in scientific literature.

Under this arrangement, the Goldstone radar developed rapidly. A Rubidium Frequency Standard and improved closed-loop ranging system were added in 1962. High-power transmitter development for the Network led to a 100-kW S-band transmitter for the radar in 1963. Continuing improvement in antenna feed systems and low-noise amplifiers gave the radar a system noise temperature of 35 kelvin in 1964. Signal processing and ranging improved again in 1966 with the addition of a 9-channel correlator and narrower and more stable range gate.

With these improvements, the capability of the radar was extended to include planets at greater distances from Earth. Mars, Mercury, and even Jupiter were soon detected, and scientifically significant observations were reported.[59] Venus was revisited in 1963, and its rotation and large bright surface features were observed.[60]

Within those first few years, half of the planets in the solar system were probed by the Goldstone radar.

The Goldstone radar, like all other science in the DSN, was affected in a major way by the availability of the first DSN 64-m antenna at Goldstone in 1966. Goldstein lost no time in taking advantage of the ten-fold increase in capability provided by the 64-m antenna, compared with the 26-m antenna he had used previously at the Venus site.

The rationale for the DSN adoption of the 64-m Az-El antenna design as the second generation of large antennas for the Network has been discussed in earlier chapters. While none of the rationale was concerned with the application of the antenna for planetary radar, the same argument applied insofar as the application of radar technology to Network problems was concerned. What was good for the radar was (in most cases) good for the Network. So the radar program continued on the 64-m antenna on the same basis as before, with two very significant exceptions. First, the radar had to compete with the flight projects for observing time on the antenna. The second exception was that all the radar-related experimental hardware and software had to be completely separate from the tightly controlled operational equipment used to support the inflight missions. Time on the antenna generally had to be justified on the grounds of relevance of the radar data to one or more of the current missions. To keep the radar equipment separate from the operational systems, it was located in a suitable area of the antenna pedestal structure, with entirely separate interfaces with the antenna itself. A separate antenna feed cone, which could be illuminated by the subreflector when needed, was allocated to radar and other research and development activity. This arrangement remained substantially unchanged through the life of the radar reviewed here and is still in use today. The new 64-m antenna site became the Mars site, and in due course the radar became the Goldstone Solar System Radar (GSSR).

Using the new Mars antenna and its enhanced radar capability, which now included range-Doppler mapping, Goldstein and others observed Venus again during the 1967 conjunction. They discovered more distinct rugged sections of surface terrain and studied a region named Beta from earlier observations in more detail.

In June 1968, the Goldstone radar detected Icarus, its first of many Earth-crossing asteroids. It was difficult to detect a target of such small size. Fortunately, a developmental 450-kW S-band transmitter was then available at the Venus site, and this, together with the Mars antenna in the bistatic mode for receiving, brought the reflected signal above the detection threshold. The bistatic arrangement also circumvented the problem of rapid changeover from transmit to receive states necessitated by the short time interval between transmission and reception of the reflected signal from close-range targets like asteroids.

The results of the observations of Icarus were the first (Goldstone) radar work published in the distinguished journal for planetary science, Icarus, which subsequently became the primary publication for planetary radar research.[61]

In 1970, the first of an ambitious series of joint programs with the Lincoln Laboratory's Haystack antenna was carried out. It was estimated that with Haystack as the transmitting antenna and Mars as the receiving antenna, a very significant increase (up to ten times) over the capability of either one could be achieved. An X-band maser tuned to the frequency of the Haystack radar transmitter was installed on the Mars antenna, and, after several frustrating delays due to 64-m-antenna scheduling problems, Goldstone receivers made a successful detection of the Jupiter satellites Ganymede and Callisto in May and June 1970. The gaseous structure of Jupiter prevented the return of a detectable signal from the planet itself.

The Deep Space Network as a Scientific Instrument

By 1969, the 400-kW S-band capability had been installed at the Mars site and was used for further studies of the Alpha and Beta regions of Venus. In December 1972, and again in January 1973, Goldstein and Morris detected the rings of Saturn with this powerful new capability. The returned signals were unexpectedly strong, 5 to 10 times stronger than those received from Venus and Mercury. This surprising result triggered an upsurge of interest in Saturn rings amongst astronomers and led to many and varied interpretations of the data.[62]

In August 1974, Richard Goldstein and George Morris succeeded in detecting Ganymede on six occasions with the Mars radar alone, using its 400-kW S-band transmitter and 35-K receiving system. They interpreted their results to indicate a considerable degree of roughness on the surface, possibly due to rocky material from meteorites embedded in a matrix of ice.

Progress on the development of a high-power X-band transmitter for the Goldstone radar was reported by C. P. Wiggins[63] in 1972. Operation at X-band would provide an increase of 6.5 dB in antenna gain, resulting in a reduction in observation time by a factor of 20. At the same time, the wider bandwidth of an X-band system would permit the use of faster range codes for more precise range measurement. The radar transmitter, receiver, microwave components and the antenna feed were to be assembled into a new Cassegrain cone that would be mounted in the Research and Development (R&D) position of the tricone on the Mars antenna. Excessive transmission losses would be avoided by operating the transmitter close to the antenna feed. The transmitter would operate at a center frequency of 8.495 MHz and be fixed tuned with a bandwidth of 50 MHz at the -1 dB points. It would consist of two 250-kW klystrons with a four-port hybrid to combine the two outputs into one waveguide port, and it would use the high-voltage power supply and cooling system that normally powered the operational DSN S-band 400-kW transmitter.

This new development was justified on the basis of a study of the rings of Saturn as a possible hazard to spacecraft navigation for the future Voyager missions to Jupiter and Saturn. The first opportunity to probe the Saturnian system with radar would start in December 1974, just three years before the two *Voyagers* were scheduled to launch and six years before the first *Voyager* would arrive at Saturn.

Despite delays caused by unavailability of 64-m antenna time for installation work, the X-band transmitter was completed and began observations on Saturn's rings in December 1974. The data from these observations confirmed the previous Goldstone results and also more recent results from the Arecibo radar, regarding the high radar cross-section (reflectivity for radar signals) and the high linear and polarization ratios. Again, the interpretation of these results was the subject of a great deal of discussion in the scientific community. Various models, from a thick cloud consisting of irregularly shaped, centimeter-size chunks of water-ice to large, meter-size metallic boulders covered with frost, were proposed. Voyager opted for the former model and continued planning its mission. The subject was not closed, however, and the rings of Saturn were observed again in 1977, 1978, and 1979.

The question of the composition and structure of the Saturn rings and the role of radar astronomy in addressing it led Andrew Butrica to observe, very astutely,

> The case of Saturn's rings resulted in radar astronomers contributing to planetary science, in contrast to their studies of the Galilean moons. Those studies . . . had been limited to epistemological studies, namely, what caused the Galilean moons' strange radar signatures? Radar contributed to Saturn science, on the other hand, by focussing less on such questions of radar technique and more on scientific questions, such as the size of the ring particles and the number and thickness of the ring layers. Although the solution of technical problems was a prerequisite for any radar astronomy problem solving, the lack of obvious relevance to planetary science was a serious matter; the ability to solve scientific problems, especially those relating to NASA space missions, was the basis on which scientists judged the value of radar astronomy and on which funding decisions were made.[64]

As subsequent events would show, the issue raised by Butrica in the preceding paragraph—justification, and its effect on funding for the continued operation of the Goldstone radar—would be a growing concern for the DSN in the years ahead.

However, in the immediate future, the Viking mission to Mars was concerned with the selection of a suitable landing site for its two Landers. Surface roughness and slope were key parameters in the landing site selection process, and these were ideally suited to radar observation. Although Mars observations had been made in 1969, they covered latitudes that were of no interest to Viking, and in any case, the results were too noisy to be of much use for landing site selection. New data was needed, particularly of the Syrtis Major and Sinus Meridiana regions which included potential Viking landing sites.

The radar group at JPL had expanded by then and included Richard R. Green, Howard C. Rumsey, George S. Downs, Paul E. Reichley, and Ray F. Jurgens, in addition to Goldstein and Morris. Using funds provided by the Viking Project office, members of

The Deep Space Network as a Scientific Instrument

this team used the Goldstone radar to make X-band observations of the Syrtis Major and Sinus Meridiana regions from October 1975 through April 1976.[65] The extent to which the Goldstone results, along with similar observations from the Arecibo and Stanford radars, influenced the final landing site decision remains an open question. It is clear, however, that the radar data, presented in the form of degrees of rms slope, was not readily understood by the scientists making the decision, who lacked an adequate understanding of the new radar techniques. For them, the high-resolution images from the Viking Orbiter were a more persuasive source of information.

Whatever the role of radar in the final decision, the site that was eventually selected proved to be very satisfactory, as was shown by the subsequent landings and the Lander's long period of successful operation on the Mars surface.

Following the radar detection of Icarus in 1968, interest in asteroids increased amongst radar astronomers, at first in simply detecting them, but later in inferring their shapes and nature of the surface material. In 1972, Goldstein observed 1685 Toro, and in 1975 the asteroid 433 Eros was probed at S-band and X-band and with left and right circular polarization. These data allowed Goldstein and Jurgens to characterize the size, shape, and surface properties of the asteroid with better resolution than had been possible to that time. This was an important step toward future progress in the techniques of asteroid characterization based on dual-frequency, dual-polarization, and radar data. Although it was not apparent at the time, asteroid research was eventually to become the principal activity for the Goldstone radar, but that situation lay twenty years into the future.

Continuing the support for the Voyager mission now approaching Jupiter, Goldstein and Green observed Ganymede and Callisto in December 1977 at X-band, and with left and right circular polarization. The unexpected radar signature they obtained confirmed earlier Arecibo results but led to an interpretation of the surface of the satellites which were at variance with the photographic images returned from *Voyager* during the actual Jupiter Encounter. The reason for this apparent contradiction remained unclear. Several theories were advanced to account for it, some of which involved a better understanding of the radar properties of planetary surfaces rather than more general scientific issues. The questions of scientific worth versus technological value were again pertinent to the continued interest of NASA in the radar.

For the next few years, the Goldstone radar received only sporadic use. The resources of the DSN were oversubscribed with flight projects like *Viking Orbiters 1* and *2*, *Viking Lander 1*, *Helios 1* and *2*, *Pioneers 6* through *9*, and *Voyagers 1* and *2*. These missions were reluctant to relinquish 64-m antenna time at Goldstone for radar experiments, no matter how scientifically interesting they claimed to be. Furthermore, Rechtin, Victor, and Goldstein had by then moved on, and within the DSN there was no longer the strong constituency for radar support as there had been in former years. Budget constraints in the TDA Office eliminated funding for maintenance of the Goldstone radar equipment, so that even when astronomers were able to obtain observing time on the antenna, precious time was often taken up with simply getting the radar to work. At NASA Headquarters, OTDA was the source of funding for the DSN. ODTA declared that all further planetary radar support by the DSN should be conducted in response to a formal requirement from OSSA in the same manner as the DSN responded to OSSA requirements for tracking and data acquisition support for flight missions. The necessary NASA Management Instruction was issued (NMI 8430.1B) to this effect, and the TDA Office at JPL developed the lower level documentation to cover radar requirements and support agreements.

While these changes were slowly evolving, and despite the difficulties of working with the radar, Jurgens and his team were able to refurbish the radar in time to make their first successful radar observation of a comet, IRAS-Araki-Alcock, in May 1982. Attempts to observe comets on subsequent occasions were not successful until 1996, when J. K. Harmon and others succeeded in detecting Comet Hyakutake in 1996.

In 1984, in response to persistent efforts by Jurgens and Downs, the NASA Office of Space Science made funds available for the procurement of a new Digital Equipment Corporation VAX 11/780 computer to be used for data acquisition and processing functions in the Goldstone radar. The computer formed the heart of a new High Speed Data Acquisition System (HSDAS), whose primary mode of operation was the collection and processing of radar data. It supported CW and ranging radar experiments, which would be required for the terrestrial and outer planet targets planned in the future years. These included Mars, Mercury, Venus, asteroids, comets, Galilean satellites, and the rings of Saturn.

As a subsystem of the planetary radar, the HSDAS collected the 15-MHz bandwidth RF signal from the radar receivers and performed all of the necessary data rate compression (filtering, phase correction, and correlation) functions necessary to get the radar data on to a general-purpose computer for further processing and analysis. These functions had been performed in the past by an ever-increasing array of hardware devices that were inadequate for the high-bandwidth radar experiments planned for the future. The extreme flexibility designed into the HSDAS and the ability to control it from simple high-level programming languages made it a useful tool for many other, nonradar purposes as well, such as radio astronomy, advanced receiver development, real-time

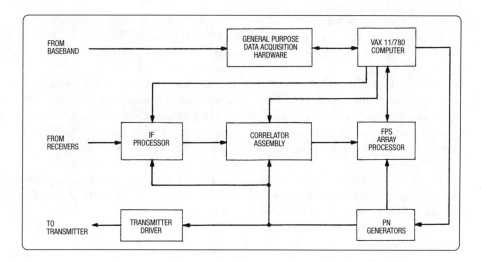

Figure 8-11. High Speed Data Acquisition System for the Goldstone Radar, 1984.

arraying, and SETI. It was installed with the other radar equipment in the DSS 14 antenna pedestal area and, with periodic improvements, served the radar well for the next decade. An overall block diagram of the High Speed Data Acquisition System is shown in Figure 8-11.[66]

In 1983, Peter T. Lyman, the Assistant Laboratory Director for Tracking and Data Acquisition, placed responsibility for direction of the Goldstone radar in the TDA Science Office under the management of Nicholas A. Renzetti. It was Renzetti's goal to convince NASA's Office of Space Science that the Goldstone radar warranted funding on the basis of its value to NASA as a scientific instrument. With the support of Steven J. Ostro, who had recently arrived at JPL from Cornell, where he had worked with Arecibo radar, they redefined the Goldstone Solar System Radar (GSSR) in terms of a "scientific instrument" on which planetary radar science could be performed. This closely paralleled the approach that had been used successfully to support radio astronomy, SETI, and crustal dynamics in the DSN. To implement this approach, and to provide an interface between the radar and the scientific community, Renzetti also created the concept of "Friend of the Radar," a position filled first by Thomas W. Thompson. Martin A. Slade filled the position in 1988.

It was expected that requests for GSSR support would be received from OSS and funding to meet the requirements would come from OTDA, just as was done with the flight projects. But this never happened, and despite the endeavors of Renzetti and others to sell the idea of the GSSR as a national resource for planetary radar research, the key issue of funding for the GSSR remained unsolved.

In 1987, the Director of the Space Science and Instruments Office at JPL obtained a modest level of funding from NASA for planetary radar research and appointed Ostro to manage its disposition in support of an observing program with the GSSR. Under Ostro's influence, science finally supplanted technology as the justification for the GSSR's existence, and excellence in its unique field of science became its goal.

Then, as had happened frequently in the past, events external to its own program overtook planetary radar and resulted in a very significant advance in the capabilities of the instrument. Over the period October 1987 through May 1988, the DSS 14 antenna was upgraded and extended from 64 meters to 70 meters to meet the requirements of the expanding planetary spacecraft program.

Driven initially by the tracking and data acquisition requirements of the Voyager Neptune Encounter, described in an earlier chapter, the 64-meter to 70-meter upgrade was a Network-wide implementation task which had been several years in the planning. DSS 14 was the last of the three antennas to be completed.

As Butrica viewed it, "the Voyager upgrade had a profound impact on the practice of Radar Astronomy at JPL; it provided the GSSR with the sensitivity needed to carry out research on a whole new set of targets (and to begin solving new sets of problems). Not only did the GSSR gain the ability to undertake asteroid research but, when linked to the Very Large Array in New Mexico, it became a new radar research tool."

The immediate benefit to the GSSR is illustrated in Figure 8-12 and shows the relative sensitivity of the 64-m and 70-m antennas for radar observations at X-band as a function of declination. The improvement of about 2 dB corresponds to the increase in antenna gain resulting from the increase in antenna diameter and other related improvements.[67]

The figure also shows the improvement that would have resulted from replacing the 400-kW transmitter with a super-high-power, 1,000-kW transmitter. This latter initiative was promoted by Renzetti in 1989 as part of a further upgrade to the GSSR, which included changes to the radar's transmit/receive system to improve its capability for tracking asteroids. Although the megawatt radar transmitter proposal was not approved, the asteroid modification to the radar feed system was eventually implemented. From 1992 through

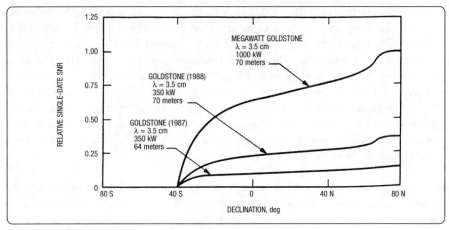

Figure 8-12. Relative sensitivity for the Goldstone Radar Antenna.

1993, a low-noise waveguide switch was developed and installed. This device enabled transmit and receive through the same microwave horn, with a switch-over time of two seconds. It thereby obviated the cumbersome sub-reflector switching procedure, which, with a switch-over time of 25 to 30 seconds, had always limited the GSSR ability to observe targets close to Earth. In 1996, the waveguide switch was superseded by an elegant arrangement of two movable mirrors which, mounted external to the feed horns, did not compromise either the radiated power or the sensitivity of the low-noise receiver.

By a fortuitous circumstance (for planetary radar), the so-called Voyager upgrade in the DSN included the addition, at DSN expense, of low-noise X-band amplifiers to all 27 of the 25-meter antennas of the Very Large Array radio telescope operated by the NRAO in New Mexico. Long affiliated with activities in the DSN, the VLA had participated in the Voyager Neptune Encounter as an element of a Goldstone-VLA array for recovery of downlink telemetry. The Voyager images of Saturn's moon Titan revealed a body covered with an opaque layer of cloud, and the instruments suggested there might be a global ocean of liquid ethane beneath the cloud cover.

Prior to the encounter, Caltech professor Duane O. Muhleman had suggested that the new VLA capability might be used effectively in the bistatic mode with GSSR for planetary radar to further study Titan, with the objective of characterizing its surface features and composition. Previous attempts to probe Titan with either the VLA or the GSSR had not been successful, but the enhanced sensitivity of both combined offered the best chance of success.

As a first demonstration of the GSSR/VLA bistatic system, and with GSSR transmitting and VLA receiving, the rings of Saturn were successfully observed in early 1988. The following year, Muhleman, Butler, and Slade observed Titan for the first time on the nights of 3–6 June. The echoes from Titan were similar to those that had been received from the Galilean satellites and suggested an icy surface, but more observations were needed. Further observations were made jointly with VLA in 1992 and again in 1993.[68] The results of these experiments, although somewhat at variance with current models of the Titan surface, were of great interest to the Cassini Saturn mission planners and were factored into the design for the Huygens Probe radar, scheduled to descend to the surface of Titan in 2004.

With minor changes to the VLA configuration, and using different polarizations, the VLA-Goldstone bistatic arrangement was used by the same experimenters to observe Mars, first in 1988 and again, several times, during the 1992–93 opposition, and to further investigate the presence of ice on the surface of Mercury in 1991.[69]

Amongst planetary radar astronomers, interest in Venus had diminished during the late 1980s for a number of reasons, some of them having to do with availability of resources and the attraction of more distant targets. With the launch of *Magellan* and the extraordinary success of its orbital radar mission in 1990, scientific investigation of Venus by planetary radar essentially came to an end. There were, however, several investigations of anomalous radar signatures from Venus conducted in subsequent years.

Asteroids

The Voyager upgrade to the Goldstone radar antenna, along with the addition of the new HSDAS, gave the GSSR the sensitivity and resolution needed for the detection of asteroids. The arrival of Steven J. Ostro provided the TDA Science Office with the experience and motivation required to engage in this new field of research in a major way. An opportunity to combine the new capability with the addition of Ostro's experience in a way that would change the course of the GSSR forever was not long in coming.

Three-dimensional modeling of asteroid shapes based on range-Doppler radar images was still a relatively new technique when the Earth-crossing asteroid 4179 Toutatis was discovered in January 1989. Toutatis was predicted to pass very close (9.4 lunar distances) to Earth on 8 December 1992. This would make it an ideal target for range-Doppler imaging with the GSSR. Ostro proposed an elaborate observing program to Renzetti that involved the GSSR, Arecibo, VLA, and the new DSS 13 BWG antenna working as an interferometer with the GSSR. The outcome would be the highest resolution three-

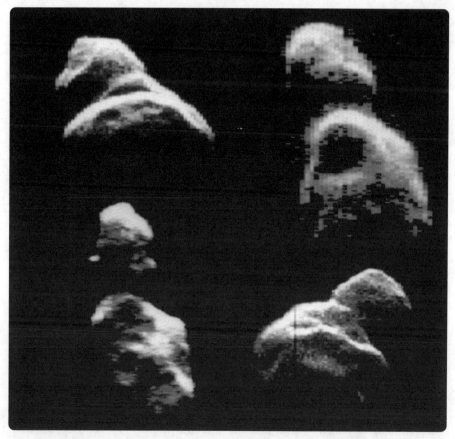

Figure 8-13. Goldstone Solar System Radar images of Asteroid Toutatis, 1992.

dimensional images of an Earth-crossing asteroid ever produced. Renzetti accepted, and the planning went forward.

Observations were carried out at the various sites over the period from 27 November through 19 December 1992, and a good data return was obtained.[70] Although the subsequent reconstruction of the images was complicated by the slow tumbling rotational state of the asteroid, they revealed the general shape and surface topography of Toutatis at the highest resolution ever obtained of an asteroid. Typical images from this experiment are shown in Figure 8-13 and reveal the presence of several craters on the two (apparently connected) components of the irregularly shaped main body of the asteroid.

Using similar techniques, Ostro and Hudson successfully observed the next Earth-crossing asteroid, 1620 Geographos, in 1994 and produced equally impressive imaging of the asteroid body. In 1995, observations of 6489 Golevka included delay-Doppler imaging, GSSR-VLA astrometry, and the first intercontinental planetary radar experiments from Goldstone to Russia and Japan. Less than 600 meters across, Golevka was the smallest solar system object imaged to that time.[71]

As scientific interest in radar observation of asteroids increased, it gave rise a heightened awareness of the hazards posed by possible Earth encounters with asteroids and raised questions of how such threats might be handled in the best interests of the public. The role that the GSSR might play in continued evolution of this intriguing subject is discussed in the relevant scientific literature.[72]

Radar Astrometry

Astrometry is a branch of astronomy that deals with the measurement of the positions, motions, and distances of planets, stars, and other celestial bodies, including asteroids. In recent times, when Ostro and others employed radar techniques to study and predict the motions of asteroids, the term "radar astrometry" was coined to describe this aspect of radar astronomy.

Thus, data generated by the GSSR were used, not only for science data for imaging and surface characterization experiments, but also to refine the orbits of Earth-crossing asteroids. This extension of the GSSR capability was proposed originally by Jurgens and carried forward by Ostro, who enlisted the support of two of JPL's orbit determination specialists, Donald K. Yeomans and Paul W. Chodas. Supported by this powerful group of experts, radar astrometry gained prominence in the period 1990–92 using data from the GSSR and the Arecibo radars.[73] Together with like advances in radar imaging, radar data were recognized by the International Astronomical Union (IAU) as a data source in its circulars to announce the discovery and orbit of newly discovered asteroids.

It seemed most likely that postdiscovery IAU circulars for new asteroids in the near future would be composed of computer files containing high-resolution video images of the objects and highly precise ephemerides, based on data taken by the GSSR and Arecibo planetary radars.

Renaissance

The renaissance of the Goldstone Solar System Radar dates from about the early 1990s. By that time, the radar was no longer perceived primarily as a test bed for new DSN technology, but was recognized in its own right as a scientific instrument for solar system research. The remarkable success in the first detection of the Earth-crossing asteroids did much to popularize the new approach to planetary radar. Over the period from 1993 to 1997, most GSSR research was conducted under the aegis of NASA's mission to the solar system. Research topics lay within the scope of the Solar System Exploration or Structure and Evolution of the Universe programs as shown in the table below.

NASA Theme: Structure and Evolution of the Universe
    GSSR Experiment:
        Mercury Tests of Gravitational Theories—
            Ongoing experiments

NASA Theme: Formation and Dynamics of Earth-like Planets
    GSSR Experiments:
        Mercury Surface Properties—
            Ongoing analysis of new and existing data
        Mercury Polar Studies—
            X-band full-disk imaging planned for 1998/1999
        Venus Surface Properties—
            Observations planned for 1998/1999
        Venus Geophysics—
            Observations planned for 1998/1999
        Mars Surface Properties—
            Observations planned for 1999 Mars opposition
        Mars Polar Studies—
            Observations planned for 1999 Mars opposition
        Mars Geophysics—
            Analysis of existing topographic maps in progress
        Galilean satellites—
            No new observations currently planned

NASA Theme: Mission to Planet Earth
    GSSR Experiments:
        Earth Stratospheric Studies—
            No new observations currently planned
        Lunar Polar Observations—
            Data from 1997 observations being analyzed

NASA Theme: Mars Exploration Support
    GSSR Experiment:
        Mars Landing Site Certification—
            Radar data aided landing site certification for 1997
            Mars Pathfinder landing[74]
            Observations in 1999 will aid future 2001 landings

NASA Theme: Building Blocks and Our Chemical Origins
    GSSR Experiments:
        Asteroids, Imaging and Orbit Refinement—
            Several asteroid observations planned each year
        Radar Data Crucial for Orbit Determination of Near-Earth
            Asteroids
        Titan Surface Properties—
            Bistatic observations with VLA planned for
                October 1998 to aid Cassini Huygens-Probe
                mission
        Comet studies—
            Comet Hyakutake observed in 1996[75]
            Observations to continue as opportunities arise

The outcome of all this activity was a plethora of publications in the scientific literature. In the ten-year period from 1988 to 1997, nearly 100 papers based on data from the GSSR were published in highly respected journals such as *Science, Icarus, Bulletin of the American Astronomical Society,* and *Lunar and Planetary Science.* The productivity of the Goldstone Radar, in terms of yearly scientific publications for the most recent ten- year period, is shown in Figure 8-14.

The data clearly show the revival of interest in radar astronomy following the Voyager upgrade and the start of the asteroid detection program. The data also suggest that the level of scientific interest, when constrained by the amount of antenna time available for radar observations, results in the publication of 10 to 12 papers per year. It does not lead to any conclusion regarding the scientific value of the papers so produced.[76, 77]

As we have seen, the constant efforts of a relatively small group of dedicated scientists produced a steady stream of scientific knowledge throughout the lifespan of the Goldstone radar reviewed here. Sometimes the stream slowed to a trickle and sometimes it was in full flood, but it never dried up. The number of publications that resulted from these efforts attests to the productivity of the radar.

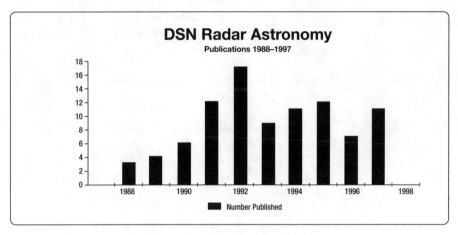

Figure 8-14. Goldstone Radar astronomy publications, 1988–97.

Significant Events

To better appreciate the overall scope of the work done with the radar over the first 35 years of its active life, the reader may wish to consult the following table of significant events in the scientific life of the radar.

Significant Events in the DSN Planetary Radar Program

1961	Venus	First detection, determined value for AU to accuracy of 500 km, refined ephemerides
1962	Venus	Determined slow retrograde motion
1963	Mars	First detection, measured albedo vs. longitude at 13 degrees latitude
1963	Mercury	First detection, measured albedo and range
1963	Jupiter	Statistically significant detection, not repeated
1964	Venus	Discovered large bright surface features

Year	Target	Description
1965	Mars	Measured albedo and roughness vs. longitude, Ma IV camera directed to image Trivium Charontis, a very bright smooth radar target
1966	Venus	Located of a number of new surface features, rotation found to be approx. 243.0 days
1967	Venus	Accurate range measurements led to improved knowledge of Venus orbit and radius, and also AU
1968	Icarus	First detection of asteroid, verified ephemeris and estimated size and roughness
1969	Venus	Range measurements, large map with highest resolution, north-south ambiguity resolved
1969	Mars	Determined accurate (±0.2 km) altitude profile of equatorial region
1969	Mercury	Surface features detected (not Venus-like)
1970	Mars	Detailed surface mapping
1970	Jupiter	Observed satellites Ganymede, Callisto for Voyager with Haystack in the bistatic mode
1971	Mars	Detailed surface mapping
1972	Toro	Asteroid 1685 Toro detected
1973	Saturn	Saturn rings observed
1974	Ganymede	Observed surface of Ganymede
1975	Mars	Surface roughness studies for Viking Lander site selection
1975	Eros	Asteroid 433 Eros detected
1977	Ganymede	Ganymede observed right and left circular polarization returns

1978	Saturn	Saturn rings observed
1979	Saturn	Saturn rings observed
1982	Comet	Comet IRAS-Araki Alcock detected
1987	Jupiter	Determined polarization ratios and radar cross sections of Galilean satellites, noted strange radar signatures
1989	Saturn	Observed satellite Titan for Cassini with Haystack in bistatic mode
1991	Mercury	Observed Mercury with Haystack in bistatic mode
1991	Earth	Observed Earth orbital debris
1992	Mars	Observed Mars with VLA in bistatic mode
1992	Asteroid	Observed first Earth-Crossing Asteroid (ECA) 4179 Toutatis with DSS 13 in bistatic mode
1993	Mars	Observed Mars with VLA in bistatic mode
1993	Mercury	Full-disk radar images, detection of ice at North Polar region
1994	Asteroid	Three-dimensional modeling of Earth-Crossing asteroid 1620 Geographos
1995	Mars	Pathfinder landing site characterization
1996	Comet	Successful observation of Comet Hyakutake
1997	Moon	Search for ice in shadowed areas of the Polar Regions

Endnotes

1. M. S. Reid, R. C. Clauss, D. A. Bathker, and C. T. Stelzried, "Low-Noise Microwave Receiving Systems in a Worldwide Network of Large Antennas," *Proceedings of the IEEE*, vol. 61, no. 9 (September 1973), p. 1330–35.

2. Andrew J. Butrica, *To See the Unseen: A History of Planetary Radar*, The NASA History Series, NASA SP-4218 (Washington, DC: NASA History Office, 1996).

3. M. J. Klein, oral interview with author, 9 April 1998.

4. N. A. Renzetti, "Oral History," J. Alonso, 17 April 1992.

5. N. A. Renzetti and A. L. Berman, "The Deep Space Network as an Instrument for Radio Science Research," JPL Publication 80-93 (15 February 1981).

6. S. W. Asmar and N. A. Renzetti, "The Deep Space Network as an Instrument for Radio Science Research," JPL Publication 80-93, Rev. 1. (15 April 1993).

7. C. T. Stelzreid, "A Faraday Rotation Measurement of a 13cm Signal in the Solar Corona," JPL Technical Report TR 32-1401 (1970), p. 83.

8. R. Woo, and R. Schwenn, "Comparison of Doppler Scintillation and In-Situ Spacecraft Plasma Measurements of Interplanetary Disturbances," *Journal of Geophysical Research* 96 (1991): 21227–44.

9. A. Kliore, D. L. Cain, and T. W. Hamilton, "Determination of Some Physical Properties of the Atmosphere of Mars from Changes in the Doppler Signal of a Spacecraft on an Earth-Occultation Trajectory," JPL Technical Report TR 32-674 (1964).

10. G. L. Tyler, "Radio Propagation Experiments in the Outer Solar System with Voyager," *Proceedings of the IEEE*, vol. 75, no. 10 (1987).

11. J. D. Anderson, P. B. Esposito, W. Martin, C. L. Thornton, and D. O. Muhleman, "Experimental Test of General Relativity Using Time-Delay Data from Mariner 6 and Mariner 7," *Astrophysical Journal* 200 (1975): 221–33.

12. F. B. Estabrook and H. D. Walhquist, "Response of Doppler Spacecraft Tracking to Gravitational Radiation," *General Relativity and Gravitation* 6 (1975): 439–47.

13. N. A. Renzetti et al., "The Deep Space Network—An Instrument for Radio Astronomy Research," JPL Publication 82-68, Rev. 1 (1 September 1988).

14. K. W. Linnes, "Radio Science Support," JPL Technical Report TR 32-1526, Vol. III, The Deep Space Network, Progress Report, March-April 1971 (15 June 1971): 46–51.

15. M. J. Batty, D. L. Jauncey, S. Gulkis, and M. J. Yerbury, "The Tidbinbilla Interferometer," PR 42-58, May-June 1980 (15 August 1980): 6–8.

16. M. J. Batty, D. L. Jauncey, P. T. Rayner, and S. Gulkis, "Accurate Radio Positions with the Tidbinbilla Interferometer," TDA Progress Report PR 42-58, May-June 1980 (15 August 1980), pp. 1–5.

17. G. S. Levy et al., "Orbiting Very Long baseline Interferometer Demonstration Using the Tracking and Data Relay Satellite System," *Proceedings of International Astronomical Union Symposium*, no. 110, VLBI and Compact Radio Sources (Bologna, Italy, 27 June–1 July 1983), pp. 405–06.

18. T. B. H. Kuiper to N. R. Haynes, "On the Role of Radio Astronomy in the DSN," internal memorandum, TBK 94-007, 6 February 1994.

19. T. B. H. Kuiper and M. R. Wick, "Radio Astronomy Observations Program: Detailed Mission Requirements," Jet Propulsion Laboratory, JPL Document No. D-12940, DSN 870-348 (1 February 1996).

20. T. B. H. Kuiper to D. J. Mudgway, "Significant Advances in Radio Astronomy Using the DSN," personal communication, August 1999.

21. J. S. Gubbay et al., "Trans-Pacific Interferometer Measurements at 2300 MHz," *Nature*: 222, 730.

22. G. S. Downs et al., "Average Pulsar Energies at Centimeter Wavelengths," *Nature* 222 (1969): 1257.

23. P. E. Reichley and G. S. Downs, "Observed Decrease in the Period of Pulsar PSR 0833-43," *Nature* 222 (1969): 229.

24. B. G. Clark et al., "Variations in the Fine Structure of Quasars 3C279 and 3C273," *Bulletin of the American Astronomical Society* 3 (1971): 383.

25. R. L. Carpenter et al., "Search for Small-scope Anisotropy in the 2.7 Cosmic Background Radiation at a Wavelength of 3.56 Centimeters," *Astrophysical Journal* 182 (1973): L61.

26. M. J. Batty et al., "Tidbinbilla Two-element Interferometer," *Astronomical Journal* 87 (1982): 938.

27. J. S. Ulvestad, C. D. Edwards, and R. Linfield, "Very Long Baseline Interferometry Using a Radio Telescope in Earth Orbit," TDA Progress Report PR 42-88, October-December 1988 (15 February 1986): 1–10.

28. R. Linfield et al., "VLBI Observations of Three Quasars Using an Earth-orbiting Radio Telescope," *Bulletin of the American Astronomical Society* 18 (1986): 993.

29. K. M. Desai et al., "First Speckle Hologram of the Interstellar Plasma," *Astrophysical Journal* 393, no. 2 (10 July 1992): L75–L78.

30. S. J. Tingay et al., "The Subparsec-Scale Structure and Evolution of Centaurus A: The Nearest Active Radio Galaxy," *Astronomical Journal* 115, issue 3 (March 1998): 960–74.

31. T. Velusamy, T. B. H. Kuiper, and W. D. Langer, "CCS Observations of the Protostellar Envelope of B335," *Astrophysical Journal* 451 (October 1995): L75.

32. T. B. H. Kuiper, W. D. Langer, and T. Velusamy, "Evolutionary Status of the Pre-protostellar Core L1498," *Astrophysical Journal* 468 (September 1996): 761.

33. C. R. Gwinn et al., "Size of the Vela Pulsar's Radio Emission Region: 500 Kilometers," *Astrophysical Journal* 483 (July 1997): L53.

34. J. C. Guirado et al., "Astrometric Detection of a Low-Mass Companion Orbiting the Star AB Doradus," *Astrophysical Journal* 490 (December 1997): 835.

35. R. E. Edelson and G. S. Levy, "The Search for Extraterrestrial Intelligence: Telecommunications Technology," TDA Progress Report PR 42-57, March-April 1980 (15 June 1980), pp. 1–8.

36. A. L. Berman, "The SETI Observational Plan," TDA Progress Report PR 42-57, March-April 1980 (15 June 1980), pp. 9–15.

37. B. K. Levitt, "SETI Pulse Detection Algorithm: Analysis of False Alarm Rates," TDA Progress Report PR 42-74, April-June 1983 (15 August 1983), pp. 149–58; S. Gulkis and E. T. Olsen, "Gain Stability Measurements at S-band and X-band," TDA Progress Report PR 42-74, April-June 1983 (15 August 1983), pp. 159–68; B. Crow, "SETI Downconverter," TDA Progress Report PR 42-67, November-December 1981 (15 February 1982), pp. 1–5; E. T. Olsen and A. Lokshin, "The SETI Interpreter Program—A Software Package for the SETI Field Tests," TDA Progress Report PR 42-74, April-June 1983 (15 August 1983), pp. 169–82.

38. B. Crow, A. Lokshin, M. Marina, and L. Ching, "SETI Radio Spectrum Surveillance System," TDA Progress Report PR 42-82, April-June 1985 (15 August 1985), pp. 173–84.

39. S. Gulkis, M. J. Klein, E. T. Olsen, R. B. Crow, R. M. Gosline, G. S. Downs, M. P. Quirk, A. Lokshin, and J. Solomon, "Objectives and First Results of the NASA SETI Sky Survey Field Tests at Goldstone," TDA Progress Report PR 42-86, April-June 1986 (15 August 1986), pp. 284–93.

40. J. Solomon, W. Lawton, M. P. Quirk, and E. T. Olsen, "A Signal Detection Strategy for the SETI All Sky Survey," TDA Progress Report PR 42-83, July-September 1985 (15 November 1985), pp. 191–201; M. P. Quirk, H. S. Wilk, and M. J. Grimm, "A Wide-Band, High Resolution Spectrum Analyzer," TDA Progress Report PR 42-83, July-September 1985 (15 November 1985), pp. 180–90.

41. M. J. Klein, S. Gulkis, E. T. Olsen, and N. A. Renzetti, "The NASA SETI Sky Survey: Recent Developments," TDA Progress Report PR 42-98, April-June 1989 (15 August 1989), pp. 218-2-26.

42. G. R. Coulter, M. J. Klein, P. R. Backus, and J. D. Rummel, "Searching for Intelligent Life in the Universe: NASA's High Resolution Microwave Survey," in Sjoerd L. Bonting, ed., *Advances in Space Biology and Medicine*, Vol. 4 (Greenwich, Connecticut: JAI Press, 1992), pp. 189–224.

43. K. M. Ong, P. F. MacDoran, J. B. Thomas, H. F. Fliegel, L. J. Skjerve, D. J. Spitzmesser, P. D. Batelaan, S. R. Paine, and M. G. Newsted, "A Demonstration of Radio Interferometric Surveying Using DSS 14 and the Project Aries Transportable Antenna," DSN Progress Report PR 42-26, January-February 1975 (15 April 1975), pp. 41–53.

44. D. L. Brun, S. C. Wu, E. H. Thom, F. D. McLaughlin, and B. M. Sweetser, "Orion Mobile Unit Design," TDA Progress Report PR 42-60, September-October 1980 (December 1980), pp. 6–32.

45. C. G. Koscielski, "Mojave Base Station Implementation," TDA Progress Report PR 42-78, April-June 1984 (15 August 1984), pp. 216–24.

46. J. M. Davidson and D. W. Trask, "Utilization of Mobile VLBI for Geodetic Measurements," TDA Progress Report PR 42-80, October-December 1984 (15 February 1985), pp. 248–66.

47. L. A. Buennagel, P. F. MacDoran, R. E. Neilan, D. J. Spitzmesser, and L. E. Young, "Satellite Emission Range Inferred Earth Survey (SERIES) Project: Final Report on Research and Development Phase, 1979 to 1983, JPL Publication 84-16 (Pasadena, California: JPL, 1984).

48. J. M. Davidson, C. L. Thornton, S. A. Stephens, S. C. Wu, S. M. Lichten, J. S. Border, O. J. Sovers, T. H. Dixon, and B. G. Williams, "Demonstration of the Fiducial Concept Using Data From the March 1985 GPS Field Test," TDA Progress Report PR 42-86 (April-June 1985), pp. 301–06.

49. J. M. Davidson, C. L. Thornton, T. H. Dixon, C. J. Vegos, L. E. Young, and T. P. Yunck, "The March 1985 Demonstration of the Fiducial Network Concept for GPS Geodesy: A Preliminary Report," TDA Progress Report PR 42-85, January-March 1986 (15 May 1986), pp. 212–18.

50. G. Lanyi, "Total Ionospheric Content Calibration Using Series GPS Satellite Data," TDA Progress Report PR 42-85, January-March 1986 (15 May 1986), pp. 1–12.

51. *GPS Flight Demonstration on Topex*, JPL proposal, July 1984.

52. DSCC Media Calibration Subsystem, DMD 1987–1992, Functional Design Review, November 1986.

53. Catherine L. Thornton, interview with Douglas J. Mudgway (April 1998).

54. W. K. Victor and R. S. Stevens, "Exploration of Venus by Radar," *Science* 134 (1961): 46.

55. Andrew J. Butrica, *To See the Unseen: A History of Planetary Radar*, The NASA History Series, NASA SP-4218 (Washington, DC: NASA History Office, 1996).

56. M. H. Brockman, L. R. Malling, and H. R. Buchanan, "Venus Radar Experiment," JPL Research Summary No. 36-8, Vol. 1 (February-April 1961), pp. 65–73.

57. W. Higa, "A Maser System for Radar Astronomy," JPL Technical Report TR 32-103 (1 March 1961).

58. R. M. Goldstein, R. Stevens, and W. K. Victor, "Goldstone Observatory Report for October-December 1962," JPL Technical Report TR 32-396 (1 March 1965).

59. R. M. Goldstein, "Mars: Radar Observations," *Science* CL (24 December 1965): 1715–17.

60. R. M. Goldstein and R. L. Carpenter, "Rotation of Venus: Period Estimated from Radar Measurement," *Science* CXXIX (8 March 1963), pp. 910–11.

61. R. M. Goldstein, "Radar Observations of Icarus," *Icarus* no. 10 (1969): 430–31.

62. R. M. Goldstein and G. A. Morris, "Radar Observations of the Rings of Saturn," *Icarus* no. 20 (1973): 260–62.

63. C. P. Wiggins, "X-band Radar Development," JPL Technical Report TR 32-1526, Vol. XII, September-October 1972 (15 December 1972), pp. 19–21.

64. See endnote 2, p. 215.

65. G. S. Downs, R. R. Green, and P. E. Reichley, "Radar Studies of the Martian Surface at Centimeter Wavelengths: The 1975 Opposition," *Icarus* no. 33 (1978): 441–53.

66. L. J. Deutsch, R. F. Jurgens, and S. S. Brokl, "The Goldstone Research and Development High-Speed Data Acquisition System," TDA Progress Report PR 42-77, January-March 1984 (15 May 1984), pp. 87–96.

67. N. A. Renzetti, T. W. Thompson, and M. A. Slade, "Relative Planetary Radar Sensitivities: Arecibo and Goldstone," TDA Progress Report PR 42-94, April-June 1988 (15 August 1988), pp. 287–93.

68. B. J. Butler, D. O. Muhleman, and M. A. Slade, "Results From 1992 and 1993 VLA/Goldstone 3.5 cm Radar Data," *Bulletin of the American Astronomical Society* 25 (1993): 1040.

69. M. A. Slade, B. J. Butler, and D. O. Muhleman, "Mercury Goldstone-VLA Radar: Part I," *Bulletin of the American Astronomical Society* 23 (1991): 1197.

70. S. J. Ostro et al., "Radar Imaging of Asteroid 4179 Toutatis," *Bulletin of the American Astronomical Society* 25 (1993): 1126.

71. S. J. Ostro, "Goldstone Radar Research of Near-Earth Asteroids," *DSN Technology and Science Program News* issue 7 (February 1997): 13.

72. S. J. Ostro, "The Role of Ground-Based Radar in Near-Earth Object Hazard Identification and Mitigation," in T. Gehrels and M. S. Matthews, eds., *Hazards Due to Comets and Asteroids* (Tucson, Arizona: University of Arizona Press, 1994), pp. 259–82.

73. D. K. Yeomans et al., "Asteroid and Comet Orbits Using Radar Data," *Astronomical Journal* 103 (1992): 303–17.

74. M. A. Slade, R. F. Jurgens, A. F. C. Haldeman, and D. L. Mitchell, "Hazard Evaluation for the Mars Pathfinder Prime Landing Site from Synthesis of Radar Ranging and Continuous Wave Observations During the 1995 Mars Opposition," *Bulletin of the American Astronomical Society* 27, no. 3 (1995): 1103.

75. J. K. Harmon et al., "Radar Detection of the Nucleus and Coma of Comet Hyakutake," (C/1996 B2), *Science* 278 (1997): 1921–24.

76. Radar astronomy online Web sites: A comprehensive listing of published scientific work based on all aspects of GSSR observations is available at *http://wireless.jpl.nasa.gov/RADAR/*.

77. Papers dealing with radar observations of asteroids are listed at *http://echo.jpl.nasa.gov/publications/pubs.html*.

CHAPTER 9

THE DEEP SPACE NETWORK
AS AN ORGANIZATION IN CHANGE

IN THE BEGINNING (1958 TO 1963)

Background

Following the ARPA decision to launch several lunar probes by the end of 1958 using the Army JUNO-II launch vehicle, the Army turned to its long-time contractor, JPL, for the payloads and the tracking and data acquisition facilities required for its portion of the new Pioneer Program.[1]

The flight paths of the lunar probes required a tracking station at Cape Canaveral to cover the launch phase, another in the vicinity of Puerto Rico for the near-Earth phase, and a deep space station in the general vicinity of JPL to track the probe on its way to the Moon. Time was short. There were only eight months to select a site, build an antenna, and set up the requisite receivers, recorders, and communication circuits. Fortunately, the Army owned a considerable amount of property in the Mojave Desert, and it was on this property at Goldstone Dry Lake, about 100 air miles from Los Angeles, that a suitable site for building the deep space antenna was found.

Construction of the site began at Goldstone in June 1958; assembly of the antenna commenced in August; and by December, the antenna, together with electronics, receivers, and communications, was complete. *Pioneer 3* was launched on 6 December 1958. This station, appropriately named Pioneer, was the first station of what was to become the Deep Space Instrumentation Facility (DSIF), and eventually the Deep Space Network (DSN) described in earlier chapters.

As these events began to unfold in 1958, the engineering organization within JPL reflected the technologies that had been appropriate to the missile guidance and telemetry support provided as a contractor to the U.S. Army. The three principal engineering departments were Aerodynamics and Propellants, Design and Power Plants, and Guidance and Electronics. The Guidance and Electronics Department, headed by Robert J. Parks, consisted of three Technical Divisions: Guidance Analysis, led by Clarence R. Gates; Guidance Development, led by Walker E. Giberson; and Guidance Research, led by Eberhardt Rechtin. Rechtin's Guidance Research Division consisted of an Electronics Research Section under Walter K. Victor and a Guidance Techniques Research Section led by Robertson Stevens.[2] Design and implementation of the Pioneer DSIF station was carried out under the technical management of engineers from these two sections, with overall responsibility shared by Rechtin, Victor, and Stevens.

The Deep Space Network as an Organization in Change

In addition to its work for the Army, JPL also supported a substantial aerodynamics research facility known as the Southern California Cooperative Wind Tunnel. This facility was funded and used, on a cooperative basis, by the many aerospace companies flourishing in the southern California region at that time. Although the Wind Tunnel was in no way associated with the DSIF, its Chief of Facilities and Mechanical Equipment would soon play a prominent role in the future history of the DSN. His name was William H. Bayley.

A mechanical engineering graduate (1952) from the University of Southern California (USC) in Los Angeles, Bill Bayley worked on facilities engineering for the Lockheed Aircraft Corporation in Burbank, California, before coming to JPL in 1956 to manage the facilities of the southern California Cooperative Wind Tunnel. At that time, and for several years afterward, this major research facility reflected JPL's association with Caltech and aeronautical research.

After JPL's transfer to NASA, Rechtin's network of tracking stations, communications, and data-processing systems, as well as the staff needed to run it, expanded rapidly. He needed help. The addition of an assistant to interact with NASA on issues of staff and budget and to deal with financial and administrative matters at JPL allowed Rechtin to address the more technical issues associated with the future world network. He turned to Bill Bayley, a perfect choice as it turned out, and made him General Manager in 1960. When Rechtin left JPL in 1967, Bayley assumed Rechtin's position as Assistant Laboratory Director for Tracking and Data Acquisition, a position he held with distinction until he retired in 1980. During his term of office, the Deep Space Network developed into a major NASA facility of 26-m, 34-m, and 64-m antennas in three countries around the globe. That achievement was due, in no small measure, to his skill in dealing with both NASA Headquarters and JPL and his ability to optimize the relationship between them in the best interests of the DSN.

A pleasant man to talk to and a gentle man to deal with, both at work and outside the work environment, Bill Bayley was popular with his colleagues at JPL and NASA Headquarters, as well as his counterparts at the government agencies in Spain, Australia, and South Africa, on whom NASA depended for support of its tracking stations in those countries. His quiet sense of humor—he was a master of the "bon mot" and "double entendre"—served to enhance his connectivity with the people he met. Bill Bayley introduced "all hands" meetings to the DSN as a way of bringing top management into direct contact with all members of the organization for a frank and open discussion of problems, policies, and procedures. Such meetings, led by Bayley himself, were always held away from the normal workplace to create a less formal and more productive atmosphere for the free expression of ideas and opinions.

He was a devotee of healthy living and exercise, so it was ironic that Bayley's untimely death in 1981 was due to a heart-related problem.

The Rechtin Years

In 1958, while the Pioneer antenna was being rushed to completion at Goldstone to meet the deadline for the first Pioneer Lunar Probe launch in December, momentous events were occurring elsewhere in the U.S. space effort. Later that year, Congress passed the National Aeronautics and Space Act to create the National Aeronautics and Space Administration (NASA). President Eisenhower immediately signed the new Act into law, and NASA began operating under the leadership of its first Administrator, T. Keith Glennan, on 1 October 1958. The transfer of JPL from the Army to NASA followed almost immediately. Instead of working under a U.S. Army contract, JPL now worked under a NASA contract. NASA inherited not only JPL's experienced personnel and its facilities, including Goldstone, but also JPL's vision of a worldwide tracking network for deep space probes.[3] NASA's plans for JPL were in keeping with this vision, and there was no interruption to the work in progress as a result of this essentially political change.

However, within NASA and JPL, the change raised serious questions regarding the status of JPL. NASA wanted to operate JPL like other Agency Centers, rather than, as JPL desired, an outside contractor managed by Caltech. Describing this situation, historian Clayton Koppes wrote, "Insider or outsider. Vast amounts of time and energy would be consumed in resolving the question of insider or outsider throughout the next decade."[4] Despite these internal disturbances, essential work moved rapidly forward.

When the new NASA Headquarters organization first came into being, responsibility for NASA-wide tracking and data acquisition services and facilities rested in Abe Silverstein's Office of Space Flight Development. Within that office, the position of Assistant Director for Space Flight Operations was filled initially by Edmond C. Buckley.

In this capacity, Buckley interacted directly with JPL on all matters related to NASA policy, guidelines, and budget for the original construction and operation of the DSIF.

Referring to Buckley's early association with the DSN, historian Cargill Hall said, "A graduate of Rensselaer Polytechnic Institute, personable and articulate, he brought to NASA many years of [NACA] experience in the development of the Wallops Island

The Deep Space Network as an Organization in Change

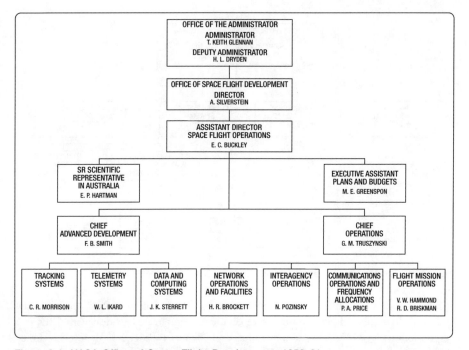

Figure 9-1. NASA Office of Space Flight Development, 1959–61.

Launch Range, where he had been responsible for the development of tracking and instrumentation associated with free flight research."

Fortunately for the fledgling DSN, Ed Buckley at NASA Headquarters and Eb Rechtin at JPL enjoyed a great deal of mutual respect and worked well together to further their common goal, the establishment of a worldwide network for NASA. Observed Cargill Hall, "No major disagreements (between NASA and JPL) marred the planning and creation of the Deep Space Network."

Responsibility for tracking and data acquisition in the NASA Office of Space Flight Development at this time is shown in Figure 9-1.

Within a very short time of its transfer to NASA, the organization of JPL was restructured to better meet its new responsibilities in the changing world. No longer was its primary business that of missiles and rockets and the tracking and data acquisition technology to support them as a contractor to the U.S. Army, but lunar and planetary

585

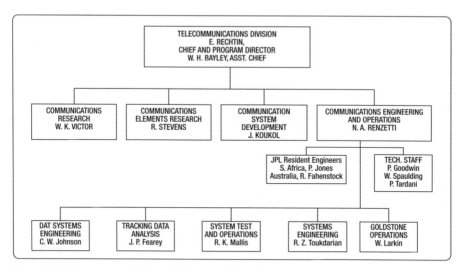

Figure 9-2. Telecommunications Division at JPL, 1961–62.

spacecraft and the technology to support them as a contractor to NASA. Seven new technical divisions and six administrative divisions were created to deal with JPL's new responsibilities in NASA's expanding Lunar and Planetary Program. All seven technical divisions were directly responsible to JPL Director William H. Pickering, while the administrative divisions reported to the Assistant Laboratory Director (ALD) of the Business Administration Office, Victor C. Larsen. Amongst the technical divisions, the Telecommunications Division (Division 33), with Rechtin named as Chief and Bayley as Assistant Chief, contained the essential technological expertise that was to become the foundation stone of the emerging worldwide DSIF. In addition to Chief of the Telecommunications Division, Rechtin was also designated as Program Director for the DSIF, an indication of the growing importance of the DSIF in the JPL organization, planning, and budget processes.

The Telecommunications Division, under Rechtin's leadership, comprised four sections to provide the state-of-the-art technical and operational resources needed to transform the DSIF from a single antenna in California to a worldwide network with tracking stations on three continents. The organization of the DSIF during this period, 1961–62, is shown in Figure 9-2.

Working within this structure, Nicholas A. Renzetti led the engineering, implementation, and operations activity supported by an aggressive and very successful research and

The Deep Space Network as an Organization in Change

development program led by Victor, Stevens, and Koukol. In 1961 and 1962, the total engineering and technical staff for both functions totaled about 200 to 250 persons, with a budget of 15 to 20 million dollars. Three years later, reflecting the demands of an expanding network, the staff had increased to 300 to 400, while the budget had increased to 50 to 60 million dollars. In addition to developing the DSIF, these same technical resources were heavily engaged in supporting the first of the NASA lunar and planetary missions, the Rangers and Mariners.

Although the *Mariner 2* mission to Venus was very successful and attracted a great deal of scientific attention, the first five Ranger missions were not. In the aftermath of a NASA inquiry into problems with the Ranger program, JPL restructured some parts of its organization that were associated with the lunar and planetary programs.[5] These changes would have a far-reaching effect on the relationship between the organization of the DSIF and that of the flight projects. The JPL structure which followed is shown in Figure 9-3.

Of most significance to the future development of the DSIF was the creation of two new offices headed by assistant laboratory directors who reported directly to the JPL director. These were a Lunar and Planetary (Flight) Projects Office headed by Robert J. Parks and a DSIF Office headed by Rechtin. These two offices would provide the programmatic direction for 1) the Mariner, Surveyor, and remaining Ranger missions and 2) the DSIF, respectively. Technical and administrative support would be provided by the seven technical divisions and the six administrative divisions. Although initially the Technical Divisions were in direct line to the JPL director, an additional Office for Technical Divisions was established within the next year with Fred H. Felberg as ALD. This office coordinated and directed the activities of all seven technical divisions at the program level. With the exception of moving Victor to the Chief position, with Joseph F. Koukol as deputy, the Telecommunications Division remained intact.

JPL had arrived at an organization within which the DSIF, later to become the DSN, would coexist with the flight projects and the technical divisions for the next several decades. As we shall see, fairly major modifications were made from time to time, but with these basic offices in place, the DSN moved rapidly forward to meet the demands for tracking and data acquisition support for the expanding NASA Lunar and Planetary program.

Equally important to the evolution of the DSN in these early years were the changes taking place in the newly formed NASA organization. As William R. Corliss explained,

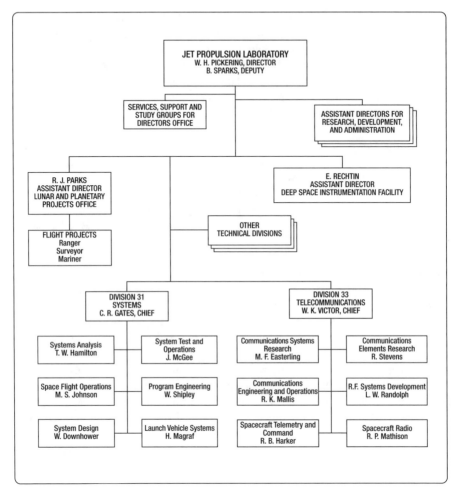

Figure 9-3. JPL organization, 1964.

Rather unexpectedly in the early 1960s, the tracking and data acquisition function was assuming more and more importance in NASA's budget and, consequently, in its organizational structure. Space flight turned out to be not just all launch rockets and spacecraft, it depended very heavily upon ground facilities for testing, launching, and, of course, tracking and communication. The early literature of space flight does not foresee these developments at all. Management practicalities soon forced NASA to recog-

The Deep Space Network as an Organization in Change

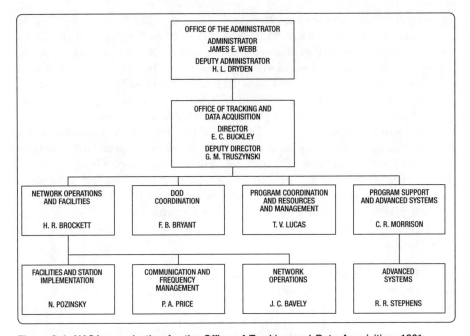

Figure 9-4. NASA organization for the Office of Tracking and Data Acquisition, 1961.

nize the importance of tracking and data acquisition by placing this function on a par with space science, manned space flight, etc. On 1 November 1961, a new Office of Tracking and Data Acquisition (OTDA) was created at NASA Headquarters. Edmond C. Buckley, who had been in charge of Space Flight Operations, was named director of the new office. In effect, the entire tracking and data acquisition function was elevated a notch in the NASA Headquarters hierarchy.

The new Office of Tracking and Data Acquisition in NASA Headquarters at the end of 1961 is shown in Figure 9-4.

The organization remained in this form until OTDA moved up another step within the NASA hierarchy when the position of Director, previously filled by Ed Buckley, was elevated to Associate Administrator status in 1966. Buckley held that position until he retired in 1968. He was replaced by Gerald M. Truszynski.

Truszynski, a native of Jersey City, New Jersey, came to NASA Headquarters in 1960 from Edwards Air Force Base in California, where he had been chief of the instrumentation division involved in developing tracking systems for jet- and rocket-powered aircraft. A year later, he was named deputy in the Office of Tracking and Data Acquisition and eventually followed Edmond Buckley to become the second Associate Administrator of that office in 1968. In that position, he was responsible, at the NASA Headquarters level, for the planning, development, and operation of global tracking networks, including the DSN, facilities for NASA Communications (NASCOM), and data acquisition and processing for all NASA spaceflight programs. His term of office covered the Mariner and Viking Eras and paralleled the Bayley period in the Tracking and Data Acquisition Office at JPL. This was a period of enormous growth and change in the DSN. It was, perhaps, the period when the Network passed from youth to maturity. Gerry Truszynski and Bill Bayley worked well together as heads of the teams at Headquarters and JPL that brought about those changes.

In addition to his involvement with NASA's Deep Space Network, Truszynski played a major role in the provision of tracking, data, and supporting services for the Apollo 8 and Apollo 11 lunar missions.

Although the DSIF Office at JPL was not made an official program office until 1963, interactions between it and OTDA had been conducted as if it were. This situation changed in October 1963, when Rechtin's title was elevated from Program Director for the DSIF to Assistant Laboratory Director for Tracking and Data Acquisition (ALD/TDA) and the former DSIF Office became the official TDA Program Office, a title it retained for the next thirty years. At last, the organization at JPL paralleled that at NASA Headquarters, as far at TDA was concerned. There remained one more event to finish the story of the formation of the modern DSN. That final event occurred in December 1963 with the formal establishment of the Deep Space Network by the JPL director. The historic interoffice Memo 218 that made the announcement, dated 24 December 1963, is reproduced in Figure 9-5.

In addition to the former DSIF, the tracking stations at Goldstone, California; Woomera, Australia; Johannesburg, South Africa; and Cape Canaveral, Florida, the official DSN included the intersite communications now called the Ground Communications Facility (GCF) and the mission-independent portions of the new Space Flight Operations Facility (SFOF) at JPL.

The Deep Space Network as an Organization in Change

JET PROPULSION LABORATORY

OFFICE OF THE DIRECTOR
Interoffice Memo 218
December 24, 1963

To: Senior Staff
Section Chiefs
Section Managers

From: W. H. Pickering

Subject: Establishment of the Deep Space Network

Effective immediately, the Deep Space Network is established by combining the Deep Space Instrumentation Facility, Interstation Communications, and the mission-independent portion of the Space Flight Operations Facility. Development and operation of this network is the responsibility of the Assistant Laboratory Director for Tracking and Data Acquisition by extension of the role statement for this Assistant Director (Office of the Director IOM 200, October 2, 1963).

Funding sources are unchanged for Fiscal 1964 and for the budget submitted by JPL for Fiscal 1965. However, JPL will endeavor to have OTDA and OSSA agree on a single source of funding as quickly as possible.

The interface with mission peculiar facilities and organizations will be worked out between the Assistant Laboratory Director for Tracking and Data Acquisition and the Assistant Laboratory Director for Lunar and Planetary Projects using as a guideline the definition of "mission-independent" as:

1. Required for two or more flight projects.

2. Best handled by JPL and not outside flight project organizations (ARC Pioneer, GSFC-MSFN, LeRc Centaur, etc.).

This change is made in order to accommodate efficiently the increasing number of outside flight projects for which the Jet Propulsion Laboratory has been tasked to supply tracking and data acquisition support. The change should also assist in closer integration of the previously separate facilities.

W. H. Pickering
Director

WHP:mc

Figure 9-5. The Deep Space Network established, December 1963.

The events just described were of fundamental importance to the shaping the DSN during its formative stages. These and subsequent events are summarized in the table below.

Top-Level Management of the DSN at JPL, 1960–97

Year	Event or Head of DSN: Title	Deputy/Program Manager	JPL Director
1958	NASA established (October); JPL moves to NASA (December)		Pickering
1960–62	Rechtin: Chief, Div. 33[1] and DSIF Director (January 1961)	Bayley	Pickering
1963	Rechtin: ALD[2]/DSIF (March) Victor: Chief, Div. 33, DSN established (December)	Bayley	Pickering
1964	Rechtin: ALD/TDA[3]	Bayley	Pickering
1967	Bayley: ALD/TDA	Victor	Pickering
1980	Lyman: ALD/TDA	Johnson[4]	Murray
1987	Dumas: ALD/TDA	Johnson	Allen
1992	Haynes: ALD/TDA	Westmoreland	Stone
1994	TMOD[5] Established		Stone
1996	Westmoreland: Director, TMOD (June)	Coffin[4]	Stone
1997	Squibb: Director, TMOD (June)	Coffin	Stone

1 Telecommunications Division.
2 Assistant Laboratory Director.
3 Tracking and Data Acquisition; changed to Telecommunications and Data Acquisition in 1982 to avoid confusion with NASA's new Space Network using the Tracking and Data Relay Satellite System (TDRSS), planned for operation the following year.
4 The title of Deputy was superseded by Program Manager in 1980 and again in 1996.
5 Telecommunications and Mission Operations Directorate.

The Deep Space Network as an Organization in Change

After the DSN was formally established at the end of 1963, the basic organizations involving the DSN at NASA Headquarters and at JPL changed very little until 1994, except for personnel and some vacillation concerning responsibility for the SFOF. Within the DSN organization, however, many changes were made in the interest of improving working relationships with OTDA and with the Flight Projects. The growth and development of the DSN organization during this thirty-year period, 1964–94, is the subject of our next discussion.

THE FORMATIVE YEARS (1964 TO 1994)

A DSN Manager for Flight Projects

Within a few short months of the formalization of the DSN, further changes were made to the organizational structure under which the DSN operated at JPL. A JPL organization chart from 1964 showing the relative positions of the TDA and flight projects offices, including their supporting technical divisions, is reproduced in Figure 9-6.

All of the technical divisions were directed by a single ALD, Fred Felberg. A technical staff element was added to the TDA Office, and Richard K. Mallis took over the Communications Engineering and Operations Section formerly managed by Renzetti. Renzetti moved to a Technical Staff position in the TDA Office as the first DSN Manager for the Ranger, Mariner, and Apollo missions. He was soon joined by DSN Managers for the Lunar Orbiter, Surveyor, and Pioneer missions. In the Systems Division, Thomas S. Bilbo replaced Marshall S. Johnson as Chief of Space Flight Operations. Thus, Mallis and Bilbo were independently responsible to their division managers for the operation of the SFOF and the DSN, respectively. They received technical support from other sections within the same divisions, but in a programmatic sense, their authority derived from completely separate sources. The SFOF traced its funding and program direction, via the Lunar and Planetary Projects Office at JPL, to OSSA at NASA Headquarters, while the DSN traced its funding and program direction, via the Tracking and Data Acquisition Office at JPL, to OTDA at NASA Headquarters.

The appearance of the DSN Manager function in the TDA organization needs some explanation at this point, since it became a key factor in determining the future working relationships between the DSN and its client users, the flight projects.

As an inevitable consequence of the creation of separate offices at NASA Headquarters for flight projects and tracking and data acquisition, a separate system of accountability for the resources required by the one to support the other came into being. Based somewhat on a long-existing system used at the military test ranges for receiving and responding to range users' requirements for instrumentation to cover their tests, the system was introduced to the DSN by a formal NASA Management Instruction (NMI 2310.1) dated March 1965. In this system, flight project requirements for DSN tracking and data acquisition services were identified in a Support Instrumentation Requirements Document (SIRD). Upon receipt of a SIRD, the DSN would respond with a NASA Support Plan that committed its resources as it deemed necessary to meet the requirements in the SIRD. The SIRD and NSP were formal documents signed by

The Deep Space Network as an Organization in Change

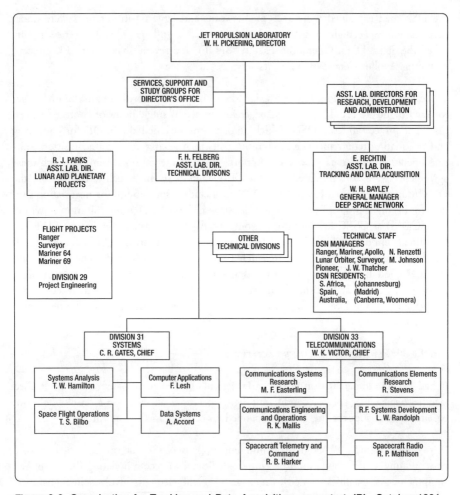

Figure 9-6. Organization for Tracking and Data Acquisition support at JPL, October 1964.

the program managers at OSSA and OTDA, respectively, and were intended to signify approval for the expenditure of the TDA resources involved.

In earlier times, when one or two tracking stations and simple facilities for control of flight operations were all managed more or less by a single entity, verbal agreements or internal memos between participating groups were all that was needed to conduct missions for a single JPL-managed spacecraft. No longer was that the case. By the mid-1960s, the DSN had expanded to two 26-m Networks and large 64-m antennas were under construction. Spaceflight missions were conducted from a large, completely new facility called the Space Flight Operations Facility (SFOF), and several major, non-JPL projects, including Apollo, were demanding its services.

To deal with this situation, Rechtin assigned a technical staff position to a DSN Manager for each flight project. The DSN manager functioned as single point of contact between his flight project and the DSN, which at that time included the SFOF, for conveying and negotiating requirements and commitments for tracking and data acquisition, mission operations, and data-processing support. The documentation system by which these agreements were consummated was the SIRD and NSP. In actuality, the role of the DSN Manager extended beyond the DSN to include responsibility for negotiating support for his flight project from other entities such as the NASA Communications Division (NASCOM) at GSFC for communications, the Air Force Eastern Test Range (AFETR) for launch support tracking coverage from land stations, Department of Defense (DOD) for ships and aircraft, the Lewis Research Center (LeRC) for launch vehicle data, and so on. Although the DSN manager did not have an inline operational role in flight operations, he was at all times held accountable to the flight project manager for the performance of the DSN in support of mission operations, according to the negotiated agreements.

The Space Flight Operations Facility

It had become apparent during the early part of the Ranger Program that the limited facilities available at JPL for conducting flight operations and mission control would not be adequate to support the greatly expanded space programs of the future.

With this in mind, JPL made a recommendation to NASA for the construction of a new building at JPL that would be entirely dedicated to accommodating the personnel, equipment, and facilities needed to manage all of the elements involved in flight operations on a continuous, round-the-clock basis. These elements included the DSIF, ground communications, data processing and display, and mission control, with all the internal communications needed to make them work together. NASA approved the recommendation in mid-1961. Under the aegis of the NASA Office of Space Science and Applications (OSSA), construction of what was called the Space Flight Operations Facility (SFOF) began shortly afterward. The SFOF was completed in October 1963. The IBM 7094 and 7040 computers and 1301 disk files were moved in immediately and began operating just before the launch of the *Mariner 4* mission to Mars in November 1964. Thus,

The Deep Space Network as an Organization in Change

Mariner 4 was the first flight project to be supported by the new SFOF operating with a combination of real-time and non-real-time computing systems.

In December 1964, however, programmatic responsibility for the SFOF was transferred from OSSA to OTDA on the basis of the interrelated functions of mission control in the SFOF and Network control in the DSN. This responsibility had been anticipated in Pickering's announcement establishing the DSN in December 1963, and it remained a major element of the DSN for almost the next decade.[6] While the TDA Office assumed programmatic responsibility for the SFOF, its operational and technical support was provided by the Space Flight Operations and Data Systems Sections of the Systems Division 31, in a somewhat analogous way to that in which the Telecommunications Division 33 supported the DSN. (See Figure 9-6.)

The years following the transfer of the SFOF to the DSN were marked by a great increase in the number and complexity of the flight missions that the DSN was called upon to support. Many of these missions were managed by institutions other than JPL. The Lunar Orbiter and Pioneer missions (and, to a lesser extent, the Surveyor and Apollo missions) were typical of this period.

At the same time, the DSN itself was expanding. The second 26-m subnet and the L/S-band conversion were completed in 1965; the first 64-m antenna became operational in 1966; and the multi-mission concept was introduced in 1967.

Toward the end of the 1960s, improvements in communication control, data distribution and display, and computer-based switching of intersite circuits were made in the SFOF. Intersite communications were carried on high-speed and wideband data lines at speeds of 4,800 bps and 50 kbps, respectively, by the DSN Ground Communications Facility, which was also part of the SFOF. Paralleling the increased data-handling capability of the Mark III DSN, the data-handling capability of the SFOF data processors was also expanded by the replacement of the old IBM computers with two IBM 360-75 and two Univac 1108 machines with appropriate interface switches for flexibility and redundancy.

While all of this development was taking place in the DSN, GCF, and SFOF, all three facilities were simultaneously conducting an intense program of lunar and planetary missions operations. The success of all these missions attests to, amongst other things, the adequacy of the organization under which they were performed. With the exception of several personnel changes, notably the resignation of Rechtin and the appointment of Bayley to replace him as ALD/TDA, the organization remained as shown in Figure 9-6.

Nevertheless, despite the apparent suitability of the existing organization to the task it was being called upon to perform, all was not well at the higher levels of management. As William Corliss explains,

> The deep space missions of the late 1960s and early 1970s brought with them substantial increases in the SFOF's data[-]processing load. Much of this data processing was strictly scientific and unrelated to DSN operations. Yet, the tracking and data acquisition function was obligated to provide for this computer time without the authority to review requirements. The flight projects were, in essence, requesting and getting large blocks of computer time and were neither financially nor managerially accountable for them. It was a bad managerial situation. NASA Headquarters recognized the situation and, in October 1971, Gerald Truszynski (OTDA) and John Naugle (OSS) reviewed the problem and decided to transfer the SFOF functions from OTDA back to OSSA. In this way the responsibility for review and validation of requirements and the associated costs of scientific data processing would be borne by the flight projects themselves.

The change at Headquarters was immediately reflected at JPL, where responsibility for the SFOF was moved from the TDA Office to a newly created Office of Computing and Information Systems (OCIS). This new organization reported not to OTDA, but to OSS. The transfer became effective in July 1972, and the organization of the technical divisions at JPL which resulted from or gave rise to (depending on the point of view) the separation of the SFOF from the DSN is shown in Figure 9-7.

The Office of Computing and Information Systems, with Allan Finerman as manager, directed the Data Systems Division, which, under the management of Glen E. Lairmore, provided technical support and operational direction for the SFOF and the GCF.

The Telecommunications Division and Mission Analysis Division provided engineering development and navigation-related analysis for the independent DSN.

The TDA Office, under the ALD (Bayley), assigned a manager to each of the three major areas of effort in the Network for which it was responsible. Both implementation of new hardware and software in the Network and Network Operations functions were carried out by Kurt Heftman, the DSN Engineering and Operations manager. Interfaces between the DSN and all flight projects for the negotiation of requirements and commitment of DSN resources were provided by Mission Support Manager Nicholas A. Renzetti and his staff of five DSN managers. The integration of the various elements of

The Deep Space Network as an Organization in Change

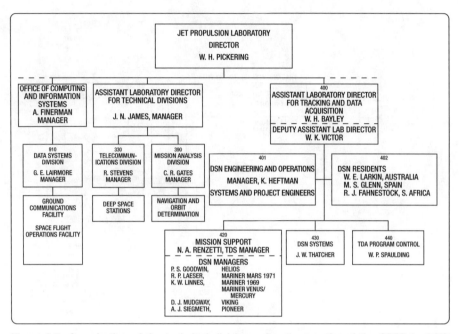

Figure 9-7. Organization of the technical divisions after the transfer of the SFOF to OCIS, 1972.

the DSN into coherent unified systems was led by John W. Thatcher. The TDA Office also maintained a representative at each overseas site to deal directly with local government officials on matters pertinent to the operation of the tracking stations in their countries. Program Control (budgetary matters) were handled by Wallace P. Spaulding.

Although the separation of the SFOF from the DSN may have appeared to be merely a "paper exercise," it was not accomplished without considerable disruption to the carefully crafted interface agreements already in place between the DSN and the Pioneer and Viking flight projects.[7, 8] Schedules, interface agreements, and capabilities had been negotiated with various elements of the flight projects and had been formally documented and approved, in accordance with current practices. These schedules, agreements, and capabilities, of course, included the SFOF as well as the DSIF. When the separation took place, new interfaces between the DSN and the flight project, and the DSN and OCIS, had to be developed and documented.

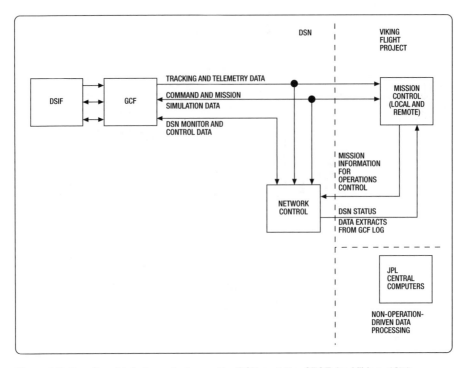

Figure 9-8. Functional interfaces between the DSN and the SFOF for Viking, 1972.

A typical example of the functional interfaces between the DSN and the SFOF after the separation of responsibility is shown in Figure 9-8.

The Viking Mission Control Center (VMCC), which included the SFOF central computing system, the mission support areas, and the Viking mission simulation system, was the joint responsibility of the OCIS and the Viking Mission Operations System. The DSN was responsible for the deep space stations, which included the 64-m and 26-m subnets, and transport of data to and from the VMCC via the high-speed and wide-band data lines of the GCF. Control and monitoring of Network performance and validation of the data streams flowing between the VMCC and the deep space stations was to be accomplished by a separate data-processing capability that would be independent of the mission-related computers in the SFOF. These functions would be accommodated in a new Network Operations Control Center (NOCC), which was being designed at the time (1972).

The Deep Space Network as an Organization in Change

As the SFOF came into use for multi-mission support in the early 1970s, common terminology was changed to better reflect the way in which the facility was actually being used. The building, JPL Building 230, and its support facilities retained the title SFOF. The SFOF was eventually designated as a Historical Landmark by the U.S. Department of the Interior in 1986. The SFOF housed two facilities that were controlled by the DSN, namely, the GCF (in the basement) and the NOCC (on the first floor). The remainder of the building accommodated the flight controllers and spacecraft analysts for the various flight projects and the data-processing equipment required to support their missions. That ensemble of accommodations and equipment was called the Mission Control and Computing Center (MCCC). It was occupied in turn by one or more Flight Projects for the duration of each mission, after which it was vacated and reconfigured for the next flight project. In some cases, non-JPL flight projects elected to conduct flight operations from Mission Control Centers at their own institutions rather than from the SFOF at JPL.

The first flight project to use the MCCC in this way was Pioneer, managed by the Ames Research Center (ARC) in Moffett Field. Similar arrangements for remote Mission Control Centers came into greater use in later years, as the DSN began supporting greater numbers of non-NASA spacecraft.

The Bayley/Lyman Years

The Tracking and Data Acquisition Office at JPL (Office 400 in the JPL hierarchy) was renamed Telecommunications and Data Acquisition Office and reached maturity under the direction of Bill Bayley over the span of his tenure as ALD between 1967–80. From a relatively small organization of fourteen in 1972 (Figure 9-7), the TDA Office had grown to an organization of seventy-four people, as shown in Figure 9-9, by 1978. Two offices—one for technology development, the other for long-range planning—had been added. Because of the tremendous growth in DSN flight operations and the expansion of the Network needed to meet the increasing number of flight missions, it was necessary to split the former Engineering and Operations Section into separate offices. One of these new offices (430) was managed by Renzetti and focused on developing the engineering systems of the DSN. Interfacing with the flight projects and scheduling antenna time, configuration control, and all other aspects of flight mission support except the actual hands-on operations and maintenance became the responsibility of the DSN Mission Support Office (440), under Spaulding.

In 1980, W. H. Bayley retired and the extant JPL director, Bruce C. Murray, selected Peter T. Lyman to replace Bayley as ALD/TDA. Lyman retained that position until

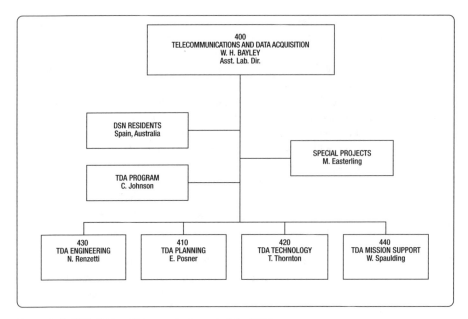

Figure 9-9. TDA Office 400 organization at JPL, 1978.

1987. Bruce Murray was succeeded as JPL Director by Lew Allen (1982), and the TDA Office continued to expand to meet the demands placed upon it by an ever-increasing number of NASA and non-NASA deep space and Earth-orbiter missions, as well as numerous nonflight, scientific programs.

In Figure 9-10, a snapshot of the TDA organization, near the end of the Lyman years in 1986, reveals the extent of the remarkable growth of the organization in that period.

Most noticeable in this organization was the addition of the TDA Science Office (450), created by Lyman in 1983 for the reasons described in the previous chapter. Renzetti managed that new office, which included a Geodynamics program, the SETI program, the Goldstone Solar System Radar program, and several other special research projects.

Paul T. Westmoreland was appointed manager of the TDA Engineering Office (430), which was by then responsible for interagency arraying, compatibility and contingency planning, and implementation of new engineering capability into the Network and GCF, in addition to its ongoing task of DSN System Engineering. The TDA Planning and Technology Development offices appeared under different managers, and they, too, had

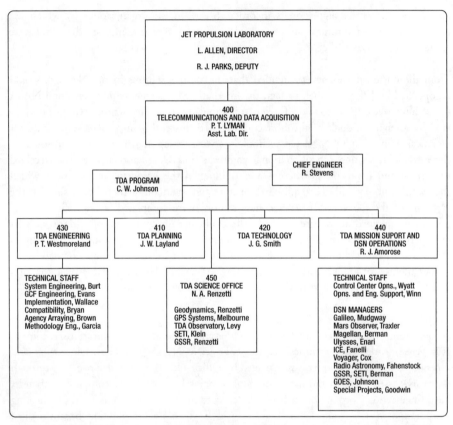

Figure 9-10. TDA Office 400 organization at JPL, 1980–87.

expanded in scope. Under the management of Raymond J. Amorose, Office 440 included management responsibility for the large Maintenance and Operations (M&O) Contract, by which an industrial aerospace contractor provided engineering and operations services to the DSN for the Goldstone Complex, the NOCC, and related activities in the Pasadena area. Three DSN engineers managed the M&O contract, the Network Control Center at JPL, and the Complex Control Center at Goldstone. Eight DSN managers attended to the interests of over twenty flight projects that were active in the DSN at that time. Finally, a program manager, chief engineer, and chief technologist provided direct support for the ALD/TDA in establishing budgetary, policy, and long-term development goals.

At this point, a brief digression is in order to explain the organizations by which the three Deep Space Communications Complexes (DSCCs) that composed the DSN were related to the TDA Office.

From the outset, it was never intended that the various stations in the Network would be operated by JPL personnel. In fact, the international agreements that permitted NASA to establish tracking stations on foreign soil stipulated that the facilities would be operated by foreign nationals of the host countries. To this end, comprehensive facilities for technical training in DSN-related technologies were set up at JPL and at Goldstone. At these facilities, visiting foreign national engineers and U.S. national engineers received theoretical and hands-on instruction and training in the hardware and software which they would subsequently be required to operate and maintain at their home sites. In the early life of the DSN (DSIF), when new stations were completed, a few JPL technical personnel remained at the site until the site could sustain continued operations with its own personnel. Later, the sites sent key personnel to the U.S. for training so that upon their return to their home bases, they could implement the new antennas or new capabilities themselves. There was a constant flow of technical personnel between the sites and Goldstone or JPL for this kind of information exchange and dissemination.

In the United States, the staff to operate the tracking stations and DSN-related services at Goldstone and the Network Control Center and Ground Communications Facility at JPL were provided by a technical services contract to NASA/JPL. The Maintenance and Operations (M&O) Contract was managed by Office 440, as discussed above, and through the years passed through the hands of various major U.S. aerospace companies—Bendix Field Engineering Corporation in 1960, Philco-Ford in 1970, and Bendix Field Engineering Corporation in 1978 and 1988. Bendix Field Engineering Corporation became Allied Signal Aerospace in 1992 without interruption to the contract.

Under the direction of JPL personnel, the M&O contractor was responsible to the DSN for the supporting services listed below (in greatly abbreviated form):

1. At JPL:
    a. Operate and maintain the Network Operations Control Center (NOCC), the Ground Communications Facility (GCF), the Compatibility Test Facility (CTA 21), and (occasionally) the DSN Launch Support Test Facility (MIL 71) at Cape Kennedy.
    b. Manage operations support programs such as antenna scheduling, discrepancy reporting, data control, radio frequency interference management, and Network performance evaluation.

c. Provide general support to all Network facilities in the areas of logistics, technical training, documentation, engineering change and configuration control, Network-level maintenance, and sustaining engineering.
d. Provide high-level technical and administrative support for specific tasks in the TDA Office and Telecommunications Division.

2. At the Goldstone Deep Space Communications Complex (GDSCC):
   a. Operate and maintain all of the tracking stations, their intrasite communications, and the research and development facility at the DSS 13 Venus site.
   b. Maintain the established Complex Maintenance Facility, Network Standards Laboratory, and other services.
   c. Maintain all of the buildings, plant equipment, roads, lighting, power, and air-conditioning, as well as fire, food, and security services.

The M&O Contract for Goldstone assigned complete responsibility for running the complex to the contractor, and over the many years of its operation, the GDSCC technical staff were always ready to respond to a call for extra effort, whether it was for an inflight mission, an engineering research and development task, or a DSN science program.

The staff and infrastructure needed to provide engineering and operations support for the Canberra Deep Space Communications Complex (CDSCC) were provided by Australian industry under a contract with the Australian government. However, all costs associated with operation of the CDSCC were borne by NASA under the terms of the intergovernmental agreement. The Complex Director remained an employee of the governmental department (agency) charged with administering the contract on behalf of the Australian government and was responsible to that department head for routine management of the Complex, its staff, and its facilities. On behalf of NASA, the ALD/TDA at JPL was responsible only for providing the operational facilities and directing their use in support of the NASA programs. The ALD/TDA retained a representative resident in Canberra to facilitate the interchange of technical and administrative information between the CDSCC and JPL. He played no part, however, in the direction or operation of the Complex, and the position was abolished in 1985.

International agreements between NASA and the Spanish government for the management, engineering, and operations support for the Madrid Deep Space Communications Complex (MDSCC) were essentially similar to those between NASA and the Australian government for management and operation of CDSCC. There, however, both the director and his staff were employees of a government agency, Instituto Nacional Tecnica Aerospacial

(INTA), until 1992, when INTA, retaining the NASA contract, delegated the maintenance and operation to its wholly owned company, Ingenieria y Servicios Aeroespaciales) (INSA).

In both situations, this seemingly awkward organizational arrangement worked well, in general terms. Some initial administrative difficulties were cleared up as time went on, but the technical staffs at both complexes were always prepared (and on many occasions were called upon) to respond unreservedly to the exigencies of deep space mission operations.

The Dumas/Haynes Years

When the position of Deputy to the Director of JPL became vacant at the retirement of Robert Parks in 1987, Lyman was appointed to fill the position, and Larry N. Dumas became ALD/TDA. Apart from the inevitable personnel changes and some structural changes within the five internal offices of the TDA (400) organization, little changed in an organizational sense in the DSN during the Dumas years, 1987 to 1992. More changes in upper management, however, were on the way.

In 1990, Edward C. Stone replaced Allen as director of JPL, and he was joined in 1992 by Dumas as deputy. The then vacant position of ALD/TDA was filled by Norman R. Haynes, who brought Westmoreland from the TDA Engineering Office to become his deputy. The TDA organization had remained essentially unchanged since 1986, and, except for changes in personnel, is well represented by Figure 9-11. This was the final form of the TDA Office. There was no hint of the very dramatic changes that would overtake the entire TDA (400) organization in the next several years.

When Haynes took over the TDA organization in 1992, it was one of several offices headed by an Assistant Laboratory Director (ALD). These offices, under the JPL director, composed the basic structure of the entire JPL organization. When he left in 1996, the JPL structure had been changed to one based upon an ensemble of directorates, each headed by its own director. The directorates were not simply renamed offices, but were restructured to reflect the "policy for change" brought to JPL by the new director, E. C. Stone.

To fully appreciate the latest changes in the TDA organization, it is necessary to understand the organizations at NASA Headquarters and in the MCCC at JPL at that time. Both influenced the shape of the future TDA organization.

The Deep Space Network as an Organization in Change

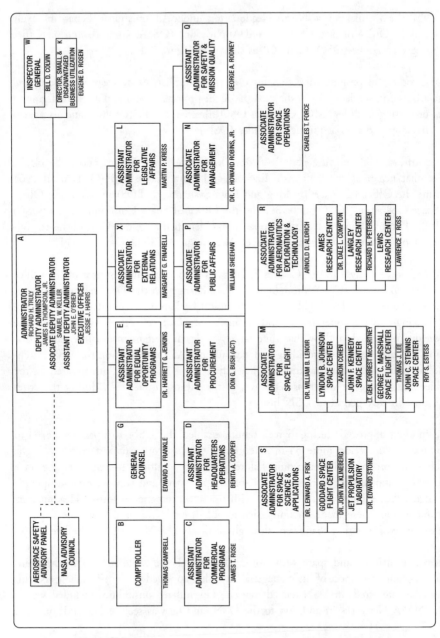

Figure 9-11. National Aeronautics and Space Administration, 1991.

By 1991, the organization at NASA Headquarters under Administrator Richard H. Truly included an Office of Space Science and Applications (OSSA, Code S) and an Office of Space Operations (OSO, Code O) as shown in Figure 9-11.

In the context of program direction for users and suppliers of deep space tracking and data-acquisition services, OSSA directed the flight projects (users) and associated ground data-processing facilities (MCCC) while OSO directed the TDA Office and its associated tracking and data-acquisition facilities (DSN).

For clarification, during the period of DSN history reviewed in this book, the NASA office that began as the Office of Tracking and Data Acquisition (OTDA) in 1961 became the Office of Space Tracking and Data Systems (OSTDS) in 1983, the Office of Space Operations (OSO) in 1988, and the Office of Space Communications (OSC) in 1992. During almost that entire period, it was identified as "Code O" and was responsible for programmatic direction of the DSN through the TDA Office at JPL.

The selection in July 1989 of Charles T. Force as Associate Administrator of the Office of Space Operations (OSO) brought immediate changes to the DSN relationship with NASA Headquarters. His highly relevant technical background, considerable experience in tracking-network operations, and strong motivation effectively streamlined the unwieldy, outmoded, and by that time largely irrelevant system of documentation that controlled agreements for providing and expending NASA resources on the DSN antennas. The time span of his position in that office, 1989–96, placed him squarely in the Galileo Era, where he was most effective, at the Headquarters level, during the crises that arose in that period of DSN history.

The funding approved during Force's tenure enabled the DSN to upgrade the 64-m antennas to 70-m, add more 34-m antennas to the Network, support recovery of the Galileo mission from almost certain failure, introduce multiple antenna arraying on a routine operational basis, move toward K-band operations, and consolidate and upgrade the Signal Processing Centers. Together with the improvements in formal Headquarters documentation, these made an impressive record of achievement that set the stage for the major reorganization that swept through all the NASA Networks in the late 1990s.

Forceful in attitude and spare with words, "Charlie" Force came straight to the point in a discussion, an aspect of his managerial style that appealed to the JPL managers with whom he interacted on DSN-related matters. His full responsibilities included several other NASA Networks in addition to the DSN, and he was seen as fair and impartial

The Deep Space Network as an Organization in Change

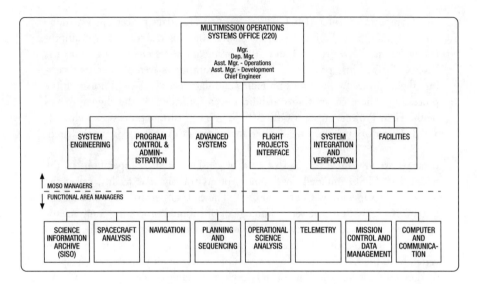

Figure 9-12. Multimission Operations Systems Office (440), 1993.

when at times his decisions regarding the allocation of limited resources were at odds with prevailing opinions at JPL.

Force joined NASA in 1965 as Director of the Guam Tracking Station and, except for returning to industry for a couple of years in the early 1980s, held increasingly responsible positions throughout his career at NASA. A native of Shoals, Indiana, and resident of the state of Maryland, he held a B.S. degree in aeronautical engineering from Purdue University. Force left NASA in 1996 to pursue commercial business interests.

By 1993, the organization at JPL formerly referred to as the MCCC had embraced the concept of multi-mission operations and had evolved into the Multimission Operations Systems Office (440) with the structure shown in Figure 9-12.

Because of its close association with the evolution of the DSN, the path by which the MCCC evolved is of historical interest and is neatly described by Alazard[9] as follows:

> From the mid-1960s through 1970, the ground data system was comprised of three IBM 7094 computers to support each of the then current (flight) projects. Between 1970 and 1977, the architecture and organization of computing changed and moved toward a multimission orientation. The Mission Control

and Computing Center (MCCC) was created and data processing was done on IBM 360-75's. The flight projects shared this single system in a multiprocessing mode. Between 1973 and 1981, minicomputers were phased into the MCCC for realtime processing. Mainframe computers were used for processing (flight) project applications programs and data records[,] i.e.[,] non-realtime processing. These computing capabilities were provided to the flight projects by the Fight Projects Support Office (FPSO). The realtime processors were usually dedicated to a given project while the non-realtime processors were shared.

In an effort to reduce the cost of data processing for flight operations in the mid-1980s, FPSO initiated development of an entirely new facility, called the Space Flight Operations Center (SFOC). The SFOC performed the functions of the MCCC with the newer technology of distributed processing using microcomputers and local area networks in a workstation environment. When it was completed in the early 1990s, existing flight project support was gradually converted from the MCCC to the new system, and new flight projects were adapted to the multimission capabilities and services. In due course the MCCC was phased out and the SFOC became the core of an even more advanced data processing system called the Advanced Multi-Mission Operations System (AMMOS).

The name FPSO was changed to Multimission Operations Systems Office (MOSO) in 1992 to better reflect the multi-mission nature of the organization. MOSO became the overall organization responsible for development, operations, and maintenance of the flight project support capabilities. AMMOS provided the direct operations support functions using the hardware and software of the Multimission Ground Data System (MGDS).

It was the task of MOSO to establish the set of multi-mission data processing capabilities and operational services that supported the flight projects in accomplishing their mission objectives. The AMMOS was the set of hardware and software tools by which this task was accomplished. Defined in this way, the MOSO task closely paralleled the DSN Operations and Mission Support Office (440) task when tracking and data-acquisition services were substituted for data-processing services and the DSN was substituted for AMMOS. The MGDS was the counterpart of the DSN.

The JPL Strategic Plan: 1994

Shortly after the Clinton Administration took office in January 1993, all federal agencies, including NASA, began to feel the effect of a groundswell of change toward a smaller, more efficient, and less costly bureaucracy. A new NASA Administrator, Daniel S. Goldin, had been appointed early in 1992, and he was quick to respond to the presidential initiative for cost reduction in government agencies. In the course of the changes that followed, the Headquarters organization was significantly reduced in size; many functions were deleted or combined; and a new, streamlined NASA organization emerged. In an environment of economic constraints and redefined national goals, the NASA budget declined in Fiscal Year 1995 for the first time since the end of the Apollo program in the late 1960s. New approaches and revised priorities for the U.S. space program were developed to ensure that space science and technology would continue to advance for the benefit of the nation. All of these factors required NASA and, consequently, JPL to rethink their ways of doing business, adopt new strategies, and create a shared understanding of how to meet customer needs in the new environment.

This state of affairs formed the driver for a Strategic Plan developed for JPL in 1994 by its new director, Edward C. Stone.[10] It defined the set of NASA programs in which

---

NASA SPACE COMMUNICATIONS AND OPERATIONS PROGRAM

- Provide affordable, world-class support to large, complex missions: Ulysses, Galileo, Cassini, and EOS.
- Advance the state of the art in low-cost support of small and moderate missions.
- Develop the capability to provide Ka-band services for future low-cost missions.
- Upgrade mission operations concepts and support systems to enable simultaneous operation of many missions at significantly reduced unit cost.
- Develop concepts for promising new services such as optical communications and non-DSN tracking terminals.
- Revolutionize ground system designs to reduce complexity and operating costs while emphasizing new technology opportunities and controlled risk.
- Support many more missions in the next five years while reducing DSN tracking operations costs by 50 percent.
- Advance the state of the art in deep space and near-Earth navigation and communications to meet expected future mission requirements.
- Seek new concepts for low-cost support of small and moderate missions.
- Increase the use and scientific impact of the DSN in radio science, astronomy, and planetary radar studies.
- Lead in developing international science agreements and cross-support systems for an affordable global space exploration program.

Figure 9-13. NASA Space Communications and Operations Program.

JPL would engage to support "NASA's strategic enterprises and other national needs." One of these programs was NASA Space Communications and Operations, and because of its importance to the future long-term development of the DSN, it is reproduced in Figure 9-13.

One of the overarching concepts embodied in the newly defined program was the combining of space communications with space operations to create a single entity of reduced complexity with lower operating costs. This was to be achieved without sacrificing performance and was concomitant with a stimulating call to further advance the state of the art in associated technologies.

Birth of the TMOD

The Telecommunications and Mission Operations Directorate (TMOD) was created in 1994 to support the NASA Space Communications and Operations program defined in the Strategic Plan described above. As such, it represented one element of the Laboratory's overall response to the NASA Strategic Plan.

The transition from the TDA organization to the TMOD organization took place during the latter half of the Haynes years (1994); although there was a considerable disruption of ongoing activities initially, by 1996, when Paul T. Westmoreland assumed the position of Director, the new arrangement had settled down and assumed the form shown in Figure 9-14.

The rearrangement and consolidation of former TDA functions, which is apparent in Figure 9-14, reflected a new and cost- effective approach to complex business management systems made popular at that time (1992–93) by Dr. Michael Hammer, co-author of the book *Reengineering the Corporation*.[11] With institutional encouragement, JPL management personnel at all levels attended Hammer's courses and returned to JPL to apply the new methodology to the workplace. Early in 1994, the TDA senior staff spent two months developing a process map for the TDA organization. Two of these processes were selected for actual reengineering, and process owners were named. The selected processes and owners were Services Fulfillment (Raymond J. Amorose) and Asset Creation (Edward L. McKinley). Leslie J. Deutsch was assigned to lead the Reengineering Team (RET) for Services Fulfillment—the process deemed most suitable for the first engineering effort. Although the RET was formed initially to reengineer only the TDA organization, the scope of its work was expanded to include the data capture and navigation functions of MOSO when the two organizations were merged to form TMOD.

The Deep Space Network as an Organization in Change

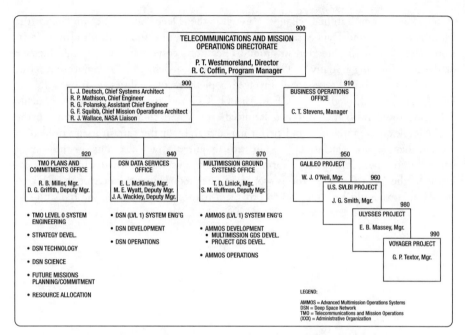

Figure 9-14. Telecommunications and Mission Operations Directorate, 1996.

The RET issued an interim report in September 1994 containing some key items for the future structure of TMOD. Among these were the three Services Fulfillment Processes recommended for reengineering—namely, Data Capture, Activity Planning, and Resources Scheduling, all defined later in this chapter. There was also a proposal that the DSN be changed from an engineering-driven organization, as it had stood in the past, to an operations-driven organization for the future.

The arrangement shown in Figure 9-14 was the first step in the overall reengineering process. Essentially, the former TDA organization was condensed into two offices: one for planning, committing, and allocating DSN resources; the other for DSN operations and system engineering. DSN science and technology were incorporated in the former, DSN development in the latter. In addition to these two offices, the Multimission Ground Systems Office, the project offices of the four inflight missions (Galileo, Space VLBI, Ulysses, and Voyager) and a new business office were added to create TMOD. There could be little doubt that the TMOD was now operations-driven rather than engineering-driven.

While these organizational changes brought the Deep Space Network and the Multimission Ground Data System together under one directorate, they did not effect, alone, the substantial reduction in ground operations costs that was the real objective of the reengineering initiative. DSN operations, now redefined in reengineering terms as the Data Capture Process, still involved a plethora of facilities, starting with the DSCCs, passing through the GCF to the NOCC and to MOSO before eventually delivering the data to the customer. Reduced cost would come from reduction in the size, complexity, redundancy, and performance margin of the infrastructure needed to operate these separate facilities. This, in turn, required a radical change in the way in which future operations were performed, and in the way in which the individual elements of the Data Capture Process would be operated. It was the task of the RET to show how this could be done and to provide a road map for making the transition to the new Data Capture Process.

REINVENTING THE FUTURE (1994 INTO THE NEW MILLENNIUM)

Reengineering TMOD

The RET completed its redesign of key TMOD subprocesses and submitted its final report to DSN Mission Operations Manager Ray Amorose in March 1995.[12] It succeeded in not only meeting but exceeding its initial cost and efficiency goals of reducing the DSN Operations budget by $9M and doubling the available tracking hours by Fiscal Year 1999. The new designs would cost about $16M to implement and were predicted to result in a cumulative savings to TMOD Services Fulfillment of approximately $35M for the period FY 1996 through FY 2000, as shown in Figure 9-15.

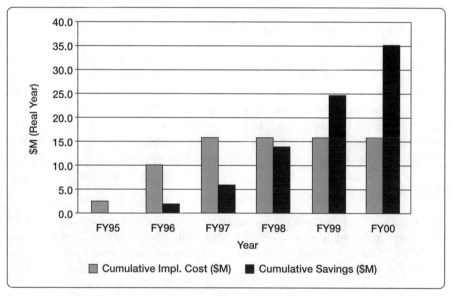

Figure 9-15. Predicted cost and savings for reengineering TMOD Services Fulfillment Processes. The three Services Fulfillment Processes include the following:
1. Data Capture—the process that provided the services of telecommunications, target tracking, and data processing to customers.
2. Activity Planning—the process that generated and consolidated the support data that was needed by the Data Capture Process to perform its functions.
3. Scheduling—the process that allocated the resources of the DSN and MGDS for use in the Data Capture Process.

Because the savings to DSN Operations were estimated to be low, the RET elected not to reengineer the Scheduling Process at that time. It was, therefore, only the Data Capture and Activity Planning Processes that, in reengineered form, contributed to the substantial projected cost savings noted above.

By basing its design on standard (rather than customized) services, the new Data Capture Process simplified mission interfaces, increased operational efficiency, and allowed full use of automation to provide these services. At the same time, a "Connection Operator" was made available to provide specially tailored services essential to the customer's needs. The parts of the NOCC, GCF, and MGDS involved in real-time operations were brought together in a single work area called Central Operations. New, more easily maintained equipment allowed them to be located in a considerably smaller work space with reduced facilities costs.

On the other hand, supporting data such as radio metric predicts, antenna pointing instructions, and schedules for Deep Space Network operations were considered an Activity Planning function and had, in past practice, been generated at the NOCC considerably in advance of the needed time. The inputs were, of necessity, immature, and they invariably led to the generation of many contingency or "just in case" versions, all of which created extra work. In the new, more rapid and efficient "just in time" system, predicts were generated at the site as needed and were based on the most up-to-date requirements and conditions. The resulting system was much more responsive to last-minute changes, more timely, more accurate, and less costly to operate.

Data Capture Process

The new Data Capture Process described above was critically dependent for its implementation on the availability of new hardware and software tools with the requisite capabilities. To a large degree, these were to be provided out of similar studies carried out by the Reengineering Team for the Asset Creation Process, which ran a parallel course to that of Customer Services Fulfillment.

In this context, it was assumed that the operations load on the 26-m/9-m subnetwork would be relieved by the introduction of a subnetwork of small unattended tracking stations to be known as Low-Earth Orbit (LEO) terminals. It was further assumed that updated electronics would allow a large degree of automation for future operation of the 26-m and 9-m antennas and would eliminate the existing need for roving operators (rovers) to configure antenna and microwave equipment in front-end areas (FEAs). At the DSCCs, the already-planned Advanced DSN Monitor and Control (ADMC) sub-

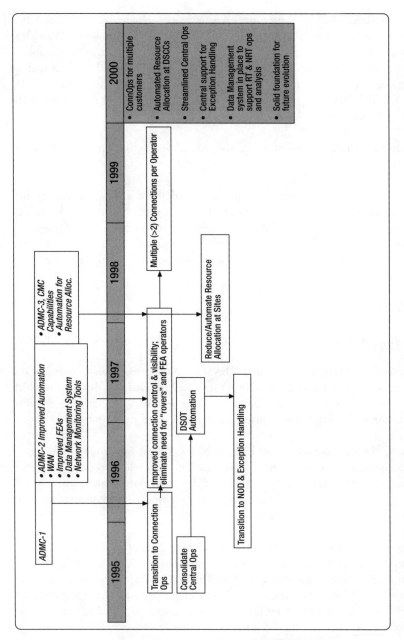

Figure 9-16. Data Capture road map.

system would provide the high degree of automatic features required for the Connection Operations concept. A new "Reliable Data Delivery System" and improved Wide Area Network were already at the advanced design stage and were assumed to be available to meet the proposed transition plan schedule. Finally, it was assumed that technology already developed in MOSO for data processing and management would be extended to incorporate the Data Systems Operations Team (DSOT) as a vital element of the new Data Capture system.

A road map for making the transition to the new Data Capture Process is shown in Figure 9-16.

The dependency on key new implementation described above is shown in the top half of the figure, while the key changes in operations processes are shown in the lower half. The plan attempted to minimize disruption to ongoing operations while it progressed rapidly toward its final realization by 2000. Near-term savings accrued from the early transition to Connection Operation in 1995, collocation of TMOD operations, and reduction in operations documentation.

Activity Plan Process

The two principal functions performed via the Activity Plan process were as follows:

1. Identifying the services to be provided. This was called Service Plan Generation and was based on a schedule showing the TMOD resources required and the time at which they were to be available, as well as a project sequence of events (PSOE) that identified what the customer would be doing during the time allocated in the schedule.

2. Generating the information, internal to the Network, that was required for operating the equipment to provide the needed services. This was called Predicts Generation and was required principally for generation of radio metric or telemetry data. It was based on spacecraft state and ephemeris and station-dependent data such as location, horizon mask, and planetary ephemeris.

At the time that the new Activity Plan Process was being designed, a standard infrastructure for the packaging and manipulation of spacecraft navigation products called the "SPICE" system had been in use for some years at JPL. The SPICE data sets were called kernels and contained the information necessary to assist users in planning and interpreting science observations from spacecraft instruments. For example, the "S" data set contained spacecraft ephemeris data, and the "P" data set contained planetary ephemerides.

In reengineering the Activity Plan Process, the RET extended this idea to create kernels containing spacecraft and planetary ephemerides data, telemetry data, station information, and time and Earth polar motion information needed to compute station locations. The data bases of these kernels would be "shadowed" or automatically updated to the tracking station's data management system and would therefore be available for use in generating telemetry and/or metric predicts on demand. At the station, computation of predicts could be initiated only minutes before the actual activity was scheduled to take place.

In addition to these innovations, the reengineered Activity Plan Process improved the way in which the customer provided the PSOE by referencing it to a standard catalog of prenegotiated services. The Activity Plan Process would then translate these service requests into the operations procedures necessary to meet them, thereby creating the actual Service Plan.

Although implementation of the new Activity Plan Process began in 1995, it depended on upgraded versions of the data management capability at the stations for full operation. Delay in providing this new capability resulted in a longer term return on the implementation costs than was the case for the Data Capture Process.

At the time of this writing, the implementation plans for the Data Capture and Activity Plan Processes were in progress along the lines discussed above. When fully completed in 1999, the reengineered TMOD Customer Services Fulfillment Process would embody the very best principles of cost-effective, process-based management, rather than the facility-based management and control principles that had been in place for the previous forty years.

Reinventing the Future

After a very short term of one year in office, Westmoreland retired in June 1997, and Gael F. Squibb was appointed director of the Telecommunications and Mission Operations Directorate (TMOD). Squibb was well qualified to guide TMOD into the future. He had served as Manager of Data Services for NASA's Space Operations Management Office since 1995 and had managed numerous scientific flight projects in thirty years of previous service with JPL.

When Squibb took over TMOD, the reengineering effort was well under way and the DSN and MGDS functions in mission operations had become recognized as contiguous elements of the Customer Services Fulfillment Process. They were, from 1994, still separate offices (940 and 970) of the original TMOD organization. It was time to bring the institutional organization more into line with the concepts of process-based management.

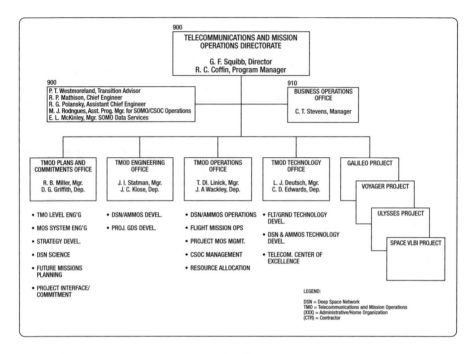

Figure 9-17. Telecommunications and Mission Operations Directorate, 1997.

The restructured Telecommunications and Mission Operations Directorate, under the direction of Gael Squibb in 1997, is shown in Figure 9-17.

The former DSN Data Services and Multimission Ground Systems Offices were combined to form the TMOD Operations Office, from which the Customer Services Fulfillment Process, including the allocation of resources, would be managed effectively. To develop the new system engineering functions for this process, a TMOD Engineering Office, which included engineering elements of both the DSN and MGDS, was also created. It would be here that the Asset Creation Process would reside. The enabling technology on which the reengineering teams based their new process-oriented designs was to be provided by the TMOD Technology Office for both the DSN and AMMOS. It was defined as a "Center of Excellence" for telecommunications to focus attention on the particular technology in which the DSN held a unique position in the NASA sphere of influence. The Plans and Commitments Office and flight project offices remained essentially the same.

The Deep Space Network as an Organization in Change

There was another feature of the TMOD organization that, although not apparent on the organization chart, affected the interaction between TMOD and its customers in a significant way—the change in the role of the DSN manager. As mentioned earlier, it was customary for the TDA Office to appoint a representative to work with each flight project in negotiating and using the tracking and data acquisition services that were required for its mission. The title of this position was "Tracking and Data System (TDS) Manager for (Name of Customer)."

Customers included those that used the DSN as a scientific instrument as well as those that used it as a tracking and data acquisition service. The TDS manager helped his designated customers to understand relevant technical and operational aspects of the DSN and acted as an authoritative single point of contact for the customers, dealing with all aspects of the service being provided. But the service rendered extended only to the interface between the DSN and the MGDS. It did not include the processing and management of the data beyond that point, which was, in reengineering terms, "facility-oriented." When TMOD was reengineered into a "process-oriented" organization, the former TDS manager positions, which were properly resident in the Plans and Commitments Office, were expanded in concept and scope to include the entire Customer Fulfillment Services Process. They became the TMOD version of the "empowered customer service representatives" advocated by Hammer and Champy and were called Telecommunications and Mission Services (TMS) Managers.

The extent to which a single person would be able to discharge this task in a meaningful way at all levels throughout the now extremely complex Services Fulfillment process remained to be seen. Hammer and Champy recognized this problem and observed, "To perform this role–that is, to be able to answer the customer's questions and solve the customer's problems–the case manager needs access to all the information systems that the people actually performing the process use and the ability to contact those people with questions and requests for further assistance when necessary." The TMS managers of the future would need full access to tools and resources such as these to properly carry out their important functions in the new process-oriented TMOD.

"Bridging the Space Frontier"

A few months after taking office, the new director of TMOD issued a statement that presented his goals and vision for the TMOD of the future. It included a plan for their realization and was called "Bridging the Space Frontier."[13] The plan traced the origin of the powerful forces throughout NASA that had influenced the organizational structure of TMOD and would determine its course for the future. The environment in which

TMOD must plan for the future, said Squibb, "is characterized by three fundamental trends: increasing demand for telecommunication and mission services and technology advancements, highly constrained resources, and organizational change."

Squibb based his plan on NASA's response to the National Space Policy, issued from the White House in 1996.[14] The National Space Policy of 1996 directed NASA to focus its research and development efforts in the four principal areas of Space Science, Earth Observation, Human Space Flight, and Space Technologies and Applications. Furthermore, it instructed NASA to "seek to privatize or commercialize its space communications operations no later than 2005" and to "examine with DoD, NOAA[,] and other appropriate federal agencies, the feasibility of consolidating ground facilities and data communications systems that cannot otherwise be provided by the private sector."

NASA began to address this latter policy by consolidating the management of all its space operations under a new Space Operations Management Office (SOMO), located at the Johnson Space Center in Houston. As the service provider, SOMO was given responsibility for implementing NASA's space operations and managing the associated space operations work process. The NASA Centers, in support of their respective implementation plans for space operations, were responsible for the execution of the space operations work process.[15]

In his plan, Squibb discussed TMOD's relationship to SOMO. "Through this cooperative arrangement with SOMO, TMOD (would) oversee operation of the Advanced Multimission Operations System (AMMOS) and the Deep Space Network (DSN). Under SOMO's leadership, efforts were underway to consolidate and streamline major support contract services." Transition to a Consolidated Space Operations Contract (CSOC) with a single, ten-year, cost-plus-award-fee contract was expected to begin in Fiscal Year 1998. In the meantime, TMOD worked to understand its future, dramatically different role as a partner with a CSOC contractor, and possibly with other United States or foreign agencies; it also tried to identify any functional activities that would be suitable for commercialization in the immediate future.

JPL's management responsibilities for the DSN and AMMOS were finally combined under Squibb's direction in 1996, and TMOD began to operate as a single process-oriented service to its customers. Joe Statman, manager of the TMOD Engineering Office, saw the new way of operating as a challenge to "create one culture—no more separate DSN/AMMOS cultures." One step remained, however, in the completion of the DSN/AMMOS unification, and that had to do with the still separate sources of programmatic direction at NASA Headquarters.

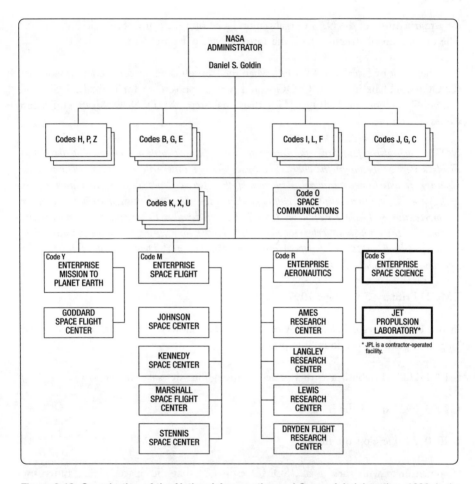

Figure 9-18. Organization of the National Aeronautics and Space Administration, 1996. In the new organization, the responsibilities of Code O (the former Office of Space Communications) were distributed among the Centers, the Office of Space Flight, and the Space Operations Management Office (SOMO). It was SOMO that provided programmatic direction for the DSN-related elements of TMOD. Code S, the home of the Space Science Enterprise, continued to provide the source of funding for the MGDS-related elements of TMOD.

The directives of the 1996 National Space Policy, and its focus on NASA research and development efforts, were eventually manifested in the NASA Strategic Plan.[16] This called for a restructuring of the NASA organization based on four grand "Strategic Enterprises." The organizational structure took the form shown in Figure 9-18.

Plans held that by Fiscal Year FY 1999, when the Consolidated Space Operations Contract (CSOC) would be in effect, CSOC would have responsibility for both the DSN and AMMOS elements of TMOD. With that final step, the DSN/AMMOS unification would be complete.

NOTE: Each of the four Strategic Enterprises was affiliated with one or more of the NASA Field Centers as shown in the chart. The Space Science Enterprise was affiliated with JPL. However, it was from endeavors in support of the Space Science, Mission to Planet Earth, and Space Flight Enterprises that requirements for the data and mission services that were the prerogative of TMOD would be derived. Two of the four Enterprises were subsequently renamed as their missions were revised. In 1997, the four Enterprises were named Space Science, Mission to Planet Earth, Human Exploration and Development of Space, and Aeronautics and Space Transportation Technology.

TMOD Primary Challenge: 1997

In the 1998 Strategic Plan, NASA established the following near-term, mid-term, and long-term goals for the four Enterprises:

- 1991–2002: Establish a virtual presence throughout the solar system

- 2003–09: Expand the horizons

- 2010–23: Develop the frontiers

Commensurate with these goals, SOMO estimated that the Space Science Enterprise would require data and mission services support for 86 missions through 2004; Mission to Planet Earth would require support for 34 missions; and Human Exploration and Development of Space would require support for 18 missions. A significant number of these missions would involve TMOD and, with more complex investigations and instruments characterizing these missions, would lead to a dramatic increase in the demand for data and mission services over the next several years. This, said Squibb, would take place against an environment of highly constrained, albeit level, budget resources. The situation and the scope of the challenge it presented to TMOD were well illustrated in the graphic shown in the 1997 Space Communications Budget Review and reproduced in Figure 9-19.

The Deep Space Network as an Organization in Change

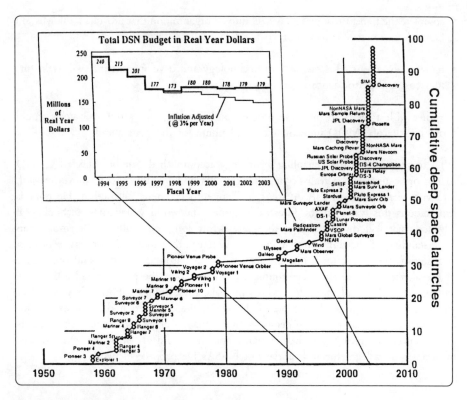

Figure 9-19. TMOD primary challenge, 1997.

As 1997 ended, TMOD was challenged to provide "world-class" data and mission support for an ever-increasing number of missions in the face of the essentially fixed budget shown in the graphic.

To meet this challenge over the following three to five years, Squibb established a set of five goals for TMOD:

- increase data return capacity by a factor of 2.6,

- accomplish a significant portion of TMOD work through interaction with at least 10 strategic partners within five years,

- improve TMOD's performance by capturing all of the data available from space missions,

- complete the transformation of TMOD into an organization that provides a full Mission Operations Services system, and

- reinvest operations costs savings into technology and development that will yield further cost and performance improvement.

Together with a metric to measure progress and a strategy for action, the goals would form the core of TMOD implementation planning for the years ahead.

With these goals and their enabling technology accomplished, Squibb expected TMOD to not only survive in a realistic new budget environment, but also to move forward, as a vital element of the NASA Strategic Plan, into the new millennium. He put his expectations this way: "In the foreseeable future, the world will witness technological advances that will dwarf those of the past. Before long, our culture will embrace an understanding of the universe that includes, among other things, its origins, evolution and destiny, the distribution and character of planets around the stars, and occurrence or prevalence of life in those environments. I fully expect the people of TMOD to be among the leaders and innovators who will make these things happen." It was a clear and ringing challenge to the people of the DSN, the engineers and scientists, technicians and administrators, and supervisors and managers at all levels who composed its vital essence. The measure of their response and the record of what they acheived was, of necessity, a task for future historians of NASA's Deep Space Network.

Endnotes

1. William R. Corliss, "A History of the Deep Space Network," NASA CR-151915 (Washington, DC: NASA, 1976).

2. J. Bluth, personal communication with author, 22 June 1998.

3. Homer E. Newell, *Beyond the Atmosphere*, NASA SP-4211 (Washington, DC: NASA, 1980).

4. Clayton R. Koppes, *JPL and the American Space Program; A History of the Jet Propulsion Laboratory* (New Haven and London: Yale University Press, 1982).

5. R. Cargill Hall, *Lunar Impact: A History of Project Ranger*, NASA SP-4210 (Washington, DC: NASA, 1977).

6. William R. Corliss, "History of the Deep Space Network," NASA CR-151915 (Washington, DC: NASA, 1976).

7. D. J. Mudgway, "Viking Mission Support," JPL Technical Report TR 32-1526, Vol. X (15 August 1972), pp. 22–26.

8. A. J. Siegmeth, "Pioneer 10 and G Mission Support," JPL Technical Report TR 32-1526, Vol. X (15 August 1972), pp. 27–34.

9. M. J. Alazard, A. I. Beers, and T. D. Linnick, "Multimission Operations Systems Office User Guide," JPL Document D-9661 (15 June 1993).

10. Jet Propulsion Laboratory, "JPL Strategic Plan," JPL Document 400-549 (April 1995); see also *http://www.jpl.nasa.gov/stratplan/*.

11. Michael Hammer and James Champy, *Reengineering the Corporation* (New York: Harper Collins, 1993).

12. R. J. Amorose, "Reengineering Team for Services Fulfillment (Operations); Final Report," TMOD internal memorandum, RJA: 95-008 (14 March 1995).

13. Gael F. Squibb, "Bridging the Space Frontier: TMOD Implementation Plan, FY 98" (Pasadena: Jet Propulsion Laboratory, 30 January 1998); see also *http://deepspace.jpl.nasa.gov/920/public/922_strat_plan/*.

14. National Space Policy 1996; see also *http://www.whitehouse.gov/WH/EOP/OSTP/NSTC/html/fs/fs-5.html*.

15. Space Operations Management Office, "Space Operations Implementation Plan-Draft E," (Houston : JSC, August 1997).

16. "NASA Strategic Plan," NASA Policy Directive (NPD)-1000.1; see also *http://www.hq.nasa.gov/office/nsp/NSPTOC.html*.

LIST OF FIGURES

Figures, Introduction

1. Figure Intro-1. *Voyager 1* outward bound for Jupiter, September 1977.

2. Figure Intro-2. *Voyager 1* views Jupiter, February 1979.

3. Figure Intro-3. Mean distances of the terrestrial planets from the Sun.

4. Figure Intro-4. Mean distances of the Jovian planets from the Sun.

5. Figure Intro-5. HA-Dec and Az-El mounted antennas of the DSN.

6. Figure Intro-6. Injection into interplanetary orbit.

7. Figure Intro-7. Cassegrain focus antenna.

8. Figure Intro-8. Improvement in DSN downlink performance during the first twenty years.

9. Figure Intro-9. An essential part of the answer; global map of the Deep Space Network, 1992.

Figures, Chapter 1

1. Figure 1-1. View of the Pioneer tracking station site, Goldstone, California, 1978.

2. Figure 1-2. Prominent personalities of the Deep Space Network unveiling the commemorative plaque at the Goldstone Pioneer Site, 28 October 1978.

3. Figure 1-3. The 26-meter antenna and tracking station (DSIF 11) at Pioneer site, Goldstone, California, 1958.

4. Figure 1-4. Essential features of the Project Echo Experiment, 1960

5. Figure 1-5. Az-El transmitting antenna for the Echo Balloon Experiment, Goldstone, 1960.

6. Figure 1-6. The 26-meter antenna and tracking station (DSIF 41), Woomera, Australia, 1961.

7. Figure 1-7. The 26-meter antenna and tracking station (DSIF 51), Hartebeestpoort, South Africa, 1961.

Figures, Chapter 2

1. Figure 2-1. World view of the Deep Space Instrumentation Facility, 1961.

2. Figure 2-2. Flight operations control center, JPL, 1961.

3. Figure 2-3. First close-up picture of Mars. Returned from *Mariner 4* on 14 July 1965.

4. Figure 2-4. Network Operations Control Center (NOCC) at JPL, 1969.

5. Figure 2-5. Side elevation of the Goldstone 64-meter azimuth-elevation antenna, 1966.

6. Figure 2-6. The 64-meter antenna at Goldstone, 1966.

7. Figure 2-7. Composition of the Deep Space Network, 1974.

Figures, Chapter 3

1. Figure 3-1. The Viking Era mission set.

2. Figure 3-2. Functional block diagram of the DSN, 1978.

3. Figure 3-3. Heliocentric orbit of *Helios 1*.

List of Figures

4. Figure 3-4. Configurations of the Viking Orbiter (top) and Viking Lander (bottom).

5. Figure 3-5. Viking Orbiter, Lander, Earth Telecommunications System.

6. Figure 3-6. First Viking close-up picture of Mars surface.

7. Figure 3-7. First panoramic picture of the surface of Mars.

8. Figure 3-8. Voyager trajectories to Jupiter, Saturn, and Uranus.

9. Figure 3-9. Downlink performance estimates for *Pioneers 10* and *11*.

10. Figure 3-10. Functional block diagram of the DSN Mark III-75 configuration for 64-meter stations.

11. Figure 3-11. Functional block diagram of the DSN Mark III-75 configuration for 26-meter stations.

12. Figure 3-12. Hydrostatic bearing and ball joint assembly.

13. Figure 3-13. Mark III-77 Network systems and subsystems.

14. Figure 3-14. General configuration of the DSN Mark III-77 Network.

15. Figure 3-15. DSN Mark III Data System implementation schedule, 1977.

16. Figure 3-16. Basic organization of the Mark III Data System.

17. Figure 3-17. Functional block diagram of the Mark III-77 Telemetry System.

18. Figure 3-18. DSN Mark III Command System, 1978.

19. Figure 3-19. DSN-GCF-NASCOM ground communications, 1976–78.

20. Figure 3-20. Structural changes to the 26-meter antenna for extension to 34-meter diameter.

21. Figure 3-21. Goldstone Station, DSS 12 antenna after conversion to 34-meter, S-X-band configuration, October 1978.

Figures, Chapter 4

1. Figure 4-1. *Voyager* Era Deep Space mission set, 1977–86.

2. Figure 4-2. *Voyager 1* image of Jupiter, 13 February 1979, at a distance of 20 million kilometers, showing Io (left) and Europa (right) against the Jupiter cloud tops.

3. Figure 4-3. *Voyager 1* image of Io, 4 March 1979.

4. Figure 4-4. *Voyager 1* trajectory through the Jupiter system, 5 March 1979.

5. Figure 4-5. *Voyager 1* image of Saturn and its rings, August 1980.

6. Figure 4-6. *Voyager 1* image of Saturn and its rings, November 1980.

7. Figure 4-7. Parkes 64-m radio astronomy antenna.

8. Figure 4-8. Enhanced link performance for *Voyager 2* at Uranus.

9. Figure 4-9. Functional block diagram of the Parkes-CDSCC telemetry array for Voyager Uranus Encounter, January 1986.

10. Figure 4-10. Pictorial diagram of the *Voyager 2* mission.

11. Figure 4-11. *Voyager 2* image of the Uranus satellite Miranda.

12. Figure 4-12. *Pioneer 10* and *11* heliocentric trajectories, September 1979.

13. Figure 4-13. DSN configuration for ICE encounter of comet Giacobini-Zinner, 11 September 1985.

14. Figure 4-14. 34-meter high efficiency (HEF) antenna: DSS 15, Goldstone.

15. Figure 4-15. Rehabilitation of the DSS 14 hydrostatic bearing and pedestal concrete, June 1983–June 1984.

16. Figure 4-16. Work in progress on the 60-m to 70-m antenna extension at DSS 14, Goldstone, December 1987.

List of Figures

17. Figure 4-17. Elements of the DSN and GSTDN, 1982.

18. Figure 4-18. Canberra Deep Space Communications Complex, January 1985.

19. Figure 4-19. Deep Space Network, Mark IVA configuration.

20. Figure 4-20. DSN antenna operations costs, 1980–2000.

21. Figure 4-21. Signal Processing Center, Mark IVA configuration.

22. Figure 4-22. Network Operations Control Center, Mark IVA.

23. Figure 4-23. Principal features of the DSCC Mark IVA Telemetry Subsystem.

24. Figure 4-24. Functional block diagram of the Mark IVA Baseband Assembly (BBA).

25. Figure 4-25. DSN Mark IVA Telemetry System key characteristics.

26. Figure 4-26. DSN Command System Mark IVA 1985.

Figures, Chapter 5

1. Figure 5-1. Galileo Era Deep Space mission set. DJM files.

2. Figure 5-2. Phobos L-band feedhorn mounted on the 64-m antenna.

3. Figure 5-3. Magellan Venus Orbit mapping profile.

4. Figure 5-4. Diagram of *Magellan* spacecraft in cruise configuration.

5. Figure 5-5. *Magellan* SAR image of an impact crater in the Atalanta Region of Venus.

6. Figure 5-6. Encounter geometry for *Voyager 2* at Neptune.

7. Figure 5-7. *Voyager 2* views three most prominent features of Neptune.

8. Figure 5-8. *Voyager 2* views south polar cap of Triton.

9. Figure 5-9. *Voyager 2* views Neptune and Triton.

10. Figure 5-10. The Galileo Venus-Earth-Earth-Gravity-Assist (VEEGA) trajectory.

11. Figure 5-11. The *Galileo* spacecraft cruise configuration.

12. Figure 5-12. The Galileo Orbiter and Probe arrival at Jupiter, 7 December 1995.

13. Figure 5-13. Network configuration for *Galileo* launch and cruise, 1989.

14. Figure 5-14. Galileo 1989 VEEGA mission: predicted X-band downlink performance at Jupiter arrival: 7 December 1995.

15. Figure 5-15. Geometry of the *Galileo* first Earth flyby.

16. Figure 5-16. *Galileo* image of Earth: Antarctica.

17. Figure 5-17. Earth and Moon from *Galileo*, December 1992.

18. Figure 5-18. Asteroid Ida and its moon.

19. Figure 5-19. The DSN in full array configuration for Galileo Jupiter orbital operations.

20. Figure 5-20. Full DSN array operations for *Galileo*: typical downlink data rate profile.

21. Figure 5-21. *Galileo* image of icebergs on the surface of Europa.

22. Figure 5-22. *Ulysses* solar polar trajectory.

23. Figure 5-23. The *Ulysses* spacecraft inflight configuration.

24. Figure 5-24. *Ulysses* second solar polar orbit.

25. Figure 5-25. Cracked bearing rollers and inner race: DSS 63, December 1989.

26. Figure 5-26. Twenty years of telemetry arraying in the DSN.

27. Figure 5-27. Parkes-CDSCC telemetry array.

List of Figures

28. Figure 5-28. Enhanced downlink performance at Uranus.

29. Figure 5-29. Enhanced downlink performance at Neptune.

30. Figure 5-30. VLA-GDSCC telemetry array.

31. Figure 5-31. The concept of DSN telemetry for the Galileo LGA mission.

32. Figure 5-32. Functional diagram of the DSN Galileo telemetry subsystem for the Canberra Deep Space Communications Complex.

33. Figure 5-33. Beam waveguide antenna in conceptual form.

34. Figure 5-34. Design for 34-meter beam waveguide antenna at DSS 13.

35. Figure 5-35. Design for operational 34-meter beam waveguide antenna.

36. Figure 5-36. DSS 34 beam waveguide antenna at Canberra Deep Space Communications Complex, Australia, 1996.

37. Figure 5-37. Cluster of three 34-m beam waveguide antennas at Goldstone, 1995.

38. Figure 5-38. Signal Processing Center configuration, 1993.

39. Figure 5-39. DSCC Telemetry Subsystem, 1997.

40. Figure 5-40. DSCC Command System, 1996.

41. Figure 5-41. DSCC Tracking Subsystem, 1995.

42. Figure 5-42. Ground Communications Facility, 1992.

Figures, Chapter 6

1. Figure 6-1. Cassini Era mission set, 1997.

2. Figure 6-2. *Mars Global Surveyor* spacecraft in the mapping configuration.

3. Figure 6-3. DSN configuration for *Mars Global Surveyor*.

4. Figure 6-4. *Mars Global Surveyor* image of Olympus Mons.

5. Figure 6-5. Pathfinder Lander deployed on surface of Mars.

6. Figure 6-6. The MPF Lander camera views the Sojourner Rover in operation near Yogi Rock on the Surface of Mars, July 1997.

7. Figure 6-7. *Cassini* interplanetary trajectory.

8. Figure 6-8. *Cassini* spacecraft with Huygens Probe.

9. Figure 6-9. Location identifiers for DSN and related sites.

10. Figure 6-10. Goldstone Deep Space Communications Complex, 1997.

11. Figure 6-11. Canberra Deep Space Communications Complex, 1997.

12. Figure 6-12. Madrid Deep Space Communications Complex, 1997.

13. Figure 6-13. OVLBI Subnetwork functional block diagram.

14. Figure 6-14. DSS 23: SVLBI 11-meter antenna at Goldstone, 1996.

Figures, Chapter 7

1. Figure 7-1. Deep Space Communications Complex in Canberra, Australia.

2. Figure 7-2. An early parabolic design of a 26-meter antenna.

3. Figure 7-3. The quasi-parabolic 34-meter high-efficiency antenna, DSS 15.

4. Figure 7-4. A 70-meter antenna with dual-shape reflector design.

5. Figure 7-5. A 34-meter beam waveguide antenna at Goldstone Technology Development Site, DSS 13.

List of Figures

6. Figure 7-6. Stationary equipment room below BWG antenna.

7. Figure 7-7. Dichroic (frequency-selective) reflector developed under Advanced Systems Program.

8. Figure 7-8. Ultra-low noise amplifier (ULNA) at DSS 13.

9. Figure 7-9. High-electron mobility transistor (HEMT) low-noise microwave amplifier.

10. Figure 7-10. Advanced Receiver (ARX) used to track the *Pioneer 10* spacecraft.

11. Figure 7-11. Telemetry communication channel performance for various coding schemes.

12. Figure 7-12. The 64-m antenna of the Radio Astronomy Observatory, Parkes, Australia.

13. Figure 7-13. The Very Large Array (VLA) of the National Radio Astronomy Observatory (NRAO) at Socorro, New Mexico.

14. Figure 7-14. The new linear ion trap (LIT) frequency standard.

15. Figure 7-15. First high-resolution radar image of Venus.

16. Figure 7-16. Profile of deep space telecommunications capability, 1960–2020.

Figures, Chapter 8

1. Figure 8-1. DSN science-related publications, 1962–97.

2. Figure 8-2. Real-time bandwidth reduction in the radio science Occultation Data Assembly.

3. Figure 8-3. Characteristic three-pulse signature of a gravitational wave in two-way Doppler data.

4. Figure 8-4. Detection strategy and system configuration for the SETI sky survey field tests, 1985.

5. Figure 8-5. SETI detection of spacecraft signals, 1985.

6. Figure 8-6. The first HRMS sky survey, Goldstone, 12 October 1992.

7. Figure 8-7. Project ARIES transportable antenna interferometer system.

8. Figure 8-8. Global Positioning System, 1986.

9. Figure 8-9. Schematic view of GPS-based ionospheric calibration.

10. Figure 8-10. Functional block diagram of the Venus radar, 1962.

11. Figure 8-11. High Speed Data Acquisition System for the Goldstone Radar, 1984.

12. Figure 8-12. Relative sensitivity for the Goldstone Radar antenna.

13. Figure 8-13. Goldstone Solar System Radar images of Asteroid Toutatis, 1992.

14. Figure 8-14. Goldstone Radar astronomy publications, 1988–97.

Figures, Chapter 9

1. Figure 9-1. NASA Office of Space Flight Development, 1959–61.

2. Figure 9-2. Telecommunications Division at JPL, 1961–62.

3. Figure 9-3. JPL organization, 1964.

4. Figure 9-4. NASA organization for the Office of Tracking and Data Acquisition, 1961.

5. Figure 9-5. The Deep Space Network established, December 1963.

List of Figures

6. Figure 9-6. Organization for Tracking and Data Acquisition support at JPL, October 1964.

7. Figure 9-7. Organization of the technical divisions after the transfer of the SFOF to OCIS, 1972.

8. Figure 9-8. Functional interfaces between the DSN and the SFOF for Viking, 1972.

9. Figure 9-9. TDA Office 400 organization at JPL, 1978.

10. Figure 9-10. TDA Office 400 organization at JPL, 1980–87.

11. Figure 9-11. National Aeronautics and Space Administration, 1991

12. Figure 9-12. Multimission Operations Systems Office (440), 1993.

13. Figure 9-13. NASA Space Communications and Operations Program.

14. Figure 9-14. Telecommunications and Mission Operations Directorate, 1996.

15. Figure 9-15. Predicted cost and savings for reengineering TMOD Services Fulfillment Processes.

16. Figure 9-16. Data Capture road map.

17. Figure 9-17. Telecommunications and Mission Operations Directorate, 1997.

18. Figure 9-18. Organization of the National Aeronautics and Space Administration, 1996.

19. Figure 9-19. TMOD primary challenge, 1997.

APPENDIX

Further Reading for Specialists

The history of technology in the Deep Space Network is well documented in a continuous series of technical articles published in the JPL Space Programs Summaries and its successor journals. Most of these articles describe work that was funded exclusively by the DSN Advanced Systems Program or was funded in cooperation with other DSN programs related to engineering implementation or mission operations. This appendix will provide the specialist reader with an overview of key developments in areas of specific interest.

Most of the citations that follow were extracted from the online index to "The Deep Space Network Progress Report" and its subsequent names, for issues from 1970 and later, or from the author index to earlier reports. The list also includes a number of citations from refereed external technical journals and several case studies of advanced technology in the DSN.

The early JPL Space Programs Summary (SPS) was published as a five-volume set designated as SPS 37-nn. One of these volumes, Volume IV, covered all Supporting Research and Advanced Development activities at JPL, including the Deep Space Instrumentation Facility (DSIF) Advanced Systems Program. Implementation and Operations activities were included in Volume III.

The Space Programs Summary became a four-volume set with SPS- 37-47. All Supporting Research and Advanced Development activities are contained in Volume III, and all DSN activity is described in Volume II.

Publication of the Space Program Summaries ceased in 1970 with SPS 37-66. However, the publication of progress reports for the DSN (including the DSN Advanced Systems Program) were continued in a series of JPL Technical Reports beginning with JPL-TR 32-1526, Volume I (15 February 1971). The DSN continued to report its progress in this form through TR 32-1526, Volume XIX (15 February 1974), when the name became "The Deep Space Network Progress Report," DSN-PR 42-nn.

DSN progress reports under the new title began with DSN-PR 42-20 in April 1974, and, except for the substitution of "Telecommunications and Data Acquisition" for "Deep Space Network" in June 1980 (TDA PR 42-57), followed by "Telecommunications and Mission Operations" in 1998, the series continued without further change through the end of the century.

For ease of identification, the appendix refers to these various publications as JPL-SPS, JPL-TR, and TDA-PR. Copies of these documents may be obtained from the Jet Propulsion Laboratory, Pasadena, California. Articles that are referenced in issues TDA-PR 42-66 and later may be accessed online at *http://tmo.jpl.nasa.gov/progress_report*.

1. **The evolution of technology in the Deep Space Network.**

Layland, J. W., and L. L. Rauch. "The Evolution of Technology in the Deep Space Network, A History of the Advanced Systems Program," TDA Progress Report PR 42-130, April-June 1997, 15 August 1997.

2. **The great antennas of the DSN.**

Potter, P. D. "Improved Dichroic Reflector Design for the 64-m Antenna S- and X-band Feed Systems," JPL-TR 32-1526, Vol. XIX: November and December 1973, 15 February 1974, pp. 55–62.

Potter, P. D. "Shaped Antenna Designs and Performance for 64-m Class DSN Antennas," DSN-PR 42-20: January-February 1974, 15 April 1974, pp. 92–111.

Galindo-Israel, V., N. Imbriale, Y. Rahmat-Samii, and T. Veruttipong. "Interpolation Methods for GTD Analysis of Shaped Reflectors," TDA Progress Report PR 42-80: October-December 1984, 15 February 1985, pp. 62–67.

Goodwin, J. P. "Usuda Deep Space Center Support for ICE," TDA Progress Report PR 42-84: October-December 1985, 15 February 1986, pp. 186–96.

Fanelli, N. A., J. P. Goodwin, S. M. Petty, T. Hayashi, T. Nishimura, and T. Takano. "Utilization of the Usuda Deep Space Center for the United States International Cometary Explorer (ICE)." Proceedings of the Fifteenth International Symposium on Space Technology and Science, Tokyo, Japan, 1986.

Otoshi, T. Y., and M. M. Franco. "Dual Passband Dichroic Plate for X-band," TDA Progress Report PR 42-94: April-June 1988, 15 August 1988, pp. 110–34.

Chen, J. C., P. H. Stanton, and H. F. Reilly. "Performance of the X-/Ka-/KABLE-band Dichroic Plate in the DSS 13 Beam Waveguide Antenna," TDA Progress Report PR 42-115: July-September 1993, 15 November 1994, pp. 54–64.

APPENDIX

Rafferty, William, Stephen D. Slobin, and Charles T. Stelzried. "Ground Antennas in NASA's Deep Space Telecommunications." Proceedings of the Institute of Electrical and Electronics Engineers, Vol. 82, No. 5, May 1994.

3. Forward command/data link (uplink).

Benjauthrit, B., and T. K. Truong. "Encoding and Decoding a Telecommunication Standard Command Code," DSN Progress Report PR 42-38: January and February 1977, 15 April 1977, pp. 115–19.

4. Return telemetry/data link (downlink).

Clauss, R. C., and R. B. Quinn. "Low Noise Receivers: Microwave Maser Development," JPL Technical Report TR 32-1526, Vol. IX: March and April 1972, 15 June 1972, pp. 128–36.

Clauss, R. C., and E. Wiebe. "Low-Noise Receivers: Microwave Maser Development," JPL Technical Report TR 32-1526, Vol. XIX: November and December 1973, 15 February 1974, pp. 93–99.

Ulvestad, J. S., G. M. Resch, and W. D. Brundage (National Radio Astronomy Observatory, New Mexico). "X-band System Performance of the Very Large Array," TDA Progress Report PR 42-92: October-December 1987, 15 February 1988, pp. 123–37.

Shell, J., and D. Neff. "A 32-GHz Reflected-Wave Maser Amplifier With Wide Instantaneous Bandwidth," TDA Progress Report PR 42-94: April-June 1988, 15 August 1988, pp. 145–62.

Tanida, L. "An 8.4-GHz Cryogenically Cooled HEMT Amplifier for DSS 13," TDA Progress Report PR 42-94: April-June 1988, 15 August 1988, pp. 163–69.

Bautista, J. J., G. G. Ortiz, K. H. G. Duh (GE Electronics Laboratory, New York), W. F. Kopp (GE Electronics Laboratory, New York), P. Ho (GE Electronics Laboratory, New York), P. C. Chao (GE Electronics Laboratory, New York), M. Y. Kao (GE Electronics Laboratory, New York), P. M. Smith (GE Electronics Laboratory, New York), and J. M. Ballingall (GE Electronics Laboratory, New York). "32-GHz Cryogenically Cooled HEMT Low-Noise Amplifiers," TDA Progress Report PR 42-95: July-September 1988, 15 November 1988, pp. 71–81.

Glass, G. W., G. G. Ortiz, and D. L. Johnson. "X-band Ultralow-Noise Maser Amplifier Performance," TDA Progress Report PR 42-116, October-December 1993, 15 February 1994, pp. 246–53.

Shell, J., and R. B. Quinn. "A Dual-Cavity Ruby Maser for the Ka-band Link Experiment," TDA Progress Report PR 42-116, October-December 1993, 15 February 1994, pp. 53–70.

5. Phase-lock tracking.

Jaffe, R., and E. Rechtin. "Design and Performance of Phase-Lock Circuits Capable of Near-Optimum Performance over a Wide Range of Input Signal and Noise Level," IRE Trans. Information Theory, Vol. IT-1, March 1955, pp. 66–76.

Tausworthe, R. C. "Another Look at the Optimum Design of Tracking Loops," JPL-SPS 37-32, Vol. IV, 30 April 1965, pp. 281–83.

Lindsey, W. C. "The Effect of RF Timing Noise in Two-Way Communications Systems," JPL-SPS 37-32, Vol. IV, 30 April 1965, pp. 284–88.

Lindsey, W. C. "Optimum Modulation Indices for Single Channel One Way & Two-Way Coherent Communications," JPL-SPS 37-37, Vol. IV, 28 February 1966, pp. 287–90.

Lindsey, W. C., and R. C. Tausworthe. "Digital Data Transition Tracking Loops," JPL-SPS 37-50, Vol. III, 30 April 1968, pp. 272–76.

Tausworthe, R. C. "Efficiency of Noisy Reference Detection," JPL-SPS 37-54, Vol. III, 31 December 1968, pp. 195–201.

Tausworthe, R. C. "Theory and Practical Design of Phase-Locked Receivers," JPL Technical Report TR 32-819, 27 April 1971.

Layland, J. W. "Noisy Reference Effects on Multiple-Antenna Reception," DSN Progress Report PR 42-25, November and December 1974, 15 February 1975, pp. 60–64.

6. Synchronization and detection.

Baumgartner, W. S., W. Frey, M. H. Brockman, R. W. Burt, J. W. Layland, G. M. Munson, N. A. Burow, L. Couvillon, A. Vaisnys, R. G. Petrie, C. T. Stelzried, and J. K. Woo. "Multiple-Mission Telemetry System," JPL-SPS 37-46, Vol. III, July 1967, pp. 175–243.

Appendix

Tausworthe, R. C., et al. "High Rate Telemetry Project," JPL-SPS 37-54, Vol. II, 30 November 1968, pp. 71–81.

Simon, M., and A. Mileant. "DSA's Subcarrier Demodulation Losses," TDA Progress Report PR 42-85: January-March 1986, 15 May 1986, pp. 111–17.

Doerksen, I., and L. Howard. "Baseband Assembly Analog-to-Digital Converter Board," TDA Progress Report PR 42-93: January-March 1988, 15 May 1988, pp. 257–64.

7. **Digital receivers.**

Kumar, R., and W. J. Hurd. "A Class of Optimum Digital Phase Locked Loops for the DSN Advanced Receiver," TDA Progress Report PR 42-83: July-September 1985, 15 November 1985, pp. 63–80.

Brown, D. H., and W. J. Hurd. "DSN Advanced Receiver: Breadboard Description and Test Results," TDA Progress Report PR 42-89: January-March 1987, 15 May 1987, pp. 48–66.

Hinedi, S. "A Functional Description of the Advanced Receiver," TDA Progress Report PR 42-100: October-December 1989, 15 February 1990, pp. 131–49.

Hinedi, S., R. Bevan, and M. Marina. "The Advanced Receiver II: Telemetry Test Results in CTA 21," TDA Progress Report PR 42-104: October-December 1990, 15 February 1991, pp. 140–56.

Sadr, R., R. Bevan, and S. Hinedi. "The Advanced Receiver II Telemetry Test Results at Goldstone," TDA Progress Report PR 42-106: April-June 1991, 15 August 1991, pp. 119–31.

8. **Encoding and decoding.**

Stiffler, J. J. "Instantly Synchronizable Block Code Dictionaries," JPL-SPS 37-32, Vol. IV, 30 April 1965, pp. 268–72.

Stiffler, J. J., and A. J. Viterbi. "Performance of a Class of Q-Orthogonal Signals for Communication over the Gaussian Channel," JPL-SPS 37-32, Vol. IV, 30 April 1965, pp. 277–81.

Green, R. R. "A Serial Orthogonal Decoder," JPL-SPS 37-39, Vol. IV, 30 June 1966, pp. 247–52.

Lushbaugh, W. A., and J. W. Layland. "System Design of a Sequential Decoding Machine," JPL-SPS 37-50, Vol. II, 31 March 1968, pp. 71–78.

McEleice, R. J., and J. W. Layland. "A Upper Bound to the Free Distance of a Tree Code," JPL-SPS 37-62, Vol. III, 30 April 1970, pp. 63–64.

Layland, J. W. "Performance of Short Constraint Length Convolutional Codes and a Heuristic Code Construction Algorithm," JPL-SPS 37-64, Vol. II, August 1970.

Lushbaugh, W. A. "Information Systems: Hardware Version of an Optimal Convolutional Decoder," JPL Technical Report TR 32-1526, Vol. II: January and February 1971, 15 April 1971, pp. 49–55.

Butman, S. A., L. J. Deutsch, and R. L. Miller. "Performance of Concatenated Codes for Deep Space Missions," TDA Progress Report PR 42-63: March and April 1981, 15 June 1981, pp. 33–39.

Divsalar, D., and J. H. Yuen. "Performance of Concatenated Reed-Solomon/Viterbi Channel Coding," TDA Progress Report PR 42-71: July-September 1982, 15 November 1982, pp. 81–94.

Yuen, J. H., and Q. D. Vo. "In Search of a 2-dB Coding Gain," TDA Progress Report PR 42-83: July-September 1985, 15 November 1985, pp. 26–33.

Dolinar, S. J. "A New Code for Galileo," TDA Progress Report PR 42-93: January-March 1988, 15 May 1988, pp. 83–96.

Dolinar, S. J. "VLA Telemetry Performance With Concatenated Coding for Voyager at Neptune," TDA Progress Report PR 42-95: July-September 1988, 15 November 1988, pp. 112–33.

Statman, J., G. Zimmerman, F. Pollara, and O. Collins. "A Long Constraint Length VLSI Viterbi Decoder for the DSN," TDA Progress Report PR 42-95: July-September 1988, 15 November 1988, pp. 134–42.

APPENDIX

Divsalar, D., M. K. Simon, and J. H. Yuen. "The Use of Interleaving for Reducing Radio Loss: Convolutionally Coded Systems," TDA Progress Report PR 42-96: October-December 1988, 15 February 1989, pp. 21–39.

Collins, O., F. Pollara, S. J. Dolinar, and J. Statman. "Wiring Viterbi Decoders (Splitting deBruijn Graphs)," TDA Progress Report PR 42-96: October-December 1988, 15 February 1989, pp. 93–103.

Berrou, C., A. Glavieux, and P. Thitimajshima. "Near Shannon Limit Error-Correcting Coding and Decoding: Turbo Codes," ICC, 1993, pp. 1064–70.

Divsalar, D. and F. Pollara. "On the Design of Turbo Codes," TDA Progress Report PR 42-123, July-September 1995, 15 November 1995, pp. 99–121.

Benedetto, S., Divsalar, D., et al. "Serial Concatenation of Interleaved Codes; Performance Analysis, Design, and Iterative Decoding," IEEE Transactions on Information Theory, Volume 44, 1998, p. 909.

9. Data compression.

Cheung, K. M., F. Pollara, and M. Shahshahani. "Integer Cosine Transform for Image Compression," TDA Progress Report PR 42-105: January-March 1991, 15 May 1991, pp. 45–53.

Ekroot, L., S. J. Dolinar, and K. M. Cheung. "Integer Cosine Transform Compression for Galileo at Jupiter: A Preliminary Look," TDA Progress Report PR 42-115, July-September 1993, 15 November 1994, pp. 110–23.

10. Arraying of antennas.

Urech, J. M. "Telemetry Improvement Proposal for the 85-foot Antenna Network," Space Programs Summary 37-63, Vol. II, Jet Propulsion Laboratory, Pasadena, California, 31 May 1970.

Urech, J. M. "Processed Data Combination for Telemetry Improvement at DSS 62," JPL Technical Report TR 32-1526, Vol. II: January and February 1971, 15 April 1971, pp. 169–76.

Wilck, H. "A Signal Combiner for Antenna Arraying," DSN Progress Report PR 42-25, November and December 1974, 15 February 1975, pp. 111–17.

Winkelstein, R. A. "Analysis of the Signal Combiner for Multiple Antenna Arraying," DSN Progress Report PR 42-26, January and December 1975, 15 April 1975, pp. 102–18.

Stevens, R. "Applications of Telemetry Arraying in the DSN," TDA Progress Report PR 42-72: October-December 1982, 15 February 1983, pp. 78–82.

Layland, J. W., and D. W. Brown. "Planning for VLA/DSN Arrayed Support to the Voyager at Neptune," TDA Progress Report PR 42-82: April-June 1985, 15 August 1985, pp. 125–35.

Brown, D. W., W. D. Brundage, J. S. Ulvestad, S. S. Kent, and K. P. Bartos. "Interagency Telemetry Arraying for Voyager-Neptune Encounter," TDA Progress Report PR 42-102: April-June 1990, 15 August 1990, pp. 91–118.

Bartok, C. D. "Performance of the Real-Time Array Signal Combiner During the Voyager Mission," TDA Progress Report PR 42-63: March-April 1981, 15 June 1981, pp. 191–202.

Layland, J. W. "ICE Telemetry Performance," TDA Progress Report PR 42-84: October-December 1985, 15 February 1986, pp. 203–13.

Hurd, W. J., F. Pollara, M. D. Russell, B. Siev, and P. U. Winter. "Intercontinental Antenna Arraying by Symbol Stream Combining at ICE Giacobini-Zinner Encounter," TDA Progress Report PR 42-84: October-December 1985, 15 February 1986, pp. 220–28.

Foster, C., and M. Marina. "Analysis of the ICE Combiner for Multiple Antenna Arraying," TDA Progress Report PR 42-91: July-September 1987, 15 November 1987, pp. 269–77.

Brown, D. W., W. D. Brundage (National Radio Astronomy Observatory), J. S. Ulvestad, S. S. Kent, and K. P. Bartos. "Interagency Telemetry Arraying for Voyager-Neptune Encounter," TDA Progress Report PR 42-102: April-June 1990, 15 August 1990, pp. 91–118.

Rogstad, D. H. "Suppressed Carrier Full-Spectrum Combining," TDA Progress Report PR 42-107: July-September 1991, 15 November 1991, pp. 12–20.

APPENDIX

11. Radio metrics—tools and techniques.

Titsworth, R. C. "Optimal Ranging Codes," JPL Technical Report TR 32-411, 15 April 1963.

Hamilton, T. W., and W. G. Melbourne. "Information Content of a Single Pass of Doppler Data from a Distant Spacecraft," JPL-SPS 37-39, Vol. III, 31 May 1966, pp. 18–23.

Tausworthe, R. C. "Minimizing Range Code Acquisition Time," JPL-SPS 37-42, Vol. IV, 31 December 1966, pp. 198–200.

Lushbaugh, W. L., and L. D. Rice. "Mariner Venus 67 Ranging System Digital Processing Design," JPL-SPS 37-50, Vol. II, 31 March 1968.

Goldstein, R. M. "Ranging with Sequential Components," JPL-SPS 37-52, Vol. II, 31 July 1968, pp. 46–49.

Martin, W. L. "A Binary Coded Sequential Acquisition Ranging System," JPL-SPS 37-57, Vol. II, 31 May 1969, pp. 72–81.

Martin, W. L. "Performance of the Binary Coded Sequential Acquisition Ranging System of DSS 14," JPL-SPS 37-62, Vol. II, 31 March 1970, pp. 55–61.

Otoshi, T. Y. "S/X Experiment: A Study of the Effects of Ambient Temperature on Ranging Calibrations," DSN Progress Report PR 42-23, July and August 1974, 15 October 1974, pp. 45–51.

Otoshi, T. Y. "S/X Band Experiment: A Study of the Effects of Multipath on Two-Way Range," DSN Progress Report PR 42-25, November and December 1974, 15 February 1975, pp. 69–83.

Melbourne, W. G. "Navigation Between the Planets," *Scientific American* 234, no. 6 (June 1976): 57–74.

Martin, W. L., and A. I. Zygielbaum. "Mu-II Ranging," JPL-TM 33-768, 15 May 1977.

Layland, J. W., and A. I. Zygielbaum. "On Improved Ranging-II," DSN Progress Report PR 42-50: January and February 1979, 15 April 1979, pp. 68–73.

Thurman, S. W. "Deep-Space Navigation With Differenced Data Types-Part I: Differenced Range Information Content," TDA Progress Report PR 42-103: July-September 1990, 15 November 1990, pp. 47–60.

Thurman, S. W. "Deep-Space Navigation With Differenced Data Types-Part II: Differenced Doppler Information Content," TDA Progress Report PR 42-103: July-September 1990, 15 November 1990, pp. 61–69.

Kahn, R. D., W. M. Folkner, C. D. Edwards, and A. Vijayaraghavan. "Position Determination of a Lander and Rover at Mars With Earth-Based Differential Tracking," TDA Progress Report PR 42-108: October-December 1991, 15 February 1992, pp. 279–93.

12. Timing standards.

Finnie, D. "Frequency Generation and Control: Atomic Hydrogen Maser Frequency Standard," JPL Technical Report TR 32-1526, Vol. I: November and December 1970, 15 February 1971, pp. 73–75.

Dachel, P. R., S. M. Petty, R. F. Meyer, and R. L. Syndor. "Hydrogen Maser Frequency Standards for the Deep Space Networks," DSN Progress Report PR 42-40, May and June 1977, 15 August 1977, pp. 76–83.

Prestage, J. D., G. J. Dick, and L. Maleki. "New Ion Trap for Atomic Frequency Standard Applications," TDA Progress Report PR 42-97: January-March 1989, 15 May 1989, pp. 58–63.

Prestage, J. D. "Improved Linear Ion Trap Physics Package," TDA Progress Report PR 42-113, January-March 1993, 15 May 1993, pp. 1–6.

13. Earth rotation and propagation media.

Winn, F. B. "Tropospheric Refraction Calibrations and Their Significance on Radio-Metric Doppler Reductions," JPL Technical Report TR 32-1526, Vol. VII: November and December 1971, 15 February 1972, pp. 68–73.

Von Roos, O. H., and B. D. Mulhall. "An Evaluation of Charged Particle Calibration by a Two-Way Dual-Frequency Technique and Alternatives to This Technique," JPL Technical Report TR 32-1526, Vol. XI: July and August 1972, 15 October 1972, pp. 42–52.

Appendix

Winn, F. B., S. R. Reinbold, K. W. Yip, R. E. Koch, and A. Lubeley. "Corruption of Radio Metric Doppler Due to Solar Plasma Dynamics: S/X Dual-Frequency Doppler Calibration for These Effects," DSN Progress Report PR 42-30, September and October 1975, 15 December 1975, pp. 88–101.

Slobin, S. D., and P. D. Batelaan. "DSN Water Vapor Radiometer- Tropospheric Range Delay Calibration," DSN Progress Report PR 42-49: November and December 1978, 15 February 1979, pp. 136–45.

Roth, M., and T. Yunck. "VLBI System for Weekly Measurement of UTI and Polar Motion: Preliminary Results," TDA Progress Report PR 42-58: May and June 1980, 15 August 1980, pp. 15–20.

Sovers, O. J.; J. L. Fanselow; G. H. Purcell, Jr.; D. H. Rogstad; and J. B. Thomas. "Determination of Intercontinental Baselines and Earth Orientation Using VLBI," TDA Progress Report PR 42-71: July-September 1982, 15 November 1982, pp. 1–7.

Scheid, J. A. "Comparison of the Calibration of Ionospheric Delay in VLBI Data by the Methods of Dual Frequency and Faraday Rotation," TDA Progress Report PR 42-82: April-June 1985, 15 August 1985, pp. 11–23.

Lindquister, U. J., A. P. Freedman, and G. Blewitt. "A Demonstration of Centimeter-Level Monitoring of Polar Motion With the Global Positioning System," TDA Progress Report PR 42-108: October-December 1991, 15 February 1992, pp. 1–9.

Linfield, R. P., and J. Z. Wilcox. "Radio Metric Errors Due to Mismatch and Offset Between a DSN Antenna Beam and the Beam of a Troposphere Calibration Instrument," TDA Progress Report PR 42-114, April-June 1993, 15 August 1993, pp. 1–13.

14. Radio science.

Levy, G. S., and G. E. Wood. "Voyager-Jupiter Radio Science Data Papers," TDA Progress Report PR 42-58: May and June 1980, 15 August 1980, pp. 114–15.

Peng, T. K., and F. F. Donivan. "Deep Space Network Radio Science System for Voyager Uranus and Galileo Missions," TDA Progress Report PR 42-84: October-December 1985, 15 February 1986, pp. 143–51.

Nelson, S. J., and J. W. Armstrong. "Gravitational Wave Searches Using the DSN," TDA Progress Report PR 42-94: April-June 1988, 15 August 1988, pp. 75–85.

Kursinski, E. R,. and S. W. Asmar. "Radio Science Ground Data System for the Voyager-Neptune Encounter, Part I," TDA Progress Report PR 42-105: January-March 1991, 15 May 1991, pp. 109–27.

15. VLBI and radio astronomy.

Miller, J. K. "The Application of Differential VLBI to Planetary Approach Orbit Determination," DSN Progress Report PR 42-40, May and June 1977, 15 August 1977, pp. 84–90.

Moultrie, B., T. H. Taylor, and P. J. Wolff. "The Performance of Differential VLBI Delay During Interplanetary Cruise," TDA Progress Report PR 42-79, July-September 1984, 15 November 1984, pp. 35–46.

Ulvestad, J. S., and R. P. Linfield. "The Search for Reference Sources for VLBI Navigation of the Galileo Spacecraft," TDA Progress Report PR 42-84, October-December 1985, 15 February 1986, pp. 152–63.

Satorius, E. H., M. J. Grimm, G. A. Zimmerman, and H. C. Wilck. "Finite Wordlength Implementation of a Megachannel Digital Spectrum Analyzer," TDA Progress Report PR 42-86, April- June 1986, 15 August 1986, pp. 244–54.

Quirk, M. P. (Institute for Defense Analyses, New Jersey), H. C. Wilck, M. F. Garyantes, and M. J. Grimm. "A Wideband, High-Resolution Spectrum Analyzer," TDA Progress Report PR 42-93, January-March 1988, 15 May 1988, pp. 188–98.

Treuhaft, R. N., and S. T. Lowe. "A Nanoradian Differential VLBI Tracking Demonstration," TDA Progress Report PR 42-109, January-March 1992, 15 May 1992, pp. 40–55.

16. Global Positioning System.

Lanyi, G. "Total Ionospheric Electron Content Calibration Using Series GPS Satellite Data," TDA Progress Report PR 42-85, January-March 1986, 15 May 1986, p. 112.

Lichten, S. M. "Precise Estimation of Tropospheric Path Delay GPS Techniques," TDA Progress Report PR 42-100, October-December 1989, 15 February 1990, pp. 1–12.

Appendix

Freedman, A. P. "Combining GPS and VLBI Earth-Rotation Data for Improved Universal Time," TDA Progress Report PR 42-105, January-March 1991, 15 May 1991, pp. 1-12.

Guinn, J., J. Jee, P. Wolff, F. Lagattuta, T. Drain, and V. Sierra. "TOPEX/POSEIDON Operational Orbit Determination Results Using Global Positioning Satellites," TDA Progress Report PR 42-116, October-December 1993, 15 February 1994, pp. 163–74.

17. The Goldstone Solar System Radar.

Carpenter, R. L., and R. M. Goldstein. "Preliminary Results of the 1962 Radar Astronomy Study of Venus," JPL-SPS 37-20, Vol. IV, 30 April 1963, pp. 182–84.

Goldstein, R. M. "Radar Observations of Icarus," JPL-SPS 37-53, Vol. II, 30 September 1968, pp. 45–48.

Leu, R. L. "X-band Radar System," JPL Technical Report TR 32-1526, Vol. XIX: November and December 1973, 15 February 1974, pp. 77–81.

Downs, G. S., R. R. Green, and P. E. Reichley. "A Radar Study of the Backup Martian Landing Sites," DSN Progress Report PR 42-36, September and October 1976, 15 December 1976, pp. 49–52.

Bhanji, A. M., M. Caplan (Varian Associates, Inc.), R. W. Hartop, D. J. Hoppe, W. A. Imbriale, D. Stone (Varian Associates, Inc.), and E. W. Stone. "High Power Ka-band Transmitter for Planetary Radar and Spacecraft Uplink," TDA Progress Report PR 42-78, April-June 1984, 15 August 1984, pp. 24–48.

Bhanji, A. M., D. J. Hoppe, B. L. Conroy, and A. J. Freiley. "Conceptual Design of a 1-MW CW X-band Transmitter for Planetary Radar," TDA Progress Report PR 42-95, July-September 1988, 15 November 1988, pp. 97–111.

Renzetti, N. A., T. W. Thompson, and M. A. Slade. "Relative Planetary Radar Sensitivities: Arecibo and Goldstone," TDA Progress Report PR 42-94, April-June 1988, 15 August 1988, pp. 287–93.

18. Telecommunications performance.

Yuen, J. H., ed. *Deep Space Telecommunications Engineering*. New York: Plenum Press, 1983.

Layland, J. W., and L. L. Rauch. "Case Studies of Technology in the DSN: Galileo, Voyager, Mariner," published in "Evolution of Technology in the Deep Space Network," TDA Progress Report PR 42-130, April-June 1997, August 1997.

Posner, E. C., et al. "Voyager Neptune Telemetry." Session V of ITC.USA/87, papers 87-0819 through 87-0824 of the International Telemetering Conference, Vol. XIII, cosponsored by the International Foundation for Telemetry and Instrumentation Society of America, 26–29 October 1987.

Posner, E. C., L. L. Rauch, and B. D. Madsen. "Voyager Mission Telecommunication Firsts," *IEEE Communications Magazine* 28, no. 8 (September 1990): 22–27.

19. Cost-reduction initiatives.

Jet Propulsion Laboratory, Charles T. Stelzried, ed. *TMOD Technology and Science Program News* 2, 4, 7, 8, 9 (March 1995-April 1998). http://deepspace.jpl.nasa.gov/technology.

20. Ka-band development.

Jet Propulsion Laboratory, Charles T. Stelzried, ed. *TMOD Technology and Science Program News* 1, 2, 3, 6, 7 (December 1994-February 1997). http://deepspace.jpl.nasa.gov/technology.

Morabito, D., R. C. Clauss, and M. Speranza. "Ka-band Atmospheric Noise-Temperature Measurements at Goldstone, California, Using a 34-m Beam-Waveguide Antenna," TMO Progress Report PR 42-132, February 1998.

Feria, Y., M. Belongie, T. McPheeters, and H. Tan. "Solar Scintillation Effects on Telecommunication Links at Ka-band and X-band," TMO Progress Report PR 42-129, May 1997.

Morabito, D. D. "The Efficiency Characterization of the DSS 13, 34-meter Beam Waveguide Antenna at Ka-band (32.0 GHz and 33.7 GHz) and at X-band (8.4 GHz)," TMO Progress Report PR 42-125, May 1996.

Vilnrotter, V. A., and B. Iijima. "Analysis of Array Feed Combining Performance Using Recorded Data," 42-125, May 1996.

Mysoor, N. R., et al. "Performance of a Ka-band Transponder Breadboard for Deep Space Applications," TMO Progress Report PR 42-122, August 1995.

APPENDIX

21. Other technologies.

Jet Propulsion Laboratory, Charles T. Stelzried, ed. *TMOD Technology and Science Program News* 2, 3, 4, 5, 8, 9 (March 1995-April 1998). *http://deepspace.jpl.nasa.gov/technology.*

22. Optical communications development.

Hemmati, H., et al. "Comparative Study of Optical and Radio Frequency Communication Systems for Deep Space Missions," TMO Progress Report PR 42-128, February 1997.

Lesh, J. R., L. J. Deutsch, and C. D. Edwards. "Optical Communications for Extreme Deep Space Missions," Review of Laser Engineering 24, issue 12 (December 1996).

Lesh, J. R. "Impact of Laser Communications on Future NASA Missions and Ground Tracking Systems." CRL International Symposium on Advanced Optical Communications and Sensing, Tokyo, Japan, 15–16 March 1995.

Wilson, K. E., J. R. Lesh, et al. "GOPEX: A Deep Space Optical Communications Demonstration with the Galileo Spacecraft," TDA Progress Report PR 42-103, July-September 1990, 15 November 1990, pp. 262–73.

Toyoshima, M., K. Araki, et al. "Reduction of ETS VI Laser Communication Equipment Optical-Downlink Telemetry Collected during GOLD," TDA Progress Report PR 42-128, 15 February 1997.

23. DSN science.

Jet Propulsion Laboratory, Charles T. Stelzried, ed. *TMOD Technology and Science Program News* 1–9 (December 1994-April 1998). *http://deepspace.jpl.nasa.gov/technology.*

ABOUT THE AUTHOR

Douglas J. Mudgway came to the United States in 1962 to work at NASA's Jet Propulsion Laboratory in Pasadena, California, following a fifteen-year career in the field of guided missile research in Australia. At JPL, he was involved in the development and operation of the Deep Space Network from its infancy in the early 1960s to its maturity in the early 1990s. He was the recipient of the NASA Exceptional Service Medal (1977) for his contribution to the Viking mission to Mars, and the Exceptional Achievement Medal (1991) for his contribution to the Galileo mission to Jupiter. A mathematics and science major from the University of New Zealand (1945), Mr. Mudgway retired in 1991 and lives in the wine country of Northern California.

INDEX

A

Active Magnetospheric Particle Explorer, AMPTE, 170
Activity Plan Process, 618, 619
Advanced DSN Monitor and Control, ADMC, 616
Advanced Multimission Operations System, AMMOS, 328, 610, 620
Advanced Receiver, ARX, 473, 474
Advanced Research Projects Agency, ARPA, 7, 8, 10, 12, 18, 19, 25
Air Force, U.S., 4, 7
Allen, Lew, 602
Ames Research Center, NASA ARC, 33, 49, 51, 475
Amorose, Raymond J., 603, 612, 615
Anderson, John D., 514
Antennas
   9-meter, 56, 169, 334, 336, 459, 541, 616
   26-meter, 4, 10, 20, 21, 30, 37, 40, 43, 44, 50, 51, 56, 57, 58, 65, 66, 77, 84, 89, 97, 98, 100, 155, 156, 157, 159, 166, 169, 265, 305, 307, 324, 328, 334, 335, 336, 383, 459, 495, 521, 597, 600, 616
   34-meter, xlvi, 84, 157, 158, 159, 166, 263, 265, 285, 290, 302, 304, 307, 324, 326, 328, 331, 334, 342, 344, 381, 439, 459, 462, 465, 467
   64-meter, 41, 43, 46, 50, 51, 57, 58, 69, 70, 75, 77, 84, 89, 96, 98, 156, 172, 256, 263, 265, 278, 290, 299, 300, 304, 337, 341, 343, 344, 481, 490, 513, 521, 597, 600
   70-meter, xlvii, 285, 290, 302, 303, 304, 306, 307, 313, 329, 333, 337, 338, 339, 340, 341, 342, 343, 459, 462, 463, 467, 523, 525
Apollo, 4, 35, 36, 43, 52, 54, 56, 57, 58, 73, 74, 597
Az-El, antenna, xvi, xx, xxiv, xxv, xxxiii, xxxiv, xxxv, xxxviii, xxxv, 17
Army, U.S., 4, 5, 7, 20
Asmar, Sami W., 507, 513
Astronomical Radio Interferometric Earth Surveying, ARIES, 541, 542
Astronomical Unit, AU, xxix, xxx, xxxii, 15, 16, 87, 88, 289, 306, 325, 330, 332, 333, 491, 550, 551
Atlantis, 286
Atlas-Agena, 38
Automated Real-time Spacecraft Navigation, ARTSN, 495
Australian Department of Supply (DOS)/Weapons Research Establishment (WRE), 19, 20, 22, 64
Australian Government's Commonwealth Scientific and Industrial Research Organisation, CSIRO, 343, 526

B

Baker-Nunn, Smithsonian tracking camera, 20, 23
Baseband Assembly, BBA, 256, 385
Bayley, William H., 3, 583, 598
Bayley/Lyman Years, 601
Beggs, James, 276
Bell Telephone Laboratories, BTL, 12, 13, 15
Bendix Field Engineering Corporation, 21
Berman, Allen L., 507, 514
Big Viterbi Decoder, BVD, 303, 309, 386, 387
Bilbo, Thomas S., 594

Blaw-Knox Company, 9, 18, 19, 20
Block V Receiver, BVR, 314, 315, 316, 358, 474
Buffered Telemetry Demodulator, BTD, 368
Butcher, Lou, 3
Butrica, Andrew J., 173, 551

C

C-band, 281
California Institute of Technology, CIT, Caltech, 6, 8, 11, 25, 171, 487, 488, 523
California, University of, Berkeley, 341
California, University, Los Angeles, 583
Canberra, Australia, CDSCC, xxxiv, xlvi, 18, 19, 21, 30, 39, 56, 66, 75, 77, 84, 102, 173, 290, 291, 303, 307, 313, 317, 324, 325, 328, 333, 337, 343, 368, 371, 379, 383, 437, 438, 459, 460, 482, 524, 526, 605
Cape Canaveral, Florida, Cape Kennedy, xxxv, 4, 5, 18, 23, 36, 37, 88, 96, 305, 328, 582
Carnegie Institute, 9
Carter, Jimmy, U.S. President, 173
Casani, John, 295, 302, 303
Cassegrain focus antenna, xxxviii
Cassini, 274, 299, 398, 408, 409, 428, 429, 430, 437, 439, 453, 497, 512
Cebreros, 158
Celestial Mechanics, 511
Central Communications Terminal, 86
Centre National d'Etudes Spatiales, CNES, 280, 327, 329, 336
Challenger, 265, 283, 295, 301, 322
Clinton, William Jefferson, U.S. President, Administration, 610
Collins Radio Company, 20, 77
Comet Giacobini-Zinner, 276, 331
Comet Halley, encounter, 274, 275, 276, 277, 278
Command and Data Console, CDC, 53
Command modulator assembly, CMA, 258, 263, 388
Command processor assembly, CPA, 258, 388
Command switching assembly, CSA, 388
Commonwealth Scientific and Industrial Research Organisation, CSIRO, 526
Compatibility Test Area 21, CTA21, 96, 97
Congress, U.S., 171
Consolidated Space and Test Center, CSTC, Air Force, 305
Corliss, William R., 42, 43, 43, 44, 45, 56, 73
Cross Support Agreements, 526
Crustal Dynamics, 541, 542, 548, 549

D

Data Capture Process, 616, 619
Data Compression, 478
Data Systems Operations Team, DSOT, 618
Declination, xxxiii
Deep Space Communications Complex, DSCC, 77, 256, 258, 264, 313, 315, 386, 389, 391, 422, 525, 604, 616

Index

Deep Space Instrumentation Facility, DSIF, 11, 15, 16, 21, 30, 31, 32, 33, 34, 36, 37, 38, 39, 40, 50, 61, 71, 507, 511, 521, 582, 591, 596, 599, 603
Deep Space Network, DSN, xxviii, xxx, xxxi, xxxii, xxxiii, xxxiv, xxxv, 3, 4, 5, 6, 8, 16, 17, 18, 25, 30, 33, 35, 36, 37, 38, 41, 42, 43, 44, 45, 46, 47, 48, 49, 50, 51, 54, 56, 57, 58, 61, 62, 63, 64, 65, 66, 69, 70, 71, 72, 73, 74, 75, 77, 78, 82, 83, 84, 87, 89, 90, 91, 92, 93, 96, 97, 99, 100, 101, 159, 166, 167, 168, 169, 170, 172, 173, 174, 176, 258, 263, 264, 265, 274, 275, 276, 277, 278, 279, 280, 281, 284, 286, 288, 289, 290, 291, 292, 293, 294, 295, 296, 297, 299, 301, 302, 303, 304, 305, 306, 307, 309, 310, 311, 312, 313, 315, 316, 317, 318, 319, 323, 324, 325, 326, 328, 329, 330, 331, 332, 333, 337, 338, 341, 343, 368, 371, 383, 398, 434, 437, 445, 450, 458, 467, 471, 472, 478, 483, 487, 488, 489, 493, 494, 498, 499, 502, 507, 508, 511, 515, 517, 521, 522, 526, 528, 550, 567, 591, 594, 597, 599, 600, 602, 603, 604, 606, 615, 620
Deep Space Network Monitor and Control, DMC, 447
Deep Space Station, DSS, 3
    DSS 16, 336
    DSS 11, Madrid, 3, 4, 56, 84
    DSS 12, 47, 84, 89, 155, 157, 158, 278, 305
    DSS 13, 47, 328, 375, 458, 494, 497, 534, 542
    DSS 14, 45, 47, 57, 58, 77, 84, 89, 96, 98, 306, 314, 315, 317, 330, 337, 340, 341, 342, 513, 514, 542
    DSS 15, 309, 314
    DSS 23, 449
    DSS 27, 336
    DSS 34, 379
    DSS 42, 56, 84, 89, 90, 158, 278
    DSS 43, 77, 84, 98, 102, 238, 314, 315, 317, 337, 338, 340, 341, 342
    DSS 44, 84, 89, 90, 158
    DSS 46, 336
    DSS 61, 56, 84, 158, 278
    DSS 62, 84, 89, 158
    DSS 63, 77, 84, 98, 238, 306, 317, 338, 339, 340, 341
    DSS 65, 288
Deep Space Tracking System, DSTS, ESA, 279
Defense, U.S. Department of, 4, 7, 8, 11, 18, 19
Deimos, 46
Delta Differential One-way Ranging, DDOR, 307, 512
Deutsch, Leslie J., 612
Differenced Range versus Integrated Doppler, DRVID, 71, 96,
Digital Equipment Corporation, DEC, 264
Digitally Controller Oscillator, DCO, 263
Doppler, xxxviii, xliv, xlvi, xlvii, 5, 9, 42, 62, 70, 74, 96, 101, 174, 278, 281, 284, 287, 304, 306, 329, 330, 338, 422, 472, 484, 485, 486, 507, 508, 511, 514, 518
DSCC Galileo Telemetry, DGT, 312, 313, 314, 315, 316, 317, 318, 359
Dryden, Hugh, 21
Dubridge, Lee A., 6
Dumas, Larry N., 526, 606
Dumas/Haynes Years, 606

E

Earth Crossing Asteroid, ECA, 499
Earth Orbiter, EO, 169
Echo, balloon experiment, 13, 14, 15, 16, 17, 42, 44, 66
Edwards, Chad, 525
Effron, Leonard, 276
Eisenhower, Dwight David, U.S. President, 7, 10, 15
Ellis, Jordan, 276
Eros, asteroid, 274, 275
European Space Agency, ESA, 168, 276, 277, 278, 279, 319, 321, 325
European Space Operation Center, ESOC, 278, 279
Explorer, missions, 5
Extra-Galactic Radio Sources, EGRS, xlvi, 525

F

Faraday Rotation, FR, 513
Federal Aviation Administration, FAA, 498
Federal Republic of Germany, 170
Feedback Concatenated Decoder, FCD, 368
Felberg, Fred, 594
Finerman, Allan, 598
Flight Operations Control Center, FOCC, 34
Flight Project Mission Operations Teams, PROJ, 447
Force, Charles T., 608
Ford, Gerald, U.S. President, 102
France, French, 169
Frequency and Timing System, FTS, 447
Full Spectrum Recorders, FSR, 368

G

Galileo, 159, 166, 168, 265, 274, 275, 283, 286, 288, 294, 295, 296, 297, 298, 299, 300, 301, 302, 303, 303, 304, 305, 306, 307, 308, 309, 310, 311, 312, 313, 314, 315, 316, 317, 318, 319, 322, 323, 329, 331, 337, 338, 339, 340, 342, 368, 398, 408, 410, 450, 477, 478, 498, 511, 512, 608
Gates, Clarence R., 582
German Space Operations Center, GSOC, 89, 90, 305
Giberson, Walker E., 582
Giotto, 276, 277, 278, 279
Global Positioning System, GPS, 483, 486, 489, 494, 547
Goddard Space Flight Center, GSFC, 33, 56, 62, 278, 335, 591
Goldin, Daniel S., NASA Administrator, 610
Goldstein, Richard M., 16, 504, 513, 528
Goldstone, California, GDSCC, xxxii, xxxiv, xl, xlvi, 2, 10, 15, 18, 35, 37, 39, 41, 42, 43, 44, 45, 47, 52, 57, 66, 69, 70, 71, 75, 77, 84, 90, 101, 157, 173, 290, 291, 302, 303, 306, 307, 316, 325, 328, 333, 336, 337, 338, 341, 342, 381, 437, 438, 450, 458, 481, 482, 490, 513, 521, 523, 530, 534, 541, 542, 603, 605
Goldstone Solar System Radar, GSSR, 16, 489, 491, 499, 522, 563, 566, 567
Grand Tour, 171, 173

Index

Greenwich Mean Time, GMT, xxxv, 102
Gravitational Wave Experiment, GWE, 323, 518
Ground Communications Facility, GCF, 39, 75, 84, 86, 368, 394, 396, 447, 597, 600, 602, 604, 614, 615
Ground Operational Equipment, GOE, 50
Ground Reconstruction Equipment, GRE, 55
Gulkis, Samuel, 524, 528

H

Helios, 86, 87, 88, 89, 90, 91, 159, 513, 514
Hall, R. Cargill, 6, 36, 73
Halley, Edmond, ii
Halley Intercept Mission, HIM, 276
Hammer, Michael, 612
Haynes, Norman R., 606, 612
Heacock, Raymond L., 171
Heftman, Kurt, 598
Higa, Walter, 551
High-Earth orbiting, HEO, 169, 256, 275, 334
High-Efficiency antenna, HEF, 299, 304
High-electron mobility transistor, HEMT, 470, 471
High-Gain antenna, HGA, 172, 284, 285, 296, 297, 307, 308, 309, 310, 311, 312, 313, 322, 323, 325, 328, 332
High-Rate Telemetry System, HRT, 44, 45
High-Resolution Microwave Survey, HRMS, 537, 538, 539
Host Country Programs, 525
Hour-Angle Declination, HA-Dec, xxxiii, xxxiv, 14
Hubble Space Telescope, HST, 288, 335, 336
Hume, Alan, Australian Minister of Supply, 21

I

Inertial Upper Stage, IUS, 168, 295, 305, 324
Ingenieria y Servicios Aerospaciales, INSA, 606
Institute for Space and Astronautical Science, ISAS, 290, 445
Instituto Geografico National, Spain, 526
Instituto Nacional Tecnica Aeroespacial, INTA, Spain, 526, 605
Interface Control Documents, ICDs, 336
Interior, U.S. Department of, 4
Intercontinental Ballistic Missile, ICBM, 4
Intermediate Range Ballistic Missile, IRBM, 7
International Astronomical Union, IAU, 566
International Cometary Explorer, ICE, 169, 170, 275, 276, 277, 300, 331
International Geophysical Year, IGY, 4, 5, 7, 13, 20
International Solar Polar Mission, ISPM, 319, 320, 321
International Solar Terrestrial Physics, ISTP, 383, 408, 434
Iowa, University of, 5

J

Jaffe, Richard, 5, 8
Jauncey, David L., 524, 528
Jet Propulsion Laboratory, JPL, xlvii, xlviii, 3, 4, 5, 6, 7, 8, 9, 10, 11, 12, 13, 14, 15, 16, 17, 18, 19, 20, 21, 23, 25, 30, 31, 33, 35, 36, 37, 38, 39, 41, 44, 45, 46, 47, 48, 49, 50, 52, 56, 60, 61, 63, 64, 65, 66, 69, 70, 72, 75, 77, 78, 83, 86, 89, 91, 93, 96, 98, 102, 171, 172, 173, 264, 276, 278, 279, 281, 295, 300, 303, 305, 306, 316, 317, 323, 329, 335, 338, 339, 368, 385, 393, 414, 450, 458, 475, 485, 502, 504, 522, 533, 582, 591, 594, 596, 597, 598, 601, 603, 605, 606, 610, 612, 618
Johannesburg, South Africa, DSIF 41, 20, 24, 25, 30, 37, 38, 39, 44, 62
   Hartebeestpoort dam site, 23
Johnson, Lyndon B., U.S. President, 35
Johnson, Lyndon B., Space Center, NASA JSC, 299, 305
Johnson, Marshall S., 594
Jupiter, 173
   Amalthea, 173
   Io, 173, 175
   Europa, 173, 175, 299
   Ganymede, 173, 175, 299
   Callisto, 173, 175, 299
Jupiter Orbit Injection, JOI, 299, 315, 316
Jupiter Orbiter Probe, JOP, 294, 295

K

Ka-band, xxxviii, 440, 496, 497, 498, 512
Ka-band Link Experiment, KABLE, 328, 496, 497, 498
Kennedy Space Center, NASA KSC, Cape Kennedy, xxvii, 265, 295, 305
Klein, M. J., 504, 528
Koscielski, Charles, 3
Kuiper, Thomas B., 504, 528, 529

L

L-band, 61, 62, 281, 316
Laeser, Richard P., 172
Laika, 4
Lairmore, Glen E., 598
Langley Aeronautical Laboratory, 13
Langley Research Center, NASA LaRC, 33, 54, 83, 87, 92
Layland, James W., 303
Leslie, Robert A., 64, 65, 66
Levy, Gerry, 525, 533
Lewis Research Center, NASA LeRC, 299
Lincoln Laboratory, 15
linear ion trap, LIT, 485, 486
Linfield, Roger, 525
Links
   One-way, 483
   Two-way, 483

INDEX

Downlink, xxxix, xli, xlvi, xlviii, 23, 468, 477, 493, 551
Uplink, xxxix, xli, xlvi, 467, 468, 497, 514, 515, 517, 551
Linnes, Karl W., 73
Lockheed Martin Astronautics, 414
Low-Earth orbiter, LEO, 334, 495, 616
Low-Gain Antenna, LGA, 297, 312, 313, 323, 330
Lunik, 34
Lyman, Peter T., 502, 561, 601

M

Madrid, Spain, MDSCC, xxxiv, 18, 19, 23, 56, 75, 77, 90, 288, 291, 302, 303, 307, 325, 337, 338, 368, 437, 439
Magellan, 159, 274, 283, 285, 286, 288, 299, 322, 323, 488, 513
Mallis, Richard K., 3, 20, 21, 594
Manchester, University of, 15
Mariner, xli, 16, 17, 25, 32, 33, 36, 37, 38, 39, 40, 41, 42, 43, 44, 45, 46, 47, 48, 50, 51, 55, 61, 62, 64, 66, 70, 71, 73, 75, 78, 82, 89, 93, 96, 172, 279, 475, 490, 491, 511, 512, 514, 515, 596
Mariner Jupiter-Saturn, 171, 172
Mark, 1A, III-77, IV-A, 71, 155, 159, 166, 167, 256, 258, 264, 285, 299, 303, 304, 323, 334, 383, 386
Mars Global Surveyor, MGS, 274, 408, 413, 414, 418, 420, 439, 453, 495, 497, 511
Mars Observer, 326, 327, 328, 329, 414
Mars Orbit Insertion, MOI, 329
Mars Pathfinder, MPF, 274, 408, 421, 422, 423, 424, 425, 439, 453
Martin, James S., Jr., 92
Martin, Warren L., 70, 169
Massachusetts Institute of Technology, MIT, 15
Mathison, Richard P., 342, 343
McElroy, Neil H., 5
McKee, Dick, 24
McKinley, Edward L., 612
Melbourne, University of, 65
Merrick, William, xx, 8, 9, 69, 70
Merritt Island, Florida, 90, 96
Messerschmicht-Bolkow-Blohm, MBB, 316
Metric Data Assembly, MDA, 263
Meyer, Don, 24
Milikan, Robert A., 6
Mission Control and Computing Center, MCCC, 84, 86, 305, 601, 606, 608
Mobile Tracking Station, MTS, 37
Moffet, Allen T., 522, 528
Moon, xxxv, 6, 7, 9, 23, 31, 33, 34, 35, 52, 54, 55, 56, 57, 78
Mu, 70, 71, 90
Mudgway, Douglas J., xvi, 73, 599
Muhleman, Duane O., 514, 563
Multimission Operations Systems Office, MOSO, 328, 610, 614
Murray, Bruce, 276, 602
Mutch, Thomas A., 176

N

NASCOM, NASA ground communications system, 52, 62, 77, 87, 90, 278, 305
National Academy of Sciences, NAS, 171
National Aeronautics and Space Administration, NASA, xxvii, xliv, xlvi, xlviii, 2, 3, 8, 11, 15, 17, 18, 19, 20, 22, 23, 25, 33, 35, 36, 38, 39, 41, 42, 43, 46, 49, 50, 54, 56, 63, 64, 65, 66, 72, 77, 78, 89, 93, 102, 155, 159, 168, 169, 171, 265, 275, 276, 277, 283, 286, 301, 303, 310, 319, 320, 321, 327, 328, 329, 334, 335, 336, 343, 502, 521, 525, 594, 602, 605
National Bureau of Standards, NBS, 71
National Geographic, 293
National Radio Astronomy Observatory, NRAO, 290, 303, 342, 343, 435, 481, 482, 523, 526, 533
National Space Development Agency, NASDA, Japan, 277, 336
Naugle, John, 598
Naval Research Laboratory, NRL, 9
Navy, U.S., 4, 20
Near Earth Asteroid Rendezvous, NEAR, 274, 412
Neher, Victor, 6
Neptune, planet, xxix, xxxi, xxxii, 289, 290, 291, 292, 293, 294, 300, 303, 332, 333, 343, 349
Network Operations Control Center, NOCC, 86, 256, 263, 264, 323, 388, 393, 434, 447, 450, 600, 604, 614, 615
New Zealand, Wellington, 6
Nigeria, South Africa, 5, 18, 19, 23
NOCC Monitor and Control, NMC, 447
NOCC Support System, NSS, 263, 264, 447
NOCC Navigation, NAV, 447

O

Occultation Data Assembly, 510
Office of Space Sciences, NASA HQ, 533
Office of Tracking and Data Acquisition, OTDA, 521, 597, 598
Olsen, Howard, 24
O'Neill, William J., 311
Optical Navigation, OPNAV, 309, 310
Orbiting Very Long Baseline Interferometry, OVLBI, 445, 446, 447, 525
Ostro, Steven J., 504, 561, 562, 564, 566

P

Parkes Radio Astronomy Observatory, 256, 290, 291, 292, 313, 343, 346, 410, 481, 482, 521
Parkes-CDSCC Telemetry Array, PCTA, 343, 344, 526
Parks, Robert J., 582
Payload Assist Module, PAM, 324
Phase-lock tracking, 472
Phobos, 46, 279, 280, 281, 282, 288
Pickering, William H., xvii, 5, 6, 35, 61, 591, 599
Piloted Space Flight Network, MSFN, 56, 57
Pioneer, 2, 3, 4, 7, 8, 10, 11, 12, 14, 20, 43, 44, 48, 50, 51, 52, 53, 62, 66, 73, 78, 82, 84, 86, 89, 91, 155, 159, 166, 167, 168, 171, 275, 276, 277, 283, 293, 294, 300, 330, 331, 341, 408, 410, 412, 473, 475, 511, 513, 514, 518, 582, 597

Index

Planetary Radar, 550
Pluto, xxix, xxxii, xxxiii
Programmed oscillator, POCA, 263
Project Operations Control Centers, POCC, 86
Project Sequence of Events, PSOE, 618, 619
Puerto Rico, 18, 23

R

Radar Astronomy, 566
Radio Astronomy, 521
Radio Frequency Interference, RFI, 534
Radio Frequency (RF) Radiation, xxxvii, xxxix
Radio Propagation and Occultation, 514
Radio Science, 507, 508, 509, 510, 513, 514
Radioisotope Thermal Generator, RTG, 51, 305, 332
Ranger, 17, 21, 23, 24, 25, 32, 33, 34, 35, 36, 38, 60, 61, 62, 64, 78, 596
Reagan, Ronald, U.S. President, 276
Rechtin, Eberhardt, 3, 5, 7, 8, 9, 30, 61, 502, 582, 584
Redstone, arsenal, 5
Relativistic Time Delay, 517
Renaissance, 567
Renzetti, Nicholas A., iii, xix, 21, 73, 155, 458, 502, 598, 601
Robledo, 84
Roentgen, satellite, 408
Rohr Corporation, 66
Round-trip light time, RTLT, 483

S

S-band, xxxvii, xlii, xliii, xlvii, 16, 39, 47, 50, 61, 62, 63, 84, 87, 97, 98, 155, 156, 157, 166, 256, 263, 264, 277, 285, 286, 290, 297, 299, 301, 304, 306, 308, 312, 313, 315, 323, 324, 325, 330, 368, 388, 467, 513, 518
Sakigake, 276
San Diego, California, 5
Saturn V, 57
Search for Extraterrestrial Intelligence, SETI, 488, 502, 504, 533, 535, 536
Shoemaker-Levy Comet, 313, 316
Siegmeth, Alfred J., 73
Signal Processing Centers, SPC, 383, 384, 385, 388, 390, 394, 447
Signal-to-Noise Ratio, SNR, xxxvii, xlvii, 478, 483
Significant Advances, 529
Significant Events, 569
   Venus, xxx, xxxi, 6, 15, 16, 31, 37, 38, 39, 42, 43, 46, 47, 48, 61, 78, 82, 155, 159, 167, 274, 277, 283, 285, 286, 305, 306, 331, 339, 490, 511, 550, 569
   Mars, Red Planet, xxvii, xxix, xxxi, xxxii, xxxv, xxxvi, 6, 31, 39, 41, 43, 44, 45, 51, 57, 62, 66, 75, 78, 82, 91, 92, 93, 96, 97, 101, 102, 103, 274, 280, 281, 282, 294, 327, 328, 329, 330, 475, 490, 511
   Mercury, xxix, xxxi, 78, 82, 569
   Jupiter, xxvii, xxix, xxxi, xxxii, 31, 46, 50, 51, 52, 74, 78, 82, 166, 167, 169, 171, 172, 173, 175, 265, 294, 295, 296, 299, 300, 301, 304, 310, 312, 313, 316, 317, 333, 478, 569

Saturn, xxvii, xxix, xxxi, 31, 52, 166, 171, 172, 173, 176, 276, 333, 571
Earth, xxix, xxxi, xxxiii, xxxv, xliii, xlvi, 4, 5, 7, 8, 12, 13, 15, 16, 33, 37, 40, 47, 54, 57, 70, 74, 86, 87, 91, 92, 100, 170, 275, 277, 284, 288, 290, 296, 325, 331, 483, 511, 517
Singapore, 5
Small Deep Space Transponder, SDST, 497
Solar Corona, wind, 513
Solar Heliospheric Observatory, Soho, 434
South African Council for Scientific and Industrial Research (CSIR)/National Institute of Telecommunications Research (NITR), 19, 23, 25
South African Department of the Postmaster General, 24
Soviet Union, 169, 283
Space Flight Operations Facility, SFOF, 39, 41, 42, 52, 60, 61, 62, 63, 72, 75, 84, 86, 596, 596, 597, 598, 599
Space Flight Operations Center, SFOC, 610
Space Operations Management Operations, SOMO, 335, 610, 612, 614, 619
Space Very Long Baseline Interferometry, SVLBI, 445
Spaceflight Tracking and Data Network, STDN, 90, 91
Spaulding, Wallace P., 599
Search for Extraterrestrial Intelligence, SETI, 488, 502, 504, 533, 535, 536
Sputnik, 4, 6, 78
Standard Format Data Unit, SFDU, 327
Stelzried, Charles, 276, 513
Stevens, Robertson S., 8, 9, 337, 582
Stoller, Floyd W., 20
Stone, Edward C., 171, 606, 611
Sun, xxix, xxxi, xxxii, xxxiii, xxxv, xxxvi, xxxvii, 16, 3, 35, 41, 74, 78, 87, 100, 169, 280, 325, 330, 331, 507, 514
Sun-Earth-Probe, SEP, 89, 90
Support Instrumentation Requirements Document, SIRD, 72, 594
SURF, satellite, 408
Surveyor, 4, 32, 35, 36, 52, 53, 55, 64, 78, 597
Synchronization and Detection, data, 472
Synthetic Aperture Radar, SAR, 284
System Noise Temperature, xliii

T

Tau, 41, 42, 71
Tausworthe, Robert C., 70
Telecommunications and Data Acquisition, TDA, 502, 503, 506, 526, 533, 541, 549, 594, 597, 598, 601, 602, 603, 605, 606, 612
Telecommunications and Mission Operations Directorate, TMOD, 334, 335, 451, 452, 458, 458, 492, 613, 618, 619, 620
  Birth of the TMOD, 612
  Reengineering TMOD, 615
Telemetry Modulation Unit, TMU, 303
Telemetry Processor Assembly, TPA, 257, 258, 385
Thatcher, John W., 73, 599
Thor-Able, 7

Index

Tidbinbilla, Australia, 21, 39, 44, 64, 65, 66, 84, 522, 524, 530
Titan-Centaur, xxvii, 88, 92, 93, 100, 172, 294, 295
TOMS-EP, 408
TOPEX-POSEIDON, 489
Toutatis, asteroid, 565
Tracking and Communications Extraterrestrial, TRACE, 18
Tracking and Data Acquisition, TDA, 96
Tracking and Data Relay Satellite System, TDRSS, 169, 275, 288, 310, 335, 336, 525, 530
Trajectory correction maneuvers, TCMs, 324
Transfer Orbit Stage, TOS, 328
Travelling Wave Tube, TWT, 330
Traveling wave masers, TWM, 469, 470, 471
Triton, 290, 292
Truly, Richard H., 608
Truszynski, Gerald, 589, 598

U

Ultra-stable Oscillator, USO, xli, xlviii, 515
Ulvestad, James, 525
Ulysses, 168, 169, 265, 285, 300, 316, 321, 322, 323, 324, 325, 329, 408, 410, 499, 613
United Kingdom, 170
United States, U.S., 78, 169, 170, 276, 283
Universal Time Coordinated, UTC, xxxv, 288
Uranus, xxix, xxxi, xxxii, 166, 171, 172, 265, 290, 333, 344
Usuda Observatory, Japan, 290, 291, 292
USSR, 283
USSR Space Research Institute, IKI, Russia, 279
Urech, Jose M., 514

V

Van Allen, James A., 5, 6, 12
Vanguard, 4, 5
Vega 1, 2, 274, 277, 279, 280, 281
Venus-Earth-Earth-Gravity Assist, VEEGA, 295, 299, 301
Venus Orbiting Imaging Radar, VOIR, 283
Venus Radar Experiment, VRE, 15, 16
Venus Radar Mapper, VRM, 283
Very Large Array, VLA, 290, 291, 303, 304, 348, 351, 471, 481, 482
Very Long Baseline Interferometry, VLBI, 263, 277, 281, 304, 341, 434, 434, 435, 445, 483, 486, 487, 488, 525, 526, 530, 542
Very Long Baseline Interferometry Space Observatory Program, VSOP, 434, 436, 445
Victor, Walter K., 8, 9, 550, 582
Viking, xli, 46, 51, 75, 78, 82, 83, 87, 89, 91, 92, 93, 96, 97, 98, 100, 101, 159, 166, 167, 172, 265, 274, 279, 294, 327, 511, 514
Viking Mission Control Center, VMCC, 600
von Braun, Wernher, 6

Voyager, xxvii, xxxv, xli 52, 78, 83, 157, 159, 166, 167, 168, 170, 171, 172, 173, 174, 175, 263, 264, 265, 275, 276, 278, 285, 286, 288, 290, 291, 292, 293, 294, 300, 315, 331, 332, 333, 343, 344, 408, 410, 475, 475, 477, 478, 488, 511, 512, 536, 613
Voyager Interstellar Mission, VIM, 331, 332

W

Wackley, Joseph A., 514
Washington, DC, 18
Weilheim, 91
Westmoreland, Paul T., 606, 612
White House, 35, 173
White Sands Missile Range, New Mexico, 4, 7, 69, 335
Woomera, Australia, DSIF 41, 20, 21, 22, 23, 25, 37, 39, 42, 66, 77, 522, 550
  Island Lagoon, 20, 21
World Net, 19

X

X-band, xxxvii, xxxviii, xlii, xliii, xlvii, 47, 84, 97, 98, 99, 100, 101, 155, 156, 157, 159, 166, 172, 173, 263, 264, 277, 285, 287, 288, 290, 296, 297, 299, 301, 304, 306, 307, 309, 310, 313, 323, 325, 327, 328, 329, 342, 383, 439, 445, 490, 512, 513, 518

Y

Yerbury, Michael J., 524

THE NASA HISTORY SERIES

Reference Works, NASA SP-4000:

Grimwood, James M. *Project Mercury: A Chronology.* NASA SP-4001, 1963.

Grimwood, James M., and C. Barton Hacker, with Peter J. Vorzimmer. *Project Gemini Technology and Operations: A Chronology.* NASA SP-4002, 1969.

Link, Mae Mills. *Space Medicine in Project Mercury.* NASA SP-4003, 1965.

Astronautics and Aeronautics, 1963: Chronology of Science, Technology, and Policy. NASA SP-4004, 1964.

Astronautics and Aeronautics, 1964: Chronology of Science, Technology, and Policy. NASA SP-4005, 1965.

Astronautics and Aeronautics, 1965: Chronology of Science, Technology, and Policy. NASA SP-4006, 1966.

Astronautics and Aeronautics, 1966: Chronology of Science, Technology, and Policy. NASA SP-4007, 1967.

Astronautics and Aeronautics, 1967: Chronology of Science, Technology, and Policy. NASA SP-4008, 1968.

Ertel, Ivan D., and Mary Louise Morse. *The Apollo Spacecraft: A Chronology, Volume I, Through November 7, 1962.* NASA SP-4009, 1969.

Morse, Mary Louise, and Jean Kernahan Bays. *The Apollo Spacecraft: A Chronology, Volume II, November 8, 1962-September 30, 1964.* NASA SP-4009, 1973.

Brooks, Courtney G., and Ivan D. Ertel. *The Apollo Spacecraft: A Chronology, Volume III, October 1, 1964-January 20, 1966.* NASA SP-4009, 1973.

Ertel, Ivan D., and Roland W. Newkirk, with Courtney G. Brooks. *The Apollo Spacecraft: A Chronology, Volume IV, January 21, 1966-July 13, 1974.* NASA SP-4009, 1978.

Astronautics and Aeronautics, 1968: Chronology of Science, Technology, and Policy. NASA SP-4010, 1969.

Newkirk, Roland W., and Ivan D. Ertel, with Courtney G. Brooks. *Skylab: A Chronology.* NASA SP-4011, 1977.

Van Nimmen, Jane, and Leonard C. Bruno, with Robert L. Rosholt. *NASA Historical Data Book, Volume I: NASA Resources, 1958-1968.* NASA SP-4012, 1976, rep. ed. 1988.

Ezell, Linda Neuman. *NASA Historical Data Book, Volume II: Programs and Projects, 1958-1968.* NASA SP-4012, 1988.

Ezell, Linda Neuman. *NASA Historical Data Book, Volume III: Programs and Projects, 1969-1978.* NASA SP-4012, 1988.

Gawdiak, Ihor Y., with Helen Fedor, compilers. *NASA Historical Data Book, Volume IV: NASA Resources, 1969-1978.* NASA SP-4012, 1994.

Rumerman, Judy A., compiler. *NASA Historical Data Book, 1979-1988: Volume V, NASA Launch Systems, Space Transportation, Human Spaceflight, and Space Science.* NASA SP-4012, 1999.

Rumerman, Judy A., compiler. *NASA Historical Data Book, Volume VI: NASA Space Applications, Aeronautics and Space Research and Technology, Tracking and Data Acquisition/Space Operations, Commercial Programs, and Resources, 1979-1988.* NASA SP-2000-4012, 2000.

Astronautics and Aeronautics, 1969: Chronology of Science, Technology, and Policy. NASA SP-4014, 1970.

Astronautics and Aeronautics, 1970: Chronology of Science, Technology, and Policy. NASA SP-4015, 1972.

Astronautics and Aeronautics, 1971: Chronology of Science, Technology, and Policy. NASA SP-4016, 1972.

Astronautics and Aeronautics, 1972: Chronology of Science, Technology, and Policy. NASA SP-4017, 1974.

Astronautics and Aeronautics, 1973: Chronology of Science, Technology, and Policy. NASA SP-4018, 1975.

Astronautics and Aeronautics, 1974: Chronology of Science, Technology, and Policy. NASA SP-4019, 1977.

Astronautics and Aeronautics, 1975: Chronology of Science, Technology, and Policy. NASA SP-4020, 1979.

Astronautics and Aeronautics, 1976: Chronology of Science, Technology, and Policy. NASA SP-4021, 1984.

Astronautics and Aeronautics, 1977: Chronology of Science, Technology, and Policy. NASA SP-4022, 1986.

Astronautics and Aeronautics, 1978: Chronology of Science, Technology, and Policy. NASA SP-4023, 1986.

Astronautics and Aeronautics, 1979-1984: Chronology of Science, Technology, and Policy. NASA SP-4024, 1988.

Astronautics and Aeronautics, 1985: Chronology of Science, Technology, and Policy. NASA SP-4025, 1990.

Noordung, Hermann. *The Problem of Space Travel: The Rocket Motor.* Edited by Ernst Stuhlinger and J. D. Hunley, with Jennifer Garland. NASA SP-4026, 1995.

Astronautics and Aeronautics, 1986-1990: A Chronology. NASA SP-4027, 1997.

Astronautics and Aeronautics, 1990-1995: A Chronology. NASA SP-2000-4028, 2000.

Management Histories, NASA SP-4100:

Rosholt, Robert L. *An Administrative History of NASA, 1958-1963.* NASA SP-4101, 1966.

Levine, Arnold S. *Managing NASA in the Apollo Era.* NASA SP-4102, 1982.

The NASA History Series

Roland, Alex. *Model Research: The National Advisory Committee for Aeronautics, 1915-1958.* NASA SP-4103, 1985.

Fries, Sylvia D. *NASA Engineers and the Age of Apollo.* NASA SP-4104, 199.

Glennan, T. Keith. *The Birth of NASA: The Diary of T. Keith Glennan.* J. D. Hunley, editor. NASA SP-4105, 1993.

Seamans, Robert C., Jr. *Aiming at Targets: The Autobiography of Robert C. Seamans, Jr.* NASA SP-4106, 1996.

Project Histories, NASA SP-4200:

Swenson, Loyd S., Jr., James M. Grimwood, and Charles C. Alexander. *This New Ocean: A History of Project Mercury.* NASA SP-4201, 1966; rep. ed. 1998.

Green, Constance McLaughlin, and Milton Lomask. *Vanguard: A History.* NASA SP-4202, 1970; rep. ed. Smithsonian Institution Press, 1971.

Hacker, Barton C., and James M. Grimwood. *On Shoulders of Titans: A History of Project Gemini.* NASA SP-4203, 1977.

Benson, Charles D., and William Barnaby Faherty. *Moonport: A History of Apollo Launch Facilities and Operations.* NASA SP-4204, 1978.

Brooks, Courtney G., James M. Grimwood, and Loyd S. Swenson, Jr. *Chariots for Apollo: A History of Manned Lunar Spacecraft.* NASA SP-4205, 1979.

Bilstein, Roger E. *Stages to Saturn: A Technological History of the Apollo/Saturn Launch Vehicles.* NASA SP-4206, 1980, rep. ed. 1997.

SP-4207 not published.

Compton, W. David, and Charles D. Benson. *Living and Working in Space: A History of Skylab.* NASA SP-4208, 1983.

Ezell, Edward Clinton, and Linda Neuman Ezell. *The Partnership: A History of the Apollo-Soyuz Test Project.* NASA SP-4209, 1978.

Hall, R. Cargill. *Lunar Impact: A History of Project Ranger.* NASA SP-4210, 1977.

Newell, Homer E. *Beyond the Atmosphere: Early Years of Space Science.* NASA SP-4211, 1980.

Ezell, Edward Clinton, and Linda Neuman Ezell. *On Mars: Exploration of the Red Planet, 1958-1978.* NASA SP-4212, 1984.

Pitts, John A. *The Human Factor: Biomedicine in the Manned Space Program to 1980.* NASA SP-4213, 1985.

Compton, W. David. *Where No Man Has Gone Before: A History of Apollo Lunar Exploration Missions.* NASA SP-4214, 1989.

Naugle, John E. *First Among Equals: The Selection of NASA Space Science Experiments.* NASA SP-4215, 1991.

Wallace, Lane E. *Airborne Trailblazer: Two Decades with NASA Langley's Boeing 737 Flying Laboratory.* NASA SP-4216, 1994.

Butrica, Andrew J., editor. *Beyond the Ionosphere: Fifty Years of Satellite Communication.* NASA SP-4217, 1997.

Butrica, Andrew J. *To See the Unseen: A History of Planetary Radar Astronomy.* NASA SP-4218, 1996.

Mack, Pamela E., editor. *From Engineering Science to Big Science: The NACA and NASA Collier Trophy Research Project Winners.* NASA SP-4219, 1998.

Reed, R. Dale, with Darlene Lister. *Wingless Flight: The Lifting Body Story.* NASA SP-4220, 1997.

Heppenheimer, T. A. *The Space Shuttle Decision: NASA's Search for a Reusable Space Vehicle.* NASA SP-4221, 1999.

Hunley, J. D., editor. *Toward Mach 2: The Douglas D-558 Program.* NASA SP-4222, 1999.

Swanson, Glen E., editor. *"Before this Decade is Out . . .": Personal Reflections on the Apollo Program* NASA SP-4223, 1999.

Tomayko, James E. *Computers Take Flight: A History of NASA's Pioneering Digital Fly-by-Wire Project.* NASA SP-2000-4224, 2000.

Morgan, Clay. *Shuttle-Mir: The U.S. and Russia Share History's Highest Stage.* NASA SP-2001-4225, 2001.

Center Histories, NASA SP-4300:

Rosenthal, Alfred. *Venture into Space: Early Years of Goddard Space Flight Center.* NASA SP-4301, 1985.

Hartman, Edwin P. *Adventures in Research: A History of Ames Research Center, 1940-1965.* NASA SP-4302, 1970.

Hallion, Richard P. *On the Frontier: Flight Research at Dryden, 1946-1981.* NASA SP- 4303, 1984.

Muenger, Elizabeth A. *Searching the Horizon: A History of Ames Research Center, 1940-1976.* NASA SP-4304, 1985.

Hansen, James R. *Engineer in Charge: A History of the Langley Aeronautical Laboratory, 1917-1958.* NASA SP-4305, 1987.

Dawson, Virginia P. *Engines and Innovation: Lewis Laboratory and American Propulsion Technology.* NASA SP-4306, 1991.

The NASA History Series

Dethloff, Henry C. *"Suddenly Tomorrow Came . . .": A History of the Johnson Space Center.* NASA SP-4307, 1993.

Hansen, James R. *Spaceflight Revolution: NASA Langley Research Center from Sputnik to Apollo.* NASA SP-4308, 1995.

Wallace, Lane E. *Flights of Discovery: 50 Years at the NASA Dryden Flight Research Center.* NASA SP-4309, 1996.

Herring, Mack R. *Way Station to Space: A History of the John C. Stennis Space Center.* NASA SP-4310, 1997.

Wallace, Harold D., Jr. *Wallops Station and the Creation of the American Space Program.* NASA SP-4311, 1997.

Wallace, Lane E. *Dreams, Hopes, Realities: NASA's Goddard Space Flight Center, The First Forty Years.* NASA SP-4312, 1999.

Dunar, Andrew J., and Stephen P. Waring. *Power to Explore: A History of the Marshall Space Flight Center.* NASA SP-4313, 1999.

Bugos, Glenn E. *Atmosphere of Freedom: Sixty Years at the NASA Ames Research Center Astronautics and Aeronautics, 1986-1990: A Chronology.* NASA SP-2000-4314, 2000.

General Histories, NASA SP-4400:

Corliss, William R. *NASA Sounding Rockets, 1958-1968: A Historical Summary.* NASA SP-4401, 1971.

Wells, Helen T., Susan H. Whiteley, and Carrie Karegeannes. *Origins of NASA Names.* NASA SP-4402, 1976.

Anderson, Frank W., Jr. *Orders of Magnitude: A History of NACA and NASA, 1915-1980.* NASA SP-4403, 1981.

Sloop, John L. *Liquid Hydrogen as a Propulsion Fuel, 1945-1959.* NASA SP-4404, 1978.

Roland, Alex. *A Spacefaring People: Perspectives on Early Spaceflight.* NASA SP-4405, 1985.

Bilstein, Roger E. *Orders of Magnitude: A History of the NACA and NASA, 1915-1990.* NASA SP-4406, 1989.

Logsdon, John M., editor, with Linda J. Lear, Jannelle Warren-Findley, Ray A. Williamson, and Dwayne A. Day. *Exploring the Unknown: Selected Documents in the History of the U.S. Civil Space Program, Volume I, Organizing for Exploration.* NASA SP-4407, 1995.

Logsdon, John M., editor., with Dwayne A. Day and Roger D. Launius. *Exploring the Unknown: Selected Documents in the History of the U.S. Civil Space Program, Volume II, Relations with Other Organizations.* NASA SP-4407, 1996.

Logsdon, John M., editor, with Roger D. Launius, David H. Onkst, and Stephen J. Garber. *Exploring the Unknown: Selected Documents in the History of the U.S. Civil Space Program, Volume III, Using Space.* NASA SP-4407, 1998.

Logsdon, John M., general editor, with Ray A. Williamson, Roger D. Launius, Russell J. Acker, Stephen J. Garber, and Jonathan L. Friedman. *Exploring the Unknown: Selected Documents in the History of the U.S. Civil Space Program, Volume IV, Accessing Space.* NASA SP-4407, 1999.

Logsdon, John M., general editor, with Amy Paige Snyder, Roger D. Launius, Stephen J. Garber, and Regan Anne Newport. *Exploring the Unknown: Selected Documents in the History of the U.S. Civil Space Program, Volume V, Exploring the Cosmos.* NASA SP-2001-4407, 2001.

Siddiqi, Asif A. *Challenge to Apollo: The Soviet Union and the Space Race, 1945-1974.* NASA SP-2000-4408, 2000.

Monographs in Aerospace History, NASA SP-4500:

Maisel, Martin D., Demo J. Giulianetti, and Daniel C. Dugan. *The History of the XV-15 Tilt Rotor Research Aircraft: From Concept to Flight.* NASA SP-2000-4517, 2000.

Jenkins, Dennis R. *Hypersonics Before the Shuttle: A Concise History of the X-15 Research Airplane.* NASA SP-2000-4518, 2000.

Chambers, Joseph R. *Partners in Freedom: Contributions of the Langley Research Center to U.S. Military Aircraft in the 1990s.* NASA SP-2000-4519, 2000.

Waltman, Gene L. *Black Magic and Gremlins: Analog Flight Simulations at NASA's Flight Research Center.* NASA SP-2000-4520, 2000.

Portree, David S. F. *Humans to Mars: Fifty Years of Mission Planning, 1950-2000.* NASA SP-2001-4521, 2001.

Thompson, Milton O., with J. D. Hunley. *Flight Research: Problems Encountered and What They Should Teach Us.* NASA SP-2000-4522, 2000.

Tucker, Tom. *The Eclipse Project.* NASA SP-2000-4523, 2000.